Lecture Notes in Mathematics

Edited by A. Dold and B. Eckmann

741

Algebraic Topology Waterloo 1978

Proceedings of a Conference Sponsored by the Canadian Mathematical Society, NSERC (Canada), and the University of Waterloo, June 1978

Edited by
Peter Hoffman and Victor Snaith

Springer-Verlag
Berlin Heidelberg New York 1979

Editors

Peter Hoffman
Department of Pure Mathematics
University of Waterloo
Waterloo, Ontario
Canada N2L 3G1

Victor Snaith
Department of Mathematics
University of Western Ontario
London, Ontario
Canada N6A 5B9

AMS Subject Classifications (1980): 10C05, 13Dxx, 16A54, 18Fxx, 18Gxx, 55-XX, 57-XX, 58Axx

ISBN 3-540-09545-4 Springer-Verlag Berlin Heidelberg New York
ISBN 0-387-09545-4 Springer-Verlag New York Heidelberg Berlin

Library of Congress Cataloging in Publication Data
Main entry under title:
Algebraic topology, Waterloo 1978.
(Lecture notes in mathematics ; 741)
Bibliography: p.
Includes index.
1. Algebraic topology--Congresses. I. Hoffman, Peter, 1941- II. Snaith, Victor Percy, 1944- III. Canadian Mathematical Congress (Society) IV. University of Waterloo. V. Series: Lecture notes in mathematics (Berlin) ; 741.
QA3.L28 no. 741 [QA612] 510'.8s [514'.2] 79-20052
ISBN 0-387-09545-4

2141/3140-543210

FOREWORD

The present volume contains the Proceedings of a conference held at the University of Waterloo in June of 1978. The conference was part of a Summer Research Institute of the Canadian Mathematical Society, and was held simultaneously with a conference in Ring Theory whose proceedings also appear in this series.

The main emphasis was on L-theory and on topological and algebraic K-theory. The volume is divided into four sections as indicated in the table of contents.

We would like to express our gratitude to all the participants for their contributions to the conference, and particularly to Ian Hambleton, Richard Kane, Stan Kochman, Reinhardt Schultz and Rick Sharpe for help with this volume. We are grateful to Sue Embro for her work with the manuscripts, and to the sponsoring organizations for financial assistance.

Waterloo, Canada
January 1979

Peter Hoffman
Victor Snaith

CONTENTS

A. L-THEORY AND ALGEBRAIC K-THEORY

B. GROUP ACTIONS

C. HOMOTOPY AND COHOMOLOGY

D. LOOP AND H-SPACES

WATERLOO TOPOLOGY SRI

List of Talks

May 15	R. Kane	:	Torsion in $BP_*(X)$
16	V. Snaith	:	On Algebraic Vector Bundles over Number Fields.
17	R. Sharpe	:	Introduction to L-theory
--	R. Steiner	:	Infinite loop constructions I
18	K. Murasugi	:	On Representations of Knot Groups
19	R. Steiner	:	Infinite loop constructions II
May 23	I. Hambleton	:	Hermitian Forms and Manifolds
24	Doug Anderson	:	Immersion Theory
25	V. Snaith	:	Introduction to the + construction and higher K-theory
31	Doug Anderson	:	The Immersion Approach to Triangulations
June 1	I. Hambleton	:	The + construction and homology spheres
2	T. Petrie	:	Background for Smooth Lie Group Actions
5	T. Petrie	:	G surgery and semiclassical applications I
--	R. Kane	:	BP theory and Finite H-spaces
--	J. Allard	:	Sums of Stably Trivial Vector Bundles
6	W. Pardon	:	Examples and Applications of Surgery Obstructions
--	R. Lashof	:	Stable G-Smoothing
--	H. Munkholm	:	Finiteness Obstructions, Whitehead Torsion and Transfer in Algebraic K-theory
--	R. Kulkarni	:	Proper Actions of non-compact Groups and Relativistic Space Forms
--	B. Williams	:	Formal Surgery Theory
7	P. May	:	Segal's maps, Kahn's maps and Mahowald's Theorem
--	I. Madsen	:	Spherical Space Forms
--	T. Petrie	:	G surgery and semiclassical applications II
--	H. Dovermann	:	Structure of the Set (Group) of Equivariant Surgery Obstructions and Applications
--	J. Harper	:	Construction of mod p H-spaces

List of Talks (con'd)

June 8	M. Rothenberg	:	Equivariant Homotopy Type of Spheres
	-- R. Lee	:	Unstable cohomology of $SL_n(Z)$
	-- W. Hsiang	:	π_i Diff (M^n)
	-- I. Hambleton	:	Decompositions of Semi-Free Actions on Homotopy Spheres
	-- J. Ewing	:	Realizing Witt classes
9	P. Zvengrowski	:	Skewness of r-fields on Spheres and Projective Stiefel Manifolds
	-- J. Milgram	:	The Swan Finiteness Obstruction
	-- T. Lada	:	A Counterexample to the Transfer Conjecture
	-- K. Murasugi	:	PSL (2,p) Coverings of Knot Manifolds
	-- S. Thomeier	:	Join Constructions and Whitehead Products
	-- G. Carlsson	:	Equivariant Embeddings in Homotopy Spheres
June 12	D. Kraines	:	Applications of the Delooping Spectral Sequence
	-- I. Madsen	:	Tangential Homotopy Equivalences
	-- W. Hsiang	:	The Topological Euclidean Space Form Problem
	-- R. Schultz	:	Compact fibrings of minimal homogeneous Spaces
	-- D. Gottlieb	:	Lefschetz Numbers of Equivariant Maps
13	A. Ranicki	:	Equivariant Wu classes
	-- J. Milgram	:	Odd Wall Groups
	-- Doug Anderson	:	Triangulations of Locally Triangulable Spaces
	-- R. Sharpe	:	Strong Euclidean Rings and K_2 (Z[Z/5])
	-- S. Lomonaco	:	Homotopy Groups of Knots
14	A. Bak	:	Surgery Groups; a survey of computations with emphasis on $\ker\left(L_{2n}^h(\pi) \to L_{2n}^p(\pi)\right)$

List of Talks (con'd)

June 15	A. Liulevicius	:	Linear Actions on Friendly Spaces
	-- A. Ranicki	:	Poincaré complexes & Surgery
	-- M. Boardman	:	Ye Olde Original Brown-Peterson Spectra
	-- R. Patterson	:	Square Preserving Algebra Endomorphisms of $H^*(BO;Z/2)$
	-- K. Varadarajan	:	Finiteness Obstruction for Nilpotent Spaces
16	K. Ehrlich	:	A Transfer Map in Algebraic K-theory
	-- V. Snaith	:	C*-extensions of Compact Lie Groups
	-- D. Sjerve	:	Homology Isomorphisms and the failure of Whitehead's Theorem
	-- D. Davis	:	BP operations and maps of stunted complex projective spaces
June 19	P. Landweber	:	New Applications of Commutative Algebra to MU and BP
20	V. Snaith	:	On $K_3(Z/4)$
	-- P. Landweber	:	The Signature of Symplectic and Self-Conjugate Manifolds
21	Don Anderson	:	Homotopy Coherence of Functors
	-- S. Kochman	:	The Symplectic Cobordism Ring I
22	S. Kochman	:	The Symplectic Cobordism Ring II

THE STRUCTURE OF ODD L-GROUPS

by
G. Carlsson
R. James Milgram*

In this paper we apply the exact sequence of [C-M] or [P] to obtain extensive information about odd surgery groups. In particular, we completely determine the proper surgery groups $L_3^p(\pi,1)$ [Ma] for π a finite 2-group. Using these groups we then obtain a system of generators for the Wall surgery groups $L_3^h(\pi,1)$ for π as above from the exact sequence

$$\text{A.}\quad H_{odd}(Z/2, \tilde{K}_o(Z(\pi))) \xrightarrow{\partial} L_3^h(\pi,1) \longrightarrow L_3^p(\pi,1) \xrightarrow{j} H_{ev}(Z/2,\tilde{K}_o(Z(\pi))).$$

For π a 2-group the 2 primary part of $\tilde{K}_o(Z(\pi))$ is determined as the quotient of a finite group depending only on the rational representation ring of π, $\mathbb{Q}(\pi)$. Then $\mathrm{im}(j)$ is easily calculated, and $\mathrm{Ker}(\partial)$ is the only datum which we presently lack in studying A.

Our main result on finite 2-groups (Theorem 4.9), is

<u>Theorem B</u>: <u>Let</u> $n(\pi)$ <u>be the number of summands</u> $M_{n_i}(F)$ <u>contained in</u> $\mathbb{Q}(\pi)$ <u>where</u> F <u>is a real field, then</u>

$$L_3^p(\pi,1) = (Z/2)^{n(\pi)-1}$$

For example, let π be the dihedral group $Z/_{2^i} \times_T Z/2(t^2 = tgtg = g^{2^i} = e)$

---

*Research supported in part by NSF MCS77-01623 and NSF MCS76-0146-A01

then

$$L_3^h(\pi,1) = L_3^p(\pi,1) = (Z/2)^{i+2} ,$$

If π is a generalized Quaternion group $Z/_{2^i} \times_T Z/2 (t^2 = (tgtg) = g^{2^{i-1}})$

$$L_3^p(\pi,1) = (Z/2)^{i+1}$$

and there is an exact sequence

$$Z/2 \xrightarrow{\partial} L_3^h(\pi,1) \longrightarrow L_3^p(\pi,1) \xrightarrow{j} Z/2$$

with j surjective.

Remark: In [Wa] it is pointed out that the surgery obstructions for surgery problems over closed manifolds factor through bordism and hence are determined by restriction to the 2-Sylow subgroup if π is finite. Thus, one would expect that combining our results with those of [Ra] will lead to extensive results along these lines. We hope to consider this in forthcoming work.

Our work here differs from previous work on these questions, for example [B-S1], in that we first use local $\Rightarrow$ global techniques for arbitrary semi-simple algebras and orders rather than only matrix rings over fields as was attempted previously. Our basic local result here is theorem 2.6.

Next, we study the Witt rings of matrix algebras over Division rings with center a finite extension of $\mathbb{Q}$. By [F - H] and [A] those Division rings which occur in $\mathbb{Q}(\pi)$ for π a finite group always have an involution. By [B-S2] the local Brauer invariants of such a Division algebra are very restrictive. (See e.g., Theorems 2.10, 7.5 and 7.6.) In particular, for algebras of this type we are able to give a generalization of the Hasse-Schilling norm theorem [S-E] , to characterize the reduced norms of elements invariant under the involution (Theorem 3.12).

These results, together with our local-global results give us enough information to make calculations in the exact sequences.

However, a restriction in the local part of the local-global theory, at this point forces us to work with the bilinear theory, rather than the quadratic theory. This only affects matters when localized at 2 , and for finite 2-groups we are able to circumvent this difficulty. But our general results relate to the structure of the odd bilinear Wall groups

$$L_1^-(Z(\pi)) ,$$

which are closely related to the Wall groups, and seem to be of interest in their own right.

We begin by showing for a general finite group π that each irreducible algebra $M_{n_i}(D_i)$ in $\mathbb{C}(\pi)$ contributes a certain amount to $L_1^-(Z\pi))$ and, indeed these contributions surject onto it.

If the involution of D fixes the center then D is type I and is either a real field or a Quaternion algebra over a real field. Otherwise, D is type II , and we have

<u>Theorem C:</u> <u>Each type</u> II <u>algebra in</u> $\mathbb{C}(\pi)$ <u>contributes at most a single</u> Z/2 <u>to</u> $L_1^-(Z(\pi))$.

(Corollary 3.13. This is a nice application of Artin reciprocity.)

The Quaternion algebras and type I fields also each give only limited contributions to $L_1^-(Z(\pi))$, but the exact results are dependent on the particular algebras involved.

The paper is organized as follows. In §1 we review the exact sequences of [C-M] and [P], and in §2 we give the basic calculations for the term

$$L_o^{\varepsilon,tor}(\mathbb{Q}(\pi),\ Z-\{o\})$$

in the exact sequence. In §3 we study the group $L_{o,f}^{\varepsilon}(\mathbb{Q}(\pi))$ and the ∂ map

D. $$\partial : L_{o,f}^{\varepsilon}(\mathbb{Q}(\pi)) \longrightarrow L_o^{\varepsilon,tor}(\mathbb{Q}(\pi),\ Z-\{o\}) .$$

This gives an effective determination of $L_1^-(Z(\pi))$ since $L_1^-(Z(\pi))$ = coker(∂) in D.

In §4 we apply the results of §2, §3 together with some facts on $\mathbb{Q}(\pi)$, for π a 2-group, coming from [M1], and some standard facts on the units in $Z(\rho_{2^i})$ to obtain our results on finite 2-groups. Then the remainder of the paper proves the basic results alluded to previously, which are used in §2, §3 to obtain our calculational results.

We would like to thank W. Pardon for some useful conversations and correspondence.

§1. Definitions; The Exact Sequence

We recall from [C-M] the basic objects of study. We let Λ be a ring with involution $\bar{\ }$, so $\overline{\lambda_1\lambda_2} = \bar{\lambda}_2\bar{\lambda}_1$. We assume that Λ is free and finitely generated over a commutative integral domain A, with A central in Λ.

Definition 1.1: An ε-symmetric Hermitian form space $(\varepsilon = \pm 1)$ is a pair (H,β), where

(a) H is a projective left Λ - module

(b) $\beta : H \times H \longrightarrow \Lambda$ is a pairing satisfying

$$\beta(\lambda_1 h_1 + \lambda_2 h_2, h_3) = \lambda_1\beta(h_1,h_3) + \lambda_2\beta(h_2,h_3)$$
$$\beta(h_1,\lambda h_2) = \beta(h_1,h_2)\bar{\lambda}$$
$$\beta(h_1,h_2) = \overline{\beta(h_2,h_1)},$$

and such that the map $ad(\beta) : H \longrightarrow H^*$ (setting $H^* = \mathrm{Hom}_\Lambda(H,\Lambda)$) defined by $ad(\beta)(h_2)(h_1) = \beta(h_1,h_2)$ is an isomorphism of Λ-modules. (The Λ- action on H^* is given by $(\lambda\phi)(h) = (\phi(h))\bar{\lambda}$, $h \in H$, $\lambda \in \Lambda$).

Given such a Hermitian form space, and K a submodule of H, we define

$$K^\perp = \{h \in H \mid \beta(h,k) = 0 \ \forall\, k \in K\}$$

If K is a direct summand of H, then so is $K^\perp$. If $K = K^\perp$, we say that K is a kernel of H. If the space (H,β) admits a kernel, we say it is split or hyperbolic.

We now let S be a multiplicative subset of the central integral domain $A \subset \Lambda$, so that $s = \bar{s}$ for all $s \in S$. We denote by M_S the module M localized at S.

Definition 1.2: An S-torsion module over Λ of projective length 1 is a Λ-module M so that $M_S = (0)$, which admits a Λ-projective resolution

$$0 \longrightarrow P_1 \longrightarrow P_2 \longrightarrow M \longrightarrow 0$$

We refer to such a module as being PL1. If the image of $[P_2] - [P_1]$ in $K_0(\Lambda)$ is zero, we say that the module is of free projective length 1, or FPL1. (Note that the "Euler Characteristic" of $M, [P_2] - [P_1]$, is an invariant of M, independent of the resolution).

Given $M_1 \subset M_2$, where M_1 and M_2 are both PL1, we say M_1 is semisplit in M_2 if M_2/M_1 is also PL1 and we write that the pair (M_1, M_2) is SPL1. Similarly, if M_1 and M_2 are FPL1, and M_2/M_1 is also FPL1, we say that the pair (M_1, M_2) is free semisplit or SFPL1.

We recall from [C-M]

Lemma 1.3: Given $M_1 \subset M_2 \subset M_3$ with (M_1, M_2) and (M_2, M_3) both SPL1, then (M_1, M_3) is also SPL1. Also, if $0 \to A \to B \to C \to 0$ is exact, and A and C are PL1, then B is PL1. (Similarly for SFPL1 and FPL1.)

Lemma 1.4: Let $0 \to A \to B \to C \to 0$ be exact and suppose that B and C are PL1. Then so is A. (Similarly for FPL1.)

We define, for M an S-torsion PL1 module, $M^* = \mathrm{Hom}_\Lambda(M, \Lambda_S/\Lambda)$. Here the Λ-structure on M^* is given by $(\lambda\phi)(x) = \phi(x)\bar{\lambda}$.

Lemma 1.5: If M is PL1, so is M^*. (Again, similarly for FPL1)

Definition 1.6: An S-torsion ε-symmetric Hermitian form space is a pair (G, β), where

(a) G is an S-torsion PL1 module

(b) $\beta : G \times G \to \Lambda_S/\Lambda$ posseses the same properties relative to the module action as the β defined in Definition 1 and is non-singular in the sense that $ad(\beta) : G \to G^*$ is an isomorphism

(c) For every $x \in G$ there is $\alpha \in \Lambda_S$ such that $\alpha = \beta(x,x) (\mathrm{mod}\ \Lambda)$ and $\alpha = \varepsilon\bar{\alpha}$.

As before, let $K \subseteq G$ and $K^{\perp} = \{g \in G | \beta(g,k) = 0 \ \forall k \in K\}$. We say that (G,β) is split if there is a PL1 submodule $K \subseteq G$, with the pair (K,G) SPL1, and $K^{\perp} = K$.

Definition 1.7: $L_o^{\varepsilon}(\Lambda)$ is the quotient of the monoid of isomorphism classes of ε-symmetric Hermitian form spaces under direct sum by split spaces. We note, as in [C-M], that this is a group.

We say that a projective Λ_S- module is extended from Λ if there is a projective Λ-module $\hat{P}$ so that $P \cong \Lambda_S \otimes_{\Lambda} \hat{P}$.

Definition 1.8: $L_o^{\varepsilon}(\Lambda, S)$ is the quotient of the monoid of ε-symmetric Hermitian form spaces (H,β) over Λ, under direct sum by spaces which admit a kernel K, also extended from Λ^n for some n. This is again a group.

Definition 1.9: $L_o^{\varepsilon,tor}(\Lambda, S)$ is the quotient of the monoid of S-torsion ε-symmetric Hermitian form spaces under direct sum by split spaces.

This is a group.

We recall the variant groups $L_{o,f}^{\varepsilon}(\Lambda)$, $L_{o,f}^{\varepsilon}(\Lambda, S)$, and $L_{o,f}^{\varepsilon,tor}(\Lambda, S)$, where the first two groups require the modules and kernels to be free, and the last that modules and kernels be FPL1. We will sometimes write $W(\Lambda)$, Witt group of Λ, for $L_{o,f}^{+}(\Lambda)$, and $W_{proj}(\Lambda)$, projective Witt group, for $L_o^{+}(\Lambda)$.

We have, for the sake of brevity, suppressed the definition of L_1^{ε} and $L_{1,f}^{\varepsilon}$, and refer the reader to [C-M].

Recall also the existence of groups $L_i^{\varepsilon,q}$, $L_{i,f}^{\varepsilon,q}$, $L_{i,f}^{\varepsilon,q,tor}$, quadratic L-groups, whose definitions we also suppress.

The following theorem is the basis of our calculations

Theorem 1.10: [C-M], [Pardon], [Ranički] . There exist exact sequences

(A) $L_{tor}^{\varepsilon}(\Lambda) \to L_o^{\varepsilon}(\Lambda,S) \to L_o^{\varepsilon,tor}(\Lambda,S) \to L_1^{-\varepsilon}(\Lambda) \to L_1^{-\varepsilon}(\Lambda,S)$

(B) $L_{o,f}^{\varepsilon}(\Lambda) \to L_{o,f}^{\varepsilon}(\Lambda,S) \to L_{o,f}^{\varepsilon,tor}(\Lambda,S) \to L_{1,f}^{-\varepsilon}(\Lambda) \to L_{1,f}^{-\varepsilon}(\Lambda,S)$

(C) $L_o^{\varepsilon,q}(\Lambda) \to L_o^{\varepsilon,q}(\Lambda,S) \to L_o^{\varepsilon,q,tor}(\Lambda,S) \to L_1^{-\varepsilon,q}(\Lambda) \to L_1^{-\varepsilon,q}(\Lambda,S)$

(D) $L_{o,f}^{\varepsilon,q}(\Lambda) \to L_{o,f}^{\varepsilon,q}(\Lambda,S) \to L_{o,f}^{\varepsilon,q,tor}(\Lambda,S) \to L_{1,f}^{-\varepsilon,q}(\Lambda) \to L_{1,f}^{-\varepsilon,q}(\Lambda,S)$

The groups $L_{1,f}^{-,q}(\mathbb{Z}\pi)$ may be identified with the surgery obstruction groups $L_3^{(h)}(\pi)$.

§2. The Groups $L^{\varepsilon,tor}_{o,p}(Z(\pi))$.

Let G be (as in §1.6) an S torsion module over $Z(\pi)$. Then we can write

$$G = \coprod_p G_p$$

where G_p is the additive p-Sylow subgroup, and we have

Theorem 2.1: (a) G is PL1 over $Z(\pi)$ if and only if G_p is PL1 for each p

(b) If p does not divide $|\pi|$ then G_p is always PL1.

Proof: (b) is Nakayama's lemma. To prove (a), from [S] we note that it is possible to find a torsion module H with $H_p = 0$ if $G_p \neq 0$ so that $L = H \oplus G$ is FPL1 . Thus $L_p = G_p$ if $G_p \neq 0$, and it suffices to consider a resolution

$$0 \longrightarrow (Z(\pi))^n \xrightarrow{m} (Z(\pi))^n \rightarrow L \longrightarrow 0$$

where m is an $n \times n$ matrix with entries in $Z(\pi)$.

Choose p so $G_p \neq 0$, and $p \,\big|\, |\pi|$. By strong approximation [B-S-1] it is possible to find $\tilde{m}$ arbitrarily close to m at p and arbitrarily close to 1 at $q \neq p$; $q \,\big|\, |\pi|$. Then

2.2 $$0 \longrightarrow (Z(\pi))^n \xrightarrow{\tilde{m}} (Z(\pi))^n \longrightarrow L_{\tilde{m}} \longrightarrow 0$$

satisfies $(L_{\tilde{m}})_p = G_p$, $(L_{\tilde{m}})_q = 0$, $p \neq q$ and $q \,\big|\, |\pi|$.

Now, consider the exact sequence

$$0 \longrightarrow G_p \longrightarrow L_{\tilde{m}} \longrightarrow L_{\tilde{m}}/G_p \longrightarrow 0 .$$

By 2.1(b) $L_{\tilde{m}}/G_p$ is PL1, and so, by lemma 1.4, is G_p. The converse is evident.

<u>Definition</u> 2.3: $X^{\varepsilon}{}_p(Z(\pi))$ <u>is the quotient of the monoid of ε-symmetric p-torsion Hermitian form spaces under direct sum by split spaces.</u>

Then we have directly from 2.1

<u>Corollary</u> 2.4: $L^{\varepsilon,tor}_{o,proj}(Z(\pi)) = \coprod_p X^{\varepsilon}{}_p(Z(\pi))$.

Localizing we now have

<u>Theorem</u> 2.5: $X^{\varepsilon}_p(Z(\pi)) = L^{\varepsilon,tor}_{o,f}(\hat{\mathbb{Q}}_p(\pi), \hat{Z}_p - \{o\})$.

(If we tensor 2.2 with $\hat{Z}_p$ we see that any PL1 , $Z(\pi)$ p-torsion module is FPL1 as a $\hat{Z}_p(\pi)$ module, and using strong approximation and 2.1(b), the converse is also true, hence the result.)

In §8 we will prove

<u>Theorem</u> 2.6: $\partial : L^{\varepsilon}_{o,f}(\hat{\mathbb{Q}}_p(\pi)) \longrightarrow L^{\varepsilon,tor}_{o,f}(\hat{\mathbb{Q}}_p(\pi), \hat{Z}_p - \{o\})$ <u>is surjective</u>.

Hence, we have

2.7 $$X^{\varepsilon}_p(Z(\pi)) = L^{\varepsilon}_{o,f}(\hat{\mathbb{Q}}_p(\pi), \hat{Z}_p-\{o\}) \Big/ \mathrm{im}(L^{\varepsilon}_{o,f}(\hat{Z}_p(\pi)))$$

To illustrate the utility of 2.7 we now turn to a discussion of $L^{\varepsilon}_{o}(\hat{\mathbb{Q}}_p(\pi))$ since $L^{\varepsilon}_{o,f}(\hat{\mathbb{Q}}_p(\pi))$ embeds in it with finite index.

Consider $\hat{\mathbb{Q}}_p(\pi) = \coprod M_{n_i}(D_i)$ where the D_i are Division algebras over finite extensions of $\hat{\mathbb{Q}}_p$. We may describe this in another way. Write

$$\mathbb{Q}(\pi) = \coprod M_{m_j}(D_j) .$$

Then, see e.g. [F-H], τ acts non-trivially on each simple algebra summand separately. Moreover, the center of D_j is a finite cyclotomic extension of $\mathbb{Q}$, say F_j. Let $\mathscr{P}_j$ be a prime over (p) in F_j, then [L. p. 39]

$$\hat{\mathbb{Q}}_p \otimes_{\mathbb{Q}} F_j = \coprod_{\mathscr{P}_j} \hat{F}_{j,(\mathscr{P}_j)} .$$

τ acts on F_j and either leaves the $\mathscr{P}_j$ invariant or permutes them in pairs. Then

2.8 $$\hat{\mathbb{Q}}_p(\pi) = \coprod_j \coprod_{\mathscr{P}_j} M_{n_j}(D_j) \otimes_{F_j} \hat{F}_{j,(\mathscr{P}_j)}$$

and τ acts according to its behavior on the primes $\mathscr{P}_j$. Moreover,

2.9 $$M_{n_j}(D_j) \otimes \hat{F}_{j,(\mathscr{P}_j)} = M_{n_j s}(\tilde{D}_{j,(\mathscr{P}_j)})$$

is again a simple algebra. In 7.5 and 7.6 we shall prove

<u>Theorem</u> 2.10: <u>In 2.9, if $\tau(\mathscr{P}_j) = \mathscr{P}_j$ 2 cases occur.</u>

(a) <u>If τ is the identify on F_j, then $\tilde{D}_{j,(\mathscr{P}_j)} = \hat{F}_{j,(\mathscr{P}_j)}$ or is the quaternion algebra over $\hat{F}_{j,(\mathscr{P}_j)}$.</u>

(b) <u>If τ is non-trivial on F_j, then $\tilde{D}_{j,(\mathscr{P}_j)} = \hat{F}_{j,(\mathscr{P}_j)}$.</u>

Since $L(A \oplus A^*,\tau) = 0$ where τ interchanges A, A^*, 2.10 reduces us to only 2 cases.

In both cases, by the Morita theory, we may make a preliminary reduction

$$2.11 \qquad L^{\varepsilon}_{o,f}(M_n(D),\tau) = L^{\varepsilon'}_{o,f}(D,\tau')$$

(For details see §5 and §6.)

We now consider the case $\tilde{D}_{j,(\mathscr{P}_j)} = \hat{F}_{j,(\mathscr{P}_j)}$. If $\tau \neq id$ on $\hat{F}_{j,(\mathscr{P}_j)}$, let

$$K \subset \hat{F}_{j,(\mathscr{P}_j)}$$

be the fixed field of τ. Then a set of generators of

$$L^+_o(\hat{F}_{j,(\mathscr{P}_j)}, \tau)$$

are the $\langle 1 \rangle$ and $\langle \alpha \rangle$ where α represents the non-trivial element in

$$2.12 \qquad Z/2 = \dot{K}/\text{Norms}(\dot{\hat{F}}_{j,(\mathscr{P}_j)})$$

<u>Theorem</u> 2.13: (a) <u>If</u> $\hat{F}_{\mathscr{P}} - K$ <u>is an unramified extension then</u> $L^+_o(\hat{F}_{\mathscr{P}},\tau) = Z/2 \oplus Z/2$

(b) <u>If</u> $\hat{F}_{\mathscr{P}} - K$ <u>is ramified and</u> $\alpha \sim -1$ <u>then</u> $L^+_o(\hat{F},\tau) = Z/4$, <u>otherwise</u> $L^+_o(\hat{F},\tau) = Z/2 \oplus Z/2$.

<u>Proof</u>: In case -1 is a norm we have $\langle -1 \rangle \sim \langle 1 \rangle$ so $2\langle 1 \rangle = 0$ and $2\langle \alpha \rangle = 0$. If -1 is not a norm then $L^+_o(\hat{F},\tau)$ is cyclic and we use the results of [M - H pp. 114-119] to obtain our result.

<u>Remark</u> 2.14: There is an element λ in $\hat{F}$ satisfying $\tau(\lambda) = -\lambda$, and using multiplication by λ we construct an isomorphism

$$L_o^+(\hat{F}_{\mathscr{P}},\tau) \cong L_o^-(\hat{F}_{\mathscr{P}},\tau)$$

which completes our analysis of this part of the first case.

<u>Theorem</u> 2.15: <u>Let</u> $\tau = \text{id}$ <u>on</u> $\hat{F}_{\mathscr{P}}$ <u>then</u> $L_o^+(\hat{F}_{\mathscr{P}})$ <u>has generators the elements of</u>

$$\dot{\hat{F}}_{\mathscr{P}}/(\dot{\hat{F}}_{\mathscr{P}})^2 .$$

<u>If</u> $p \neq 2$, <u>then</u> $\dot{\hat{F}}_{\mathscr{P}}/\dot{\hat{F}}_{\mathscr{P}}^{2}$ <u>has</u> 2 <u>generators</u> α, π, <u>where</u> π <u>is a uniformizing parameter and</u>

$$L_o^+(\hat{F}_{\mathscr{P}}) = \begin{cases} (Z/2)^4, & -1 \neq \alpha \\ (Z/4)^2, & -1 = \alpha \end{cases}$$

<u>Remark</u> 2.16: When $p = 2$ there are a large number of elements in $\dot{\hat{F}}_{\mathscr{P}}/\dot{\hat{F}}_{\mathscr{P}}^{2}$. They are studied in [O-M , pp. 150 160, 170]. Using this and the results of [M - H, pp. 76-83, especially 8.1] we can read off $L_o^+(\hat{F}_{\mathscr{P}})$ in all cases.

Also, in all cases where $\tau = \text{id}$ on $\hat{F}_{\mathscr{P}}$ we have

$$L_o^-(\hat{F}_{\mathscr{P}}) = 0 .$$

We turn now to the case when D is the quaternion algebra over $\hat{F}_{\mathscr{P}}$.

Then $D = \hat{F}_{\mathscr{P}}(1) + \hat{F}_{\mathscr{P}}(i) + \hat{F}_{\mathscr{P}}(j) + \hat{F}_{\mathscr{P}}(ij)$ with $i^2 = a$, $j^2 = b$, $ij = -ji$. (Actually, we can be more explicit, let $\hat{K}$ be the degree 2 non-ramified extension of $\hat{F}_{\mathscr{P}}$, and $\lambda \in \hat{K}$ satisfies $\psi(\lambda) = -\lambda$ where ψ is the Galois automorphism, then we can choose i , j so $i^2 = \lambda^2 = a$ and $j^2 = \pi$ where π is a uniformizing parameter for $\hat{F}_{\mathscr{P}}$.) The usual involution τ on D is given by $1 \leftrightarrow 1$, $i \leftrightarrow -i$, $j \leftrightarrow -j$, $ij \leftrightarrow -ij$,and we have

<u>Theorem</u> 2.17: (a) $L_0^+(D,\tau) = Z/2$ <u>generator</u> $<1>$

(b) $L_0^-(D,\tau)$ <u>is a</u> $Z/2$<u>-vector space generated by the elements of</u> $(\dot{\hat{F}}_{\mathscr{P}}/\dot{\hat{F}}_{\mathscr{P}}^2 - (1))$, <u>with relations</u> $<\nu> \perp <\omega> \sim <\nu+\omega> \perp <\nu/\omega(\nu+\omega)>$

<u>Proof</u>: To begin we may assume our form diagonalized (by Lemma 5.4). Then in case (a) the elements all lie in $\dot{\hat{F}}_{\mathscr{P}}$. But $<\alpha> \sim <s\alpha\tau(s)> \sim <\alpha N(s)>$ for any $s \in D$, and given $\lambda \in \dot{\hat{F}}_{\mathscr{P}}$ there is some $s \in D$ so that $N(s) = \lambda$. Hence (a) follows.

To prove (b), we again assume the form diagonal. Now the elements $\nu \in \hat{F}_{\mathscr{P}}(i,j,ij)$.

<u>Lemma</u> 2.18: <u>Let</u> ν , $\omega \in \hat{F}_{\mathscr{P}}(i,j,ij)$ <u>and suppose</u> $N(\nu) = N(\omega)f^2$ <u>for some</u> $f \in \dot{\hat{F}}_{\mathscr{P}}$. <u>Then there is an</u> $s \in D$ <u>so that</u>

$$s\nu\tau(s) = \omega .$$

<u>Proof</u>: If $N(\nu) = N(\nu)f^2$ then the degree 2 extension of $\hat{F}_{\mathscr{P}}$, $\hat{F}_{\mathscr{P}}$ is isomorphic to $\hat{F}_{\mathscr{P}}(\omega)$ and there is an s so that

$$s\nu s^{-1} = \omega/f .$$

But $s^{-1} = \tau(s)/N(s)$, and we have

2.19 $$s\nu\tau(s) = (N(s)/f)w .$$

Now, if $\lambda = a + b\omega$ then $\lambda\omega\tau(\lambda) = N(\lambda)w$. Also, if we write $D = \hat{F}_{\mathscr{P}} + \hat{F}_{\mathscr{P}}\omega + \hat{F}_{\mathscr{P}}\mu + \hat{F}_{\mathscr{P}}\mu\omega$ with $\mu\omega = -\omega\mu$, $\tau(\mu) = -\mu$, then $\mu^2 \in \dot{\hat{F}}_{\mathscr{P}}$ and cannot be $N(\lambda)$ for any $\lambda = a + b\omega$. Indeed, <u>if we could solve the equation</u> $\mu^2 = N(\lambda)$ <u>then</u>

$$(\lambda^{-1}\mu)^2 = \lambda^{-1}\tau(\lambda^{-1})\mu^2 = 1$$

and $(\lambda^{-1}\mu - 1)(\lambda^{-1}\mu + 1) = 0$ which is impossible.

Returning to 2.19, if $N(s)/f = N(\lambda)$ then $\lambda^{-1}s\nu\tau(\bar{\lambda}^{-1}s) = w$. If not then $N(s)/f = N(\lambda\mu)$ for some λ and $(\lambda\mu)^{-1}s\nu\tau(\lambda u)^{-1}s) = w$. In either case 2.18 follows.

Now we check that $2 < \omega > = 0$. By the arguments in the proof of lemma 2.18 there is some element θ

$$\theta = \lambda\mu \quad \text{or} \quad \lambda'$$

so that $\theta\omega\tau(\theta) = -(\omega)$, and $2 < \omega > = <\omega> \perp < \omega > = 0$.

Finally, we verify the relation. Let ν , w satisfy $\tau(\nu) = (-\nu)$ $\tau(\omega) = (-\omega)$. We consider the space $< \nu > \perp < \omega >$ with basis elements e, f . Choose a new basis $e + f = e'$

$$e - \nu\omega^{-1}f = f' .$$

We easily verify these are orthogonal and have the indicated lengths.

Remark 2.20: Taken together, 2.7, 2.10, 2.13, 2.15, and 2.17 give us a complete set of generators for $X_p^{\varepsilon}(Z(\pi))$. The main problem then remaining in its calculation is the determination of $\mathrm{im}[L_{o,f}^{\varepsilon}(\hat{Z}_p(\pi))]$. For $p \nmid |\pi|$ this will reduce to "classical" calculations, while when $p \mid |\pi|$, the group will be slightly larger than classical calculations would lead us to expect. These deviations lie at the heart of our analysis of $L_1^{\varepsilon}(Z(\pi))$.

§3. Calculations

Maximal Orders. Let $\mathcal{M}_\pi \subset \mathbb{Q}(\pi)$ be a maximal order containing $Z(\pi)$ so if

$$\mathbb{Q}(\pi) = \coprod M_{n_i}(D_i)$$

then

$$\mathcal{M}_\pi = \coprod M_{n_i}(\mathcal{N}_i)$$

where $\mathcal{N}_i$ is a maximal order in D_i. An order $\mathcal{M}$ is maximal $\leftrightarrow \hat{\mathcal{M}}_{\mathcal{P}}$ is maximal for each finite prime $\mathcal{P}$ in the center. For an exposition of the theory of maximal orders see e.g. [R] or [S-E]. The simplest case is when D_i is a field, so a cyclotomic extension of $\mathbb{Q}$. Then $\mathcal{N}_i$ is the ring of algebraic integers in D_i. Again, this splits into two cases, [I] that when D_i is fixed under τ, and [II] when $\tau|D_i$ is complex conjugation.

We consider case II first. Let $\mathcal{P}$ be a prime of F over p in K. Then either $\tau(\mathcal{P}) = \mathcal{P} = (p)$ or $\tau(\mathcal{P})\cdot\mathcal{P} = (p)$. In the second case $\hat{F}_{(p)} = \hat{F}_{\mathcal{P}} \oplus \hat{F}_{\tau(\mathcal{P})}$ and $L_o^{\pm}(F_{(p)}) = 0$. We now consider the case $\tau(\mathcal{P}) = \mathcal{P}$, so $L_o^{+}(\hat{F},\tau) = L_o^{-}(\hat{F}_{\mathcal{P}},\tau)$ is given by Theorem 2.13.

Theorem 3.1:

Consider the sequence 1:10(B) for $(\hat{F}_{\mathcal{P}}, \hat{Z}_{\mathcal{P}}, \tau)$ then

(a) If $\mathcal{P} = (p)$ is ramified

$$L_{o,f}^{+\,tor} = 0$$

(b) If $\mathcal{P} = (p)$ is non-ramified then

$$L_{o,f}^{\varepsilon,tor} = Z/2\,.$$

Proof: If $\mathcal{P} = (p)$ then π, the uniformizing parameter represents $\hat{K}^{\bullet}_{(p)}/N(\hat{F}^{\bullet}_{\mathcal{P}}) = Z/2$. Clearly, $\langle 1 \rangle$ is in the image of $L^+_o(\Lambda)$, and just as clearly $\langle \pi \rangle$ is not in this image.

If $\mathcal{P} - (p)$ is ramified then either $\langle 1 \rangle$ generates $L^+_o(\hat{F}_{\mathcal{P}}, \tau)$ in which case we are done or $\langle 1 \rangle, \langle \alpha \rangle$ generate and we are still done.

We now consider the L^- groups. In the non-ramified case there is always a unit λ so $\tau(\lambda) = -\lambda$ and $(\cdot\lambda) : L^{\varepsilon} \leftrightarrow L^{-\varepsilon}$ gives an isomorphism of the $+$, $-$ sections of 1.10(B).

Remark 3.2: If $\mathcal{P} - (p)$ is ramified then 2 cases occur, either the uniformizing parameter π' of $\hat{F}_{\mathcal{P}}$ can be chosen so $\tau(\pi') = -\pi'$ or there is a unit μ so $\tau(\mu) = -\mu$. (The latter case occurs only if $\mathcal{P}$ lies over (2), for example $\hat{F}_{\mathcal{P}} = \hat{\mathbb{Q}}_2(i)$.) We then have

Theorem 3.3: *If $\mathcal{P} - (p)$ is ramified then in the sequence* 1.10(B)

(a) *If $\mathcal{P}$ lies over (2) then*

$$L^{-tor}_{o,f} = \begin{cases} 0, & \text{if there is a unit } \mu \text{ so } \tau(\mu) \equiv -\mu \\ Z/2 & \text{otherwise} \end{cases}$$

(b) *If $\mathcal{P}$ does not lie over 2 then*

$$L^{-,tor}_{o,f} = \begin{cases} Z/2 \oplus Z/2 & \text{if } \alpha \neq -1 \\ Z/2 & \text{if } \alpha = -1 \end{cases}$$

Proof: One checks easily in case $\tau(\pi') = -\pi'$, that the generators of

$$L^-_{o,f}(\hat{Z}_{\mathcal{P}}, \tau)$$

are given by the 2×2 matrices

3.4 $$m = \begin{pmatrix} r\pi' & \lambda \\ -\tau(\lambda) & s\pi' \end{pmatrix}$$

where λ is a unit, $r,s \in \hat{Z}_{(p)}$. Then the discriminant of m is $rs(\pi')^2 + N(\lambda)$. In case (a) any unit has this form, hence we can obtain

$$<\pi'> \perp <\pi'(N(\lambda) + r(\pi')^2)>$$

in the image and (a) follows, since the image elements all have even rank.

However, in case (b), since $\hat{Z}_{(p)}/(m)$ is not perfect, we obtain exactly the units in $N(U(\hat{Z}_{\mathscr{P}}))$, and so the image is $2<\pi'>$. 3.3 follows.

<u>Theorem</u> 3.5: <u>Let</u> F <u>be a cyclotomic extension of</u> $\mathbb{Q}$ <u>with involution</u> τ (= complex conjugation) <u>and fixed field</u> K. <u>Then</u>, <u>in the situation of</u> 1.10(A) <u>or</u> (B)

$$\partial : L_o^+(F,\tau) \longrightarrow L_o^{+\ tor}(F,\mathcal{O}_F,\tau)$$

<u>is onto where</u> $\mathcal{O}_F$ <u>is the ring of algebraic integers in</u> F.

<u>Proof</u>: We begin by ignoring the ramified primes. Then

$$L_o^{+\ tor}(F,\mathcal{O}_F,\tau) = I_{K,c}/N(I_{F,c})$$

where c is all the ramified primes, and $I_{F,c}$ is the group of fractional ideals in $\mathcal{O}_F$ prime to c, so $I_{K,c} = \coprod_{\mathscr{P}\text{ non ramified}} Z(e_{\mathscr{P}})$.

Now, let $\mathcal{O}_{K,c}$ be the set of elements in $\mathcal{O}_K$ which are congruent to 1 modulo a power of c (the conductor). Then given $\alpha\varepsilon\mathcal{O}_{K,c}$, $\partial<\alpha>$ in $L_o^{+\,tor}(F,\mathcal{O}_F,\tau)$ is easily seen to be equal to $\{\alpha\}$ in $I_{K,c}/N(I_{F,c})$.

Thus, ignoring ramified primes at which $L_o^{+\,tor} \equiv 0$ anyway, the image of ∂ is a quotient of

3.6 $$I_{K,c}/\dot{K}_c N(I_{F,c}) = Z/2$$

by the Artin reciprocity theorem (see e.g. [L , Theorem 3, p. 205]) .

<u>Lemma</u> 3.7: <u>Let</u> $F = \mathbb{C}(\rho_n)$, $K = \mathbb{C}(\rho_n + \rho_n^{-1})$, <u>and</u> τ <u>complex conjugation</u>. <u>Then there is an element</u> α <u>in</u> $\dot{K}$ <u>with</u> $\partial<\alpha>$ = <u>non-trivial class in</u> $I_{K,c}/K_c N(I_{F,c})$.

<u>Proof</u>: Recall that $c = n^+$, which means $\dot{K}_c$ is all elements of K congruent to 1(n) and positive at all infinite places. Now there is some element β in $\mathcal{O}_K$ with $N_{\mathbb{C}}^K(\beta) \equiv -1(n)$. Since $<\beta , F/K> = 1$ we must have $<\beta, F/K>_\infty = -1$, $<\beta, F/K>_{\text{finite}} = -1$, and, in particular since we can ignore infinite places the result follows.

Now we can complete the proof of 3.5. Indeed $F \subset \mathbb{C}(\rho_n)$ for some n , so $K \subset \mathbb{C}(\rho_n + \rho_n^{-1})$, then $N_K^{\mathbb{C}(\rho_n+\rho_n^{-1})}(\beta) = \alpha\varepsilon K$ satisfies $\partial<\alpha>$ represents the non-trivial class in 3.6.

In [M - H Appendix 2] a map re is defined so the diagram

3.8 $$\begin{array}{ccc} L_o^{\varepsilon}(F,\tau) & \xrightarrow{re} & L_o^{\varepsilon}(K) \\ \downarrow{\scriptstyle \partial_F} & & \downarrow{\scriptstyle \partial_K} \\ L_o^{\varepsilon,tor}(F,\mathcal{O}_F,\tau) & \xrightarrow{\overline{re}} & L_o^{\varepsilon,tor}(K,\mathcal{O}_K) \end{array}$$

commutes, and $im(\overline{re})$ is the fundamental ideal $I_K = \coprod_p I_p$ where I_p is generated by $<\alpha> - <1>$.

<u>Lemma</u> 3.9: <u>Given</u> K <u>a finite extension of</u> $\mathbb{C}$ <u>and an element</u> $\theta \in I_K$, <u>there is an element</u> $\coprod <\beta_i>$ <u>in</u> $L_o^{\varepsilon}(K)$ <u>with</u> $\partial \coprod <\beta_i> = \theta$.

<u>Proof</u>: We show it for $\beta_i \in I_{p_i}$. Let p' represent $[p]^{-1}$ in $CL(\mathcal{O}_K)$ so $p \cdot p' = (c)$. Choose $\alpha \equiv \begin{cases} \text{non square mod } p \\ \text{square mod } p' \\ 1 \text{ mod sufficiently high powers of all ramified primes} \end{cases}$

then set $F = K(\sqrt{\alpha})$. Clearly, $\partial_F(c) = <1>_{p_i}$, and

$$\overline{re}\, \partial_F(c) = \beta_i \quad . \quad 3.9 \text{ follows.}$$

<u>Corollary</u> 3.10: $L_o^{\varepsilon,tor}(K,\mathcal{O}_K)/im\partial_K \cong H_2(Z/2, CL(\mathcal{O}_K))$.

(The proof now parallels [M - H , pg 93-94].)

Now we consider the general case.

By 2.6 and weak approximation we see that coker $\partial : L_{o,f}^{\varepsilon}(\mathbb{Q}(\pi)) \rightarrow L_o^{\varepsilon,tor}$, is the image of all but a finite number of the $X_p^{\varepsilon}(Z(\pi))$. In particular

3.11 $$p : \coprod_{p \nmid |\pi|} X_p^{\varepsilon}(Z(\pi)) \longrightarrow \bar{L}_{1\ proj}(Z(\pi)) \longrightarrow 0$$

is exact. On the other hand $Z(\frac{1}{|\pi|})(\pi)$ is a maximal order, so

$$X_p^{\varepsilon}(Z(\pi)) = X_p^{\varepsilon}(\mathcal{M}(Z(\pi))) = \coprod X_p^{\varepsilon}(M_{n_i}(\mathcal{O}_{D_i})) = \coprod X_p^{\varepsilon}(\mathcal{O}_{D_i},\tau_i) .$$

This allows us to consider the various summands $M_{n_i}(D_i)$ separately.

Theorem 3.12 (Norm) theorem): Let (D_i, τ) be a type II algebra, with center F and fixed field $K \subset F$ occurring in $M_{n_i}(D_i)$ for some group ring $\mathbb{Q}(\pi)$. Then given any element $\alpha \in K$ there are two elements $s, t \in \mathrm{fix}(\tau) \subset D_i$ with

$$\tilde{N}(s)\tilde{N}(t) = \alpha$$

(The proof will be given in §9)

Remark: In 3.12 it could well be that one of s, t is 1.

Corollary 3.13: Each type II algebra in $\mathbb{Q}(\pi)$ contributes at most a single $Z/2$ to $L_{1,proj}(Z(\pi))$.

Proof: Let c be divisible by all primes dividing $|\pi|$, then $I_c(\mathcal{O}_K)/N(\dot{F}_c)NI_c(\mathcal{O}_F) = Z/2$.

In particular from 3.12 given $n \in N(\dot{F}_c)$ there are s, t with $\tilde{N}(s)\tilde{N}(t) = n$. Consequently, away from primes dividing $|\pi|$

$$\partial(\langle s \rangle \perp \langle t \rangle) = \{n\} \in I_c(\mathcal{O}_K)/N(I_c(\mathcal{O}_F))$$

At $p \,\big|\, |\pi|$, $\tilde{N}(s)\tilde{N}(t) \equiv 1$, so from Theorem 2.13

$$(\langle s \rangle \perp \langle t \rangle)_p = \begin{cases} 0 & \text{in case (a) or (b)}, \ \alpha \not\sim -1 \\ 2\langle 1 \rangle_p & \text{in case (b)} \ \ \alpha \sim -1 \end{cases}.$$

Thus $(\langle s \rangle \perp \langle t \rangle - 2\langle 1 \rangle)_p \equiv 0$ for $p \nmid |\pi|$, and $\partial(\langle s \rangle \perp \langle t \rangle - 2\langle 1 \rangle) = \{n\}$. 3.13 follows.

Remark 3.14: As in the proof of 3.7, it is often possible to show that the contribution of D_i is actually 0 . For example, this happens when there is a unit ε in $\mathcal{O}_K$ with $N^K_{\mathbb{C}}(\varepsilon) = -1$.

Remark 3.15: In the type I case either D_i is a Quaternion algebra or a real subfield of a cyclotomic field. In the latter case, it is convenient to reverse the viewpoint taken for the type II algebras above, and instead measure the deviation between the image of the $\coprod_{p||\pi|} X_p(K,\mathcal{O}_K)$ and the entire image. Thus, let $\bar{\partial}$ be the composite

$$L^+_o(K,O_K) \xrightarrow{\partial} \coprod X_p \xrightarrow{p} \coprod_{p \nmid |\pi|} X_p \, .$$

Then we have, as in 3.10

Theorem 3.16: coker $\bar{\partial} \cong H_2(Z/2, CL(\mathcal{O}_K))$.

It remains to discuss the case when D_i is a type I Quaternion algebra. We assume the involution τ is the usual one $i \leftrightarrow -i$, $j \leftrightarrow -j$, $ij \leftrightarrow -ij$. Then we have

Theorem 3.17: (a) $L^{!}_o(D_i,\tau)$ has generators the elements of $\dot{K}/\dot{K}^{+}$ where $\dot{K}^+$ is the set of elements positive at all infinite places of K , if D_i is a Quaternion algebra at ∞ , otherwise $L^+_o(D_i,\tau) = Z/2$.

(b) At a prime $\mathcal{P}$ of K at which $D_i \otimes_K \hat{K}_{\mathcal{P}}$ is not a Quaternion algebra we have

$$L^+_o(D_i \otimes_K \hat{K}_{\mathcal{P}}, \mathcal{O}_{D_i} \otimes_{\mathcal{O}_K} \mathcal{O}_{\mathcal{P}}, \tau) = 0 \, .$$

<u>Proof</u>: If D_i is a quaternion algebra at any infinite prime it is such as at all ∞ places by the Benard-Schacher theorem [B-S2]. Then (a) follows from the Hasse-Schilling norm theorem [S-E].

To show (b), note that the fixed set of τ is one dimensional. Hence, since $D_i \otimes_K \hat{K}_{\mathcal{P}} = M_2(\hat{K}_{\mathcal{P}})$ by assumption, we must have that $\tau \approx$ the "symplectic" involution:

$$\tau\begin{pmatrix} a & b \\ c & d \end{pmatrix} = \begin{pmatrix} d & -b \\ -c & a \end{pmatrix}$$

But $L^+(M_2(F), \tau_{sp}) = L^-(M_2(F), \tau_{ord}) = 0$ where τ_{ord} is the usual transpose and (b) follows.

To study the situation $L_o^-(D_i,\tau) \xrightarrow{\partial} \coprod (X_p)$ we content ourselves with the observation that Theorem 2.17 (b) should be modified for global fields only by pointing out that

3.18 $L_o^-(D,\tau)$ <u>is a</u> Z/2 <u>vector space generated by the elements of</u> $K^{\cdot +}/(K^{\cdot +})^2 - (1)$ <u>with the same relation as occurs in</u> 2.17 (b).

§4. An application: $L_1^{-,q}(Z(\pi))$ for π a finite 2-group.

The representation ring for a finite 2-group. Recall that any irreducible representation of a finite p-group is monomial. This says, given any irreducible representation f on π there is a subgroup $H \subset \pi$, a projection $p : H \to E$ where E is an extension of $Z/_{p^r}$ by $Z/_{p^\ell} \subset \mathrm{Aut}(Z/_{p^r})$, the irreducible representation

$$r : E \longrightarrow \mathbb{Q}(\rho_{p^n}) \otimes_{Z(Z/p^n} Z(E)$$

and $f = \mathrm{tr}^H{}_\pi(r \circ p)$.

Theorem 4.1: If π is a 2-group then E above can be assumed to be

(i) $Z/_{2^i} \times_T Z/2$; $t^2 = g^{2^{i-1}}$, $t^{-1}gt = g^{-1}$

(ii) $Z/_{2^i} \times_T Z/2$; $t^2 = 1$, $t^{-1}gt = g^{-1}$

(iii) $Z/_{2^i} \times_T Z/2$; $t^2 = 1$, $t^{-1}gt = g^{2^{i-1}+1}$

(iv) $Z/_{2^i}$

(This follows from the calculations in [M.1] on $H^*(Z/2^r, Z/_{2^i})$, and the observation that there are exactly 3 distinct subfields between

$$\mathbb{Q}(\rho_{2^{i-1}} + \rho_{2^{i-1}}^{-1}) \text{ ——— } \mathbb{Q}(\rho_{2^i})$$

which are

4.2 $$\mathbb{Q}(\rho_{2^i} + \rho_{2^i}^{-1}),\ \mathbb{Q}(\rho_{2^i} - \rho_{2^i}^{-1}),\ \mathbb{Q}(\rho_{2^{i-1}}).$$

If the center of (f) is of the last type then $Z/_{2^i} = E$. If the center is of the second type then $E \cong 4.1(iii)$. If the center is of the first type then 4.1(i) or 4.1(ii) occurs.)

More exactly, we have

Theorem 4.3: The simple algebra

$$\mathbb{Q}(\rho_{2^i}) \otimes_{Z(Z/_{2^i})} Z(E) \text{ in } 4.1 \text{ is}$$

(i) D for 4.1 (i) where D is the Quaternion algebra with center $\mathbb{Q}(\rho_{2^i} + \rho_{2^i}^{-1})$, invariants $\frac{1}{2}$ at all infinite primes and

(a) 0 at all finite primes $i > 2$

(b) $\frac{1}{2}$ at 2, 0 otherwise $i = 2$

(ii) $M_2(\mathbb{Q}(\rho_{2^i} + \rho_{2^i}^{-1}))$ for 4.1 (ii)

(iii) $M_2(\mathbb{Q}(\rho_{2^i} - \rho_{2^i}^{-1}))$ for 4.1(iii)

(iv) $\mathbb{Q}(\rho_{2^i})$ for 4.1 (iv)

(For the definitions of the local invariants of a division algebra see e.g. [Se]. What (i) means is D is the ordinary Quaternion algebra at ∞ primes and $M_2(\hat{F}_p)$ in case (a) for $\mathscr{P}$ over (2), while in case (b) it is $\hat{\mathbb{Q}}_2(i,j)$ the non-trivial Quaternion algebra at 2 and $M_2(\hat{\mathbb{Q}}_p)$ otherwise. The result follows directly from [M.1].)

The units in the subfields of $\mathbb{Q}(\rho_{2^i})$: Before we can proceed with the calculations we need some information on units. Recall, from [H] that $U(\mathbb{Q}(\rho_{2^i})) = U(\mathbb{Q}(\rho_{2^i} + \rho_{2^i}^{-1})) \times_{Z/2} (Z/_{2^i})$

Now, let $\lambda_i = \rho_{2^i} + \rho_{2^i}^{-1}$, then inductively we have

4.4 $$\lambda_i = \sqrt{2 + \lambda_{i-1}}$$

(Indeed $\lambda_i^2 = (\rho_{2^i} + \rho_{2^i}^{-1})^2 = 2 + \lambda_{i-1}$.) Thus

<u>Lemma</u> 4.5: $\varepsilon_i = 1 + \lambda_i$ <u>is a unit and</u> $N_{\mathbb{Q}}^{D(\lambda_i)}(\varepsilon_i) = -1$.

<u>Proof:</u> Let $g \in \mathrm{Gal}\,(\mathbb{Q}(\rho_{2^i}) - \mathbb{Q})$ send $\rho_{2^i} \longrightarrow \rho_{2^i}^{2^{i-1}+1} = -\rho_{2^i}$. Then $\mathrm{fix}(g) \cap \mathbb{Q}(\lambda_i) = \mathbb{Q}(\lambda_{i-1})$ and $g(\lambda_i) = -\lambda_i$. Hence

$$N_{\mathbb{Q}(\lambda_{i-1})}^{\mathbb{Q}(\lambda_i)}(\varepsilon_i) = (1 + \lambda_i)(1 - \lambda_i) = 1 - (2 + \lambda_{i-1}) = -\varepsilon_{i-1} .$$

Now proceed by induction, using

$$N_{\mathbb{Q}}^{\mathbb{Q}(\lambda_i)}(-x) = N_{\mathbb{Q}}^{\mathbb{Q}(\lambda_{i-1})} N_{\mathbb{Q}(\lambda_{i-1})}^{\mathbb{Q}(\lambda_i)}(x) .$$

<u>Theorem</u> 4.6 (Weber): <u>Let h generate $\mathrm{Gal}(\mathbb{Q}(\lambda_i) - \mathbb{Q})$ then the units</u> $-1, \varepsilon_i, h(\varepsilon_i)\, h^2(\varepsilon_i) \cdots h^{2^{i-2}-1}(\varepsilon_i)$ <u>are linearly independent and the</u> <u>subgroup spanned by them has odd index in</u> $U(\mathbb{Q}(\lambda_i))$.

(See e.g., [H].)

We also need the following fairly standard result which is in any case a key step in the proof of 4.6.

<u>Lemma</u> 4.7: <u>In</u> $Z_2(\lambda_i)$ <u>the group of units</u>

$$(\hat{Z}_2^+)^{2^{i-2}} \times Z/2$$

<u>has generators</u>

$$<-1>, <5>, <\varepsilon_i>, <h(\varepsilon_i)> \cdots, <h^{2^{i-2}-1}(\varepsilon_i)>.$$

Finally, we quote another well known theorem of Weber

Theorem 4.8: $CL(Z(\lambda_i))$, $CL(Z(\rho_{2^i} - \rho_{2^i}^{-1}))$, $CL(Z(\rho_{2^i}))$ are all of odd order.

We are now ready to apply the results of §3. We begin with the sequence 1.10(C), note that $L_{1,f}^{-1,q}(\mathbb{Q}(\pi)) = 0$ for π a finite group, and calculate

Theorem 4.9: Let π be a finite two group, write

$$\mathbb{Q}(\pi) = \coprod_k tr_\pi^{H_k}(rop)[\mathbb{Q}(\rho_{2^j}) \otimes_{Z(Z/2^j)} Z(E_k)]$$

and let $n(\pi)$ be the number of E_k occurring above of type 4.1(ii), then

$$L_1^{-1,q}(Z(\pi)) = (Z/2)^{n(\pi)-1}$$

For example, if $D_i = Z/2^i \times_T Z/2$ is the group of 4.1(ii), then $n(D_i) = i + 3$ and

$$L_1^{-1,q}(Z(D_i)) = (Z/2)^{i+2}.$$

If $Q_i = Z/2^i \times_T Z/2$ is the group of 4.1(i) then

$$n(Q_i) = i + 2,$$

and

$$L_1^{-1,q}(Z(Q_i)) = (Z/2)^{i+1}.$$

If V_i is the group of 4.1(iii) then $n(V_i) = 4$ and

$$L_1^{-1,q}(Z(V_i)) = (Z/2)^3$$

Finally, for $Z/_2 i$, $n(Z/_2 i) = 2$ and

$$L_1^{-1,q}(Z(Z/_2 i)) = Z/2 .$$

Proof: From the fact that all the class numbers are odd and the fact that there is a unit ε_i in $(\mathbb{C}(\lambda_i))$ with $N(\varepsilon_i) = -1$ it follows that no type II algebra [4.1(iii) or 4.1(iv)] contributes anything to the above group, and that the type I 4.1(ii) algebra has image coming entirely from $X_2(\pi)$.

Now, we check also for the type 4.1(i) algebras. Note first that the involution on each of these is the usual involution, and transferring up does not change this. Hence, we are dealing with $L_o^+(D,\tau)$. In all cases of primes away from 2 this is $M_2(\hat{F}_p(\lambda_i))$ with symplectic involution, hence away from 2 these contribute nothing. Again, at 2, only in case 4.1(i) (b) are we dealing with an example in which $L_o^{|}(D \otimes \hat{\mathbb{C}}_2, \tau) \neq 0$. But clearly here this group = Z/2 and the generator is the image of $< 1 > \in L_o^+(D,\tau)$, so no type 4.1(i) algebra contributes anything.

It remains to look more closely at the type 4.1(ii) algebras. Here we once more obtain

$$\hat{\mathbb{C}}_2 \otimes_{Z(Z/_2 i)} Z(E) = M_2(\hat{\mathbb{C}}_2(\lambda_i))$$

with representation

$$g \longrightarrow \begin{pmatrix} 0 & 1 \\ -1 & \lambda_i \end{pmatrix}, \; t \longrightarrow \begin{pmatrix} 0 & 1 \\ 1 & 0 \end{pmatrix}$$

and the involution is given by $\tau(x) = s(x^t)s^{-1}$ where $s^t = s$. Under these circumstances we will see that

$$L_o^+(M_2(\hat{\mathbb{C}}_2(\lambda)),\tau) = L_o^+(\hat{\mathbb{C}}_2(\lambda_i))$$

which is generated by

$$<1>, <-1>, <5>, <\varepsilon_i> \cdots <h^r\varepsilon_i> \cdots$$

$$<\lambda_i> <-\lambda_i>, <5\lambda_i> \cdots \text{ etc.}$$

Now, all of these but $<5u>$, $<\varepsilon 5\lambda_i>$ u, ε units, are in the image of elements in $L_o^+(\mathbb{C}(\lambda_i))$ which are either units or have norm 2 . In either case they cancel out. Now away from 2 all these remaining elements only have ∂_p parts at 5 .

<u>Lemma</u> 4.10: (5) <u>is prime in</u> $Z(\lambda_i)$

<u>Proof</u>: $5^{2^{i-2}} \equiv 1(2^i)$, but $5^{2^{i-3}} \not\equiv 1(2^i)$. Hence (5) splits in $Z(\rho_{2^i})$ into 2 primes interchanged by complex conjugation. 4.10 follows.

Now, using the technique of 3.9 we see that $I \subset X_5$ is in the image of ∂ exactly. Hence, all the elements $<\varepsilon 5\lambda_i>$, $<5u>$ all have the same image in $L_1^-(Z(\pi))$, and each type 4.1(ii) representation contributes at most a $Z/2$ to $L_1^-(Z(\pi))$, $L_1^{-,q}(Z(\pi))$.

In order to verify the remainder of 4.9 we must first show a result analogous to Theorem 2.6 for $X_2^q(Z(\pi))$.

<u>Lemma</u> 4.11: $L_1^{-,q}(\hat{Z}_2(\pi)) = L_1^-(\hat{Z}_2(\pi)) = 0$ <u>for</u> π <u>a finite 2-group</u>.

We defer the proof for the moment. Note that 4.11 implies

$$\partial : L_{o,f}^+(\hat{Q}_2(\pi)) \longrightarrow L_{o,f}^{+,tor,q}(\hat{Z}_2(\pi)) \quad \text{is onto.}$$

We now study its kernel. This breaks up into 2 parts, when we are looking at L_o^+ or $L_o^{+,q}$. In the former case the group $L_o^+(\hat{Z}_2(\pi))$ is entirely calculated in [C] and, consequently we defer the complete determination of

$$L_1^-(\hat{Z}_2(\pi))$$

to [C] . However, in the quadratic case we have

<u>Lemma</u> 4.12: $L_o^{+,q}(\hat{Z}_2(\pi)) = Z/2$ <u>with generator</u>

$$\begin{pmatrix} -2 & 1 \\ 1 & 2 \end{pmatrix} .$$

<u>Remark</u> 4.13: We thank W. Pardon for correspondence and comments which led to 4.12.

The image of $\begin{pmatrix} -2 & 1 \\ 1 & 2 \end{pmatrix}$ corresponds to the diagonal $\langle -5 \rangle \perp \langle 1 \rangle$ at all 1 dimensional representations, and 4.9 now follows.

<u>Remark</u> 4.14: $L_1^-(Z(\pi))$ is a quotient of $L_1^{-,q}(Z(\pi))$ so that 4.9 also gives us representative generators for this group.

Remark 4.15: The actual surgery obstruction group $L^h_3(Z(\pi)) = L^{-,q}_{1,f}(Z(\pi))$ is connected with $L^{-,q}_1(Z(\pi))$ by the exact sequence

4.16:
$$\cdots \longrightarrow H_{odd}(Z/2, \tilde{K}_o(Z(\pi))) \xrightarrow{\partial} L^{-,q}_{1,f}(Z(\pi)) \longrightarrow L^{-,q}_1(Z(\pi)) \xrightarrow{j} H_{ev}(Z/2, \tilde{K}_o(Z(\pi))) \longrightarrow \cdots$$

and in certain cases this gives good control of $L^{-,q}_{1,f}(Z(\pi))$. For example, from [F-K-W], we have

$$\tilde{K}_o(Z(\pi))_2 = 0$$

if π is a group of type 4.1(ii) so in this case

$$L^h_3(Z(\pi)) = (Z/2)^{i+2} .$$

Again, if π is a group of type 4.1(i) then $\tilde{K}_o(Z(\pi))_2 = Z/2$. In this case it may be verified that j is onto and so either

$$L^h_3(Z(\pi)) = (Z/2)^i$$

or there is an extension

$$Z/2 \longrightarrow L^h_3(Z(\pi)) \longrightarrow (Z/2)^i \longrightarrow 0 .$$

Remark 4.17: More general techniques for calculating $\tilde{K}_o(Z(\pi))$ are given in [M,2]. Applied to finite 2-groups using 4.1, 7.3, 4.6, and

4.8 we can get good bounds on the 2-torsion, by bounding the contributions from each irreducible subalgebra separately.

See the appendix for more details on the structure of $K_o(Z(\pi))$.

Using the results of the appendix it is direct to calculate a generating set for $H_1(Z/2, \tilde{K}_o(Z(\pi)))$. These are represented by PL1 torsion modules which, when direct summed with their duals are FPL1. Such a module defines a projectively trivial form, the hyperbolic form on

$$M = A \oplus A^*$$

but since $A \oplus A^*$ is FPL1, M represents an element in $L^{*,tor}_{o,f}(\mathbb{Q}(\pi), Z(\pi))$ which is in the image of ∂ in 4.16. These M's, together with the elements j in the kernel of j (see A in the introduction) now span $L^h_3(Z(\pi))$.

Moreover, again using the description of $\tilde{K}_o(Z(\pi))_2$ given in the appendix, the image of j is obvious on the generators in Theorem 4.9.

The proof of Lemma 4.11:

Lemma 4.18: Let Λ satisfy the following. Let $W = W_1 \oplus W_2$ be any f.g. free Λ-module and e a basis vector for W. If

$$p_i(e_i) \in W_i$$

is the projection, then one of the $p_i(e_i)$ is a basis vector for W_i. In this case

$$L^{\varepsilon}_{1,f}(\Lambda) = 0 , \quad L^{\varepsilon,q}_{1,f}(\Lambda) = 0 .$$

Proof: Let $(H : K_1, K_2)$ represent an element of $L^{\varepsilon}_{1,f}(\Lambda)$. Write H as $K_1 \oplus K_1^{\#}$ and choose a Λ-basis for $K_2 : f_{1,\ldots,} f_n$. Then $f_1 = (p_1(f_1), p_2(f_1))$ and one of these is a basis element. Say it is $p_1(f_1)$. [If not $(H,K_1,K_2) \sim (H,K_1^{\#},K_2)$ so replace K_1 by $K_1^{\#}$.] Then we have K_1 spanned by $p_1(f_1) = e_1, e_2,\ldots,e_n$, and $K_1^{\#}$ is spanned by $e_1^{\#}, e_2^{\#},\ldots e_n^{\#}$, the dual basis.

Write $p_2(f_1) = \beta_1 e_1^{\#} + \sum_{j\geq 2} \beta_j e_j^{\#}$.

Since K_2 is a kernel

$$0 = \langle f_1, f_1 \rangle = \beta_1 + \varepsilon\tau(\beta_1)$$

in the $L^{\varepsilon}_{1,f}$ case, and in the $L^{\varepsilon,q}_{1,f}$ case, also

$$0 = u(f_1) = [\beta_1] \in \Lambda/\{\lambda - \varepsilon\tau\lambda \mid \lambda\varepsilon\Lambda\}.$$

Hence $\beta_1 = -\varepsilon\tau(\beta_1)$ and in the $L^{\varepsilon,q}_{1,f}$ case $\beta_1 = \lambda - \varepsilon\tau(\lambda)$. In either case set

$$M = \begin{pmatrix} \beta_1 & -\tau(\beta_2) & -\tau(\beta_3) \ldots, & -\tau(\beta_n) \\ \beta_2 & 0 & 0 & 0 \\ \cdot & \cdot & \cdot & \cdot \\ \beta_n & 0 & 0 & 0 \end{pmatrix}$$

then $(H,K_{1,M},K_2) \sim (H,K_1,K_2)$ where $K_{1,M}$ is the graph formation, and a basis for $K_{1,M}$ is

$$(f_1, e_2', e_3', \ldots e_n').$$

Now choose

$$K^{\#}_{1,M} = K^{\#}_{1}\,.$$

with new dual basis $f^{\#}_1, e'^{\#}_1, \ldots, e'^{\#}_n$.

We split H as $< f_1, f^{\#}_1 > \oplus < f_1, f^{\#}_1 >^{\perp}$ and clearly

4.19) $(H, K_{1,M}, K_2) \sim (< f_1, f^{\#}_1 >, < f_1 >, < f_1 >) \perp (< f_1, f^{\#}_1 >^{\perp}_{;} K'_{1,M}, K'_2)$.

The first summand in 4.19 is trivial, the second summand has smaller rank and we can iterate the argument to finish the proof of 4.18.

Now 4.11 follows since for $\hat{Z}_2(\pi)$, π a 2 group, if J is the Jacobson radical then $\hat{Z}_2(\pi)/J = Z/2$ and an element of $\hat{Z}_2(\pi)$ is a unit iff $\alpha = 1 + \ell$ with $\ell \varepsilon J$. Hence the condition of 4.18 is satisfied for $\Lambda = \hat{Z}_2(\pi)$.

The proof of 4.12:

We first observe that every generator of $L^{+,q}_{o,f}(\hat{Z}_2(\pi))$ has the form

4.20 $$\begin{pmatrix} \alpha & 1 \\ 1 & \beta \end{pmatrix}$$

with $\alpha = \xi + \tau(\xi)$, $\beta = \eta + \tau(\eta)$. Indeed, choose a basis for (V,A) $e_1 \ldots e_n$. Then by non-singularity there is an $e^{\#}_1$ so $<e_1, e^{\#}_1> = 1$ and, since $< e_i, e_i >$ is even it belongs to J so the matrix of the form on the subspace $< e_1, e^{\#}_1 >$ is of the form 4.20 and is non-singular. Hence V splits as $<e_1, e^{\#}_1> \perp (<e_1, e^{\#}_1>^{\perp})$.

Suppose $\xi \varepsilon J = \ker(\hat{Z}_2(\pi) \longrightarrow \hat{Z}_2(\pi)/J = Z/2)$. Conjugate by $\begin{pmatrix} 1 & -\xi \\ 0 & 1 \end{pmatrix}$ to obtain

$$\begin{pmatrix} \alpha & 1 \\ 1 & \beta \end{pmatrix} \sim \begin{pmatrix} \xi\eta\tau(\xi) + \xi\tau\eta\tau(\xi) , & 1 \\ 1 & \beta' \end{pmatrix}$$

but the upper left hand term now has the form $\xi' + \tau(\xi')$ with $\xi' \in J^2$. Proceeding inductively $\xi' \in J^{2^r}$ for any r and $\begin{pmatrix} \alpha & 1 \\ 1 & \beta \end{pmatrix}$ is split.

Lemma 4.21: Suppose $\alpha \neq \xi + \tau(\xi)$ with $\xi \varepsilon J$, then

$$\begin{pmatrix} \alpha & 1 \\ 1 & \beta \end{pmatrix} \perp \begin{pmatrix} 2 & 1 \\ 1 & \beta \end{pmatrix} \quad \text{is split.}$$

Proof: Give basis vectors e_1, f_1, e_2, f_2 , with $\langle e_1, f_1 \rangle = 1$, $\langle e_1, e_1 \rangle = \alpha$ $\langle e_2, e_2 \rangle = 2$ etc. Then

$$L = \langle e_1 + e_2, f_1 \rangle$$

is a direct summand on which the pairing is given by

$$\begin{pmatrix} \alpha + 2 & 1 \\ 1 & \beta \end{pmatrix}$$

which is split since $\alpha + 2 = (\xi + 1) + \tau(\xi + 1)$ and $\xi + 1 \varepsilon J$. A basis for $L^{\perp}$ is given by $e_3 = (e_2 - 2f_2)$, $(-\beta e_1 + f_1 + e_2 + (\beta\alpha - 3)f_2) = f_3$.

But now

$$\langle f_3,f_3\rangle = \beta\alpha\beta - 3\beta - 4 + \beta\alpha + \alpha\beta + (\beta\alpha-3) \quad \beta \quad (\beta\alpha-3) ,$$

and $\beta\alpha\beta = \beta\xi\beta + \beta\tau(\xi)\beta$, with $\beta\varepsilon J$, $-4 = -2 + (-2)$, $2\varepsilon J$,

$\beta\alpha + \alpha\beta = \beta\alpha + \tau(\beta\alpha)$ with $\beta\alpha\varepsilon J$, finally $-3\beta + (\beta\alpha - 3)\beta(\beta\alpha - 3)$ $= (-3\eta + (\beta\alpha - 3)\eta(\beta\alpha - 3)) + \tau(-\eta + (\beta\alpha -3)\eta(\beta\alpha -3))$ and $-3\eta + (\beta\alpha-3)\eta(\beta\alpha -3)\varepsilon J$

Hence $\langle f_3,f_3\rangle = \gamma + \tau(\gamma)$ with $\gamma\varepsilon J$ so $L^{\perp}$ is split, and the lemma follows.

Thus $L^{+,q}_{1,f}(\hat{Z}_2(\pi))$ has generator $\begin{pmatrix} 2 & 1 \\ 1 & 2 \end{pmatrix}$ and is either $Z/2$ or zero. However, projecting $\hat{Z}_2(\pi)$ onto $\hat{Z}_2$ shows that it is $Z/2$. 4.12 follows.

§5. Involutions on Matrix Rings

We collect the necessary basic facts about involutions on the $n \times n$ matrix ring $M_n(D)$, where D is a division algebra. The results are proved in [A] and [J]. Let $C_D \subseteq D$ denote the center of D.

Definition: An involution τ on $M_n(D)$ is an antiautomorphism of $M_n(D)$ so that $\tau^2 = id$. We say that a matrix M is ε-τ-symmetric if $\tau(M) = \varepsilon(M)$, $\varepsilon = \pm 1$.

Theorem 5.1: $M_n(D)$ admits an involution if and only if D does.

Let τ denote an involution on $M_n(D)$. C_D is included in $M_n(D)$ as the diagonal matrices, and τ preserves C_D, hence acts as an automorphism of order 1 or 2 on C_D. Denote the fixed field of $\tau|C_D$ by F_τ. Given a non-singular ε-τ-symmetric matrix A, we define an autiautormorphism τ_A of $M_n(D)$ by $\tau_A(M) = A\tau(M)A^{-1}$. τ_A is in fact an involution, for

$$\tau_A^2(M) = \tau_A(A\tau(M)A^{-1}) = A\tau(A)^{-1}\tau^2(M)\tau(A)A^{-1} = \varepsilon^2 M = M .$$

Theorem 5.2: Given any involution of $M_n(D)$ with $F_\sigma = F_\tau$, there is an ε-symmetric matrix A so that $\sigma = \tau_A$.

If $\alpha = \tau_A$, and A is +1 or -1 -τ-symmetric, we say that σ is of positive or negative type respectively.

We say that two involutions τ and σ of $M_n(D)$ are equivalent if there is an automorphism α of $M_n(D)$ so that $\tau \circ \alpha = \alpha \circ \sigma$. (This means that $(M_n(D), \tau)$ and $(M_n(D), \sigma)$ are isomorphic as rings with involution.)

Fixing a particular involution τ of $M_n(D)$, we note

Lemma 5.3: If $A' = BA\tau(B)$, where B is a non-singular matrix over D , then τ_A is equivalent to $\tau_{A'}$.

Proof: Define $\alpha(M) = B^{-1}MB$. Then we have

$$\tau_A \circ \alpha(M) = A\tau(B^{-1}MB)A^{-1} = A\tau(B)\tau(M)\tau(B^{-1})A^{-1} =$$

$$B^{-1}BA\tau(B)\tau(M)\tau(B^{-1})A^{-1}B^{-1}B = B^{-1}A'\tau(M)A'^{-1}B = \alpha \circ \tau_A(M) .$$

Remark: Thus, the classification of involutions τ on $M_n(D)$ with F_τ equal to a given subfield of index 1 or 2 in C_D up to equivalence is equivalent to the classification of non-singular ε-σ-symmetric matrices A under the equivalence relation $\underset{\sigma}{\sim}$, where $A \underset{\sigma}{\sim} A' \iff A = BA'\sigma(B)$ for some non-singular matrix B , and σ is a particular involution with $F_\sigma = F$. Letting σ be the conjugate transpose involution on $M_n(D)$ associated to an involution $\tilde{\sigma}$ of D so that $F_{\tilde{\sigma}} = F$, we find that the classification of involutions τ with $F_\tau = F$ is equivalent to the classification of ε-symmetric Hermitian forms of rank n with respect to $\tilde{\sigma}$.

Lemma 5.4: If $\tilde{\sigma}$ is a non-trivial involution of D , any ε-$\tilde{\sigma}$-symmetric Hermitian form may be diagonalized.

If D is a field, char$(D) \neq 2$, and $\tilde{\sigma}$ is trivial, any $+\tilde{\sigma}$-symmetric form may be diagonalized, and any $-\tilde{\sigma}$-symmetric form may be put in the form

$$\begin{pmatrix} 0 & I \\ -I & 0 \end{pmatrix}$$

If D is a field of characteristic 2, and $\tilde{\sigma}$ is trivial, we may put any $\pm\tilde{\sigma}$-symmetric form in the form

$$\begin{pmatrix} 0 & I & 0 \\ I & 0 & 0 \\ 0 & 0 & D \end{pmatrix},$$

where D is a diagonal matrix.

Combining the remark and lemma 5.4, we have

Theorem 5.5: Let τ be an involution on $M_n(D)$ with $F_\tau = F_\sigma$. Then

(i) If $\tilde{\sigma}$ is nontrivial, τ is equivalent to σ_A, where A is diagonal and ε-σ-symmetric.

(ii) If D is a field, char(D) $\neq 2$, and τ is of $+$ type, τ is equivalent to σ_A, where A is diagonal and ε-σ-symmetric.

(iii) If D is a field, char(D) $\neq 2$, $\tilde{\sigma}$ is trivial, and τ is of $-$ type, then τ is equivalent to σ_A, where $A = \begin{pmatrix} 0 & I \\ -I & 0 \end{pmatrix}$.

(iv) If D is a field of characteristic 2, and $\tilde{\sigma}$ is trivial, τ is equivalent to σ_A, where

$$A = \begin{pmatrix} 0 & I & 0 \\ I & 0 & 0 \\ 0 & 0 & D \end{pmatrix}, \quad D \text{ a diagonal matrix.}$$

We make some remarks about the L-groups associated to these involutions. We consider $M_n(D)$, endowed with an involution τ, D is a division algebra. As before, the notation $A \underset{\tau}{\sim} A'$ will mean that $A = BA'\tau(B)$ for some non-singular matrix B.

Lemma 5.6: Let $A = \delta\tau(A)$, $\delta = \pm 1$. Then M is ε-τ_A-symmetric if and only if MA is $\varepsilon\delta$-τ-symmetric.

Proof: $M = \varepsilon\tau_A M \Leftrightarrow M = \varepsilon A\tau(M)A^{-1} \Leftrightarrow MA = \varepsilon A\tau(M) = \varepsilon\delta\tau(MA)$.

Lemma 5.7: $M \underset{\tau A}{\sim} M'$ if and only if $MA \underset{\tau}{\sim} M'A$.

Proof: $M \underset{\tau A}{\sim} M' \Leftrightarrow M = BM'\tau_A(B) = BM'A\tau(B)A^{-1} \Leftrightarrow MA = BM'A\tau(B) \Leftrightarrow MA \underset{\tau}{\sim} M'A$.

Putting these two lemmas together, we obtain

Lemma 5.8: Let R_τ denote a ring $M_n(D)$ with involution τ, and let σ denote τ_A, for some ε-τ-symmetric matrix A. Then

$$L^{\varepsilon}_{o,f}(R_\sigma) \cong L^{\varepsilon}_{o,f}(R_\tau) \quad \text{if} \quad A = \tau(A)$$

$$L^{\varepsilon}_{o,f}(R_\sigma) \cong L^{-\varepsilon}_{o,f}(R_\tau) \quad \text{if} \quad A = -\tau(A)$$

Proof: Lemmas 5.6 and 5.7 set up bijective correspondences between ε-symmetric Hermitian form spaces over R_σ and R_τ if $A = \tau(A)$ and between ε-symmetric Hermitian form spaces over R_σ and $-\varepsilon$-symmetric Hermitian form spaces over R_τ if $A = -\tau(A)$, which give the isomorphisms.

§6. Involutions on Quaternion Algebras

Let $k\langle a,b\rangle$ denote the algebra over a field k with generators i and j and relations $ij = -ji$, $i^2 = a$, $j^2 = b$. The standard involution on $k\langle a,b\rangle$ is given by

$$\tau = \text{id} \quad \text{on } k$$

$$\tau(i) = -i \ , \ \tau(j) = -j \ .$$

A short calculation shows $x\tau(x) \in k$ for every $x \in k\langle a,b\rangle$ and the reduced norm $N(x)$ is defined to be $x\tau(x)$. Note that $N(xy) = N(x)N(y)$, since $xy\tau(xy) = xy\tau(y)\tau(x) = x\tau(x)y\tau(y)$, $y\tau(y)$ being central. $N(x)$ is non-zero if and only if x is invertible, and then $x^{-1} = \frac{1}{N(x)} \cdot x$.

Also, note that $x + \tau(x) \in k$. $x + \tau(x)$ is the trace of x , denoted $tr(x)$. Clearly, x satisfies the polynomial equation $f_x(y) = y^2 - tr(x)y + N(x) = 0$ which generates a degree 1 or 2 extension field of k . If $f_x(y)$ is reducible over k and $x \notin k$ then $v = x - \frac{1}{2}tr(x)$ satisfies $f_v(y) = y^2 - N(v) = 0$, and $f_v(y)$ is also reducible over k so $N(v) = w^2$ for some $w \in k$. Thus, $(v-w)(v+w) = 0$ and $k\langle a,b\rangle$ is not a division algebra. (It is the matrix algebra $M_2(k)$).

Thus we have

Lemma 6.1: *Let $k\langle a,b\rangle$ be a division algebra D , and $x \notin k$ belonging to D , then the characteristic polynomial $f_x(y)$ of x is irreducible and $k(x)$ is a degree 2 extension of k .* $k(x) \cong k(\sqrt{N(x-\frac{1}{2}tr(x))})$.

Note also that $\tau(x - \frac{1}{2}tr(x)) = -(x - \frac{1}{2}tr(x))$. We now make some observations on $N(x)$ regarded as a quadratic form on $k\langle a,b\rangle$. Clearly, if

$x = \alpha + \beta i + \gamma j + \delta ij$ then

$$N(x) = \alpha^2 + \beta^2 a + \gamma^2 b + \delta^2 ab$$

so as a quadratic form space

$$k < a,b > = < 1 > \perp < a > \perp < b > \perp < ab >$$

and $k < a,b >$ is a division algebra if and only if $< 1 > \perp < a > \perp < b > \perp < ab >$ has no isotropy vectors.

Now assume that $k < a,b >$ is a division algebra, then the quadratic form $N|k(x)$ is non-singular and $k(x)^{\perp}$ is two-dimensional.

Since $N(v + w) = N(v) + N(w) + v\tau(w) + w\tau(v)$ we see that the associated bilinear form $< v,w >$ is $\frac{1}{2}(v\tau(w) + w\tau(v))$. Thus, if $w \in (k(x))^{\perp}$ then

(1) $\tau(w) = -w$

(2) if $v = x - \frac{1}{2}tr(x)$, then $wv = -vw$, and we have

Lemma 6.2: <u>Let</u> $\alpha = N(x - \frac{1}{2}tr(x))$, $\beta = N(w)$ <u>for any</u> $w \in k(x)^{\perp}$, <u>then</u>

$$k < a,b > \cong k < \alpha,\beta >$$

We say an involution on $k < a,b >$ is of type I if it leaves the center k fixed.

<u>Lemma</u> 6.3 : <u>Let</u> τ' <u>be a type</u> I <u>involution on</u> $k < a,b >$, <u>then</u> $\tau = \tau'$ <u>or there is an</u> s <u>such that</u> $\tau(s) = -s$ <u>and</u> $\tau'(x) = s^{-1}\tau(s)s$ <u>for every</u> $x \in k < a,b >$.

Proof: $\tau'\tau(x) = x'$ is an automorphism over k of $k < a,b >$. Hence $x' = s^{-1}xs$ for some $s \in k < a,b >$ and $(\tau'\tau)(\tau(x)) = \tau'(x) = s^{-1}\tau(x)s$.

Definition: Let τ' be a type I involution of $k < a,b>$ associated with s so $\tau(s) = -s$ as above. Then τ' is said to be of type I(B).

Suppose we have a type I(B) involution associated to s. Then write $k < a,b > = k(s) \oplus k(s)w = k < \alpha, \beta >$, and assume $s = i$, $w = j$, then $\tau'(i) = -i$ and $\tau'(j) = j$, $\tau'(ij) = ij$ and the fixed set of τ' is associated with the quadratic form

$$< 1 > \perp < \beta > \perp < \alpha\beta >$$

Corollary 6.4: Let $x \in k(a,b)$ then there is a type I(B) involution fixing x.

Proof: Let $w \in k(x)^{\perp}$ then if $\tau'(y) = w^{-1}\tau(y)w$ we have $x \in k(x) \subset$ fix set τ' so 4 follows.

Proposition 6.5: Let τ be a type II involution on $K < a,b >$, so τ acts non-trivially on K, with fixed field $k \subseteq K$. Then $K < a,b > = k < \alpha,\beta > \otimes_k K$.

Proof: Let F be the fixed set of τ. Then F is a 4-dimensional vector space over k. Let $\lambda \in F$, $\lambda \notin K$ then $K < a,b > = M_2(K) = M_2(k) \otimes_k K$ or $f_{\lambda}(y) = y^2 - tr(\lambda)y + N(\lambda)$ is irreducible over K. I claim $tr(\lambda)$, $N(\lambda)$ both belong to k (since they belong to K and are invariant under τ). Now, take $\cup \in K(\lambda)^{\perp}$ so $\cup\lambda = -\lambda\cup$, $\cup^2 = \gamma \in K$. Note that

$$\tau(\cup)\lambda = \tau(\lambda\cup) = -\tau(\cup\lambda) = -\lambda\tau(\cup)$$

so

$$(\cup + \tau(\cup) = -\lambda(\cup + \tau(\cup)).$$

If

$$U \neq \tau U \text{, then } (U + \tau U)^2 \in k \text{ and}$$

$k < \lambda, U + \tau U > \subset k < a,b >$, so $K < a,b > = k < \lambda, U + \tau U > \otimes_k K$.

Otherwise, $U = -\tau U$ so $U\tau(U) = -\gamma \in k$ and $k < \lambda, U > \subset k < a,b >$.

In either case, 6.5 follows.

§7. Involutions and Division Algebras

One of the terms in our exact sequence is $L^+_{o,f}(\mathbb{Q}(\pi))$, where $\mathbb{Q}(\pi)$ is given an involution τ by $\tau(g) = g^{-1}$. Recall that $\mathbb{Q}(\pi) \cong \oplus M_{n_i}(D_i)$, where each D_i is a division algebra over $\mathbb{C}$. Each summand corresponds to a central idempotent e_i in $\mathbb{Q}(\pi)$, so

$$M_{n_i}(D_i) = \mathbb{C}(\pi)\cdot e_i$$

Moreover, as was pointed out to us by I . Herstein [F-H], we have

<u>Lemma</u> 7.1: $\tau(e_i) = e_i$.

<u>Proof</u>: We write $e_i = \frac{d_i}{|\pi|} \sum_{g \in \pi} \chi(g^{-1})g$, where χ is the character corresponding to the i-th representation. Now, $\chi(g^{-1}) = \overline{\chi(g)}$, $^-$ denoting complex conjugation, but $\chi(g)$ is real-valued, hence $\chi(g^{-1}) = \chi(g)$. Now

$$\tau(e_i) = \frac{d_i}{|\pi|} \sum_{g \in \pi} \chi(g^{-1})g^{-1} = \frac{d_i}{|\pi|} \sum_{g \in \pi} \chi(g)g^{-1} = e_i .$$

Thus, $\mathbb{C}(\pi)$splits as a direct sum of central simple algebras with involution. We have already analyzed the involutions on matrix rings over division algebras in terms of the involutions on the division algebras themselves. We now study division algebras with involution.

We first recall some facts about division algebras over local fields. Let K_p be a finite extension of $\hat{\mathbb{Q}}_p$ for some prime p, with valuation v and uniformizing parameter π . Given a cyclic degree n extension of K_p , say L_p , ψ an element in the Galois group, and γ an element of K_p , we define a central simple algebra (L,ψ,γ) as the algebra over L_p generated by an indeterminate y subject to the relations

$$y\ell = \ell^{\psi} y,\ y^n = \gamma .$$

(L,ψ,γ) is a central simple algebra. We have

Theorem 7.2: Theorem 23]: Any central simple algebra A over K_p of dimension n^2 is of the form (L_p,ψ,π^{μ}), $0 \le \mu < n$, where L_p is the unramified degree n cyclic extension of K_p and ψ in the Frobenius generator of its Galois group. Furthermore, A is a division algebra if and only if $(\mu,n) = 1$. If $(\mu,n) = \ell > 1$, then $A = M_n(D)$, where $D = (L^1,\psi^1,\pi^{\mu/\ell})$ and L^1 is unramified of degree n/ℓ and cyclic over K_p.

Since the residue field $\hat{k}$ is finite there is a unique unramified extension of degree n of K_p for each n, so we obtain an invariant $\mu/n \in \mathbb{Q}/\mathbb{Z}$ which determines the division algebra associated to A. We write $\mathrm{inv}_{K_p}(A) = -\mu/n$.

Theorem 7.3: $\mathrm{inv}_{K_p}(A \otimes_{K_p} B) = \mathrm{inv}_{K_p}(A) + \mathrm{inv}_{K_p}(B)$

$\mathrm{inv}_{K_p}(A) = 0$ if and only if $A \cong M_n(K_p)$.

Moreover, let W_p be a degree m extension of K_p. Then

Theorem 7.4: $\mathrm{inv}_{W_p}(A \otimes_{K_p} W_p) = m(\mathrm{inv}_{K_p}(A))$.

We have a complete description of all division algebras over K_p. For the case of the archimedean primes, we note that all central simple algebras over $\mathbb{C}$ are matrix rings $M_n(\mathbb{C})$, and that any central simple algebra over $\mathbb{R}$ is either isomorphic to $M_n(\mathbb{R})$ or to $M_n(\mathbb{H})$, where

$\mathbb{H}$ denotes the division ring of real quaternions. $\mathbb{H} \otimes_{\mathbb{R}} \mathbb{H} \cong M_4(\mathbb{R})$.

Having completed a description for all local fields, we ask to what extent do the algebras $A \otimes_K K_v$ determine A , where A is a central simple algebra over K , a finite extension of the rationals, and v is any valuation of K . The theorem is

<u>Theorem</u> 7.5: (Albert-Brauer-Hasse-Noether): <u>Let</u> A <u>be a central simple</u> K-<u>algebra, then</u> $A_v = A \otimes_K K_v$ <u>is a central simple</u> K_v <u>algebra and</u> A <u>is a matrix ring over</u> K <u>if and only if</u> $\text{inv}_v(A) = \text{inv}_{K_v}(A \otimes_K K_v) = 0$ <u>for all valuations</u> v <u>of</u> K .

For a given algebra A , only finitely many of the numbers $\text{inv}_v(A)$ are non-zero. These are referred to as the <u>ramified primes</u> of A . Also, if B is central simple over K then $A \otimes_K B$ is again central simple over K , and $M_r(K) \otimes_K A \cong M_s(K) \otimes_K B$ if and only if $\text{inv}_v(A) = \text{inv}_v(B)$ for all valuations v .

We now specialize to consider the local invariants of a simple algebra A appearing in the direct sum decomposition of $\mathbb{C}(\pi)$ for some finite group π .

<u>Remark</u>: The set of isomorphism classes of central simple finite dimensional K-algebras factored by the equivalence relation $A \otimes_K M_n(K) \sim A$ forms a group under tensor product $\otimes_K$ called the <u>Brauer group</u> $B(K)$. In our case the invariant map

$$\coprod \text{inv}_v : B(K) \longrightarrow \coprod_{v \text{ real}} \mathbb{Z}/2 \times \coprod_{v \text{ finite}} \mathbb{Q}/\mathbb{Z}$$

gives an injection, and a basic result of class field theory says that the cokernel is isomorphic to $\mathbb{Q}/\mathbb{Z}$. Clearly, each equivalence class contains a single division algebra central over K , and, as a set, $B(K)$ is isomorphic to the set of distinct isomorphism classes of such division algebras.

Similarly, the set of equivalence classes of those A , central over K , which occur in the direct sum decomposition of $\mathbb{Q}(\pi)$ for some finite group π form a subgroup $S(K) \subseteq B(K)$ called the Schur subgroup of B(K) . See e.g., [M-1], [Y] . We should note that K is a subfield of a cyclotomic extension $\mathbb{Q}(\zeta_n)$, where ζ_n is s primitive n-th root of unity and $n \mid |\pi|$.

We say that an involution on a central simple algebra A over K is of type I if it is trivial when restricted to K , and type II if it is not. Note that the involution restricts to an automorphism of the center.

We have

Theorem 7.6 :(Brauer-Speiser): Let $\tau : M_{n_i}(D_i) \longrightarrow M_{n_i}(D_i)$ be a type I involution. Then D_i is either a field or a Quaternion algebra $k\langle a,b\rangle$ for some field k .

Here, $k\langle a,b\rangle$ is the k-algebra with generators i, j and relations $ij = -ji$, $i^2 = a$, $j^2 = b$.

Theorem 7.7: Let $A \in S(K)$, and suppose A has an involution of type II induced by the involution of $\mathbb{C}(\pi)$. (So, we may take the fixed field under the involution to be the totally real subfield of K.) If σ is the involution restricted to K , then

(1) $\mathrm{inv}_v(A) = -\mathrm{inv}_{\sigma(v)}(A)$ for all v with $\sigma(v) \neq v$ (Note that σ is the non-trivial element in the Galois group $G(K/K^{re})$, so σ acts on the valuations of K).

(2) $\mathrm{inv}_v(A) = 0$ for v infinite.

(3) $\underset{\sim}{\text{inv}}_v$ (A) = 0 <u>if</u> $\sigma(v) = v$ <u>and</u> v <u>is finite</u>.

<u>Proof</u>: Benard and Schacher [B-S-2] prove that if $\theta \in G(K/\mathbb{Q})$, and θ extended to $\mathbb{Q}(\zeta_n)$ satisfies $\theta(\zeta_n) = \zeta_n^d$, then

$$d\ \underset{\sim}{\text{inv}}_v(A) = \underset{\sim}{\text{inv}}_{\theta(v)}(A)$$

Since σ extends to the automorphism $x \to \bar{x}$ we have $\zeta_n \to \zeta_n^{-1}$, and this gives (1). K is normal over $\mathbb{Q}$ since K is cyclotomic and K is of degree 2 over K^{re}. Hence, K has no real embeddings so (2) follows. It remains to show (3). Suppose not. By (1),

$\underset{\sim}{\text{inv}}_v(A) = 1/2$ for some v such that $\sigma(v) = v$. But then $A \underset{K}{\otimes} K_v$ is a matrix ring over a Quaternion algebra D_v. But since $\sigma(v) = v$, we have that K_v is a degree 2 extension of K_v^{re}, so $D_v = D_{K_v^{re}} \otimes_{K_v^{re}} K_v = M_2(K_v)$, since $\underset{\sim}{\text{inv}}_{K_v}(D \otimes_{K_v^{re}} K_v) = 2\ \underset{\sim}{\text{inv}}_{K_v^{re}}(D_{K_v^{re}})$. But this is a contradiction, completing the proof.

<u>Remark</u>: Actually 7.7.1 and 7.7.3 hold for general type II involutions on central simple algebras over a finite extension K of $\mathbb{Q}$. Specifically (compare [A, p. 162, Theorem 22]) by the Gruenwald-Wang theorem [W1] there is a degree n cyclic extension L of k, (the fixed field of the involution) so that $L \cap K = k$ and so $L \cdot K \underset{K}{\otimes} A = M_n(LK)$. Then $A \cong (LK, \psi, \gamma)$ and Albert proves A has a type II involution if and only if $\tau(\gamma) \cdot \gamma$ is the norm of an element in L. But $(KL,\psi,\gamma) \otimes (KL,\psi,\tau(\gamma)) = \{(KL,\psi,\gamma\tau(\gamma))\}$ and if $\gamma\bar{\gamma}$ is a norm from KL

then $(KL,\psi,\gamma\tau(\gamma)) = M_n(KL)$. Consequently, $\underset{\sim}{\text{inv}}_v(KL,\psi,\gamma) = -\underset{\sim}{\text{inv}}_v(KL,\psi,\tau(\gamma))$, but $\underset{\sim}{\text{inv}}_v(KL,\psi,\tau(\gamma)) = \underset{\sim}{\text{inv}}_{\tau(v)}(KL,\psi,\gamma)$. Finally, the proof of 6.3 does not have to be changed to apply to this more general situation.

We need one further result on the local invariants of a division algebra in $S(K)$

Theorem 7.8 (Yamada): Let $M_n(D)$ be a direct summand of $\mathbb{C}(\pi)$ for π of order ℓ and suppose K = center (D) . Then

$$\underset{\sim}{\text{inv}}_v(D) \neq 0$$

implies v divides ℓ . (i.e., the only ramified primes of D lie over the primes dividing ℓ .) Finally, we have

Proposition 7.9: Let A be a central simple k-algebra and τ, τ' two type II involutions, or A a Quaternion algebra and τ, τ' two type I(B) involutions ($\tau|k = \tau'|k$ if τ is of type II), then $\tau'(x) = s^{-1}\tau(x)s$ for some s with $\tau(s) = \pm s$. (The sign can be chosen + if τ has type II .) Let $\text{Fix}(\tau)$, $\text{Fix}(\tau')$ be the fixed sets of τ, τ', then $\text{Fix}(\tau') = \text{Fix}(\tau) \cdot s$.

Proof: If τ, τ' have type I(B), then we can assume $\text{Fix}(\tau) = k(1) + k(j) + k(ij)$ and $\tau(x) = -x$ if and only if $x = ki$. But $(ki)^{-1}\tau(x)(ki)$ is then the usual involution on $k < a,b >$. Now assume $\tau'(x) = s^{-1}\tau(x)s$, and $\tau(s) = s$, then $\tau'(xs) = s^{-1}s\tau(x)s = xs$ if and only if $x \in \text{Fix}(\tau)$ and conversely. To complete the proof it suffices to note in the case of a type II involution that if $\tau(s) = -s$ then there is an $\alpha \in k$ so $\tau(\alpha) = -\alpha$ and $(\alpha s)^{-1}\tau(\alpha s) = s^{-1}\tau(x)s$ while $\tau(\alpha s) = \alpha s$.

§8. The Local Lifting Theorem

We recall from [C-M] the definition of the boundary map

$$\partial : L^{\varepsilon}_{o,f}(\Lambda, S) \longrightarrow L^{\varepsilon,tor}_{o,f}(\Lambda,S)$$

Given a Hermitian space over Λ_S, (H,β) , where the module H is free, we choose a Λ-free submodule $L \subset H$ so that $\beta|L \times L$ takes values in Λ. Let $L^{\#} = \{h \in H | \beta(h,\ell) \in \Lambda \; \forall \; \ell \in L\}$, and define $M = L^{\#}/L$. Now, a pairing $\tilde{\beta} : M \times M \longrightarrow \Lambda_S/\Lambda$ is defined by $\tilde{\beta}(m_1,m_2) = \beta(\ell_1,\ell_2)$, where ℓ_1 and ℓ_2 are representatives for m_1 and m_2 in $L^{\#}$. The torsion Hermitian space $(M,\tilde{\beta})$, which is of the same symmetry as (H,β) , is now defined to be the image under ∂ of the class of (H,β) in $L^{\varepsilon}_{o,f}(\Lambda,S)$.

We specialize to the case $\Lambda = \mathbb{Z}_{(p)}(\pi)$, $\Lambda_S = \mathbb{Q}(\pi)$, where $\mathbb{Z}_{(p)}$ denotes $\mathbb{Z}$ localized at the prime p , and π is a finite group. The involution on both these rings is specified by $g \to g^{-1}$ for $g \in \pi$. Thus, we have the localization sequence

$$L^{\varepsilon}_{o,f}(\mathbb{Z}_{(p)}(\pi)) \longrightarrow L^{\varepsilon}_{o,f}(\mathbb{Q}(\pi)) \xrightarrow{\partial} L^{\varepsilon,tor}_{o,f}(\mathbb{Z}_{(p)}(\pi)) \longrightarrow L^{-\varepsilon}_{1,f}(\mathbb{Z}_{(p)}(\pi))$$

We state some preliminary lemmas

<u>Lemma</u> 8.1: <u>A matrix with coefficients in</u> $\mathbb{Z}_{(p)}(\pi)$ <u>is invertible if and only if its mod</u> p <u>reduction is invertible.</u>

<u>Proof</u>: Regard M simply as a matrix over $\mathbb{Z}_{(p)}$. Then M invertible $\Leftrightarrow \det(M) \not\equiv 0 \bmod (p) \Leftrightarrow \det(\rho(M)) \neq 0 \Leftrightarrow \rho(M)$ invertible, where $\rho : \hat{\mathbb{Z}}_{(p)}(\pi) \longrightarrow Z/p(\pi)$ is reduction, mod (p) .

Lemma 8.2: *Let* R *be a finite dimensinal algebra over a field* k , $J(R)$ *its Jacobson radical. Then a matrix* M *over* R *is invertible if and only if* $\varepsilon(M)$ *is invertible, where* $\varepsilon : R \to R/J(R)$ *is the projection.*

Proof: Since R is a finite dimensional algebra, it is Artinian, consequently its Jacobson radical is nilpotent. Thus, given M , such that $\varepsilon(M)$ is invertible, we find a matrix N over R so that $MN = I + L$, where I denotes the identify matrix, and L has entries in the $J(R)$. $I + L$ is invertible, the inverse given by the sum.

$$K = \sum_{i=0}^{\infty} (-1)^i L^i$$

which is finite by the nilpotence of $J(R)$. Thus, $M^{-1} = NK$, which proves the lemma.

We note that the Jacobson radical of $\mathbb{Z}/p(\pi)$ is involution invariant since it is characterized both as the intersection of all maximal right ideals and as the intersection of all maximal left ideals. Consequently, the involution $g \to g^{-1}$ on $\mathbb{Z}/p(\pi)$ induces an involution τ on $\mathbb{Z}/p(\pi)/J(\mathbb{Z}/p(\pi))$ which we now denote by $\tilde{\Lambda}$. Also let $\Lambda = \mathbb{Z}_{(p)}(\pi)$. By the Wedderburn-Artin-Albert theory, $(\tilde{\Lambda}, \tau)$ is a semisimple ring with involution, hence

$$(\tilde{\Lambda}, \tau) \cong \bigoplus_i (\tilde{\Lambda}_i, \tau_i) ,$$

where

$$(\tilde{\Lambda}_i, \tau_i)$$

denotes a simple ring with involution.

It is known (see e.g.[F-M]) that a simple ring with involution is either a simple ring equipped with an involution, or of the form $(\Sigma \oplus \Sigma, \tau)$, where Σ is a simple ring, and $i(\sigma_1,\sigma_2) = (\alpha^{-1}(\sigma_2), \alpha(\sigma_1))$, where α is an antiautomorphism of Σ.

Given a ring Σ with involution $\bar{\ }$, we let

$$\Sigma_+ = \{\sigma \in \Sigma | \sigma = \bar{\sigma}\}$$

$$\Sigma_- = \{\sigma \in \Sigma | \sigma = -\bar{\sigma}\}$$

We wish to examine the maps $\Lambda_+ \to (\tilde{\Lambda}_i)_+$ and $\Lambda_- \to (\tilde{\Lambda}_i)_-$. Either (i) $\tilde{\Lambda}_i \cong M_n(\mathbb{F}_{p^r})$ or (ii) $\tilde{\Lambda}_i \cong M_n(\mathbb{F}_{p^r}) \oplus M_n(\mathbb{F}_{p^r})$, where $\mathbb{F}_{p^r}$ denotes the finite field of order p^r.

<u>Lemma</u> 8.3: (a) $\Lambda_+ \to (\tilde{\Lambda}_i)_+$ <u>is always surjective</u>

(b) <u>If</u> $p \neq 2$, <u>or</u> $p = 2$ <u>and we are in case</u> (i) <u>with the involution acting non-trivially on the center of</u> $M_n(\mathbb{F}_{2^r})$, <u>or</u> $p = 2$ <u>and we are in case</u> (ii), <u>then</u>

$$\Lambda_- \to (\tilde{\Lambda}_i)_- \quad \underline{\text{is surjective.}}$$

<u>Proof</u>: (a) We note that if we are in case (ii), or in case (i) with $p \neq 2$, $\lambda = \bar{\lambda} \Rightarrow \lambda = \nu + \bar{\nu}$ for some $\nu \in \tilde{\Lambda}_i$. Thus, since $\pi : \Lambda \to \tilde{\Lambda}_i$ is surjective, we may pick ν' with $\pi(\nu') = \nu$, and find $\pi(\nu' + \bar{\nu}') = \lambda$. If we are in case (i), $p = 2$, and the involution acts non-trivially on the center of $M_n(\mathbb{F}_{2^r})$ it is a well-known fact that $H^1(G, \mathbb{F}_{2^r}) = 0$,

where $G = \mathbb{Z}/2\mathbb{Z}$ is the Galois group of $\mathbb{F}_{2^r}$ over the fixed field of the involution. But this means that $\lambda = \bar{\lambda} \Rightarrow \lambda = \nu + \bar{\nu}$, and we may argue as above. Finally, if the involution acts trivially, we note that if the image of Λ_+ contains all the diagonal matrices, then $\pi_+ : \Lambda_+ \to (\tilde{\Lambda}_i)_+$ is surjective. But in $\mathbb{F}_{2^r}$, every element is a square, hence we may write any diagonal matrix D as E^2, where E is a diagonal matrix. Since $\pi : \Lambda \to \tilde{\Lambda}_i$ is surjective, we may pick $E' \varepsilon \Lambda$ so that $\pi(E') = E$, and so $\pi(E'\bar{E}') = \pi(E)\pi(E') = E^2 = D$, and $(E'\bar{E}') = E'\bar{E}'$, hence the result

(b) The proof of (a) shows that in these cases, we may write $\bar{\lambda} = -\lambda$ if and only if $\lambda = \nu - \bar{\nu}$, hence we may argue as in (a).

We now prove the main theorem in this section.

<u>Theorem</u> 8.4: <u>The map</u> $L^{\varepsilon}_{o,f}(\mathbb{Q}(\pi)) \xrightarrow{\partial} L^{\varepsilon,tor}_{o,f}(\mathbb{Z}_{(p)}(\pi))$ <u>is surjective</u>.

<u>Proof</u>: Given a torsion FPL1 Hermitian space (M,β) over $\mathbb{Z}_{(p)}(\pi)$, we form a short free resolution of M

$$0 \longrightarrow F_1 \longrightarrow F_2 \longrightarrow M \longrightarrow 0$$

(For later simplicity assume that F_1 and F_2 are of even rank.)

We form the dual resolution

$$0 \longrightarrow F_2^* \longrightarrow F_1^* \longrightarrow M_* \longrightarrow 0$$

where the map $F_2^* \longrightarrow F_1^*$ is restriction, and $F_1^* \longrightarrow M_*$ is obtained by extending an element in F_1^* to an element in $\mathrm{Hom}_{\mathbb{Z}_{(p)}(\pi)}(F_2, \mathbb{Q}(\pi))$. Now

β induces a map ad(β) , which we will denote by $\theta : M \longrightarrow M_*$, with $\theta = \varepsilon\theta_*$. Thus, we have the diagram

$$\begin{array}{ccc}
0 & & 0 \\
\downarrow & & \downarrow \\
F_1 & & F_2^* \\
\iota \downarrow & & \downarrow \iota^* \\
F_2 & \xrightarrow{\tilde{\theta}} & F_1^* \\
\pi \downarrow & & \downarrow \\
M & \xrightarrow{\theta} & M_* \\
\downarrow & & \downarrow \\
0 & & 0
\end{array}$$

We may lift θ to a map $\tilde{\theta} : F_2 \rightarrow F_1^*$ by standard arguments, and condition (c) in Definition (1.6) guarantees that we may lift θ to a $\tilde{\theta}$ so that $(\tilde{\theta} \circ \iota)^* = \varepsilon(\tilde{\theta} \circ \iota)$. (For, choosing $\tilde{\theta}$ simply amounts to choosing a $\mathbb{Q}(\pi)$valued pairing $\tilde{\beta}$ on F_2 so that $\tilde{\beta}(x,y) = \beta(\pi(x),\pi(y))(\text{mod } \mathbb{Z}_{(p)}(\pi))$ and condition (c) guarantees that we may make this pairing ε-symmetric.)

The condition that (M,β) should be the image of the element in $L^{\varepsilon}_{o,f}(\mathbb{Q}(\pi))$obtained from the pairing $\tilde{\beta}$ is that the map $\tilde{\theta}$ should be an isomorphism, as is seen by examining the definition of the boundary map ∂ .

Now, we know that θ is an isomorphism, but not that $\tilde{\theta}$ is. If $\alpha : F_2 \longrightarrow F_2^*$ satisfies $\alpha = \varepsilon\alpha^*$, then $\tilde{\tilde{\theta}} = \tilde{\theta} + i^*\alpha$ is also a lift of θ satisfying $(\tilde{\tilde{\theta}} \circ \tilde{\theta})^* = \tilde{\tilde{\theta}} \circ i$, and all liftings satisfying this condition are of that form. Note that the map $\tilde{\theta} + i^* : F_2 \oplus F_2^* \longrightarrow F_1^*$ is surjective, since θ is an isomorphism. Thus, it will suffice to show that for any

$n \times n$ matrices M and Θ over $\mathbb{Z}_{(p)}(\pi)$, so that $(M\Theta)^t = M\Theta$, and $\mathrm{im}(\Theta) + \mathrm{im}(M^t) = F^n$ (where all matrices are regarded as endomorphisms of the free module of rank n over $\mathbb{Z}_{(p)}(\pi), F^n$), there is a matrix N, with $N = \varepsilon N^t$, so that $\Theta + M^t N$ is an isomorphism. (Here A^t denotes the conjugate transpose of A with respect to the involution on $\mathbb{Z}_{(p)}(\pi)$). But by lemmas 1 and 2, this is possible precisely if it is possible to find N, with $N = \varepsilon N^t$, so that $\rho(\Theta + M^t N)$ is an isomorphism, where $\rho : \Lambda \to \tilde{\Lambda}$ is the reduction.

Thus, we consider matrices over $\tilde{\Lambda}$, which is a direct sum of matrix rings. We may split the matrices according to the splitting of $\tilde{\Lambda}$ into simple rings with involution, say as

$$\tilde{\Lambda} \cong \bigoplus_i \tilde{\Lambda}_i .$$

We attempt to solve the problem within each summand. Thus, suppose that we are in case (i), i.e., $\tilde{\Lambda}_i \cong M_n(\mathbb{F}_{p^r})$ with involution σ. We recall from §5 that $\sigma = \tau_A$, where τ is the conjugate transpose involution with respect to some involution on the center, with $A = \delta\tau(A)$, $\delta = \pm 1$. Given an involution σ on $\mathbb{Z}_{(p)}(\pi)$, $\tilde{\sigma}$ the induced involution on $\tilde{\Lambda}_j^{(m)}$, we say that an ε-$\tilde{\sigma}$-symmetric matrix N is <u>ε-σ-liftable</u> if there is a matrix N' over $\mathbb{Z}_{(p)}(\pi)$ with $\sigma(N') = \varepsilon N'$, and $\rho(N') = N$, where $\rho : \mathbb{Z}_{(p)}(\pi) \to \tilde{\Lambda}^{(m)}$ is the reduction. Note that if α is an invertible element of $\mathbb{Z}_{(p)}(\pi)$, with $\alpha = \pm\sigma(\alpha)$ for an involution σ, then $x \to \alpha\tau(x)\alpha^{-1}$ defines an involution on $\mathbb{Z}_{(p)}(\pi)$, which we denote by σ_α. We now claim that we may assume that $\tilde{\sigma}$, the induced involution on the matrix ring, is actually τ, the conjugate transpose involution with respect to some involution on the center. For given the diagram

(I)
$$\begin{array}{ccc} F & & F \\ {\scriptstyle M}\Big\downarrow & \overset{N}{\nearrow} & \Big\downarrow \tilde{\sigma}(M) = A\tau(M)A^{-1} \\ F & \xrightarrow{\quad \Theta \quad} & F \end{array}$$

with $\tilde{\sigma}(\Theta M) = \varepsilon\Theta M$ and $\Theta \oplus \sigma(M)$ surjective, the existence of an ε-σ- liftable N with $\Theta + \sigma(M)N$ an isomorphism is equivalent to the existence of an $\varepsilon\delta$ - $\tilde{\sigma}_{\alpha}$- liftable $\tilde{N}$ in

(II)
$$\begin{array}{ccc} F & & F \\ {\scriptstyle M}\Big\downarrow & \overset{\tilde{N}}{\nearrow} & \Big\downarrow {\scriptstyle \tau(M)} \\ F & \xrightarrow{\quad A^{-1}\Theta \quad} & F \end{array}$$

with $A^{-1}\Theta + \tau(M)\tilde{N}$ an isomorphism, provided that A is ± 1 - σ - liftable to a matrix α . But Lemma 3 guarantees that any $\pm$ -symmetric matrix is either $+ 1$ - σ - liftable or -1 - $\tilde{\sigma}$ liftable. To see that the two problems are equivalent, given such an $\tilde{N}$, we note that $A\tilde{N}$ is a solution for N in (I). Noting that $\tau(A^{-1}\Theta M) = \delta\tau(\Theta M)A^{-1} = \delta A^{-1}\sigma(\Theta M)$ $= \varepsilon\delta A^{-1}\Theta M$, we see that we have a problem of the same type as (I), but now with respect to the involution τ . Thus, we assume that $\tilde{\sigma}$ is the conjugate transpose involution, and henceforth write $\sigma(M) = M^t$ for clarity.

We now attempt to solve problem (I). By multiplying M on the right by L and M^t on the left by L^t , where L is an appropriately chosen invertible matrix, we may assume that $M = \begin{pmatrix} I & 0 \\ 0 & 0 \end{pmatrix}$, $M^t = \begin{pmatrix} I & 0 \\ 0 & 0 \end{pmatrix}$.

If we rewrite Θ as $\begin{pmatrix}\theta_{11} & \theta_{12}\\ \theta_{21} & \theta_{22}\end{pmatrix}$, the condition $(\Theta M)^t = \varepsilon\Theta M$

becomes $\varepsilon\begin{pmatrix}\theta_{11} & 0\\ \theta_{21} & 0\end{pmatrix} = \begin{pmatrix}\theta_{11}^t & \theta_{21}^t\\ 0 & 0\end{pmatrix} \Rightarrow \theta_{11}^t = \varepsilon\theta_{11}\ ,\ \theta_{21} = 0\ ,$

so $\Theta = \begin{pmatrix}\theta_{11} & \theta_{12}\\ 0 & \theta_{22}\end{pmatrix}$.

The condition that $\Theta \oplus M^t$ is surjective forces θ_{22} to be non-singular. We now attempt to find

$$N = \begin{pmatrix}\nu_{11} & \nu_{12}\\ \nu_{21} & \nu_{22}\end{pmatrix} \quad \text{with}$$

$\nu_{11}^t = \varepsilon\nu_{11},\ \nu_{22}^t = \varepsilon\nu_{22},\ \nu_{12}^t = \varepsilon\nu_{21}$ so that $\Theta + M^tN$ is an isomorphism.

But $\Theta + M^tN = \begin{pmatrix}\theta_{11}+\nu_{11} & \theta_{12}+\nu_{12}\\ 0 & \theta_{22}\end{pmatrix}$

Thus, we must simply choose ν_{11} so that $\theta_{11}+\nu_{11}$ is an isomorphism. If $\varepsilon = 1$, Lemma 3(a) guarantees that $\nu_{11} = I - \theta_{11}$ is the image of a symmetric matrix, so we are done, since I is non-singular. Suppose that $\varepsilon = -1$, and there is a unit $d\varepsilon\tilde{\Lambda}_-^{(i)}$ which is in the image of $(\mathbb{Z}_{(p)}(\pi))_-$. Then, since by construction $\begin{pmatrix}\theta_{11} & 0\\ 0 & 0\end{pmatrix}$ is the image of

a -symmetric matrix over $\mathbb{Z}_{(p)}(\pi)$, so is $\begin{pmatrix}d\cdot I - \theta_{11} & 0\\ 0 & 0\end{pmatrix}$

Setting $\nu_{11} = dI - \theta_{11}$, we are done.

We are left with the case where there is no unit $d \in \tilde{\Lambda}_-^{(i)}$ in the image of $(\mathbb{Z}_{(p)}(\pi))_-$. It is easy to see that this means $\tilde{\Lambda}^{(i)}$ must be a matrix ring of odd rank with trivial involution on the center. But cokernel (M) admits a non-singular -symmetric Hermitian form λ, namely (cokernel (M), $\lambda) \cong (H \otimes \tilde{\Lambda}^{(i)}, \beta \otimes id)$ such that $\lambda(x,x) \in im(\mathbb{Z}_p(\pi)_-)$ for all x. (Here (H, β) denotes the original torsion Hermitian space.) Now $\lambda(x,x)$ is of the form $\alpha - \bar{\alpha}$ for all x, hence coker (M) must have even rank as a $\mathbb{F}_{p^r}$-vector space. Consequently, so must im(M), since we assumed that the original free modules in our resolution had even rank. Therefore, θ_{11} is a $2k \times 2k$ matrix for some k. Now, the matrix

$$K = \begin{pmatrix} 0 & I_k \\ -I_k & 0 \end{pmatrix} \text{ satisfies } K = X - X^t, \text{ where } X = \begin{pmatrix} 0 & I_k \\ 0 & 0 \end{pmatrix}, \text{ so } K$$

is the image of a -symmetric matrix over $\mathbb{Z}_p(\pi)$. Hence, set $\nu_{11} = K - \theta_{11}$, and we are done.

Finally, if our simple ring with involution is the direct sum of two simple rings, then every $\pm$ symmetric matrix is the image of a $\pm$ symmetric matrix of the form $X \pm X^t$. The argument is now the same as above, part (i). This completes the proof.

<u>Corollary</u> 8.5: <u>The map</u> $L^{\varepsilon}_{o,f}(\mathbb{C}(\pi)) \to L^{\varepsilon,tor}_{o,f}(\mathbb{Z}_{(p_1,\ldots,p_s)}(\pi))$ <u>is surjective for any finite set of primes</u> $\{p_1,\ldots,p_s\}$. (Here $\mathbb{Z}_{(p_1,\ldots,p_s)}$ denotes $\mathbb{Z}$ with all primes except $\{p_1,\ldots,p_s\}$ inverted.)

<u>Proof</u>: Note that a matrix over $\mathbb{Z}_{(p_1,\ldots,p_s)}(\pi)$ is invertible if and only if its reduction mod $p_1 p_2 \cdots p_s$ is invertible, and then apply the proof of Theorem 4 at each prime separately, noting that $\mathbb{Z}/(p_1,\ldots p_s)(\pi) \cong \bigoplus_{p_i} \mathbb{Z}/p_i(\pi)$.

Corollary 8.6: The map $L^{\varepsilon}_{o,f}(\hat{\mathbb{Q}}_p(\pi)) \longrightarrow L^{\varepsilon,tor}_{o,f}(\hat{\mathbb{Z}}_{(p)}(\pi))$ is surjective.

Proof: The proof for $\mathbb{Z}_{(p)}(\pi)$ applies equally well to the situation $\hat{\mathbb{Z}}_p(\pi) \longrightarrow \hat{\mathbb{Q}}_p(\pi)$.

§9. The Norm Theorem

Let A be a division algebra over K, a finite extension of $\mathbb{Q}$, with involution τ. Then there is a finite extension L of K so that $A \underset{K}{\otimes} L \cong M_n(L)$. Let $x \in A$, and let $f_x(\lambda)$ be the characteristic polynomial of x regarded as a matrix in $M_n(L)$. Then

$$f_x(\lambda) = \lambda^n - tr(x)\lambda^{n-1} + \cdots + (-1)^n N(x)$$

where $N(x) = \det(x)$ denotes the reduced norm of x. Similarly, let $\phi_x(\lambda)$ denote the minimal polynomial of x. Clearly, $N(xy) = N(x)N(y)$, and $N(x) \neq 0$ for $x \neq 0$. Moreover, the coefficients of $f_x(\lambda)$ lie in K. In particular, if $\tau(x) = x$, and $k \subset K$ is the fixed subfield of K under τ, then $f_x(\lambda) \varepsilon k(\lambda)$.

Now, let $F(A,\tau)$ be the set $\{a\varepsilon A | \tau(a) = a\}$. If τ is a type II involution, then $F(A,\tau)$ is a vector space of dimension n^2 over k. If τ is a type I involution (thus A is a Quaternion algebra), then $F(A,\tau)$ has dimension one or three depending on whether τ has type I(A) or I(B).

Definition 9.1: $N(A,\tau)$ is the multiplicative subgroup of $\dot{k}$ generated by the norms of elements in $F(A,\tau)$.

Similarly, $I(A,\tau)$ is the set of elements $\alpha\varepsilon A$ for which $\tau(\alpha) = -\alpha$, and if τ has type II, with $a\varepsilon K$ satisfying $\tau(a) = -a$, then $I(A,\tau) = aF(A,\tau)$. Also, if (A,τ) has type I(A) then $I(A,\tau)$ is the set of pure imaginary elements in the Quaternion algebra A . The group $N(I(A,\tau))$ also plays a role. For (A,τ) of type II,

$$N(I(A,\tau)) = N(a) \cdot N(A,\tau) .$$

For (A,τ) of type $I(A)$, $N(I(A,\tau)) = N(I(K < a,b >))$ is the subgroup of K generated by the numbers $m^2a + n^2b + \ell^2 ab, m,n,\ell \in K$.

We now turn to the determination of $N(A,\tau)$. To begin, we note

Lemma 9.2: Let $V \subset A$ be the set of elements v so that v is fixed under some type II involution of A with fixed field $k \subsetneq K$. Then $N(V) = N(A,\tau)$. (Here $N(V)$ is the subgroup of $\dot{k}$ generated by the $N(v)$, $v \in V$) .

Proof: $v \in V$ if and only if there is a τ' so that $\tau'(v) = v$, but $\tau'(x) = s^{-1}\tau(x)s$ for some s with $\tau(s) = \pm s$. So if $\tau(s) = -s$, then $\tau(as) = as$ works as well, where $a \in \dot{K}$ is such that $\tau(a) = -a$. Hence (by Prop. 7.8) $F(A,\tau') = F(A,\tau)s$ and $N(v) = N(y)N(s)$ where $y,s \in F(A,\tau)$.

The following result is a special case of a result proved in [A, Theorem 15, p. 157].

Lemma 9.3: Let A be a division algebra over K . If A has an involution with fixed field $k \subsetneq K$ then x is fixed under some type II involution with fixed field k if and only if the minimal polynomial $\phi_x(\lambda)$ has coefficients in k .

Proof: Let $\deg \phi_x(\lambda) = m$, then since A is a division algebra, $K(x) \subseteq A$ is a subfield of degree m over K . If $\tau(x) = x$ for some τ with fixed field k then $0 = \psi(x) = \phi_x(x) - \tau(\phi_x(x)) = \sum_{i>0} (k_i - \tau k_i)x^{m-i}$ and by the definition of ϕ_x, $\psi \equiv 0$ so $k_i = \tau(k_i)$ for each i.

Now suppose $\phi_x(\lambda)$ has coefficients in k. Then $\phi_x(\tau(x)) = \tau(\phi_x(x)) = 0$, so there is an $s \in A$ so that $s^{-1}xs = \tau(x)$, or $x = (\tau(s)s^{-1}))\,x\,(s\tau(s^{-1}))$. Hence, set $\beta = \tau(s)s^{-1}$. If $\beta = -1$, then $\tau(s) = -s$, otherwise $y = s + \tau(s) = (1 + \beta^{-1})s \neq 0$ in A, and $s^{-1}(1+\beta^{-1})^{-1}x(1+\beta^{-1})s = s^{-1}xs = \tau(x)$, so $y^{-1}xy = \tau(x)$ and $y\tau(x)y^{-1} = x$. But $y\tau(z)y^{-1} = \tau'(z)$ is an involution of the desired type.

Theorem 9.4: Let A be a division algebra with center K and involution τ of type II or type I(B) where K is a finite extension of $\mathbb{C}$, then $N(A,\tau) = \dot{k}$ if K has no real embeddings v for which $\mathrm{inv}_v(A) = 1/2$. Otherwise $N(A,\tau)$ is the subgroup of $\dot{k}$ consisting of elements positive at all the real embeddings above.

Proof: We need

Lemma 9.5: Let L be a degree n extension of K and A be central simple over K of dimension n^2. Then L embeds in A as a subalgebra if and only if $L \otimes_K A \cong M_n(L)$.

Proof: (See e.g. [S-E], p. 188)

Assume now A has a type II involution.

Using 7.7.1 and 7.7.3 which by the (remarks following 7.7) hold for general A we see that if v is a valuation on K with $\mathrm{inv}_v(A) \neq 0$, then $\tau(v) \neq v$ and $K_v = k_w$, where w is the valuation on k associated to v. Let $m \in \dot{k}$ be positive at all infinite v with $\mathrm{inv}_v(A) = 1/2$. (If such invariants occur, n is even, where n^2 is the dimension of the algebra.) From

([S-E] lemma 9.26, p. 200) it follows that there is a polynomial

$$x^n + a_1x^{n-1} + \cdots + (-1)^n m = f(x)$$

irreducible over k, with $k_v(\alpha) - k_w$ a degree n extension for each finite w with $\underset{\sim}{\text{inv}}_v(A) \neq 0$, where v is a valuation extending w, and α is a root of $f(x)$. Furthermore, $k_v(\alpha) = \mathbb{C}$ for each infinite v above. Then since $K_v = k_w$,

$$K_v(\alpha) - K_v$$

is also of degree n and so $f(x)$ is irreducible over K as well. Thus

$$A \underset{K}{\otimes} K \cdot k(\alpha)$$

is a matrix algebra and $K \cdot k(\alpha) \subset A$. Now the minimal polynomial of α = characteristic polynomial of $\alpha = f(x)$, which has coefficients in k, and $N(\alpha) = m$. Thus, this case follows from lemma 3.

We now complete the proof by assuming we are in case I(B).

From corollary 3.4 there is a type I(B) involution fixing x for any $x \in k < a,b >$. Thus, by lemma 2 $N(A,\tau) = N(k < a,b >)$. But by the Hasse-Schilling norm theorem, $N(k < a,b >) = \{x \in k \mid x > 0$ at all infinite primes with $\underset{\sim}{\text{inv}}_v(x) \neq 0\}$.

Remark: If A has a type II involution and is in $S(K)$, then $k = K^{re}$ and K has no real embeddings, consequently $N(A,\tau) = \dot{k}$ in these cases.

Appendix: $K_o(Z(\pi))$ for π a finite 2-group.

Let $\mathcal{M} \subset \mathbb{Q}(\pi)$ be a maximal order containing $Z(\pi)$. In [R-U], [F], an exact sequence is studied.

A.1: $0 \longrightarrow D(\pi) \longrightarrow K_o(\pi) \longrightarrow K_o(\mathcal{M}) \longrightarrow 0$

which defines $D(\pi)$ for any finite group π. In [F] this group $D(\pi)$ is shown to depend only on reduced norms or determinants.

Precisely, let $\mathcal{M}$ be a maximal order for a simple algebra $A = M_n(D)$ where D is a finite dimensional division algebra over a finite extension of $\mathbb{Q}$ or $\tilde{\mathbb{Q}}_p$. Then $\mathcal{M} = M_n(\mathcal{M}_D)$ where $\mathcal{M}_D$ is a maximal order in D, and set

$$K_1'(\mathcal{M}) = K_1'(\mathcal{M}_D) = \text{im } K_1(\mathcal{M}) \longrightarrow K_1(A) .$$

By [W.2][S-E], [B],

$$K_1(A) = K_1(\text{center } A) = (\text{center } A)^{\bullet}$$

if D is not a Quaternion algebra at any infinite prime, and $K_1(A) = K_1^+(\text{center } A)$ = elements in (center A) which are positive at all infinite places where D is a Quaternion algebra.

Moreover, $K_1'(\mathcal{M})$ = units in $K_1(A)$, coming from the center of $\mathcal{M}$.

Remark A.2: The isomorphisms above to the units of the center are all obtained by taking reduced norms or determinants. Also, for $\mathcal{M} \subset \mathbb{Q}(\pi)$ write $\mathbb{Q}(\pi) = \coprod_i M_{n_i}(D_i)$ and $\mathcal{M} = \coprod \mathcal{M}_i$ where $\mathcal{M}_i \subset M_{n_i}(D_i)$ is a maximal order.

Hence we can define $K_1'(\mathcal{M}) = \coprod(K_1'(\mathcal{M}_i))$.

Similarly, let $\mathcal{M}_p$ be a maximal order in $\hat{\mathbb{Q}}_p(\pi)$ containing $\hat{Z}_p(\pi)$, and we have

$$K_1'(\mathcal{M}_p) = \coprod (K_1'(\mathcal{M}_{p,i})) .$$

Then in both cases above the maps $Z(\pi) \to \mathcal{M}$, $(\hat{Z}_p(\pi) \to \mathcal{M}_p)$ induce maps

A.3 (a) $K_1(Z(\pi)) \to K_1'(\mathcal{M})$

(b) $K_1(\hat{Z}_p(\pi)) \to K_1'(\mathcal{M}_p)$

and 2.3(b) may be calculated as follows. Since $Z_p(\pi)$ is local $U(\hat{Z}_p(\pi)) \to K_1(\hat{Z}_p(\pi))$ is surjective. Now, take a unit and look at its image in the various p-adic representations of π, and take its reduced norm in each.

Then we have

Theorem 1.4 [F]: $D(\pi) = \left(\coprod_{p \mid |\pi|} K_1'(\mathcal{M}_p) / \operatorname{im} K_1(\hat{Z}_p(\pi)) \right) \Big/ \operatorname{im}(K_1'(\mathcal{M}))$.

Remark A.5: For examples of calculations with 2.4 see e.g. [F-K,W] and more particularly [M2].

From now on π is a finite 2-group.

We are interested now in the structure of the 2 primary part of $K_o(Z(\pi))$. From 4.8, $K_o(Z(\pi))_2 = D(\pi)_2$, and from A.4, 4.1, $D(\pi) = D(\pi)_2$.

Now, consider $V(\pi) = K_1^1(\mathcal{M}_2) / \operatorname{im} K_1^1(\mathcal{M})$, which has $D(\pi)$ as a quotient.

To make (the) calculation more easily stateable we also factor out by $(1 + |\pi|^{\ell}\alpha)$ where $\alpha \in$ center $\mathcal{M}_2$, so let

A.6 $$W_{\ell}(\pi) = K_1(\mathcal{M}_2) / \operatorname{im}(1 + |\pi|^{\ell}\alpha) \operatorname{im} K_1'(\mathcal{M})$$

We have

<u>Theorem</u> A.7: (a) $W_{\ell}(\pi)$ <u>is finite and for</u> ℓ <u>sufficiently large</u> $\operatorname{im}(1 + |\pi|^{\ell}\alpha) \subseteq \operatorname{im} K_1'(\hat{Z}_2(G))$ <u>so</u> $W_{\ell}(\pi)$ <u>surjects onto</u> $D(\pi)$

(b) $W_{\ell}(\pi) = \coprod W_{\ell}(E_f)$ <u>where</u> f <u>runs over all irreducible representations of</u> G.

<u>Proof</u>: (a) follows since $Z(\pi) \subset \mathcal{M}$ is of finite index, so $\hat{Z}_2(\pi) \subset \mathcal{M}_2$ is also of finite index and $2^r \mathcal{M}_2 \subset \hat{Z}_2(\pi)$ for some r.

(b) follows since $W_{\ell}(\pi) = K_1'(\mathcal{M}_2/2^{\ell s}\mathcal{M}_2)/\operatorname{im} K_1'(\mathcal{M})$ where we interpret $K_1'(\mathcal{M}_2/2^{\ell s}\mathcal{M}_2)$ as the image of $K_1(\mathcal{M}_2)$ in $K_1(\mathcal{M}_2 \otimes \hat{\mathbb{C}}_2)$ where we factor out by appropriate powers of fundamental ideals.

But $(K_1'(\mathcal{M}_2) = \coprod K_1'(\mathcal{M}_{2,i}) = \coprod K_1'(\mathcal{M}_{2,f}) = \coprod K_1'(\hat{Z}_2(\rho_{2^i}) \times_{Z(Z/2^i)} E_f)$.

We calculate $W_{\ell}(E_f)$ as follows using 4.6.

A.8 (a) For E_f of type 4.1(i)

$W_{\ell}(E_f) = Z/2^{\ell} + \oplus (Z/2)^{2^{i-2}}$ with generators $\langle 5 \rangle$ of $Z/2^{\ell}$, $\langle -1 \rangle$, $g^j(1+\lambda_i)$ for the $(Z/2)$'s.

(b) For E_f of type 4.1(ii)

$W_{\ell}(E_f) = Z/2^{\ell-1}$ with generator $\langle 5 \rangle$

(c) For E_f of type (4.1(iii)

$$W_\ell(E_f) = Z/2^{\ell-1} \quad \text{with generator } \langle 5 \rangle$$

(d) For E_f of type 4.1(iv)

$$W_\ell(E_f) = (Z/2^{\ell-1})^{2^{i-2}+1}$$

and one of the generators is $\langle 5 \rangle$. The action of the Galois automorphism $g(\rho_{2^i}) = (\rho_{2^i})^{-1}$ is $g \langle 1 + 2i \rangle = \langle 1 - 2i \rangle = \langle 5/1 + 2i \rangle$ otherwise on generators. $g(x_i) = -x_i$.

Only (d) needs any detailed argument.

We know

$$\hat{H}_1(Z/2, \hat{\mathbb{C}}_2(\rho_{2^i})^\bullet) = Z/2$$

$$\hat{H}_2(Z/2, \hat{\mathbb{C}}_2(\rho_{2^i})^\bullet) = 0$$

from local class field theory. I claim

A.9: $$\hat{H}_1(Z/2, \cup(\hat{Z}_2(\rho_{2^i}))) = \hat{H}_2(Z/2, \cup(\hat{Z}_2(\rho_{2^i}))) = Z/2 .$$

<u>Proof</u>: Consider the exact sequence

$$0 \longrightarrow \cup\, \hat{Z}_2(\rho_{2^i}) \longrightarrow \hat{\mathbb{C}}_2(\rho_{2^i})^\bullet \longrightarrow Z \longrightarrow 0$$

g is the identity on Z and so we have

$$0 \longrightarrow \hat{H}_1(Z/2, \cup) \longrightarrow \hat{H}_1(Z/2, \hat{\mathbb{C}}_2(\rho_{2^i})^\bullet) \xrightarrow{j_*} (Z/2) \xrightarrow{\partial} \hat{H}_0(Z/2, \cup) \longrightarrow 0$$

but since the generator of $\hat{H}_1(Z/2, \hat{\mathbb{C}}_2(\rho_{2^i})^{\cdot})$ can be chosen to be a unit (indeed $(1 + \lambda_i)$ will do) it follows that $j_* = 0$ and A.9 follows.

Next, $\cup(\hat{Z}_2(\rho_{2^i})) = Z/2^i \times (\hat{Z}_2^+)^{2^{i-1}}$, and this splitting is either with respect to g or not. I claim it cannot be with respect to g. Indeed, if it were then $(\hat{Z}_2^+)^{2^{i-1}}$ would give trivial homology under $Z/2$, and hence would be free. But then the invariants would all be of the form $x\,g(x)$. However, again by local class-field theory invariants/norms $= Z/2$. Thus, $\cup(\hat{Z}_2(\rho_{2^i}))/\text{Tor} = \text{Free}\ (\hat{Z}^+_{(2)}(\hat{Z})) \oplus \hat{Z}_2^+ \oplus \hat{Z}_2^-$. Moreover, the $\hat{Z}_2^+$ is generated by $(1 + \lambda_i)$, so the $\hat{Z}_2^-$ extends the trosion module non-trivially. This implies, if v generates the $\hat{Z}_2^-$, that $N(v) = -1$. Thus, in the special case of $\hat{Z}_2(i)$ the argument is slightly different. Here i, $(1 + 2i)$, $1 - 2i$, generate the units.

In either case A.10(d) follows directly.

Bibliography

[A] A. A. Albert, Structure of Algebras, Amer. Math. Soc. Colloquium Pub. Vol. XXIV, (1939)

[B-S1] A. Bak, W. Scharlau, "Grothendieck and Witt groups of orders and finite groups", Invent. Math. 23 (1974) 207-240

[B] H. Bass, Algebraic K-theory, Benjamin (1968)

[B-S2] M. Benard, M. M. Schacher, "The Schur subgroup II", J. Algebra 22 (1972) 378-385

[C] G. Carlsson, "On the Witt group of a Z-odic group ring", (to appear)

[C-M] __________, R. J. Milgram, Some exact sequences in the theory of Hermitian forms", J. Purc and Applied Algebra (to appear)

[F-H] K. L. Fields - I. N. Herstein, "On the Schur Subgroup of the Brauer group", J. Algebra 20 (1972) 70-71

[F] A. Fröhlich, "Locally free modules over arithmetic orders", J. Reine Angew Math 274/75 (1975) 112-138

[F-M] __________, A. McEvett, "Forms over rings with involution", J. Algebra 12 (1969) 79-104

[F-k-W] __________, M. E. Keating-S. M. J. Wilson, "The classgroups of quaternion and dihedral 2-groups", Mathematika 21 (1974) 64-71

[H] H. Hasse, Uber die Klassenzahl Abelscher Zahlkörper, Berlin, Akademic-Verlag (1952)

[J] N. Jacobson, Structure of Rings, Amer. Math. Soc. Colloquium Public, Vol. XXXVII (1956)

[L] S. Lang, Algebraic Number Theory, Addison-Wesley (1968)

[Ma] S. Maumary, "Proper surgery groups and Wall-Novikov groups", Springer Lecture Notes in Mathematics, Vol. 343, (1973) 526-539

[M1] R. J. Milgram, "The determination of the Schur subgroup", (Mimeo) Stanford (1977)

[M2] __________, "Evaluating the Swan obstruction for periodic groups", (Mimeo) Stanford (1978)

[M-H] J. Milnor-D. Husemoller, Symmetric Bilinear Forms, Springer-Verlag (1973)

[O-M] O. T. O'Meara, Introduction to Quadratic Forms, Springer-Verlag (1963)

[P] W. Pardon, Local Surgery and the Exact Sequence of a Localization for Wall Groups, Mem. A.M.S. #196 (1977)

[Ra] A. Ranicke, The Algebraic Theory of Surgery, (Mimeo)

[R] I. Reiner, Maximal Orders, Academic Press (1975)

[R-U] ________, S. Ullom, "A Meyer-Vietoris sequence for class groups", J. Alg. 31 (1974), 305-342

[Se] J. P. Serre, Corps Locaux, Hermann, Paris (1968)

[S] R. J. Swan, "Induced representations and projective modules", Ann. of Math. (2) 71 (1960) 552-578

[S-E] __________, E. G. Evans, K-theory of Finite Groups and Orders, Springer Lecture notes in mathematics #149 (1970)

[W1] S. Wang, "On Gruenwald's theorem", Ann. of Math. (2) 51 (1950) 471-484

[W2] _______, "On the commutator group of a simple algebra", Amer. J. of Math 72 (1950), 323-334

[Wa] C. T. C. Wall, "Formulae for the surgery obstructions", Topology 15 (1976) 189-210

[Y] T. Yamada, The Schur Subgroup of the Brauer Group, Springer Lecture Notes in Mathematics #397 (1974)

The surgery group $L_3^h(Z(G))$ for G a finite 2-group

by

Ian Hambleton
R. James Milgram

In [C-M,2] a theorem is proved which expresses $L_3^p(Z(G))$ as a simple functor of the rational representation ring $R_{\mathbb{Q}}(G)$ when G is a finite 2-group. In the appendix to [C-M,2] one of us shows that the 2-primary part of $\tilde{K}_o(Z(G))$ is the quotient of a finite group depending only on $R_{\mathbb{Q}}(G)$ and the order of G.

Here we determine the structure of $L_o^p(Z(G))$, and provide a complete determination of a factorization of the map d in the Ranicki-Rothenberg sequence

$$* \ \ldots \to L_{2i}^p(Z(G)) \xrightarrow{d} H_{od}(Z/2, \tilde{K}_o(Z(G))) \xrightarrow{\partial} L_{2i-1}^h(Z(G)) \to L_{2i-1}^p(Z(G)) \to H_{ev} \to .$$

through the group alluded to above. In particular we apply our results to obtain $L_3^h(Z(G))$, the surgery obstruction group, when G is a generalized quaternion 2-group. This in turn leads to examples of the existence of semi-free group actions on homotopy spheres which do not admit balanced splittings, (see [A-H] for definitions, and the reduction to properties of * in particular pp. 8-9).

In detail we have

Theorem A: Let G be a finite 2-group, then $L_o^p(Z(G)) = Z^{\ell(G)}$ where $\ell(G)$ is the number of irreducible real representations of G.

Theorem B: For G a finite 2-group the kernel K in the map

$$0 \to K \to L_3^h(Z(G)) \to L_3^p(Z(G))$$

is known once the map $\varphi : W_\ell(G) \to D(G)$ is known, where $W_\ell(G)$ is given in [C-M,2 Appendix, especially A.7, A.8], for ℓ sufficiently large.

Indeed in §2,3, we give all the information needed to determine K explicitly. Also, note that $W_\ell(G)$ depends only on the rational representation ring of G, while the ℓ is determined by $|G|$. We remark that even the extension is determined from the information in φ, though we don't explain this here. Finally, we point out that the map d in * $L^p_{od}(Z(G)) \to H_{ev}(Z/2,\tilde{K}_o(Z(G)))$ is already implicitly determined in [C-M,2], our techniques here can also be used to determine the map

$$L^p_2(Z(G)) \to H_{ev}(Z/2,\tilde{K}_o(Z(G)))$$

and in each case a theorem similar to B holds.

In §4, we apply these results to the generalized quaternion groups.

Theorem C: Let $Q_{2^i,2}$ be the generalized quaternion group $\{x,y|x^{2^i} = y^2 = (xy)^2\}$ then d is surjective in * for $i = 0$ and $L^h_3(Z(Q_{2^i,2})) = (Z/2)^{i+1}$ injects into $L^p_3(Z(Q_{2^i,2}))$.

The application to balanced splittings results since [F-K-W], [M] show that the Swan homomorphism T is onto the 2-torsion in $\tilde{K}_o(Z(Q_{2^i,2}))$. See also, §4.1, 4.2.

§1. <u>The proof of theorem A.</u>

Consider the diagrams of long exact sequences

1.1
$$\begin{array}{ccccccccc}
\cdots \to & H_{od}(Z/2,\tilde{K}_o(\mathbb{C}(G))) & \to & L_1^h(\mathbb{Q}(G)) & \to & L_1^p(\mathbb{Q}(G)) & \to & H_{ev}(Z/2,\tilde{K}_o)\cdots & \to \\
& \downarrow s & & \downarrow \bar{s} & & \downarrow & & \downarrow & \\
\cdots \to & H_{od}(Z/2,\tilde{K}_o(\hat{\mathbb{C}}_2(G))) & \to & L_1^h(\hat{\mathbb{C}}_2(G)) & \to & L_1^p(\hat{\mathbb{C}}_2(G)) & \to & H_{ev}(Z/2,\tilde{K}_o)\cdots & \to
\end{array}$$

1.2
$$\begin{array}{ccccccccc}
\cdots \to & L_1^h(\mathbb{Q}(G)) & \to & L_1^{p,tor}(Z(G)) & \to & L_o^p(Z(G)) & \to & L_o^h(\mathbb{Q}(G)) & \to \cdots \\
& \downarrow \bar{s} & & \downarrow r & & \downarrow & & \downarrow & \\
\cdots \to & L_1^h(\hat{\mathbb{C}}_2(G)) & \xrightarrow{\partial} & L_1^{h,tor}(\hat{Z}_2(G)) & \to & L_o^h(\hat{Z}_2(G)) & \xrightarrow{i} & L_o^h(\hat{\mathbb{C}}_2(Q)) & \to \cdots
\end{array}$$

From [C-M,2 p. 33-35] or [R] we have that

1.3
$$L_1^p(\mathbb{Q}(G)) = L_1^p(\hat{\mathbb{C}}_2(G)) = 0 .$$

Since G is a finite group $K_o(\mathbb{C}(G)) = R_{\mathbb{C}}(G)$, $= Z^{\ell}$ where ℓ is the number of irreducible $\mathbb{C}$ representations of G . Also, since G is a 2-group we have that

1.4
$$K_o(\hat{\mathbb{C}}_2(G)) \cong K_o(\mathbb{C}(G))$$

under the natural inclusion [s] . Hence in 1.1 s is an isomorphism and $\bar{s}$ is a surjection of $L_1^h(\mathbb{C}G)) \to L_1^h(\hat{\mathbb{C}}_2(G))$.

Now consider 1.2. In [C-M,2, p. 31] we have shown that $L_o^h(\hat{Z}_2(G)) = Z/2$ injects into $L_o^h(\hat{\mathbb{C}}_2(G))$. So

1.5 $$\partial : L_1^h(\hat{\mathbb{C}}_2(G)) \to L_1^{h,tor}(\hat{Z}_2(G))$$

is onto. But from [C-M,1 §2] and [C-M,2, p. 10] (or arguments totally analogous to those) we have that

$$r : L_1^{p,tor}(Z(G)) \to L_1^{h,tor}(\hat{Z}_2(G))$$

is an isomorphism. Hence from the surjectivity of ∂ and $\bar{s}$ it follows that the map

1.6 $$L_o^p(Z(G)) \to L_o^h(\mathbb{C}(G))$$

is an injection.

At this point, consider the diagram of exact sequences

$$\begin{array}{ccccccc} & & \downarrow & & \downarrow & & \\ & & L_o^p(Z(G)) & \longrightarrow & L_o^p(\mathcal{M}(G)) & & \\ & & \downarrow & & \downarrow & & \\ 0 & \longrightarrow & L_o^h(\mathbb{C}(G)) & \longrightarrow & L_o^p(\mathbb{C}(G)) & \longrightarrow & H_{od}(Z/2,\tilde{K}_o) \end{array}$$

where $\mathcal{M}$ is a Z-maximal order containing $Z(G)$ in $\mathbb{C}(G)$, which shows that

$$L_o^p(Z(G)) \hookrightarrow \text{im} : L_o^p(\mathcal{M}(G)) \hookrightarrow L_o^p(\mathbb{C}(G))$$

Now, $L_o^p(\mathbb{C}(G)) = \coprod_i L_o^p(R_i(G))$, where R_i is the ith irreducible representation algebra. These are classified as to type in [C-M,2, p. 26]. Using Morita equivalence, the results of [M-H,pp. 117-118] for the type 4.3(ii) and 4.3(iv) representations (in the notation of [C-M,2, p. 26]), [M-H, p. 95] for the type 4.3(ii) representations and a direct calculation in the 4.3(i) case we see that im $L_o^p(\mathcal{M}(G))$ in $L_o^p(\mathbb{C}(G))$ is a direct sum of Z's and the proof of theorem A is complete.

Remark 1.8: Similar techniques can be applied to calculate $L_i^p(Z(G))$ for G a finite 2-group when $i = 1,2$, as well. These results will be written down in their entirety in [C-M-P] where the general case of G a 2-hyperelementary group will also be studied.

Remark 1.9: It is not true for finite 2 groups that $L_o^h(Z(G)) = Z^\ell$, as $K_o(Z(G))_{(2)}$ tends to grow very large and $L_1^p(Z(G))$ is zero except for some Z/2's coming from the type 4.3(i) representations of [C-M,2, p. 26]. So $L_o^h(Z(G)) = Z^\ell \oplus (Z/2)^s$. The Z's may be detected via the Atiyah-Singer G-signature theorem [P], but we have no idea of what occurs with the Z/2's.

§2. Factoring the map d .

Throughout this section we assume that the reader is familiar with the appendix in [C-M,2].

Begin with the local-global pull-back diagram

2.1
$$\begin{array}{ccc} Z(G) & \longrightarrow & \mathcal{M}(G) \\ \big\downarrow & & \big\downarrow i \\ \hat{Z}_2(G) & \xrightarrow{\ j\ } & \mathcal{M}(G)\otimes_Z \hat{Z}_2 \end{array}$$

where $\mathcal{M}(G)$ is a maximal Z-order for $Z(G)$ in $\mathbb{Q}(G)$ and $\mathcal{M}(G)\otimes_Z\hat{Z}_2$ is a maximal $\hat{Z}_2$ order.

2.1 allows us to construct projective $Z(G)$ modules together with non-singular forms by mixing forms over $\hat{Z}_2(G)$ with forms over $\mathcal{M}(G)$ on $\mathcal{M}(G)\otimes_Z\hat{Z}_2$. Specifically, let $(\mathcal{M}(G)^n, A_n)$, $(\hat{Z}_2(G)^n, B_n)$ be suitable forms and assume there is a C_n in $GL_n(\mathcal{M}(G)\otimes_Z\hat{Z}_2)$ so that

2.2
$$C_n \cdot i(A_n)C_n^* = j(B_n)$$

Then on the projective module W defined by C_n,

2.3
$$\begin{array}{ccc} W & \longrightarrow & \mathcal{M}(G_T)^n \\ & & \big\downarrow i^n \\ \big\downarrow & & \mathcal{M}(G)\otimes_Z\hat{Z}_2 \\ & & \big\downarrow C_n \\ \hat{Z}_2(G)^n & \xrightarrow{\ j^n\ } & \mathcal{M}(G)\otimes_Z\hat{Z}_2 \end{array}$$

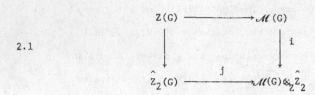

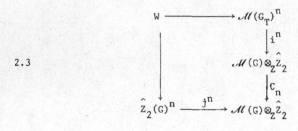

2.2 gives a form which becomes A_n when tensoring W with $\mathcal{M}(G)$, and B_n on tensoring with $\hat{Z}_2(G)$. We denote the form on W by

$$[W, A_n, B_n, C_n] .$$

In the appendix to [C-M,2], the group $D(G) \subset \tilde{K}_o(ZG)$ is described on page A.2, see in particular Theorem 1.4, as a quotient of $K_1(\mathcal{M}(G) \otimes_Z \hat{Z}_2)$.

Then the following lemma is clear.

Lemma 2.4: The image of $[C_n]$ in $D(G) \subset \tilde{K}_o(Z(G))$ represents

$$d([W, A_n, B_n, C_n]) .$$

Throughout the remainder of this section we assume $B_n = \begin{pmatrix} 0 & I \\ I & 0 \end{pmatrix}$ so that C_n makes A_2 2-locally equivalent to a hyperbolic form. (Actually, this assumption holds for every element of $L_o^p(Z(G))$.)

Lemma 2.5: Let A_n be itself hyperbolic except at a single representation $M_n(F)$ where F is a formally real field, then

$$d(W, A_n, B_n, C_n) = 1 .$$

Proof: At $M_n(F \otimes_{\mathbb{Q}} \hat{\mathbb{Q}}_2)$ we have

$$\begin{pmatrix} 0 & I \\ I & 0 \end{pmatrix} = C_n A_n C_n^*$$

and taking determinants $\pm 1 = (\det C_n)^2 \det(A_n)$ but $\det(A_n)$ is a unit in $Z(\rho_{2^i} + \rho_{2^i}^{-1})$, the ring of algebraic integers in F. Now, use the unit calculations of §4 of [C-M,2], in particular 4.6, 4.7 to see that det (C_n) is likewise a unit in $Z(\rho_{2^i} + \rho_{2^i}^{-1})$, hence in the kernel of d.

Lemma 2.6: Let A_n be hyperbolic except at a representation $M_n(\mathcal{Q}(F))$ where $\mathcal{Q}(F)$ is the type 4.3(i)(a) simple algebra of [C-M,2], then the class of A in the Witt ring is determined by its multisignature at the various real places of F, and if A_n has signature 0 except at the ith place ∞_i, where it has signature ± 2, then

$$d[W, A_n, B_n, C_n] = (\varepsilon_i^{-1})$$

where ε_i is any unit of F positive at all ∞_j, $j \neq i$ and negative at ∞_i.

Proof: The maximal order $\mathcal{M}(G)$ can be chosen to be $M_n(\mathcal{O}_{\mathcal{Q}(F)}) \oplus \mathcal{M}^1$ where $\mathcal{O}_{\mathcal{Q}(F)}$ is a maximal order in $\mathcal{Q}(F)$. Indeed we can take

2.7 $$\mathcal{O}_{\mathcal{Q}(F)} = Z(\tfrac{1+i+j+k}{2}, i, j) \otimes_Z Z(\rho_{2^\ell} + \rho_{2^\ell}^{-1}).$$

In this $\mathcal{O}_{\mathcal{Q}(F)}$, $1 = \frac{1+i+j+k}{2} + \frac{1-i-j-k}{2}$ and so all elements of the center are even. Now consider the form $A_n = \left[\begin{pmatrix} \varepsilon_i & 0 \\ 0 & -1 \end{pmatrix}, \begin{pmatrix} 0 & 1 \\ 1 & 0 \end{pmatrix}\right]$.

As $F \neq \mathbb{C}$, $\mathcal{O}_{\mathcal{Q}(F)} \otimes_Z \hat{Z}_2 = M_2(\hat{Z}_2(\rho_{2^\ell} + \rho_{2^\ell}^{-1}))$ [C-M,2, Theorem 4.3.(i)], and the involution is given up to equivalence by

$$\begin{pmatrix} a & b \\ c & d \end{pmatrix} \to \begin{pmatrix} d & -b \\ -c & a \end{pmatrix}, \begin{pmatrix} \varepsilon_i & \\ & -1 \end{pmatrix} \longmapsto \begin{pmatrix} \varepsilon_i & 0 & & \\ 0 & \varepsilon_i & 0 & \\ & 0 & -1 & 0 \\ & & 0 & -1 \end{pmatrix}$$

Now, we remark that it is sufficient to study C_n in $\mathcal{O}_{\mathcal{Q}(F)} \otimes_Z \hat{\mathbb{Q}}_2$, since, by [C-M, 2, A-4] no information is lost by using $K' = \operatorname{im} K_1(\mathcal{O} \otimes \hat{Z}_2)$ in $K_1(\mathcal{O} \otimes \hat{\mathbb{Q}}_2)$. Here we may choose

$$C_2 = C^1 \begin{pmatrix} \varepsilon_i^{-1} & & & \\ & 1 & & \\ & & 1 & \\ & & & 1 \end{pmatrix}$$

where C^1 effects the isomorphism $\begin{pmatrix} 0 & I \\ I & 0 \end{pmatrix} = C^1 \begin{pmatrix} 1 & 0 & & \\ 0 & 1 & & \\ & & -1 & 0 \\ & & 0 & -1 \end{pmatrix}$

which is valid over $\mathbb{C}$, and clearly $\det C^1 = +1$. Thus, $\det C_2 = (\varepsilon_i^{-1})$ and 2.6 follows.

The situation is slightly different at the ordinary quaternion algebra $\mathcal{Q}(\mathbb{Q})$.

Lemma 2.8: *Let A_n be hyperbolic except at $M_n(\mathcal{Q}(\mathbb{Q}))$, then A_n has signature $2i$ and*

$$d([W, A_n, B_n, C_n]) = (-1)^i_{\mathcal{Q}} .$$

Proof: We may assume $\begin{pmatrix} 1 & 0 \\ 0 & 1 \end{pmatrix} = A_n$. Now there is a $\nu \in \hat{Z}_2 \otimes_Z Z(i, j, \frac{i+j+k+1}{2})$ with norm $\nu\bar{\nu} = -1$. Set

$$C_2 = \begin{pmatrix} \frac{1}{2} & \frac{1}{2} \\ -1 & 1 \end{pmatrix} \begin{pmatrix} \nu & 0 \\ 0 & 1 \end{pmatrix}$$

and $N(C_2) = -1$.

The remaining cases are all type II algebras of the form $M_n(Z(\rho_{2^i}))$ or $M_n(Z(\rho_{2^i} - \rho_{2^i}^{-1}))$.

§3. The type II algebras and theorem B.

We begin by obtaining the structure of the units over complex conjugation t in the rings $\hat{Z}_2(\rho_{2^i})$, $\hat{Z}_2(\tau_i)$ where $\tau_i = \rho_{2^i} - \rho_{2^i}^{-1}$. For the next 2 results we assume $i \geq 3$.

Theorem 3.1: Let $\varepsilon_1 = 1 + \rho_{2^i} + \rho_{2^i}^{-1} = 1 + \lambda_i$. There is a unit $\nu(i)$ such that $\nu(i)t(\nu(i)) = -1$ and as a module over t we have $\hat{Z}_2(\rho_{2^i})^{\cdot} = \hat{Z}_{2+} \times \bar{M}_- \times \Pi\, \hat{Z}_2(t)/t^2 =$ Moreover $\hat{Z}_{2+}$ is generated by ε_1, and M_- is the module $\hat{Z}_2 \times Z/2^i$ with t action $t(a,b) = (-a,-b+2^{i-1})$. The generators of M_- are ν and ρ_{2^i}.

Proof: Using Artin reciprocity the norms in $\hat{Z}_2(\lambda_i)$ of $\hat{Z}_2(\rho_{2^i})$ have index 2 and ε_1 is not a norm since its norm in $\hat{Z}_2$ is -1. Now $(\hat{Z}_2(\rho_{2^i}))^{\cdot} = Z/2^i \times (\hat{Z}_2)^{2^{i-1}}$ and $\hat{Z}_2(\lambda_i)^{\cdot} = Z/2 \times (\hat{Z}_2)^{2^{i-2}}$. Write the generators of this latter group $-1, \varepsilon_1, n_2 \ldots n_{2^{i-2}}$ where the n_i are all norms, say $n_i = w_i \cdot t(w_i)$. Clearly, the w_i, ε_1, ν and ρ_{2^i} generate $\hat{Z}_2(\rho_{2^i})^{\cdot}$ and 3.1 follows directly.

Similarly, we have

Theorem 3.2: In $\hat{Z}_2(\tau_{i+1})$ there is a unit $\nu(i+1)$ with $\nu(i+1)t(\nu(i+1)) = -1$ and as a module over t we have

$$\hat{Z}_2(\tau_{i+1})^{\cdot} = \hat{Z}_2^{+} \times M_- \times \Pi\, \hat{Z}_2(t)/t^2 = 1$$

Moreover, $\hat{Z}_2^{+}$, M_- are given as in 3.1.

Remark 3.3: The only differences in these 2 discriptions comes on comparing the images of global units, which give all but 1 of the η_i and are the same. However, in $\hat{Z}_2(\tau_{i+1}) - \hat{Z}_2(\lambda_i)$, $\varepsilon_1 5$ is the remaining η_i while in $\hat{Z}_2(\rho_{2^i}) - \hat{Z}_2(\lambda_i)$ the remaining η_i can be taken to be 5.

Remark 3.4: The cases not covered in the above are $\hat{Z}_2(i) - \hat{Z}_2$ where $\hat{Z}_2(i)^{\cdot} = Z/4 \times \hat{Z}_2(t)/t^2{=}1$ with generators i, $i + 2i$, and $\hat{Z}_2(\sqrt{-2}) - \hat{Z}_2$ where $\hat{Z}_2(\sqrt{-2})^{\cdot} = Z/2 \times \hat{Z}_2(t)/t^2{=}1$ with generators -1, $1 + \sqrt{-2}$.

Hence, as in [C-M,2, A.8, A.9] on factoring out global units (and typos) we have

3.5 $$W_\ell(\hat{Z}_2(\rho_{2^i}),t) = (Z/2^\ell)_-^{2^{i-2}} \times Z/2^\ell(t)/t^2{=}1$$

with generators ν, w_j, $1+2i$, where $(w_j)t(w_j) = \eta_j$ a global unit, for $i \geqslant 3$

3.6 $$W_\ell(\hat{Z}_2(i),t) = Z/2^\ell(t)/t^2{=}1 \quad .$$

Also,

3.7 $$W_\ell(\hat{Z}_2(\tau_{i+1}),t) = (Z/2^\ell)^{2^{i-2}} \times Z/2^\ell(t)/t^2{=}1$$

with generators ν, w_j, w, $wt(w) = 5\varepsilon_1$, $i \geqslant 3$ and $w_j t(w_j) = \eta_j$ a global unit, while

3.8 $$W_\ell(\hat{Z}_2(\sqrt{-2}),t) = Z/2^\ell(t)/t^2{=}1$$

with generator $1 + \sqrt{-2}$.

Now we have

<u>Theorem</u> 3.9: <u>The image of</u> d <u>in the</u> W_ℓ <u>above is precisely the</u> w_j <u>with norm a global unit</u>.

<u>Proof</u>: From [M-H, p. 118, example 2] we have that ker(rank homomorphism) $r : W(\hat{\mathbb{Q}}_2(\rho_{2^i})) \to Z/2$ is $\mathbb{Z}/2$ generated by $\langle\varepsilon_1\rangle - \langle 1\rangle$, $r : W(\hat{\mathbb{Q}}_2(\tau_{i+1})) \to Z/2$ is $\langle\varepsilon_1\rangle - \langle 1\rangle$ for $i \geq 3$ and $\langle -1\rangle - \langle 1\rangle$ in the remaining cases. In particular, the forms $\langle\eta_1\rangle - \langle 1\rangle$ for η_i a global unit are all trivial. But rationally

$$\begin{pmatrix} w_i^{-1} & 0 \\ 0 & 1 \end{pmatrix}\begin{pmatrix} \eta_i & 0 \\ 0 & -1 \end{pmatrix}\begin{pmatrix} t(w_i^{-1}) & 0 \\ 0 & 1 \end{pmatrix} = \begin{pmatrix} 1 & 0 \\ 0 & -1 \end{pmatrix}$$

So 3.9 follows, on checking from §1 that the $\langle\eta_i\rangle - \langle 1\rangle$, $\langle -1\rangle - \langle 1\rangle$, $2(\langle\varepsilon_1\rangle - \langle 1\rangle)$ generate the piece of $L_o^P(Z(G))$ coming from this representation.

Finally, putting 3.9 together with 2.5-2.8, and checking, again using §1, that every element in $L_o^P(Z(G))$ goes to 0 in $L_o(\mathcal{M}(G) \otimes_Z \hat{Z}_2)$, we see that we have completely determined d so theorem B follows.

§4. The proof of theorem C.

We use the notation of [C-M,2, Appendix], and begin by observing that according to [F-K-W], $D(Q_{2^i,2}) = \tilde{K}_o(Z(Q_{2^i,2}))_{(2)} = Z/2$, and in particular, for $Q_{2,2}$, the element $(1_{\mathcal{Q}}, 1_{--}, 1_{-+}, 1_{+-}, \langle 3\rangle_{++})$ (where $\mathcal{Q}$ is the quaternion representation and $(\pm,\pm)$ are the 1 dimensional representations) represents the generator.

But the unit $1 + x + y$ in $\hat{Z}_2(Q_{2,2})$ has image $\langle 3,1,1,1,3\rangle$, and $1 + 2xy + 2y \mapsto \langle 1,1,1,3,3\rangle$ $1 + 2x \mapsto \langle -3,1,1,3,3\rangle$ so the product $(1 + x + y)(1 + 2x)(1 + 2xy + 2y)^{-1} \mapsto \langle -9,1,1,1,3\rangle$ and, on factoring out squares, we have that $(-1,1,1,1,1)$ also generates $D(Q_{2,2})$. By 2.8 this last element is in the image of d. On the other hand, by definition the elements $(1,\ldots 1,\theta_{++})$ are the image of the Swan homomorphism T. Hence we have

Theorem 4.1: For $G = Q_{2,2}, K_o(G) = D(G) = Z/2$ is in the image of both T and d.

More generally

Theorem 4.2: a) The Swan homomorphism

$$T : (Z/2^{i+2})^{\cdot} \to D(Q_{2^i,2}) = Z/2$$

is surjective.

b) For $Q_{2^i,2}$, $i > 1$ the non-trivial element in $D(Q_{2^i,2})$ is represented by ε_1 at the quaternion algebra and ones at the remaining representations.

Proof: We display the representations as $\mathcal{Q}$, $M_2(Q(\lambda_i))$, $M_2(Q(\lambda_{i-1}))$, ..., $M_2(Q)$, Q_{--}, Q_{-+}, Q_{+-}, Q_{++} where $\lambda_i = \rho_{2^i} + \rho_{2^i}^{-1}$ then in $W_\ell(Q_{2^i,2})$ we have for $i \geqslant 3$

$$1 + x^{2^{i-2}} + y \mapsto (3,1,3,3,\ldots,3,1\_\_3,1,3)$$

$$1 + 2y \mapsto (-3\ ,\ 3,3,\ldots,3,1,\ 3,1,3)$$

$$1 + 2x^{2^{i-2}} \mapsto (-3,1,9\ \ldots\ \ 9,1,\ 3,1,3)$$

$$1 + 2x^{2^{i-3}} \mapsto (-3,3,1,9\ \ldots\ 9,1,\ 3,1,3)$$

etc. provided $2^{i-j} > 1$.

Comparing successive terms and factoring out squares we have

$$(1,3,1,\ldots\qquad ,1) =$$

$$(1,1,3,1,\ldots\qquad ,1) =$$

$$(1,1,1,3,1\ldots\qquad ,1) =$$

$$(1,1,\ldots,\qquad 3,1,1,1,1) =$$

$$(-1,1,1,\ldots,\ \ldots,\qquad 1) = 1$$

Next use

$$1 + x + y^2 \mapsto (1,3\ ,\ldots,\qquad 3,1,1,3,3)$$

$$3 \mapsto (9,9\ ,\ldots,\qquad 9,3,3,3,3)$$

$$-1 + x^{2^{i-1}} - x^{-2^{i-1}} \mapsto (5,1,1,\ldots,\ \ldots\qquad 1)$$

These imply $(3,1,\ldots,\ \ldots,\qquad 1) =$

$$(1\ \ldots,\ \ldots,\qquad 1,3,1,3,1) =$$

$$(1\ \ldots,\ \ldots,\qquad 1,1,3,3,1) =$$

$$(1\ \ldots,\ \ldots,\qquad 1,1,1,3,3) = 1$$

Finally, note

$$1 + x^s + y \rightarrow \quad (3 + \lambda(s), 1, \ldots, \ldots 1, 3) \quad \delta \text{ odd}$$

$$(3 + \lambda(s), 1, \ldots, \ldots) \quad \delta \text{ even}$$

where $\lambda(s) = \rho^{s}_{2^i} + \rho^{-s}_{2^i}$.

These relations show

$$(1, \ldots, \ldots, 1, 3)$$

generates $D(Q_{2^i,2})$ and prove (a).

To prove (b) we do arithmetic in $\hat{Z}_2(\lambda_i)$ but factor out squares. doing this we can factor out by the ideal $4(\lambda_i)$. Thus

$$(1 + \lambda(s)) \sim (1 + \lambda(s))^3 = (1 + 3\lambda(s))(1 + 3\lambda(s)^2)$$

but

$$\lambda(s)^2 = \lambda(2s) + 2$$

so

$$(1 + 3\lambda(s)^2) = (-1 + 3\lambda(2s))$$

$$\sim -1(1 + 5\lambda(2s))$$

$$\sim -1(1 + \lambda(2s))$$

Whence,

$$(1 + 3\lambda(s)) \sim (-1)\ (1 + \lambda(s))\ (1 + \lambda(2s)) \ .$$

and we obtain

$$(-1)^i(1+3\lambda(s))(1+3\lambda(2s)) \cdots (1+3\lambda(2^i s)) \sim (1+\lambda(s))$$

Using the already determined relations we now have

$$(1+\lambda(s),1,\ldots 1,1) \sim (1,1,\ldots 1,3)$$

for s odd. On the other hand, $\delta_s = 1+\lambda(s)$ is a global unit, as we see from [C-M,2, p. 27, lemma 4.5], with norm -1, and -1 together with the δ_s generate the global units mod squares. Hence, ε_i is an odd product of the δ_s, -1 and squares, which completes the proof.

Bibliography

[A-H] D. Anderson, I. Hambleton, "Balanced splittings of semi-free actions on homotopy spheres", these proceedings.

[C-M,1] G. Carlsson, R. J. Milgram, "Some exact sequences in the theory of Hermitian forms", J. Pure and Applied Algebra.

[C-M,2] __________, ________, "The structure of odd L-groups", these proceedings.

[C-M-P] __________, ________, W. Pardon (to appear).

[F-K-W], A. Fröhlich, M. E. Keating, S. M. J. Wilson,"The class groups of quaternion and dihedral 2-groups", Mathematika 21 (1974) 64-71.

[M] R. J. Milgram, "Evaluating the Swan finiteness obstruction for periodic groups", Stanford (1978).

[M-H] J. Milnor, D. Husemoller, Symmetric Bilinear Forms, Springer Verlag, 1973.

[P] T. Petri, "The Atiyah-Singer invariant, the Wall groups $L_n(\pi,1)$ and the function $(te^x + 1)/(te^x - 1)$",Ann. of Math. 92 (1970) 174-187.

[R] A. Ranicki, "On the algebraic L-theory of semi-simple rings", J. of Algebra 50 (1978) 242-243.

[S] R. Swan, E. G. Evans, K-theory of Finite Groups and Orders, Springer Verlag,Lecture Notes in Mathematics #149 (1970).

[W] C. T. C. Wall, "Formulae for the surgery obstruction", Topology 15 (1976) 189-210.

Institute for Advanced Study
Stanford University

November 13, 1978

Whitehead torsion for PL fiber homotopy equivalences

by

Hans J. Munkholm*)

I. Results

Let E and B be finite, connected polyhedra and $p : E \to B$ a PL fibration in the sense of Hatcher [Ha]. Let $A \subseteq B$ be a subpolyhedron, and the inclusion a homotopy equivalence with Whitehead torsion $\tau(B,A) \in Wh(\pi_1(B))$, see e.g. [Mi]. We study the relationship between $\tau(B,A)$ and $\tau(E,E_A)$ where $E_A = p^{-1}(A)$.

Recall, [Qu, p. 1o3] and [MP], that some group homomorphisms $\varphi: \pi \to \rho$ give rise to a transfer map $\varphi^*: Wh(\rho) \to Wh(\pi)$. For example this is the case if $[\rho : \mathrm{Im}\,\varphi] < \infty$ and $\nu = \mathrm{Ker}(\varphi)$ is a group of type (FF), i.e. $\mathbb{Z}$ admits a finite resolution by finitely generated free modules over $\mathbb{Z}\nu$.

Theorem A If p has fiber $B\nu$ where ν is of type (FF) and $\pi_1(F) \to \pi_1(E)$ is injective, then

$$\tau(E,E_A) = \varphi^*(\tau(B,A))$$

where $\varphi = p_* : \pi_1(E) \to \pi_1(B)$.

Remark Note that any compact, flat riemannian or hyperbolic manifold is such a $B\nu$. Especially, the theorem covers circle- (and torus-) fibrations with $\pi_1(F) \to \pi_1(E)$ injective.

If $\varphi : \pi_1(E) \to \pi_1(B)$ is onto and has $\mathrm{Ker}(\varphi) = C_k$ = cyclic of order k (with generator t) then there is a pull back diagram of rings

$$\begin{array}{ccc} \mathbb{Z}\pi_1(E) & \xrightarrow{\varphi} & \mathbb{Z}\pi_1(B) \\ r\downarrow & & \downarrow\bar{r} \\ \mathbb{Z}\pi_1(E)/(N) & \xrightarrow[\psi]{} & (\mathbb{Z}/k)\pi_1(B) \end{array}$$

*) Princeton University and Odense University. Partially supported by the Danish Natural Science Research Council.

where (N) is the ideal generated by $N = 1+t+t^2+\dots+t^{k-1}$. The transfer ψ^* is defined, see [MP]. Also there is an automorphism β of $(\mathbb{Z}/k)\pi_1(B)$ sending $\overline{r}(\overline{\xi})$ to $\overline{r}(n\overline{\xi})$ where $\overline{\xi} = \varphi(\xi)$, $\xi \in \pi_1(E)$ and $\xi t \xi^{-1} = t^n$. Also, β induces a transfer map β^*(which equals $(\beta_*)^{-1}$).

<u>Theorem B</u> Let p have fiber $F = S^{2\ell-1}/C_k$, $\ell \geq 1$. Assume that $\pi_1(F) \to \pi_1(E)$ maps onto a copy of C_k. Then

$$r_*\tau(E,E_A) = \Sigma_{i=0}^{\ell-1} \psi^*(\beta^*)^i \, \overline{r}_* \, \tau(B,A)$$

<u>Remarks</u> 1. For $\ell = 1$ this deals with S^1-fibrations having $\pi_1(F) \to \pi_1(E)$ non-injective. For $\ell > 1$ it deals with lens space fibrations having $\pi_1(F) \to \pi_1(E)$ injective.

One can sometimes compute the transfer maps. Thus we get

<u>Corollary C</u> For any orientable S^1-fibration $p : E \to B$ with

$$\mathrm{Im}\,(\pi_1(S^1)) \cap [\pi_1(E),\pi_1(E)] = 1$$

one has

(i) $\tau(E,E_A) = 0$, if $\pi_1(S^1) \to \pi_1(E)$ is injective

(ii) $r_*\tau(E,E_A) = 0$, otherwise

<u>Remark</u> K. Ehrlich [Eh_1,Eh_2] has shown that $\tau(E,E_A) = 0$ also in the case (ii) of Corollary C.

Theorem A is a special case of

<u>Theorem D</u> Assume that $p_* : \pi_1(E) \to \pi_1(B)$ factors as $\pi_1(E) \xrightarrow{s} \pi \xrightarrow{\varphi} \pi_1(B)$ where s is onto and $\nu = \mathrm{Ker}(\varphi)$ is of type (FF). Let $\overline{F}$ be the covering of the base point compo-

nent F_o of F corresponding to the kernel of the epimorphism $\pi_1(F_o) \to \pi_1(E) \to \nu$. Assume that

(i) ν acts trivially on $H_*(\bar{F};\mathbb{Z})$

(ii) $H_*(\bar{F};\mathbb{Z})$ is finitely generated over $\mathbb{Z}$.

Then

$$s_*(\tau(E,E_A)) = \varphi^*(\tau(B,A)) \cdot \Sigma(-1)^i [H_i(\bar{F};\mathbb{Z})]$$

Remarks 1. $[H_i(\bar{F};\mathbb{Z})]$ is the element of the integral representation ring $G(\pi)$, in the sense of [PT] , represented by $H_i(F;\mathbb{Z})$. The dot indicates the right $G(\pi)$ module structure on $Wh(\pi)$.

2. If p is a PL fiber bundle, if $\pi=\pi_1(B)$ (so φ=identity), and each $H_i(\bar{F};\mathbb{Z})$ is free over $\mathbb{Z}$, then Theorem D becomes the main result of Anderson, [An_1]. In [Mu] we removed the freeness condition.

Further references. Related results are found also in [An_2,An_3,An_4].

2. Outline of proof of Theorem D

Step 1 We generalize Milnor's treatment of torsion of based chain complexes with based homology, see [Mi], to the case of finite chain complexes C_* where each C_i and each $H_i(C_*)$ has a preferred b.f.r. Here a b.f.r. for an R module M is a finite resolution

$$0 \to F_n \to \dots \to F_o \to M \to 0$$

of M together with a basis for each F_i . Given two such, say $\varepsilon : F_* \to M$ and $\varepsilon' : F'_* \to M$, one has a chain map $\varphi_*: F_* \to F'_*$ lifting 1_M . The mapping cone, C_*, of φ_* is a based acyclic chain complex.

We let $-[\varepsilon/\varepsilon'] = \tau(C_*) \in \bar{K}_1(R)$, see [Mi] . A preferred b.f.r. then means an equivalence class of b.f.r.'s where $\varepsilon \sim \varepsilon'$ if $[\varepsilon/\varepsilon'] = 0$. All of sections 2,3,4 of [Mi] now generalizes in a straightforward way. Thus we shall speak of $\tau(C_*) \in \bar{K}_1(R)$ for chain complexes where C_i and $H_i(C_*)$ have preferred b.f.r.'s over R .

Remarks 1. A similar generalization, using slightly different methods, has been given by Maumary [Ma] .

2. Details of the above, for the case where all b.f.r.'s have length ≤ 1, can be found in [Mu] .

Step 2 Let C_* be a finitely filtered, finite chain complex of R modules. Assume that each term $E^r_{s,t}$ in the resulting spectral sequence has a preferred b.f.r. $\varepsilon^r_{s,t}$, and that $\varepsilon^r_{s,t}$ is constant for r large. There results preferred b.f.r.'s

$$\gamma_n = \varepsilon^o_{0,n}\, \varepsilon^o_{1,n-1} \cdots \varepsilon^o_{n,0}$$

$$\chi_n = \varepsilon^\infty_{0,n}\, \varepsilon^\infty_{1,n-1} \cdots \varepsilon^\infty_{n,0}$$

for C_n and H_nC (notation as in [Mi]). As in [Mi] one shows that each $B^r_{**} = \mathrm{Im}(d^r : E^r_{**} \to E^r_{**})$ admits some preferred b.f.r., say $\beta^r_{s,t}$, and if one lets

$$(1) \qquad \tau(E^*_{**}) = \Sigma(-1)^{s+t} [\beta^r_{s,t}\, \varepsilon^{r+1}_{s,t}\, \beta^r_{s-r,t+r-1}/\varepsilon^r_{s,t}]$$

then this depends only on ε^o_{**} and ε^∞_{**} . Also, using the above γ_n, χ_n

$$(2) \qquad \tau(E^*_{**}) = \tau(C_*)$$

Remarks 1. Maumary, [Ma] , has a similar result, assuming that the chain complexes $E^o_{s,*}$ have $\tau(E^o_{s,*}) = 0$ (w.r.t. the b.f.r.'s $\varepsilon^o_{s,t}$ and $\varepsilon^1_{s,*}$)

2. The special case where all b.f.r.'s have length ≤ 1 is in [Mu] .

Step 3 We relate the torsion of chain complexes and the $G(\pi)$ module structure on $Wh(\pi)$ as follows.

(3) $$\tau(C_* \otimes_{\mathbb{Z}} N) = \tau(C_*) \cdot [N]$$

Here C_* is a chain complex of $\mathbb{Z}\pi$ modules; each C_i and each $H_i(C_*)$ is $\mathbb{Z}$-free and has a preferred b.f.r. Also N is a $\mathbb{Z}\pi$ module, finitely generated over $\mathbb{Z}$. The formula means that each $C_i \otimes_{\mathbb{Z}} N$ and each $H_i(C_* \otimes_{\mathbb{Z}} N) = H_i(C_*) \otimes_{\mathbb{Z}} N$ (diagonal π-action) inherit preferred b.f.r.'s and that, so equipped, $C_* \otimes_{\mathbb{Z}} N$ has torsion equal to $\tau(C_*) \cdot [N]$.

Remark The special case, where C_i and $H_i(C_*)$ are all based, is proved in [Mu] .

Step 4 Let $\varphi : \pi \to \rho$ be onto and have kernel ν of type (FF). From a b.f.r., say $\chi : P_* \to \mathbb{Z}$, of $\mathbb{Z}$ as a trivial ν-module, one can construct a b.f.r. for $\varphi^*(\mathbb{Z}\rho)$

$$1 \otimes \chi : \mathbb{Z}\pi \otimes_{\mathbb{Z}\nu} P_* \to \mathbb{Z}\pi \otimes_{\mathbb{Z}\nu} \mathbb{Z} = \varphi^*(\mathbb{Z}\rho)$$

over $\mathbb{Z}\pi$. This in turn can be modified, see lemma 7.1 of [Ba] , to a b.f.r.

$$\chi_1 : Q_* \to \varphi^*(\mathbb{Z}\rho)$$

with the property that any automorphism of $\varphi^*(\mathbb{Z}\rho)^n$ (direct sum of n copies) lifts to a map of resolutions $Q^n_* \to Q^n_*$

which is an automorphism on each Q_i^n .

If M is a $\mathbb{Z}\rho$ module with basis μ then we may view μ as an isomorphism $\varphi^*(\mathbb{Z}\rho)^n \to \varphi^*(M)$ $(n = \mathrm{rank}_{\mathbb{Z}\rho}(M))$. And we may equip $\varphi^*(M)$ with the b.f.r.

$$\mu\chi_1^n : Q_*^n \to \varphi^*(M)$$

Thus every based $\mathbb{Z}\rho$ module M gives rise to a $\mathbb{Z}\pi$ module $\varphi^*(M)$ with a preferred b.f.r.

Now, if C_* is a based complex with based homology modules over $\mathbb{Z}\rho$ then $\varphi^*(C_*)$ is a complex of $\mathbb{Z}\pi$ modules with preferred b.f.r.´s for $\varphi^*(C_i)$ and $H_i(\varphi^*(C_*))$. Therefore, $\tau(\varphi^*(C_*))$ is defined and the terminology is not misleading, i.e. one can prove the formula

$$(4) \qquad \varphi^*\tau(C_*) = \tau(\varphi^*(C_*))$$

in $\overline{K}_1(\mathbb{Z}\pi)$ (and in $\mathrm{Wh}(\pi)$) .

Note that (4) especially implies that $\tau(\varphi^*(C_*))$ is independent of the choice of χ_1 .

<u>Step 5</u> To prove Theorem D one first reduces to the case where $p_*(\pi_1(E)) = \pi_1(B)$. This is done by showing that there is a diagram of fibrations

$$\begin{array}{ccc} F & \supset & F_o \\ \downarrow & & \downarrow \\ E & = & E \\ \downarrow & & \downarrow \\ B & \longleftarrow & \tilde{B}/\overline{\pi} \end{array}$$

where F_o is the base point component of F , $\overline{\pi} = \mathrm{Im}(p_*)$, and $\tilde{B}$ is the universal covering of B .

Once p_*, and φ, is onto one considers

$$\begin{array}{ccccc} F & = & F & \leftarrow & \overline{F} \\ \downarrow & & \downarrow & & \downarrow \\ E & \leftarrow & \overline{E} & \leftarrow & \hat{E} \\ p\downarrow & & \downarrow\overline{p} & & \downarrow\hat{p} \\ B & \underset{q}{\leftarrow} & \overline{B} & = & B \end{array}$$

where $\overline{p}$ is the pull back of p via q and $\hat{E}$ is the covering of $\overline{E}$ corresponding to the epimorphism $\pi_1(\overline{E}) = \mathrm{Im}(\pi_1(F) \to \pi_1(E)) \to \nu = \mathrm{Ker}(\varphi)$. By definition $s_*\tau(E,E_A) = \tau(C_*(\hat{E},\hat{E}_A))$ where C_* is the simplicial chain functor. Assume that p, and hence $\hat{p}$, are simplicial; filter $C_*(\hat{E},\hat{E}_A)$ by inverse images of skeleta of $(\tilde{B},\tilde{A})$. The resulting spectral sequence is a spectral sequence of $\mathbb{Z}\pi$ modules and

$$E^1_{*t} = \varphi^* C_*(\tilde{B},\tilde{A}) \otimes_{\mathbb{Z}} H_t(\overline{F};\mathbb{Z}) \tag{5}$$

where π acts diagonally and the action on $H_t(\overline{F};\mathbb{Z})$ is explained e.g. in [MP]. Also $E^2_{**} = 0$ and $\tau(B,A) = \tau(C_*(\tilde{B},\tilde{A}))$, so using (2), (1), (5), (4), and (3)

$$\begin{aligned} s_*\tau(E,E_A) &= \tau(C_*(\hat{E},\hat{E}_A)) \\ &= \tau(E^*_{**}) \\ &= \Sigma(-1)^s\ \tau(E^o_{s,*}) + \Sigma(-1)^t\ \tau(E^1_{*,t}) \\ &= \Sigma(-1)^s\ \tau(E^o_{s,*}) + \varphi^*(\tau(B,A))\cdot\Sigma(-1)^t[H_t(\overline{F};\mathbb{Z})] \end{aligned}$$

Thus we only have to show that $\Sigma(-1)^s\tau(E^o_{s,*}) = 0$. This occupies the rest of the present section.

Step 6 For computing $\Sigma(-1)^s\ \tau(E^o_{s,*})$ we start by noticing that

$$(-1)^s\ \tau(E^o_{s,*}) = \tau(C_*(\hat{E}_s,\hat{E}_{s-1})) \tag{6}$$

where $\hat{E}_s = \hat{p}^{-1}(\text{s-skeleton of } (\tilde{B},\tilde{A}))$; $C_*(\hat{E}_s,\hat{E}_{s-1})$ has its standard basis, and $H_*(\hat{E}_s,\hat{E}_{s-1}) = E^1_{s,*}$ has the preferred b.f.r. coming via the isomorphism

(7) $\theta : \varphi^* C_*(\tilde{B},\tilde{A}) \otimes_{\mathbb{Z}} H_*(\bar{F};\mathbb{Z}) \to E^1_{s,*}$

(cfr. steps 3 and 4). For suitable triangulations one has a commutative diagram

$$\begin{array}{ccc} \coprod_{\sigma} \pi \times_\nu (M(\bar{f}_1^{(\sigma)},\ldots,\bar{f}_s^{(\sigma)}),\ \dot{M}(\bar{f}_1^{(\sigma)},\ldots,\bar{f}_s^{(\sigma)})) & \xrightarrow{\hat{\psi}} & (\hat{E}_s,\hat{E}_{s-1}) \\ \downarrow & & \downarrow \\ \coprod_{\sigma}(M(f_1^{(\sigma)},\ldots,f_s^{(\sigma)}),\ \dot{M}(f_1^{(\sigma)},\ldots,f_s^{(\sigma)})) & \xrightarrow{\psi} & (E_s,E_{s-1}) \end{array}$$

Here σ ranges over the s-simplices of B . ψ comes from an iterated mapping cylinder structure for $E \to B$. Thus $M(f_1^{(\sigma)},\ldots,f_s^{(\sigma)})$ is the iterated mapping cylinder for simplicial maps $f_i^{(\sigma)}:F_{\sigma(i-1)} \to F_{\sigma(i)}$ where $F_{\sigma(i)} = p^{-1}(i^{th}$ vertex $\sigma(i)$ of $\sigma)$. Moreover, for each simplex σ of B we have fixed a lifted simplex $\tilde{\sigma}$ of $\tilde{B}$; and $\bar{f}_i^{(\sigma)} : \bar{F}_{\tilde{\sigma}(i-1)} \to \bar{F}_{\tilde{\sigma}(i)}$ is a lifting of $f_i^{(\sigma)}$ to $\bar{F}_{\tilde{\sigma}(i-1)} = \hat{p}^{-1}(\tilde{\sigma}(i-1))$ which covers $F_{\sigma(i-1)}$ with covering transformation group isomorphic to ν . $\hat{\psi}$ is π-equivariant, and ψ and $\hat{\psi}$ are relative homeomorphisms. Since $\tilde{B}$ is simply connected there is a well defined isomorphism $H_*(\bar{F}) \to H_*(\bar{F}_{\tilde{\sigma}(0)})$ "induced" by any path in $\tilde{B}$ from the base point to $\tilde{\sigma}(0)$. Also $H_{*-s}(\bar{F}_{\tilde{\sigma}(0)}) \cong H_*(M^{(\sigma)}, \dot{M}^{(\sigma)})$, where $M^{(\sigma)} = M(\bar{f}_1^{(\sigma)},\ldots,\bar{f}_s^{(\sigma)})$, induced by the map $\bar{F}_{\tilde{\sigma}(0)} \to \Delta^s \times \bar{F}_{\tilde{\sigma}(0)} \xrightarrow{proj} M^{(\sigma)}$. We use the composite isomorphism to identify $H_*(M^{(\sigma)}, \dot{M}^{(\sigma)})$ with $H_{*-s}(\bar{F})$. Since $\mathbb{Z}$ has a b.f.r. over $\mathbb{Z}\nu$ so does any finitely generated abelian group (viewed as trivial module over ν). Fix a b.f.r., say χ , for $\tilde{H}_{*-s}(\bar{F})$. Use the above isomorphism to give a b.f.r., say, $\chi^{(\sigma)}$ for $H_*(M^{(\sigma)}, \dot{M}^{(\sigma)})$.
Finally equip

$$H_*(\coprod_{\sigma}\pi\times_\nu(M^{(\sigma)},\dot{M}^{(\sigma)})) = \bigoplus_{\sigma} \mathbb{Z}\pi \otimes_{\mathbb{Z}\nu} H_*(M^{(\sigma)},\dot{M}^{(\sigma)})$$

with the b.f.r. $\bigoplus_{\sigma}(\mathbb{Z}\pi \otimes_{\mathbb{Z}\nu} \chi^{(\sigma)})$ (terminology obvious). Then $\tau(C_*(\coprod_{\sigma} \pi\times_{\nu}(M^{(\sigma)},\dot{M}^{(\sigma)}))) \in \overline{K}_1(\mathbb{Z}\pi)$ is defined (use the obvious $\mathbb{Z}\pi$-basis for the chains), and Anderson's excision lemma, [An$_1$], (suitably generalized to allow b.f.r.'s rather than bases) applies to give

(8) $$\tau(C_*(\coprod_{\sigma} \pi\times_{\nu}(M^{(\sigma)},\dot{M}^{(\sigma)}))) = \tau(C_*(\hat{E}_s,\hat{E}_{s-1}))$$

provided that we prove (see step 7)

(9) $$\hat{\psi}_*: H_*(\coprod_{\sigma} \pi\times_{\nu}(M^{(\sigma)},\dot{M}^{(\sigma)})) \to H_*(\hat{E}_s,\hat{E}_{s-1}) \text{ is simple}$$

It is obvious that

(10) $$\tau(C_*(\coprod_{\sigma} \pi\times_{\nu}(M^{(\sigma)},\dot{M}^{(\sigma)}))) = \sum_{\sigma} i_*\tau(C_*(M^{(\sigma)},\dot{M}^{(\sigma)}))$$

where $i : \nu \to \pi$ is the inclusion. Also, standard excision arguments show that

(11) $$\tau(C_*(M^{(\sigma)},\dot{M}^{(\sigma)})) = (-1)^s\tau(C_*(\overline{F}_{\sigma(s)})) = (-1)^s\tau(C_*(\overline{F}))$$

Now (6), (8)-(11) show that

$$\Sigma(-1)^s\tau(E^o_{s,*}) = \chi(B,A)\tau(C_*(\overline{F}))$$

which vanishes, since the Euler characteristic does so. Note that in (11) one uses the fact, [Ha], that $\overline{F}_{\sigma(s)}$ and $\overline{F}$ are simple homotopy equivalent.

Step 7 We must still show that $\hat{\psi}_*$ is simple, cfr. (9). This is done by making θ explicit, cfr. (7). Since $\tilde{B}$ is simply connected, for each simplex $\xi\tilde{\sigma}$ of $\tilde{B}$ ($\tilde{\sigma}$ a lift of σ, $\xi \in \pi_1(B)$) there is a well defined isomorphism

$$H_*(\overline{F}) \to H_*(\overline{F}_{\sigma(0)})$$

"induced" by a path from the base point to $\xi\,\tilde{\sigma}(0)$. Note that $\hat{p}^{-1}(\xi\tilde{\sigma}(0)) = \overline{F}_{\sigma(0)}$. One may piece these together to define an isomorphism of $\mathbb{Z}\pi_1(B)$ modules

$$\theta' : C_*(\tilde{B},\tilde{A}) \otimes_{\mathbb{Z}} H_*(\bar{F}) \to C_*(\tilde{B},\tilde{A}\ ;\mathcal{H}_*(\bar{F}_{.(0)}))$$

where $\pi_1(B)$ acts diagonally on the $\otimes$ product. Now the twisted chain module $C_*(\tilde{B},\tilde{A}\ ;\mathcal{H}_*(\bar{F}_{.(0)}))$ is precisely

$$\begin{aligned} H_*(\coprod_\sigma \pi\times_\nu(M^{(\sigma)},\dot{M}^{(\sigma)})) &= \bigoplus_\sigma \mathbb{Z}\,\pi \otimes_{\mathbb{Z}\nu} H_*(M^{(\sigma)},\dot{M}^{(\sigma)}) \\ &= \bigoplus_\sigma \mathbb{Z}\,\bar{\pi} \otimes_{\mathbb{Z}} H_*(M^{(\sigma)},\dot{M}^{(\sigma)}) \end{aligned}$$

(Here we need the action of ν on $H_*(\bar{F})$ to be trivial). And θ is given by

$$\theta = \hat{\psi}_*\ \theta'$$

The way we chose b.f.r.'s this means that we must check that (the inverse of) θ' is simple. But θ' is the direct sum of isomorphisms

$$\theta'_\sigma : \varphi^*(\mathbb{Z}\,\bar{\pi} \otimes_{\mathbb{Z}} H_*(\bar{F})) \to \varphi^*(\mathbb{Z}\,\bar{\pi} \otimes_{\mathbb{Z}} H_*(\bar{F}_{\sigma(0)}))$$

where $\bar{\pi} = \pi_1(B)$ acts diagonally on the left hand side and on the left factor only on the right hand side. Each θ'_σ can be shown to be simple. Hence so is θ.

3. Outline of other proofs.

Theorem A is an immediate corollary of Theorem D. Theorem B is proved just as Theorem D, using $\mathbb{Z}\,\pi_1(E)/(N) \otimes_{\mathbb{Z}\pi_1(E)} C_*(\tilde{E},\tilde{E}_A)$, $\tilde{E}$ the universal covering of E. It is the simple homology structure of $\bar{F} = S^{2\ell-1}$ that makes it possible to compute the E^1-term. The computations are analogous to those of section 5 of [MP].

Finally we note that under the condition (i) of corollary C one has a homomorphism $\rho : \pi_1(E) = \pi \to \nu = \mathbb{Z}$ s.t. $\nu \to \pi \to \nu$ is multiplication by some $m \neq 0$. Now let $\bar{\pi} = \pi_1(B)$ and let $\bar{F}$ be a free $\mathbb{Z}\bar{\pi}$ module with an automorphism $\bar{f}$ representing the element $x \in K_1(\mathbb{Z}\bar{\pi})$. One may verify that

$$F = \rho^*(\mathbb{Z}\nu) \otimes_{\mathbb{Z}} \varphi^*(\overline{F})$$

and

$$f = \rho^*(\mathbb{Z}\nu) \otimes_{\mathbb{Z}} \varphi^*(\overline{f})$$

define a free module F over $\mathbb{Z}\pi$ and an automorphism f of F. Furthermore there is a commutative diagram

$$\begin{array}{ccccccccc} 0 & \to & F & \xrightarrow{\alpha} & F & \xrightarrow{\xi} & \overline{F} & \to & 0 \\ & & \downarrow f & & \downarrow f & & \downarrow \overline{f} & & \\ 0 & \to & F & \xrightarrow{\alpha} & F & \xrightarrow{\xi} & \overline{F} & \to & 0 \end{array}$$

where $\alpha = \rho^*(r_{t-1}) \otimes_{\mathbb{Z}} \varphi^*(\overline{F})$, $r_{t-1} : \mathbb{Z}\nu \to \mathbb{Z}\nu$ is right multiplication by $t-1$, and $t \in \nu$ is a generator. Moreover, the row is exact. Therefore, $\varphi^*(x) = [F,f] - [F,f] = 0$.

A similar argument shows that $\psi^*\overline{r}_* = 0$ under the assumption in (ii).

4. References

[An_1] D.R. Anderson, The Whitehead torsion of the total space of a fiber bundle, Topology 11 (1972), 179-194.

[An_2] D.R. Anderson, Whitehead torsions vanish in many S^1 bundles, Inventiones Math. 13 (1971), 305-324.

[An_3] D.R. Anderson, A note on the Whitehead torsion of a bundle modulo a sub-bundle, Proc. Amer. Math. Soc. 32 (1972), 593-595.

[An_4] D.R. Anderson, The Whitehead torsion of a fiber homotopy equivalence, Mich. Math. J. 21 (1974), 171-180.

[Ba] H. Bass, Introduction to some methods of algebraic K-theory, Amer. Math. Soc., 1974.

[Eh_1] K. Ehrlich, Thesis, Cornell Univ., 1977.

[Eh_2] K. Ehrlich, Finiteness obstruction of fiber spaces, preprint, Purdue University, 1978.

[Ha] A.E. Hatcher, Higher simple homotopy theory, Ann. of Math. 102 (1975), 101-137.

[Ma] S. Maumary, Contributions á la théorie du type simple d'homotopie, Comm. Math. Helv. 44 (1969), 410-437.

[Mi] J. Milnor, Whitehead torsion, Bull. Amer. Math. Soc. 72 (1966), 358-426.

[MP] H.J. Munkholm and E.K. Pedersen, On the Wall finiteness obstruction for the total space of certain fibrations, preprint, Princeton University, 1978.

[Mu] H.J. Munkholm, Whitehead torsion for spectral sequences and for PL fiber bundles, preprint, Princeton University, 1978.

[Qu] D. Quillen, Higher algebraic K-theory: I, in Springer Lecture Notes in Math. vol. 341 (1973).

[PT] E.K. Pedersen and L. Taylor, The Wall finiteness obstruction for a fibration, to appear in Amer. J. Math.

Localization in quadratic L-theory

by Andrew Ranicki, Princeton University

Introduction

Localization is an indispensable tool in the computation of the surgery obstruction groups $L_n(\pi) \equiv L_n(\mathbb{Z}[\pi])$ (n(mod 4)) of Wall [3], at least for finite groups π. The L-groups $L_n(A)$ of a ring with involution A are compared with the L-groups $L_n(S^{-1}A)$ of the localization $S^{-1}A$ inverting some multiplicative subset $S \subset A$, the difference being measured by certain L-groups $L_n(A,S)$ depending on the category of S-torsion A-modules. In particular, if $A = \mathbb{Z}[\pi]$, $S = \mathbb{Z} - \{0\} \subset A$ and π is finite then $S^{-1}A = \mathbb{Q}[\pi]$ is semi-simple, and it is comparatively easy to compute $L_n(\mathbb{Q}[\pi])$, $L_n(\mathbb{Z}[\pi],S)$ and hence $L_n(\mathbb{Z}[\pi])$.

Localization in algebraic L-theory has already been studied by many authors, including Wall [1],[2],[6], Passman and Petrie [1], Connolly [1], Milnor and Husemoller [1], Bak and Scharlau [1], Karoubi [1], Pardon [1],[2], Carlsson and Milgram [1], though not in the generality obtained here.

The behaviour of the L-groups under localization is governed by an exact sequence of the type

$$\ldots \longrightarrow L_n(A) \longrightarrow L_n(S^{-1}A) \longrightarrow L_n(A,S) \longrightarrow L_{n-1}(A) \longrightarrow L_{n-1}(S^{-1}A) \longrightarrow \ldots.$$

Our immediate aim in this paper is to obtain a precise statement of this sequence (Proposition 2.4). We shall go some way towards a proof, but the detailed account is deferred to a projected instalment of the series "The algebraic theory of surgery" (Ranicki [2]), where we shall also prove a localization exact sequence of this type for symmetric L-theory.

Apart from the localization exact sequence itself we shall discuss the following applications:

- Let $\hat{A} = \varprojlim_{s \in S} A/sA$ be the S-adic completion of A. There are defined excision isomorphisms

$$L_n(A,S) \longrightarrow L_n(\hat{A},\hat{S}) \quad (n(\text{mod } 4))$$

and a Mayer-Vietoris exact sequence of the type

$$\ldots \longrightarrow L_n(A) \longrightarrow L_n(\hat{A}) \oplus L_n(S^{-1}A) \longrightarrow L_n(\hat{S}^{-1}\hat{A}) \longrightarrow L_{n-1}(A) \longrightarrow \ldots$$

(Proposition 3.2).

- If the ring A is an R-module then the symmetric Witt group $L^0(R)$ acts on the localization exact sequence of (A,S). This $L^0(R)$-module structure is used to prove that natural maps of the type

$$L_n(\mathbb{Z}[\pi]) \longrightarrow L_n(\mathbb{Q}[\pi]) \quad (n(\mathrm{mod}\ 4))$$

are isomorphisms modulo 8-torsion, and that the L-groups $L_n(\mathbb{Z}_m[\pi])$ are of exponent 8 (Propositions 4.2,4.4).

- If the ring A is an algebra over a Dedekind ring R and $S = R - \{0\} \subset A$ there are defined natural direct sum decompositions

$$L_n(A,S,\varepsilon) = \bigoplus_{\mathcal{P}} L_n(A,\mathcal{P}^{\infty},\varepsilon) \quad (n(\mathrm{mod}\ 4))$$

with $\mathcal{P}$ ranging over the non-zero prime ideals of R such that $\bar{\mathcal{P}} = \mathcal{P}$. The L-groups $L_n(A,\mathcal{P}^{\infty},\varepsilon)$ are defined using quadratic structures on $\mathcal{P}$-primary S-torsion A-modules. (Proposition 5.1).

We shall consistently use the language of forms and formations of Ranicki [1]. We shall omit the proofs of results of the following nature:

i) some relation, invariably called "cobordism", involving forms and formations is claimed to be an equivalence relation such that the equivalence classes define an abelian group with respect to the direct sum $\oplus$

ii) some function between such cobordism groups is claimed to be an isomorphism.

The chain complex formulation of quadratic L-theory in Ranicki [2] lends itself more readily to proofs of such results, those of type i) being obtained by an algebraic mimicry of the cobordism of manifolds, and those of type ii) by identifying cobordism groups of forms and formations with cobordism groups of quadratic Poincaré complexes. From the point of view of Ranicki [2] the L-groups $L_n(A)$ are defined for $n \geqslant 0$ to be the algebraic cobordism groups of pairs (C,ψ) such that C is an n-dimensional f.g. projective A-module chain complex and ψ is a quadratic structure inducing Poincaré duality $H^{n-*}(C) = H_*(C)$. The groups $L_n(A,S,\varepsilon)$ are defined for $n \geqslant 0$ to be the algebraic cobordism groups of pairs (D,θ) such that D is an (n+1)-dimensional f.g. projective A-module chain complex which becomes chain contractible over $S^{-1}A$ and θ is a quadratic structure inducing Poincaré duality $H^{n+1-*}(D) = H_*(D)$. It is relatively easy to prove the exact sequence

$$\ldots \longrightarrow L_n(A) \longrightarrow L_n(S^{-1}A) \longrightarrow L_n(A,S) \longrightarrow L_{n-1}(A) \longrightarrow L^S_{n-1}(S^{-1}A) \longrightarrow \ldots,$$

so that to obtain a localization exact sequence for the surgery obstruction groups it remains only to identify the chain complex L-groups with the 4-periodic L-groups defined using forms and formations. Although this identification can be used to both state and prove the localization exact sequence in terms of forms and formations we find the chain complex approach more illuminating, at least as far as proofs are concerned.

* * *

§1. Quadratic L-theory

We recall some of the definitions and results of Ranicki [1],[2].

Let A be an associative ring with 1, and with an involution

$$\bar{\ } : A \longrightarrow A \ ; \ a \longmapsto \bar{a}$$

such that

$$(\overline{ab}) = \bar{b}.\bar{a} \quad , \quad (\overline{a+b}) = \bar{a} + \bar{b} \quad , \quad \bar{1} = 1 \quad , \quad \bar{\bar{a}} = a \in A \qquad (a,b \in A).$$

A-modules will always be taken to have a left A-action.

The dual of an A-module M is the A-module

$$M^* = \mathrm{Hom}_A(M,A) \ ,$$

with A acting by

$$A \times M^* \longrightarrow M^* \ ; \ (a,f) \longmapsto (x \longmapsto f(x)\bar{a}) \ .$$

The dual of an A-module morphism $f \in \mathrm{Hom}_A(M,N)$ is the A-module morphism

$$f^* : N^* \longrightarrow M^* \ ; \ g \longmapsto (x \longmapsto g(f(x))) \ .$$

If M is a f.g. projective A-module then so is the dual M^*, and there is defined a natural A-module isomorphism

$$M \longrightarrow M^{**} \ ; \ x \longmapsto (f \longmapsto \overline{f(x)})$$

which we shall use to identify $M^{**} = M$.

Let $\varepsilon \in A$ be a central unit such that

$$\bar{\varepsilon} = \varepsilon^{-1} \in A$$

(for example, $\varepsilon = \pm 1$). Given a f.g. projective A-module M define the ε-duality involution

$$T_\varepsilon : \mathrm{Hom}_A(M,M^*) \longrightarrow \mathrm{Hom}_A(M,M^*) \ ; \ \varphi \longmapsto (\varepsilon\varphi^* : x \longmapsto (y \longmapsto \varepsilon\overline{\varphi(y)(x)})) \ ,$$

let

$$Q^\varepsilon(M) = \ker(1-T_\varepsilon : \mathrm{Hom}_A(M,M^*) \longrightarrow \mathrm{Hom}_A(M,M^*))$$

$$Q_\varepsilon(M) = \mathrm{coker}(1-T_\varepsilon : \mathrm{Hom}_A(M,M^*) \longrightarrow \mathrm{Hom}_A(M,M^*)) \ ,$$

and define a morphism of abelian groups

$$1+T_\varepsilon : Q_\varepsilon(M) \longrightarrow Q^\varepsilon(M) \ ; \ \psi \longmapsto \psi + \varepsilon\psi^* \ .$$

An $\begin{cases}\underline{\varepsilon\text{-symmetric}}\\ \underline{\varepsilon\text{-quadratic}}\end{cases}$ form over A $\begin{cases}(M,\varphi)\\(M,\psi)\end{cases}$ is a f.g. projective A-module M together with an element $\begin{cases}\varphi\in Q^{\varepsilon}(M)\\ \psi\in Q_{\varepsilon}(M)\end{cases}$. A morphism (resp. isomorphism) of such forms

$$\begin{cases}f : (M,\varphi)\longrightarrow(M',\varphi')\\ f : (M,\psi)\longrightarrow(M',\psi')\end{cases}$$

is an A-module morphism (resp. isomorphism) $f\in \mathrm{Hom}_A(M,M')$ such that

$$\begin{cases}f^*\varphi' f = \varphi\in Q^{\varepsilon}(M)\\ f^*\psi' f = \psi\in Q_{\varepsilon}(M)\ .\end{cases}$$

The form $\begin{cases}(M,\varphi)\\(M,\psi)\end{cases}$ is non-singular if $\begin{cases}\varphi\in \mathrm{Hom}_A(M,M^*)\\ \psi+\varepsilon\psi^*\in \mathrm{Hom}_A(M,M^*)\end{cases}$ is an isomorphism.

A sublagrangian of a non-singular $\begin{cases}\varepsilon\text{-symmetric}\\ \varepsilon\text{-quadratic}\end{cases}$ form over A $\begin{cases}(M,\varphi)\\(M,\psi)\end{cases}$ is a direct summand L of M such that the inclusion $j\in \mathrm{Hom}_A(L,M)$ defines a morphism of forms

$$\begin{cases}j : (L,0)\longrightarrow(M,\varphi)\\ j : (L,0)\longrightarrow(M,\psi)\ .\end{cases}$$

The annihilator of a sublagrangian L is the direct summand $L^{\perp}$ of M defined by

$$\begin{cases}L^{\perp} = \ker(j^*\varphi : M\longrightarrow L^*)\\ L^{\perp} = \ker(j^*(\psi+\varepsilon\psi^*) : M\longrightarrow L^*)\ .\end{cases}$$

A lagrangian is a sublagrangian L such that

$$L^{\perp} = L\ .$$

A non-singular $\begin{cases}\varepsilon\text{-symmetric}\\ \varepsilon\text{-quadratic}\end{cases}$ form over A is hyperbolic if it admits a lagrangian, or equivalently if it is isomorphic to the standard hyperbolic form

$$\begin{cases} H^{\varepsilon}(P,\theta) = (P\oplus P^*, \begin{pmatrix} 0 & 1 \\ \varepsilon & \theta \end{pmatrix} \in Q^{\varepsilon}(P\oplus P^*)) \\ H_{\varepsilon}(P) = (P\oplus P^*, \begin{pmatrix} 0 & 1 \\ 0 & 0 \end{pmatrix} \in Q_{\varepsilon}(P\oplus P^*)) \end{cases}$$

for some $\begin{cases} \varepsilon\text{-symmetric form over } A\ (P^*,\theta\in Q^{\varepsilon}(P^*)) \\ \text{f.g. projective } A\text{-module } P \end{cases}$.

The $\begin{cases} \underline{\varepsilon\text{-symmetric}} \\ \underline{\varepsilon\text{-quadratic}} \end{cases}$ Witt group of A $\begin{cases} L^0(A,\varepsilon) \\ L_0(A,\varepsilon) \end{cases}$ is the abelian group with respect to the direct sum $\oplus$ of the equivalence classes of non-singular $\begin{cases} \varepsilon\text{-symmetric} \\ \varepsilon\text{-quadratic} \end{cases}$ forms over A $\begin{cases} (M,\varphi) \\ (M,\psi) \end{cases}$ under the equivalence relation

$\begin{cases} (M,\varphi) \sim (M',\varphi') \\ (M,\psi) \sim (M',\psi') \end{cases}$ if there exists an isomorphism of forms

$$\begin{cases} f : (M,\varphi)\oplus H^{\varepsilon}(P,\theta) \longrightarrow (M',\varphi')\oplus H^{\varepsilon}(P',\theta') \\ f : (M,\psi)\oplus H_{\varepsilon}(P) \longrightarrow (M',\psi')\oplus H_{\varepsilon}(P') \end{cases}$$

for some $\begin{cases} \varepsilon\text{-symmetric forms over } A\ (P^*,\theta),(P'^*,\theta') \\ \text{f.g. projective } A\text{-modules } P,P' \end{cases}$.

The ε-symmetrization map of Witt groups

$$1+T_{\varepsilon} : L_0(A,\varepsilon) \longrightarrow L^0(A,\varepsilon) \ ; \ (M,\psi) \longmapsto (M,(1+T_{\varepsilon})\psi)$$

is an isomorphism modulo 8-torsion.

From now on we shall restrict attention to just those aspects of symmetric L-theory which we shall use in our treatment of quadratic L-theory. We refer to Part I of Ranicki [2] for a more thorough development of symmetric L-theory.

An $\underline{\varepsilon\text{-quadratic formation over A}}$ $(M,\psi;F,G)$ is a non-singular ε-quadratic form over A (M,ψ) together with a lagrangian F and a sublagrangian G. An <u>isomorphism</u> of formations

$$f : (M,\psi;F,G) \longrightarrow (M',\psi';F',G')$$

is an isomorphism of forms $f:(M,\psi)\longrightarrow(M',\psi')$ such that

$$f(F) = F' \ , \ f(G) = G' \ .$$

A <u>stable isomorphism</u> of formations

$$[f] : (M,\psi;F,G) \longrightarrow (M',\psi';F',G')$$

is an isomorphism of formations

$$f : (M,\psi;F,G)\oplus(H_\varepsilon(P);P,P^*) \longrightarrow (M',\psi';F',G')\oplus(H_\varepsilon(P');P',P'^*)$$

for some f.g. projective A-modules P,P'.

An ε-quadratic formation $(M,\psi;F,G)$ is <u>non-singular</u> if G is a lagrangian of (M,ψ).

The <u>boundary</u> of an ε-quadratic $\left\{\begin{matrix}\text{form}\\ \text{formation}\end{matrix}\right.$ over A $\left\{\begin{matrix}(M,\psi)\\ (M,\psi;F,G)\end{matrix}\right.$ is the non-singular $\left\{\begin{matrix}(-\varepsilon)\text{-}\\ \varepsilon\text{-}\end{matrix}\right.$ quadratic $\left\{\begin{matrix}\text{formation}\\ \text{form}\end{matrix}\right.$ over A

$$\begin{cases}\partial(M,\psi) = (H_{-\varepsilon}(M);M,\{(x,(\psi+\varepsilon\psi^*)(x))\in M\oplus M^* \mid x\in M\}) \\ \partial(M,\psi;F,G) = (G^\perp/G,\psi^\perp/\psi) \quad .\end{cases}$$

An ε-quadratic $\left\{\begin{matrix}\text{form}\\ \text{formation}\end{matrix}\right.$ is non-singular if and only if its boundary $\left\{\begin{matrix}\text{formation}\\ \text{form}\end{matrix}\right.$ is $\left\{\begin{matrix}\text{stably isomorphic to } 0\\ 0\end{matrix}\right.$.

Non-singular ε-quadratic $\left\{\begin{matrix}\text{forms}\\ \text{formations}\end{matrix}\right.$ over A $\left\{\begin{matrix}(M,\psi),(M',\psi')\\ (M,\psi;F,G),(M',\psi';F',G')\end{matrix}\right.$ are <u>cobordant</u> if there exists $\left\{\begin{matrix}\text{an isomorphism}\\ \text{a stable isomorphism}\end{matrix}\right.$ of $\left\{\begin{matrix}\text{forms}\\ \text{formations}\end{matrix}\right.$

$$\begin{cases} f : (M,\psi)\oplus(M',-\psi') \longrightarrow \partial(N,\varphi;H,K) \\ [f] : (M,\psi;F,G)\oplus(M',-\psi';F',G') \longrightarrow \partial(N,\varphi)\end{cases}$$

for some $\left\{\begin{matrix}\varepsilon\text{-}\\ (-\varepsilon)\text{-}\end{matrix}\right.$ quadratic $\left\{\begin{matrix}\text{formation}\\ \text{form}\end{matrix}\right.$ over A $\left\{\begin{matrix}(N,\varphi;H,K)\\ (N,\varphi)\end{matrix}\right.$.

<u>Proposition 1.1</u> Cobordism is an equivalence relation on the set of non-singular ε-quadratic $\left\{\begin{matrix}\text{forms}\\ \text{formations}\end{matrix}\right.$ over A, such that the equivalence classes define an abelian group $\left\{\begin{matrix}L_0(A,\varepsilon)\\ L_1(A,\varepsilon)\end{matrix}\right.$ with respect to the direct sum $\oplus$.

[]

The cobordism group of forms $L_0(A,\varepsilon)$ is just the Witt group of ε-quadratic forms over A, as defined previously.

Define abelian groups $L_n(A,\varepsilon)$ for n(mod 4) by

$$L_n(A,\varepsilon) = \begin{cases} L_0(A,(-)^i\varepsilon) \\ L_1(A,(-)^i\varepsilon) \end{cases} \text{if } n = \begin{cases} 2i \\ 2i+1 \end{cases} .$$

For $\varepsilon = 1 \in A$ we shall write

$$L_n(A,1) = L_n(A) \quad , \quad L^0(A,1) = L^0(A) .$$

In the terminology of Part I of Ranicki [1]

$$L_n(A) = U_n(A) .$$

Given a subgroup $X \subseteq \widetilde{K}_0(A)$ (resp. $X \subseteq \widetilde{K}_1(A)$) which is preserved as a set by the duality involution

$$* : \widetilde{K}_0(A) \longrightarrow \widetilde{K}_0(A) \; ; \; [P] \longmapsto [P^*]$$

$$(\text{resp. } * : \widetilde{K}_1(A) \longrightarrow \widetilde{K}_1(A) \; ; \; \tau(f:P \longrightarrow Q) \longmapsto \tau(f^*:Q^* \longrightarrow P^*))$$

let $L_n^X(A,\varepsilon)$ (n(mod 4)) be the L-groups defined as in Proposition 1.1, but using only forms and formations involving f.g. projective A-modules P such that $[P] \in X \subseteq \widetilde{K}_0(A)$ (resp. based f.g. free A-modules such that all isomorphisms $f \in \mathrm{Hom}_A(P,Q)$ have torsion $\tau(f) \in X \subseteq \widetilde{K}_1(A)$). In particular, for $X = \widetilde{K}_0(A)$

$$L_n^{\widetilde{K}_0(A)}(A,\varepsilon) = L_n(A,\varepsilon) .$$

For $\varepsilon = 1 \in A$ we shall write

$$L_n^X(A,1) = L_n^X(A) .$$

In the terminology of Part III of Ranicki [1]

$$L_n^X(A) = U_n^X(A) \text{ for } X \subseteq \widetilde{K}_0(A) \quad (\text{resp. } L_n^X(A) = V_n^X(A) \text{ for } X \subseteq \widetilde{K}_1(A)) .$$

<u>Proposition 1.2</u> Given $*$-invariant subgroups $X \subseteq Y \subseteq \widetilde{K}_m(A)$ (m = 0 or 1) there is defined an exact sequence of abelian groups

$$\ldots \longrightarrow \hat{H}^{n+1}(\mathbb{Z}_2;Y/X) \longrightarrow L_n^X(A,\varepsilon) \longrightarrow L_n^Y(A,\varepsilon) \longrightarrow \hat{H}^n(\mathbb{Z}_2;Y/X) \longrightarrow L_{n-1}^X(A,\varepsilon) \rightarrow \ldots$$

with the Tate $\mathbb{Z}_2$-cohomology groups defined by

$$\hat{H}^n(\mathbb{Z}_2;Y/X) = \{g \in Y/X \,|\, g^* = (-)^n g\} / \{h + (-)^n h^* \,|\, h \in Y/X\} .$$

[]

(In dealing with based A-modules it is convenient to assume that A is such that the rank of a f.g. free A-module is well-defined and $\tau(\varepsilon:A \rightarrow A) = 0 \in \widetilde{K}_1(A)$).

In order to define even-dimensional relative L-groups we shall need the following refinement of the notion of formation.

A split ε-quadratic formation over A $(F,(\begin{pmatrix}\gamma\\ \mu\end{pmatrix},\theta)G)$ is an ε-quadratic formation over A $(H_\varepsilon(F);F,G)$, where $\begin{pmatrix}\gamma\\ \mu\end{pmatrix}:G \longrightarrow F\oplus F^*$ is the inclusion, together with a hessian $(-\varepsilon)$-quadratic form over A $(G,\theta\in Q_{-\varepsilon}(G))$ such that

$$\gamma^*\mu = \theta - \varepsilon\theta^* : G \longrightarrow G^* .$$

Such a split formation will normally be written as (F,G).

An isomorphism of split ε-quadratic formations

$$(\alpha,\beta,\psi) : (F,G) \longrightarrow (F',G')$$

is defined by A-module isomorphisms $\alpha\in \mathrm{Hom}_A(F,F')$, $\beta\in \mathrm{Hom}_A(G,G')$ together with a $(-\varepsilon)$-quadratic form $(F^*,\psi\in Q_{-\varepsilon}(F^*))$ such that

$$\text{i)}\ \alpha\gamma + (\psi-\varepsilon\psi^*)^*\mu = \gamma'\beta : G \longrightarrow F'$$

$$\text{ii)}\ \alpha^{*-1}\mu = \mu'\beta : G \longrightarrow F'^*$$

$$\text{iii)}\ \theta + \mu^*\psi\mu = \beta^*\theta'\beta \in Q_{-\varepsilon}(G) \quad .$$

A stable isomorphism of split ε-quadratic formations

$$[\alpha,\beta,\psi] : (F,G) \longrightarrow (F',G')$$

is an isomorphism of the type

$$(\alpha,\beta,\psi) : (F,G)\oplus(P,P^*) \longrightarrow (F',G')\oplus(P',P'^*) ,$$

for some f.g. projective A-modules P,P' with $(P,P^*) = (P,(\begin{pmatrix}0\\ 1\end{pmatrix},0)P^*)$.

An isomorphism of split ε-quadratic formations $(\alpha,\beta,\psi):(F,G)\longrightarrow(F',G')$ determines an isomorphism of the underlying ε-quadratic formations

$$\begin{pmatrix}\alpha & \alpha(\psi-\varepsilon\psi^*)^* \\ 0 & \alpha^{*-1}\end{pmatrix} : (H_\varepsilon(F);F,G) \longrightarrow (H_\varepsilon(F');F',G') .$$

Conversely, every isomorphism of ε-quadratic formations

$$f : (H_\varepsilon(F);F,G) \longrightarrow (H_\varepsilon(F');F',G')$$

can be refined to an isomorphism of split ε-quadratic formations $(\alpha,\beta,\psi):(F,G)\longrightarrow(F',G')$. Similarly for stable isomorphisms.

The split boundary of an ε-quadratic form over A $(M,\psi\in Q_\varepsilon(M))$ is the non-singular split $(-\varepsilon)$-quadratic formation over A

$$\partial(M,\psi) = (M,(\begin{pmatrix}1\\ \psi+\varepsilon\psi^*\end{pmatrix},\psi)M) .$$

A morphism of rings with involution is a function

$$f : A \longrightarrow B$$

such that

$$f(a_1+a_2) = f(a_1)+f(a_2)\ ,\ f(a_1a_2) = f(a_1)f(a_2)\ ,\ f(\bar{a}) = \overline{f(a)}\ ,\ f(1) = 1 \in B$$
$$(a_1,a_2,a \in A)\ .$$

Given such a morphism regard B as a (B,A)-bimodule by

$$B \times B \times A \longrightarrow B\ ;\ (b,x,a) \longmapsto b.x.f(a)\ .$$

A f.g. projective A-module M induces a f.g. projective B-module $B\otimes_A M$, and there is defined a natural B-module isomorphism

$$B\otimes_A M^* \longrightarrow (B\otimes_A M)^*\ ;\ b\otimes f \longmapsto (c\otimes x \longmapsto c.f(x).\bar{b})$$

which we shall use to identify $(B\otimes_A M)^* = B\otimes_A M^*$. Given a central unit $\varepsilon \in A$ such that $\bar{\varepsilon} = \varepsilon^{-1}$ (as above) we have that $\overline{f(\varepsilon)} = f(\varepsilon)^{-1} \in B$, and it will be assumed that $f(\varepsilon)$ is central in B. It is convenient to also denote $f(\varepsilon) \in B$ by ε. An ε-quadratic $\begin{Bmatrix}\text{form}\\ \text{formation}\end{Bmatrix}$ over A $\begin{Bmatrix}(M,\psi)\\ (M,\psi;F,G)\end{Bmatrix}$ induces an ε-quadratic $\begin{Bmatrix}\text{form}\\ \text{formation}\end{Bmatrix}$ over B

$$\begin{cases} B\otimes_A(M,\psi) = (B\otimes_A M, 1\otimes\psi) \\ B\otimes_A(M,\psi;F,G) = (B\otimes_A M, 1\otimes\psi; B\otimes_A F, B\otimes_A G)\ , \end{cases}$$

and there are induced morphisms in the L-groups

$$f : L_n(A,\varepsilon) \longrightarrow L_n(B,\varepsilon)\ ;\ x \longmapsto B\otimes_A x \quad (n(\text{mod } 4))\ .$$

We shall now define relative L-groups $L_n(f,\varepsilon)$ $(n(\text{mod } 4))$ to fit into an exact sequence

$$\ldots \longrightarrow L_n(A,\varepsilon) \xrightarrow{f} L_n(B,\varepsilon) \longrightarrow L_n(f,\varepsilon) \longrightarrow L_{n-1}(A,\varepsilon) \longrightarrow \ldots\ .$$

A <u>relative ε-quadratic form over $f:A \longrightarrow B$</u> $((F,G),(M,\psi),h)$ is a triple consisting of a non-singular split $(-\varepsilon)$-quadratic formation over A (F,G), an ε-quadratic form over B (M,ψ), and a stable isomorphism of non-singular split $(-\varepsilon)$-quadratic formations over B

$$h : B\otimes_A(F,G) \longrightarrow \partial(M,\psi)\ .$$

The relative forms $((F,G),(M,\psi),h),((F',G'),(M',\psi'),h')$ are <u>cobordant</u> if there exist a $(-\varepsilon)$-quadratic form over A (L,φ) and a stable isomorphism of non-singular split $(-\varepsilon)$-quadratic formations over A

$$k : \partial(L,\varphi) \longrightarrow (F',G')\oplus -(F,G) \quad (\text{where } -(F,G) = (F,(\begin{pmatrix}-\gamma\\ \mu\end{pmatrix},-\theta)G))$$

such that the non-singular ε-quadratic form over B obtained by glueing

$$(N,\nu) = B\otimes_A(L,\varphi)\cup_{(h'\oplus h)(1\otimes k)}((M',-\psi')\oplus(M,\psi))$$

is null-cobordant, that is

$$(N,\nu) = 0 \in L_0(B,\varepsilon) .$$

The glueing operation was introduced in the proof of Theorem 4.3 of Part I of Ranicki [1], and it has also been described in Wall [6],[7]. We shall not repeat its definition here.

A <u>relative ε-quadratic formation over $f:A\longrightarrow B$</u> $((P,\theta),Q,h)$ is a triple consisting of a non-singular ε-quadratic form over A (P,θ), a f.g. projective B-module Q, and an isomorphism of non-singular ε-quadratic forms over B

$$h : B\otimes_A(P,\theta) \longrightarrow H_\varepsilon(Q) .$$

The relative ε-quadratic formations $((P,\theta),Q,h),((P',\theta'),Q',h')$ are <u>cobordant</u> if there exist an ε-quadratic formation over A $(M,\psi;F,G)$ and an isomorphism of non-singular ε-quadratic forms over B

$$k : \partial(M,\psi;F,G) \longrightarrow (P',\theta')\oplus(P,-\theta)$$

such that the non-singular ε-quadratic formation over B

$$(N,\nu;H,K) = (B\otimes_A(M,-\psi)\oplus H_\varepsilon(Q);(B\otimes_A F)\oplus Q,$$
$$\{(x+y,(h'\oplus h)(1\otimes k)(y))\in B\otimes_A M\oplus(Q\oplus Q^*)\,|\,x\in B\otimes_A G, y\in B\otimes_A(G^\perp/G)\})$$

is null-cobordant, that is

$$(N,\nu;H,K) = 0 \in L_1(B,\varepsilon) .$$

<u>Proposition 1.3</u> Cobordism is an equivalence relation on the set of relative

ε-quadratic $\begin{cases}\text{forms}\\ \text{formations}\end{cases}$ over $f:A\longrightarrow B$, such that the equivalence classes

define an abelian group $\begin{cases}L_0(f,\varepsilon)\\ L_1(f,\varepsilon)\end{cases}$ with respect to the direct sum $\oplus$.

The L-groups defined for $n \pmod 4$ by

$$L_n(f,\varepsilon) = \begin{cases}L_0(f,(-)^i\varepsilon)\\ L_1(f,(-)^i\varepsilon)\end{cases} \text{if } n = \begin{cases}2i\\ 2i+1\end{cases}$$

fit into an exact sequence of abelian groups

$$\ldots \longrightarrow L_n(A,\varepsilon) \xrightarrow{f} L_n(B,\varepsilon) \longrightarrow L_n(f,\varepsilon) \longrightarrow L_{n-1}(A,\varepsilon) \longrightarrow \ldots,$$

with

$$L_n(B,\varepsilon) \longrightarrow L_n(f,\varepsilon) \; ; \; x \longmapsto (0,x,0)$$

$$L_n(f,\varepsilon) \longrightarrow L_{n-1}(A,\varepsilon) \; ; \; (y,x,g) \longmapsto y \quad .$$

[]

In the case $\varepsilon = 1$ we shall write

$$L_n(f,1) = L_n(f) .$$

Relative L-groups $L_n(f)$ were first defined by Wall [3] (for n odd) and Sharpe [1] (n even), in the case when all the modules involved are f.g. free.

The above definition of the relative ε-quadratic L-groups $L_n(f,\varepsilon)$ generalizes immediately to the intermediate ε-quadratic L-groups. Given $*$-invariant subgroups $X \subseteq \widetilde{K}_m(A)$, $Y \subseteq \widetilde{K}_m(B)$ ($m = 0$ or 1) such that $B \otimes_A X \subseteq Y$ there are defined L-groups $L_n^{X,Y}(f,\varepsilon)$ ($n \pmod 4$) which fit into an exact sequence of abelian groups

$$\ldots \longrightarrow L_n^X(A,\varepsilon) \xrightarrow{f} L_n^Y(B,\varepsilon) \longrightarrow L_n^{X,Y}(f,\varepsilon) \longrightarrow L_{n-1}^X(A,\varepsilon) \longrightarrow \ldots .$$

§2. Localization

In setting up the localization exact sequence for quadratic L-theory we follow the pattern established for the localization exact sequence of algebraic K-theory

$$K_1(A) \longrightarrow K_1(S^{-1}A) \longrightarrow K_1(A,S) \longrightarrow K_0(A) \longrightarrow K_0(S^{-1}A)$$

in Chapter IX of Bass [1]. (The extension of the sequence to the lower K-groups K_i $(i \leqslant -1)$ of Bass and the higher K-groups K_i $(i \geqslant 2)$ of Quillen need not concern us here). There are three stages :

I) For any ring morphism $f:A \longrightarrow B$ there is defined a relative K-group $K_1(f)$ to fit into an exact sequence

$$K_1(A) \xrightarrow{f} K_1(B) \longrightarrow K_1(f) \longrightarrow K_0(A) \xrightarrow{f} K_0(B) \quad .$$

Specifically, $K_1(f)$ is a Grothendieck group of triples (P,Q,g) consisting of f.g. projective A-modules P,Q and a B-module isomorphism $g:B\otimes_A P \longrightarrow B\otimes_A Q$.

II) For a localization map $f:A \longrightarrow S^{-1}A$ it is possible to express g as $\frac{h}{s}$ for some $h \in \mathrm{Hom}_A(P,Q)$, $s \in S$ such that h induces an isomorphism over $S^{-1}A$. Thus $K_1(A \longrightarrow S^{-1}A)$ can be expressed as a Grothendieck group of triples such as (P,Q,h).

III) Define $K_1(A,S) = K_0$(exact category of h.d. 1 S-torsion A-modules) and observe that there is a natural isomorphism of abelian groups

$$K_1(A \longrightarrow S^{-1}A) \longrightarrow K_1(A,S) \ ; \ (P,Q,h) \longmapsto [\mathrm{coker}(h:P \longrightarrow Q)] \quad .$$

We have already developed the L-theoretic analogue of I) in §1 above.

(As in the algebraic K-theory of Bass [1] we shall only consider localizations $A \longrightarrow S^{-1}A$ inverting subsets $S \subset A$ of central elements. There is some interest in the L-theory of eccentric localizations, inverting non-central elements. The work of Smith [1] considers localizations of the type $A \longrightarrow S^{-1}A$ with $S = f^{-1}(1) \subset A$ for some ring morphism $f:A \longrightarrow B$ such that a morphism $g \in \mathrm{Hom}_A(P,Q)$ of f.g. projective A-modules P,Q becomes an isomorphism $1\otimes g \in \mathrm{Hom}_B(B\otimes_A P, B\otimes_A Q)$ if and only if $\ker(g) = 0$ and $\mathrm{coker}(g)$ is an S-torsion A-module. In principle, our methods permit a generalization to quadratic L-theory of any K-theoretic eccentric localization sequence).

Let A be a ring with involution (as in §1).

A <u>multiplicative subset</u> $S \subset A$ is a subset of A such that

i) $st \in S$ for all $s,t \in S$

ii) $\bar{s} \in S$ for all $s \in S$

iii) if $sa = 0$ for some $a \in A, s \in S$ then $a = 0$

iv) $as = sa \in A$ for all $a \in A, s \in S$

v) $1 \in S$.

The <u>localization of A away from S</u> $S^{-1}A$ is the ring with involution defined by the equivalence classes of pairs $(a,s) \in A \times S$ under the relation

$$(a,s) \sim (a',s') \text{ if } s'a = sa' \in A ,$$

with addition, multiplication and involution by

$$(a,s) + (b,t) = (at+bs,st) \ , \ (a,s)(b,t) = (ab,st) \ , \ \overline{(a,s)} = (\bar{a},\bar{s}) \ .$$

As usual, the class of (a,s) is denoted by $\frac{a}{s} \in S^{-1}A$. The inclusion

$$A \longrightarrow S^{1}A \ ; \ a \longmapsto \frac{a}{1}$$

is a morphism of rings with involution. An A-module M induces an $S^{-1}A$-module

$$S^{-1}M = S^{-1}A \otimes_A M$$

which can be identified with the $S^{-1}A$-module of equivalence classes of pairs $(x,s) \in M \times S$ under the relation

$$(x,s) \sim (x',s') \text{ if } s'x = sx' \in M .$$

Again, the class of (x,s) is denoted by $\frac{x}{s} \in S^{-1}M$. Given A-modules M,N regard $\mathrm{Hom}_A(M,N)$ as an A-module by

$$A \times \mathrm{Hom}_A(M,N) \longrightarrow \mathrm{Hom}_A(M,N) \ ; \ (a,f) \longmapsto (x \longmapsto f(x)\bar{a}) \ ,$$

and use the natural $S^{-1}A$-module isomorphism

$$S^{-1}\mathrm{Hom}_A(M,N) \longrightarrow \mathrm{Hom}_{S^{-1}A}(S^{-1}M,S^{-1}N) \ ; \ \frac{f}{s} \longmapsto (\frac{x}{t} \longmapsto \frac{f(x)}{t\bar{s}})$$

as an identification. In particular, for N = A we have the identification

$$(S^{-1}M)^* = S^{-1}(M^*) \ .$$

For example, if $A = \mathbb{Z}$, $S = \mathbb{Z} - \{0\}$ then $S^{-1}A = \mathbb{Q}$.

Let $L_n^S(S^{-1}A,\varepsilon)$ $(n(\bmod 4))$ be the intermediate ε-quadratic L-groups of $S^{-1}A$ associated to the $*$-invariant subgroup $S=\mathrm{im}(\widetilde{K}_0(A)\to\widetilde{K}_0(S^{-1}A))\subseteq\widetilde{K}_0(S^{-1}A)$ of the projective classes of f.g. projective $S^{-1}A$-modules induced from f.g. projective A-modules. Let $L_n^S(A\longrightarrow S^{-1}A,\varepsilon)$ $(n(\bmod 4))$ be the relative L-groups appearing in the exact sequence

$$\ldots\longrightarrow L_n(A,\varepsilon)\longrightarrow L_n^S(S^{-1}A,\varepsilon)\longrightarrow L_n^S(A\longrightarrow S^{-1}A,\varepsilon)\longrightarrow L_{n-1}(A,\varepsilon)\to\ldots\ .$$

In the first instance we shall express $L_n^S(A\longrightarrow S^{-1}A,\varepsilon)$ for $n=\begin{cases}2i\\2i+1\end{cases}$ as the cobordism group of non-singular $\begin{cases}\text{split } (-)^{i-1}\varepsilon\text{-quadratic formations}\\(-)^i\varepsilon\text{-quadratic forms}\end{cases}$ over A which become $\begin{cases}\text{stably isomorphic to } 0\\\text{hyperbolic}\end{cases}$ over $S^{-1}A$, corresponding to stage II) of the above programme. We shall then use this expression to identify

$$L_n^S(A\longrightarrow S^{-1}A,\varepsilon)=L_n(A,S,\varepsilon)\quad(n(\bmod 4))$$

with $\begin{cases}L_{2i}(A,S,\varepsilon)\\L_{2i+1}(A,S,\varepsilon)\end{cases}$ $(i(\bmod 2))$ the Witt group of non-singular $(-)^i\varepsilon$-quadratic linking $\begin{cases}\text{forms}\\\text{formations}\end{cases}$ defined using h.d. 1 S-torsion A-modules, corresponding to stage III).

An A-module morphism $f\in\mathrm{Hom}_A(M,N)$ is an <u>S-isomorphism</u> if the induced $S^{-1}A$-module morphism

$$S^{-1}f:S^{-1}M\longrightarrow S^{-1}N\ ;\ \frac{x}{s}\longmapsto\frac{f(x)}{s}$$

is an isomorphism.

An <u>S-isomorphism</u> of ε-quadratic forms over A

$$f:(M,\psi)\longrightarrow(N,\phi)$$

is a morphism of ε-quadratic forms such that $f\in\mathrm{Hom}_A(M,N)$ is an S-isomorphism. There is induced an isomorphism of ε-quadratic forms over $S^{-1}A$

$$S^{-1}f:S^{-1}(M,\psi)\longrightarrow S^{-1}(N,\phi)\quad .$$

An ε-quadratic form over A (M,ψ) is <u>non-degenerate</u> if $\psi+\varepsilon\psi^*\in \mathrm{Hom}_A(M,M^*)$ is an S-isomorphism.

An <u>S-lagrangian</u> of a non-degenerate ε-quadratic form over A (M,ψ) is a f.g. projective submodule L of M such that the inclusion $j\in \mathrm{Hom}_A(L,M)$ defines a morphism of forms over A

$$j : (L,0) \longrightarrow (M,\psi)$$

which becomes the inclusion of a lagrangian over $S^{-1}A$. The inclusion j extends to an S-isomorphism of non-degenerate ε-quadratic forms over A

$$(j\ \ k) : (L\oplus L^*, \begin{pmatrix} 0 & s \\ 0 & 0 \end{pmatrix}) \longrightarrow (M,\psi)$$

for some $k\in \mathrm{Hom}_A(L^*,M)$, $s\in S$.

A <u>non-degenerate ε-quadratic formation over A</u> $(M,\psi;F,G)$ is a non-singular ε-quadratic form over A (M,ψ) together with a lagrangian F and an S-lagrangian G.

A non-degenerate ε-quadratic $\begin{cases}\text{form}\\ \text{formation}\end{cases}$ over A $\begin{cases}(M,\psi)\\ (M,\psi;F,G)\end{cases}$ induces a non-singular ε-quadratic $\begin{cases}\text{form}\\ \text{formation}\end{cases}$ over $S^{-1}A$ $\begin{cases}S^{-1}(M,\psi)\\ S^{-1}(M,\psi;F,G)\end{cases}$, representing an element of $\begin{cases}L_0^S(S^{-1}A,\varepsilon)\\ L_1^S(S^{-1}A,\varepsilon)\end{cases}$. Conversely, every element of $\begin{cases}L_0^S(S^{-1}A,\varepsilon)\\ L_1^S(S^{-1}A,\varepsilon)\end{cases}$ is represented by a $\begin{cases}\text{form}\\ \text{formation}\end{cases}$ of this type.

(We could achieve a more systematic terminology by calling non-degenerate objects over A 'S-non-singular'. We prefer to bow to the tradition of calling forms over $\mathbb{Z}$ which become non-singular over $\mathbb{Q}$ 'non-degenerate').

An <u>ε-quadratic S-form over A</u> $(M,\psi;L)$ is a non-degenerate ε-quadratic form over A (M,ψ) together with an S-lagrangian L. The S-form is <u>non-singular</u> if the form (M,ψ) is non-singular, in which case there is defined an <u>associated</u> relative ε-quadratic formation over $A \longrightarrow S^{-1}A$

$$((M,\psi),S^{-1}L,(j\ \frac{k}{s})^{-1} : S^{-1}(M,\psi) \longrightarrow H_\varepsilon(S^{-1}L))$$

with $j \in \mathrm{Hom}_A(L,M)$, $k \in \mathrm{Hom}_A(L^*,M)$, $s \in S$ as above.

An <u>isomorphism</u> of ε-quadratic S-forms over A

$$f : (M,\psi;L) \longrightarrow (M',\psi';L')$$

is an isomorphism of forms

$$f : (M,\psi) \longrightarrow (M',\psi')$$

such that

$$f(L) = L' .$$

A <u>stable isomorphism</u> of ε-quadratic S-forms over A

$$[f] : (M,\psi;L) \longrightarrow (M',\psi';L')$$

is an isomorphism of the type

$$f : (M,\psi;L) \oplus (H_\varepsilon(P);P) \longrightarrow (M',\psi';L') \oplus (H_\varepsilon(P');P')$$

for some f.g. projective A-modules P,P'.

An <u>ε-quadratic S-formation over A</u> $(M,\psi;F,G)$ is a non-degenerate ε-quadratic formation over A such that the A-module morphism

$$G \longrightarrow M/F ;\ x \longmapsto [x]$$

is an S-isomorphism. The S-formation is <u>non-singular</u> if G is a lagrangian of (M,ψ).

An <u>isomorphism</u> of ε-quadratic S-formations over A

$$f : (M,\psi;F,G) \longrightarrow (M',\psi';F',G')$$

is an isomorphism of ε-quadratic forms over A

$$f : (M,\psi) \longrightarrow (M',\psi')$$

such that

$$f(F) = F' ,\ f(G) = G' .$$

A <u>stable isomorphism</u> of ε-quadratic S-formations over A

$$[f] : (M,\psi;F,G) \longrightarrow (M',\psi';F',G')$$

is an isomorphism of the type

$$f : (M,\psi;F,G)\oplus(H_{\varepsilon}(P);P,P^*) \longrightarrow (M',\psi';F',G')\oplus(H_{\varepsilon}(P');P',P'^*)$$

for some f.g. projective A-modules P,P'.

A <u>split ε-quadratic S-formation over A</u> $(F,(\begin{pmatrix}\gamma\\ \mu\end{pmatrix},\theta)G)$ is an ε-quadratic S-formation over A $(H_{\varepsilon}(F);F,G)$, where $\begin{pmatrix}\gamma\\ \mu\end{pmatrix}:G\longrightarrow F\oplus F^*$ is the inclusion, together with a <u>hessian</u> $(-\varepsilon)$-quadratic form over A $(G,\theta\in Q_{-\varepsilon}(G))$ such that

$$\gamma^*\mu = \theta - \varepsilon\theta^* : G \longrightarrow G^* .$$

Such a split S-formation will normally be written as (F,G), denoting $(F,(\begin{pmatrix}-\gamma\\ \mu\end{pmatrix},-\theta)G)$ by $-(F,G)$. Note that $\mu\in \mathrm{Hom}_A(G,F^*)$ is an S-isomorphism.

A split ε-quadratic S-formation (F,G) is <u>non-singular</u> if G is a lagrangian of $H_{\varepsilon}(F)$, that is if the sequence of A-modules

$$0 \longrightarrow G \xrightarrow{\begin{pmatrix}\gamma\\ \mu\end{pmatrix}} F\oplus F^* \xrightarrow{(\varepsilon\mu^*\ \gamma^*)} G^* \longrightarrow 0$$

is exact. For non-singular (F,G) define the <u>associated</u> relative $(-\varepsilon)$-quadratic form over $A \longrightarrow S^{-1}A$ $((F,G),0,0)$.

An <u>isomorphism</u> of split ε-quadratic S-formations over A

$$(\alpha,\beta,\psi) : (F,G) \longrightarrow (F',G')$$

is defined by A-module isomorphisms $\alpha\in\mathrm{Hom}_A(F,F')$, $\beta\in\mathrm{Hom}_A(G,G')$ together with a $(-\varepsilon)$-quadratic form $(F^*,\psi\in Q_{-\varepsilon}(F^*))$ such that

i) $\alpha\gamma + (\psi-\varepsilon\psi^*)^*\mu = \gamma'\beta : G \longrightarrow F'$

ii) $\alpha^{*-1}\mu = \mu'\beta : G \longrightarrow F'^*$

iii) $\theta + \mu^*\psi\mu - \beta^*\theta'\beta \in \ker(S^{-1}:Q_{-\varepsilon}(G)\longrightarrow Q_{-\varepsilon}(S^{-1}G))$.

A <u>stable isomorphism</u> of split ε-quadratic S-formations over A

$$[\alpha,\beta,\psi] : (F,G) \longrightarrow (F',G')$$

is an isomorphism of the type

$$(\alpha,\beta,\psi) : (F,G)\oplus(P,P^*) \longrightarrow (F',G')\oplus(P',P'^*)$$

for some f.g. projective A-modules P,P'.

The boundary of a non-degenerate ε-quadratic $\begin{cases}\text{form}\\ \text{formation}\end{cases}$ over A $\begin{cases}(M,\psi)\\ (M,\psi;F,G)\end{cases}$ is the non-singular $\begin{cases}\text{split } (-\varepsilon)\text{-quadratic } S\text{-formation}\\ \varepsilon\text{-quadratic } S\text{-form}\end{cases}$ over A

$$\begin{cases}\partial(M,\psi) = (M,(\begin{pmatrix}1\\ \psi+\varepsilon\psi^*\end{pmatrix},\psi)M)\\ \partial(M,\psi;F,G) = (M,\psi;G)\end{cases}$$

Non-singular $\begin{cases}\varepsilon\text{-quadratic } S\text{-forms}\\ \text{split } \varepsilon\text{-quadratic } S\text{-formations}\end{cases}$ over A $\begin{cases}(M,\psi;L),(M',\psi';L')\\ (F,G),(F',G')\end{cases}$ are cobordant if there exists a stable isomorphism

$$\begin{cases}[f] : (M,\psi;L)\oplus(M',-\psi';L') \longrightarrow \partial(N,\varphi;H,K)\\ [\alpha,\beta,\psi] : (F,G)\oplus-(F',G') \longrightarrow \partial(N,\varphi)\end{cases}$$

for some non-degenerate $\begin{cases}\varepsilon\text{-}\\ (-\varepsilon)\text{-}\end{cases}$ quadratic $\begin{cases}\text{formation}\\ \text{form}\end{cases}$ over A $\begin{cases}(N,\varphi;H,K)\\ (N,\varphi)\end{cases}$ such that

$$\begin{cases}S^{-1}(N,\varphi;H,K) = 0 \in L_1^S(S^{-1}A,\varepsilon)\\ S^{-1}(N,\varphi) = 0 \in L_0^S(S^{-1}A,-\varepsilon)\end{cases}$$

<u>Proposition 2.1</u> Cobordism is an equivalence relation on the set of non-singular $\begin{cases}\varepsilon\text{-quadratic } S\text{-forms}\\ \text{split } \varepsilon\text{-quadratic } S\text{-formations}\end{cases}$ over A, such that the equivalence classes define an abelian group with respect to the direct sum ⊕.

The cobordism group of non-singular $\begin{cases}(-)^i\varepsilon\text{-quadratic } S\text{-forms}\\ \text{split } (-)^i\varepsilon\text{-quadratic } S\text{-formations}\end{cases}$ over A is naturally isomorphic (via the associated relative $\begin{cases}\text{formation}\\ \text{form}\end{cases}$ construction) to the relative L-group $L_n^S(A \longrightarrow S^{-1}A,\varepsilon)$ for $n = \begin{cases}2i+1\\ 2i+2\end{cases}$.

The morphisms of the exact sequence

$$\ldots \longrightarrow L_n(A,\varepsilon) \longrightarrow L_n^S(S^{-1}A,\varepsilon) \longrightarrow L_n^S(A \longrightarrow S^{-1}A,\varepsilon) \longrightarrow L_{n-1}(A,\varepsilon) \longrightarrow \ldots$$

involving $L_n^S(A \longrightarrow S^{-1}A,\varepsilon)$ are given in terms of S-forms and S-formations by

$$\begin{cases} L_{2i}^S(S^{-1}A,\varepsilon) \longrightarrow L_{2i}^S(A \longrightarrow S^{-1}A,\varepsilon) \ ; \ S^{-1}(M,\psi) \longmapsto \partial(M,\psi) \\ L_{2i+1}^S(S^{-1}A,\varepsilon) \longrightarrow L_{2i+1}^S(A \longrightarrow S^{-1}A,\varepsilon) \ ; \ S^{-1}(M,\psi;F,G) \longmapsto \partial(M,\psi;F,G) \end{cases}$$

$$\begin{cases} L_{2i}^S(A \longrightarrow S^{-1}A,\varepsilon) \longrightarrow L_{2i-1}(A,\varepsilon) \ ; \ (F,G) \longmapsto (H_{(-)^{i-1}\varepsilon}(F);F,G) \\ L_{2i+1}^S(A \longrightarrow S^{-1}A,\varepsilon) \longrightarrow L_{2i}(A,\varepsilon) \ ; \ (M,\psi;L) \longmapsto (M,\psi) \ . \end{cases}$$

[]

An A-module M is <u>S-torsion</u> if

$$S^{-1}M = 0 \ ,$$

or equivalently if for every $x \in M$ there exists $s \in S$ such that $sx = 0 \in M$.

An A-module M is <u>h.d. 1</u> (= homological dimension 1) if it admits a f.g. projective A-module resolution of length 1

$$0 \longrightarrow P_1 \xrightarrow{\ d\ } P_0 \longrightarrow M \longrightarrow 0 \ .$$

An h.d. 1 S-torsion A-module is thus an A-module which admits a f.g. projective A-module resolution of length 1 with $d \in \mathrm{Hom}_A(P_1,P_0)$ an S-isomorphism.

Regard the abelian group $S^{-1}A/A$ as an A-module by

$$A \times S^{-1}A/A \longrightarrow S^{-1}A/A \ ; \ (a,\tfrac{b}{s}) \longmapsto \frac{ab}{s} \ .$$

The <u>S-dual</u> of an A-module M is the A-module

$$M^\wedge = \mathrm{Hom}_A(M,S^{-1}A/A)$$

with A acting by

$$A \times M^\wedge \longrightarrow M^\wedge \ ; \ (a,f) \longmapsto (x \longmapsto f(x)\bar{a}) \ .$$

The S-dual of an A-module morphism $f \in \mathrm{Hom}_A(M,N)$ is the A-module morphism

$$f^\wedge : N^\wedge \longrightarrow M^\wedge \ ; \ g \longmapsto (x \longmapsto g(f(x))) \quad .$$

The S-dual of an h.d. 1 S-torsion A-module $M = \mathrm{coker}(d:P_1 \longrightarrow P_0)$ is an h.d. 1 S-torsion A-module $M^\wedge$, with resolution

$$0 \longrightarrow P_0^* \xrightarrow{\ d^*\ } P_1^* \longrightarrow M^\wedge \longrightarrow 0$$

where

$$\cdots = dy \in P_0) \ .$$

The natural A-module morphism

$$M \longrightarrow M^{\wedge\wedge} \ ; \ x \longmapsto (f \longmapsto \overline{f(x)})$$

is an isomorphism if M is an h.d. 1 S-torsion A-module, in which case we shall use it as an identification, and to define the <u>ε-duality</u> involution

$$T_\varepsilon : \mathrm{Hom}_A(M,M^\wedge) \longrightarrow \mathrm{Hom}_A(M,M^\wedge) \ ; \ \varphi \longmapsto (\varepsilon\varphi^\wedge : x \longmapsto (y \longmapsto \varepsilon\overline{\varphi(y)(x)})) \ .$$

An <u>ε-symmetric linking form over (A,S)</u> (M,λ) is an h.d. 1 S-torsion A-module M together with an element $\lambda \in \ker(1-T_\varepsilon : \mathrm{Hom}_A(M,M^\wedge) \longrightarrow \mathrm{Hom}_A(M,M^\wedge))$. Equivalently, λ is given by a pairing

$$\lambda : M \times M \longrightarrow S^{-1}A/A \ ; \ (x,y) \longmapsto \lambda(x)(y)$$

satisfying

$$\begin{aligned} &\text{i) } \lambda(x,ay) = a\lambda(x,y) \in S^{-1}A/A \\ &\text{ii) } \lambda(x,y+y') = \lambda(x,y) + \lambda(x,y') \in S^{-1}A/A \\ &\text{iii) } \lambda(y,x) = \varepsilon\overline{\lambda(x,y)} \in S^{-1}A/A \qquad (x,y,y' \in M) \ . \end{aligned}$$

Define the abelian groups

$$\begin{aligned} Q_\varepsilon(A,S) &= S^{-1}A/\{a + \varepsilon\bar{a} \mid a \in A\} \\ Q_\varepsilon(S^{-1}A/A) &= (S^{-1}A/A)/\{b - \varepsilon\bar{b} \mid b \in A\} \end{aligned}$$

and the abelian group morphism

$$1+T_\varepsilon : Q_\varepsilon(S^{-1}A/A) \longrightarrow Q_\varepsilon(A,S) \ ; \ c \longmapsto c + \varepsilon\bar{c} \ .$$

An <u>ε-quadratic linking form over (A,S)</u> (M,λ,μ) is an ε-symmetric linking form over (A,S) (M,λ) together with a function

$$\mu : M \longrightarrow Q_\varepsilon(A,S)$$

such that

$$\begin{aligned} &\text{i) } \mu(ax) = a\mu(x)\bar{a} \in Q_\varepsilon(A,S) \\ &\text{ii) } \mu(x+y) - \mu(x) - \mu(y) = \lambda(x,y) + \varepsilon\overline{\lambda(x,y)} \in Q_\varepsilon(A,S) \\ &\text{iii) } [\mu(x)] = \lambda(x)(x) \in S^{-1}A/A \qquad (x,y,y' \in M, \ a \in A). \end{aligned}$$

The linking forms appearing in the work of Wall [2], Passman and Petrie [1], Connolly [1] and Pardon [1],[2] on odd-dimensional surgery obstructions are just the ε-quadratic linking forms over $(\mathbb{Z}[\pi], \mathbb{Z} - \{0\})$, with $\varepsilon = \pm 1$ and π a finite group.

A <u>split ε-quadratic linking form over (A,S)</u> (M,λ,ν) is an ε-symmetric linking form over (A,S) (M,λ) together with a function

$$\nu : M \longrightarrow Q_\varepsilon(S^{-1}A/A)$$

such that

$$\begin{aligned} &\text{i)} \quad \nu(ax) = a\nu(x)\bar{a} \in Q_\varepsilon(S^{-1}A/A) \\ &\text{ii)} \quad \nu(x+y) - \nu(x) - \nu(y) = [\lambda(x)(y)] \in Q_\varepsilon(S^{-1}A/A) \\ &\text{iii)} \quad \nu(x) + \varepsilon\overline{\nu(x)} = \lambda(x)(x) \in S^{-1}A/A \qquad (x,y\in M,\ a\in A). \end{aligned}$$

Split ε-quadratic linking forms were introduced by Karoubi [1].

A <u>morphism</u> (resp. <u>isomorphism</u>) of $\begin{cases} \varepsilon\text{-symmetric} \\ \varepsilon\text{-quadratic} \\ \text{split } \varepsilon\text{-quadratic} \end{cases}$ linking forms over (A,S)

$$\begin{cases} f : (M,\lambda) \longrightarrow (M',\lambda') \\ f : (M,\lambda,\mu) \longrightarrow (M',\lambda',\mu') \\ f : (M,\lambda,\nu) \longrightarrow (M',\lambda',\nu') \end{cases}$$

is a morphism (resp. isomorphism) $f \in \mathrm{Hom}_A(M,M')$ such that

$$f^\wedge \lambda' f = \lambda \in \mathrm{Hom}_A(M,M^\wedge)$$

and also

$$\begin{cases} \mu : M \xrightarrow{f} M' \xrightarrow{\mu'} Q_\varepsilon(A,S) \\ \nu : M \xrightarrow{f} M' \xrightarrow{\nu'} Q_\varepsilon(S^{-1}A/A) \ . \end{cases}$$

It can be shown that the forgetful functor

$$\begin{aligned} &(\text{split } \varepsilon\text{-quadratic linking forms over } (A,S)) \\ &\qquad \longrightarrow (\varepsilon\text{-quadratic linking forms over } (A,S)) \ ; \\ &(M,\lambda,\nu) \longmapsto (M,\lambda,\mu = (1+T_\varepsilon)\nu : M \xrightarrow{\nu} Q_\varepsilon(S^{-1}A/A) \xrightarrow{1+T_\varepsilon} Q_\varepsilon(A,S)) \end{aligned}$$

defines a surjection of isomorphism classes, which is a bijection if $\frac{1}{2} \in S^{-1}A$, e.g. if $A = \mathbb{Z}[\pi]$, $S = \mathbb{Z} - \{0\}$, $S^{-1}A = \mathbb{Q}[\pi]$. (This may be deduced from Proposition 2.2 below). In §6 we shall give examples of triples (A,S,ε) for which there is a perceptible difference between split ε-quadratic and ε-quadratic linking forms over (A,S).

An $\begin{cases} \varepsilon\text{-symmetric} \\ \varepsilon\text{-quadratic} \\ \text{split } \varepsilon\text{-quadratic} \end{cases}$ linking form over (A,S) $\begin{cases} (M,\lambda) \\ (M,\lambda,\mu) \\ (M,\lambda,\nu) \end{cases}$ is

<u>non-singular</u> if $\lambda \in \mathrm{Hom}_A(M,M^\wedge)$ is an isomorphism.

As in §1 we shall concentrate on the ε-quadratic L-theory, leaving the ε-symmetric L-theory of linking forms to the relevant part of Ranicki [2].

There is a close connection between linking forms over (A,S) and S-formations over A, which was first observed by Wall [1] in the case $A = \mathbb{Z}$, $S = \mathbb{Z} - \{0\}$, $S^{-1}A = \mathbb{Q}$.

<u>Proposition 2.2</u> The isomorphism classes of (non-singular) $\begin{cases} \varepsilon\text{-quadratic} \\ \text{split } \varepsilon\text{-quadratic} \end{cases}$ linking forms over (A,S) $\begin{cases} (M,\lambda,\mu) \\ (M,\lambda,\nu) \end{cases}$ are in a natural one-one correspondence with the stable isomorphism classes of (non-singular) $\begin{cases} (-\varepsilon)\text{-quadratic} \\ \text{split } (-\varepsilon)\text{-quadratic} \end{cases}$ S-formations over A $\begin{cases} (N,\psi;F,G) \\ (F,G) \end{cases}$. The linking form $\begin{cases} (M,\lambda,\mu) \\ (M,\lambda,\nu) \end{cases}$ corresponding to the S-formation $\begin{cases} (N,\psi;F,G) \\ (F,(\binom{\gamma}{\mu},\theta)G) \end{cases}$ is defined by

$$\begin{cases} \begin{cases} M = N/(F+G)\ ,\ \lambda : M \longrightarrow M^\wedge\ ;\ x \longmapsto (y \longmapsto \frac{1}{s}(\psi-\varepsilon\psi^*)(x)(g)) \\ \mu : M \longrightarrow Q_\varepsilon(A,S)\ ;\ y \longmapsto \frac{1}{s}(\psi-\varepsilon\psi^*)(y)(g) - \psi(y)(y) \\ \qquad\qquad\qquad (x,y\in N,\ s\in S,\ g\in G,\ sy - g \in F) \end{cases} \\ \begin{cases} M = \mathrm{coker}(\mu : G \longrightarrow F^*)\ ,\ \lambda : M \longrightarrow M^\wedge\ ;\ x \longmapsto (y \longmapsto \frac{1}{s}\gamma^*(x)(g)) \\ \nu : M \longrightarrow Q_\varepsilon(S^{-1}A/A)\ ;\ y \longmapsto \frac{\theta(g)(g)}{s\bar{s}} \\ \qquad\qquad\qquad (x,y\in F^*,\ s\in S,\ g\in G,\ sy = \mu g \in F^*). \end{cases} \end{cases}$$

[]

A <u>sublagrangian</u> of a non-singular split ε-quadratic linking form over (A,S) (M,λ,ν) is a submodule L of M such that

i) L, M/L are h.d. 1 S-torsion A-modules

ii) the inclusion $j \in \mathrm{Hom}_A(L,M)$ defines a morphism of linking forms

$$j : (L,0,0) \longrightarrow (M,\lambda,\nu)$$

iii) the A-module morphism

$$[\lambda] : M/L \longrightarrow L^{\wedge} \ ; \ [x] \longmapsto (y \longmapsto \lambda(x)(y)) \quad (x \in M, y \in L)$$

is onto.

The <u>annihilator</u> of a sublagrangian L in (M,λ,ν) is the submodule $L^{\perp}$ of M defined by

$$L^{\perp} = \ker(j^{\wedge}\lambda : M \longrightarrow L^{\wedge}) \ ,$$

which is such that $L \subseteq L^{\perp}$.

A <u>lagrangian</u> of (M,λ,ν) is a sublagrangian L such that

$$L^{\perp} = L \ .$$

A non-singular split ε-quadratic linking form which admits a lagrangian is <u>hyperbolic</u>. For example, if L is a sublagrangian of (M,λ,ν) then there is defined a non-singular split ε-quadratic linking form $(L^{\perp}/L, \lambda^{\perp}/\lambda, \nu^{\perp}/\nu)$ such that $(M,\lambda,\nu) \oplus (L^{\perp}/L, -\lambda^{\perp}/\lambda, -\nu^{\perp}/\nu)$ is hyperbolic, with lagrangian

$$L' = \{(x,[x]) \in M \oplus L^{\perp}/L \mid x \in L^{\perp}\} \ .$$

Given an h.d. 1 S-torsion A-module P define the standard hyperbolic split ε-quadratic linking form over (A,S)

$$H_{\varepsilon}(P) = (P \oplus P^{\wedge}, \lambda : P \oplus P^{\wedge} \longrightarrow (P \oplus P^{\wedge})^{\wedge} ; (x,f) \longmapsto ((y,g) \longmapsto f(y) + \varepsilon\overline{g(x)}) \ ,$$
$$\nu : P \oplus P^{\wedge} \longrightarrow Q_{\varepsilon}(S^{-1}A/A) \ ; \ (x,f) \longmapsto f(x) \ ) \ .$$

A <u>split ε-quadratic linking formation over (A,S)</u> $(F,(\begin{pmatrix}\gamma\\ \mu\end{pmatrix},\theta)G)$ is defined by a sublagrangian G in a standard hyperbolic split ε-quadratic linking form over (A,S) $H_{\varepsilon}(F)$, together with a <u>hessian</u> $(-\varepsilon)$-quadratic linking form over (A,S)

$$(G, \gamma^{\wedge}\mu \in \mathrm{Hom}_A(G,G^{\wedge}), \theta : G \longrightarrow Q_{-\varepsilon}(A,S))$$

where $\begin{pmatrix}\gamma\\ \mu\end{pmatrix} : G \longrightarrow F \oplus F^{\wedge}$ is the inclusion. Such objects first appeared in the work of Pardon [1], and similar structures have been studied by Karoubi [1].

We shall normally write $(F,(\binom{\gamma}{\mu},\theta)G)$ as (F,G), denoting $(F,(\binom{-\gamma}{\mu},-\theta)G)$ by $-(F,G)$.

An <u>isomorphism</u> of split ε-quadratic linking formations over (A,S)

$$f : (F,G) \longrightarrow (F',G')$$

is an isomorphism of the hyperbolic split ε-quadratic linking forms

$$f : H_\varepsilon(F) \longrightarrow H_\varepsilon(F')$$

such that

$$f(F) = F' \ , \ f(G) = G'$$

and also

$$\theta : G \xrightarrow{f|} G' \xrightarrow{\theta'} Q_{-\varepsilon}(A,S) \ .$$

A <u>sublagrangian</u> of a split ε-quadratic linking formation over (A,S) (F,G) is a sublagrangian L of $H_\varepsilon(F)$ such that

i) $L \subseteq G$, with G/L an h.d. 1 S-torsion A-module

ii) $F \cap L = \{0\}$, $F \oplus F^\wedge = F + L^\perp$.

Such a sublagrangian determines an <u>elementary equivalence</u> of split ε-quadratic linking formations over (A,S), the transformation

$$(F,G) \longrightarrow (F',G') \quad ,$$

with (F',G') defined by

$$F' = F \cap L^\perp \quad , \quad G' = G/L$$

$$\gamma' : G' \longrightarrow F' \ ; \ [x] \longmapsto \gamma(x)$$

$$\mu' : G' \longrightarrow F'^\wedge \ ; \ [x] \longmapsto (y \longmapsto \mu(x)(y))$$

$$\theta' : G' \longrightarrow Q_{-\varepsilon}(A,S) \ ; \ [x] \longmapsto \theta(x) \qquad (x \in G, y \in F\ ).$$

Elementary equivalences and isomorphisms generate an equivalence relation on the set of split ε-quadratic linking formations over (A,S), which we shall call <u>stable equivalence</u>.

A split ε-quadratic linking formation over (A,S) (F,G) is <u>non-singular</u> if G is a lagrangian of $H_\varepsilon(F)$, or equivalently if the sequence

$$0 \longrightarrow G \xrightarrow{\binom{\gamma}{\mu}} F \oplus F^\wedge \xrightarrow{(\varepsilon\mu^\wedge\ \gamma^\wedge)} G^\wedge \longrightarrow 0$$

is exact. Any linking formation stably equivalent to a non-singular one is itself non-singular.

There is a close connection between linking formations over (A,S) and S-forms over A.

<u>Proposition 2.3</u> The stable equivalence classes of (non-singular) split ε-quadratic linking formations over (A,S) (F,G) are in a natural one-one correspondence with the stable isomorphism classes of (non-singular) ε-quadratic S-forms over A (M,ψ;L). The linking formation (F,G) corresponding to the S-form (M,ψ;L) is defined as follows: extend the inclusion $j \in \mathrm{Hom}_A(L,M)$ to an S-isomorphism of ε-quadratic forms over A

$$(j\ \ k)\ :\ (L\oplus L^*, \begin{pmatrix} 0 & s \\ 0 & 0 \end{pmatrix}) \longrightarrow (M,\psi)$$

for some $k \in \mathrm{Hom}_A(L^*,M)$, $s \in S$, set

$$F = \mathrm{coker}(\bar{s}:L \longrightarrow L)\ ,\ G = \mathrm{coker}((j\ k):L\oplus L^* \longrightarrow M)\ ,$$

define $\begin{pmatrix}\gamma\\ \mu\end{pmatrix}:G \longrightarrow F\oplus F^\wedge$ via the resolution

$$\begin{array}{ccccccccc}
0 & \longrightarrow & L\oplus L^* & \xrightarrow{(j\ k)} & M & \longrightarrow & G & \longrightarrow & 0 \\
 & & \Big\downarrow {\scriptstyle \begin{pmatrix} \varepsilon & 0 \\ 0 & 1 \end{pmatrix}} & & \Big\downarrow {\scriptstyle \begin{pmatrix} k^*(\psi+\varepsilon\psi^*) \\ j^*(\psi+\varepsilon\psi^*) \end{pmatrix}} & & \Big\downarrow {\scriptstyle \begin{pmatrix} \gamma \\ \mu \end{pmatrix}} & & \\
0 & \longrightarrow & L\oplus L^* & \xrightarrow{\begin{pmatrix} \bar{s} & 0 \\ 0 & s \end{pmatrix}} & L\oplus L^* & \longrightarrow & F\oplus F^\wedge & \longrightarrow & 0\ ,
\end{array}$$

and let $(G,\hat{\gamma\mu}\in\mathrm{Hom}_A(G,G^\wedge),\theta:G\longrightarrow Q_{-\varepsilon}(A,S))$ be the (-ε)-quadratic linking form over (A,S) corresponding to the ε-quadratic S-formation over A

$$(H_\varepsilon(M^*);M^*,\mathrm{im}(\begin{pmatrix} -\bar{\varepsilon}\psi j & \psi^* k \\ j & k \end{pmatrix}:L\oplus L^* \longrightarrow M^*\oplus M))\ .$$

[]

The <u>boundary</u> of a split ε-quadratic linking $\begin{Bmatrix}\text{form}\\ \text{formation}\end{Bmatrix}$ over (A,S) $\begin{Bmatrix}(M,\lambda,\nu)\\ (F,G)\end{Bmatrix}$ is the non-singular split $\begin{Bmatrix}(-\varepsilon)\text{-}\\ \varepsilon\text{-}\end{Bmatrix}$ quadratic linking $\begin{Bmatrix}\text{formation}\\ \text{form}\end{Bmatrix}$ over (A,S)

$$\begin{cases} \partial(M,\lambda,\nu) = (M,(\begin{pmatrix}1\\ \lambda\end{pmatrix},(1+T_\varepsilon)\nu)M) \\ \partial(F,G) = (G^\perp/G,\lambda^\perp/\lambda,\nu^\perp/\nu)\ ,\ \text{where } H_\varepsilon(F) = (F\oplus F^\wedge,\lambda,\nu)\ . \end{cases}$$

A split ε-quadratic linking $\begin{cases}\text{form}\\ \text{formation}\end{cases}$ is non-singular if and only if its boundary linking $\begin{cases}\text{formation}\\ \text{form}\end{cases}$ is $\begin{cases}\text{stably equivalent to } 0\\ 0\end{cases}$.

Non-singular split ε-quadratic linking $\begin{cases}\text{forms}\\ \text{formations}\end{cases}$ over (A,S) $\begin{cases}(M,\lambda,\nu),(M',\lambda',\nu')\\ (F,G),(F',G')\end{cases}$ are <u>cobordant</u> if there exists $\begin{cases}\text{an isomorphism}\\ \text{a stable equivalence}\end{cases}$

$$\begin{cases}(M,\lambda,\nu)\oplus(M',-\lambda',-\nu')\longrightarrow\partial(K,L)\\ (F,G)\oplus-(F',G')\longrightarrow\partial(N,\varphi,\psi)\end{cases}$$

for some split $\begin{cases}\varepsilon\text{-}\\ (-\varepsilon)\text{-}\end{cases}$ quadratic linking $\begin{cases}\text{formation}\\ \text{form}\end{cases}$ over (A,S) $\begin{cases}(K,L)\\ (N,\varphi,\psi)\end{cases}$.

<u>Proposition 2.4</u> Cobordism is an equivalence relation on the set of non-singular split ε-quadratic linking $\begin{cases}\text{forms}\\ \text{formations}\end{cases}$ over (A,S), such that the equivalence classes define an abelian group $\begin{cases}L_0(A,S,\varepsilon)\\ L_1(A,S,\varepsilon)\end{cases}$ with respect to the direct sum $\oplus$. The L-groups defined for n(mod 4) by

$$L_n(A,S,\varepsilon)=\begin{cases}L_0(A,S,(-)^i\varepsilon)\\ L_1(A,S,(-)^i\varepsilon)\end{cases}\text{ if } n=\begin{cases}2i\\ 2i+1\end{cases}$$

fit into the localization exact sequence

$$\ldots\longrightarrow L_n(A,\varepsilon)\longrightarrow L_n^S(S^{-1}A,\varepsilon)\longrightarrow L_n(A,S,\varepsilon)\longrightarrow L_{n-1}(A,\varepsilon)\longrightarrow\ldots\,.$$

The fit is achieved by natural isomorphisms

$$L_n(A,S,\varepsilon)\longrightarrow L_n^S(A\longrightarrow S^{-1}A,\varepsilon)\quad(n(\text{mod }4))\ ,$$

defined by sending a non-singular linking $\begin{cases}\text{form}\\ \text{formation}\end{cases}$ over (A,S) to the corresponding non-singular $\begin{cases}S\text{-formation}\\ S\text{-form}\end{cases}$ over A (given by Proposition $\begin{cases}2.2\\ 2.3\end{cases}$).

[]

Note that $L_0(A,S,\varepsilon)$ can also be viewed as the abelian group of equivalence classes of non-singular split ε-quadratic linking forms over (A,S) under the relation

$(M,\lambda,\nu)\sim(M',\lambda',\nu')$ if there exists an isomorphism

$$f : (M,\lambda,\nu)\oplus(N,\varphi,\psi) \longrightarrow (M',\lambda',\nu')\oplus(N',\varphi',\psi')$$

for some hyperbolic split ε-quadratic linking forms $(N,\varphi,\psi),(N',\varphi',\psi')$.

The localization exact sequence of Proposition 2.4 was first obtained by Pardon [1] in the case $A = \mathbb{Z}[\pi]$ (π finite), $S = \mathbb{Z}-\{0\}$ following on from the earlier work of Wall [1],[2], Passman and Petrie [1], Connolly [1] and his own work on rational surgery (Pardon [2]). These authors only work with f.g. free A-modules - we shall discuss the effect of this restriction in §7 below.

Karoubi [1] obtained a localization exact sequence in the context of hermitian K-theory. However, the methods of that paper are not sufficient for a localization sequence in the surgery obstruction groups, since it is frequently assumed that $1/2 \in A$, the formula for the quadratic function Q on p.366 of Part I is not well-defined in general, and the quadratic linking formations do not include the hessian θ appearing in the definition of (F,G) (introduced by Pardon [1]) which carries delicate quadratic information such as the Arf invariant.

The localization exact sequence is natural, in the following sense.

Let $f:A \longrightarrow B$ be a morphism of rings with involution such that $f(S)\subseteq T$ for some multiplicative subsets $S\subset A$, $T\subset B$. Given an h.d. 1 S-torsion A-module M with a f.g. projective A-module resolution

$$0 \longrightarrow P_1 \xrightarrow{\ d\ } P_0 \longrightarrow M \longrightarrow 0$$

we have that $d \in \mathrm{Hom}_A(P_1,P_0)$ is an S-isomorphism, and hence that $1\otimes d \in \mathrm{Hom}_B(B\otimes_A P_1, B\otimes_A P_0)$ is a T-isomorphism. Also, the functor

$$B\otimes_A - : (\text{A-modules}) \longrightarrow (\text{B-modules}) \ ; \ P \longmapsto B\otimes_A P$$

is right exact, so that we have a f.g. projective B-module resolution

$$0 \longrightarrow B\otimes_A P_1 \xrightarrow{1\otimes d} B\otimes_A P_0 \longrightarrow B\otimes_A M \longrightarrow 0$$

and $B\otimes_A M$ is an h.d. 1 T-torsion B-module. Thus f induces a functor

$$B\otimes_A - : (\text{h.d. 1 S-torsion A-modules}) \longrightarrow (\text{h.d. 1 T-torsion B-modules}) ;$$
$$M \longmapsto B\otimes_A M$$

and there are defined abelian group morphisms

$$f : L_n(A,S,\varepsilon) \longrightarrow L_n(B,T,\varepsilon) \; ; \; x \longmapsto B\otimes_A x \qquad (n(\text{mod } 4)) .$$

<u>Proposition 2.5</u> A morphism of rings with involution $f:A \longrightarrow B$ such that $f(S)\subseteq T$ for some multiplicative subsets $S\subset A$, $T\subset B$ induces a morphism of exact sequences of abelian groups

$$\begin{array}{ccccccccc}
\ldots \longrightarrow & L_n(A,\varepsilon) & \longrightarrow & L_n^S(S^{-1}A,\varepsilon) & \longrightarrow & L_n(A,S,\varepsilon) & \longrightarrow & L_{n-1}(A,\varepsilon) & \longrightarrow \ldots \\
 & \downarrow f & & \downarrow S^{-1}f & & \downarrow f & & \downarrow f & \\
\ldots \longrightarrow & L_n(B,\varepsilon) & \longrightarrow & L_n^T(T^{-1}B,\varepsilon) & \longrightarrow & L_n(B,T,\varepsilon) & \longrightarrow & L_{n-1}(B,\varepsilon) & \longrightarrow \ldots \; .
\end{array}$$

[]

Were it necessary we could define relative L-groups $L_n(f,S,\varepsilon)$ for n(mod 4) (as cobordism groups of relative linking forms and formations) to fit into exact sequences

$$\ldots \longrightarrow L_n(A,S,\varepsilon) \xrightarrow{f} L_n(B,T,\varepsilon) \longrightarrow L_n(f,S,\varepsilon) \longrightarrow L_{n-1}(A,S,\varepsilon) \longrightarrow \ldots$$
$$\ldots \longrightarrow L_n(f,\varepsilon) \longrightarrow L_n^S(S^{-1}f,\varepsilon) \longrightarrow L_n(f,S,\varepsilon) \longrightarrow L_{n-1}(f,\varepsilon) \longrightarrow \ldots \; .$$

§3. Cartesian squares

We shall now investigate the conditions under which a morphism of rings with involution and multiplicative subsets

$$f : (A,S) \longrightarrow (B,T)$$

induces excision isomorphisms

$$f : L_n(A,S,\varepsilon) \longrightarrow L_n(B,T,\varepsilon) \quad (n(\bmod 4))$$

and a Mayer-Vietoris exact sequence

$$\ldots \longrightarrow L_n(A,\varepsilon) \longrightarrow L_n^S(S^{-1}A,\varepsilon) \oplus L_n(B,\varepsilon) \longrightarrow L_n^T(T^{-1}B,\varepsilon) \longrightarrow L_{n-1}(A,\varepsilon) \longrightarrow \ldots .$$

Define a partial ordering on S by

$$s \leqslant s' \text{ if there exists } t \in S \text{ such that } s' = st \in S .$$

Define also a direct system of abelian groups $\{A/sA \mid s \in S\}$ with structure maps

$$A/sA \longrightarrow A/stA \; ; \; x \longmapsto tx .$$

The abelian group morphisms

$$A/sA \longrightarrow S^{-1}A/A \; ; \; a \longmapsto \frac{a}{s}$$

allow the identification

$$\varinjlim_{s \in S} A/sA = S^{-1}A/A .$$

The involution

$$- : S^{-1}A/A \longrightarrow S^{-1}A/A \; ; \; \frac{a}{s} \longmapsto \overline{\frac{a}{s}}$$

is identified with the involution

$$- : \varinjlim_{s \in S} A/sA \longrightarrow \varinjlim_{s \in S} A/sA \; ; \; \{a_s \in A/sA \mid s \in S\} \longmapsto \{\overline{a_{\bar{s}}} \in A/sA \mid s \in S\} .$$

A morphism of rings with involution and multiplicative subsets

$$f : (A,S) \longrightarrow (B,T)$$

is <u>cartesian</u> if $f(S) = T$ and if for every $s \in S$ the map

$$f : A/sA \longrightarrow B/tB \; ; \; x \longmapsto f(x) \quad (t = f(s) \in T)$$

is an isomorphism of abelian groups. It follows that there is induced an isomorphism of abelian groups with involution

$$f : \varinjlim_{s \in S} A/sA = S^{-1}A/A \longrightarrow \varinjlim_{t \in T} B/tB = T^{-1}B/B \; ; \; x \longmapsto f(x) ,$$

and hence that the commutative square of rings with involution

$$\begin{array}{ccc} A & \longrightarrow & S^{-1}A \\ f\downarrow & & \downarrow f \\ B & \longrightarrow & T^{-1}B \end{array}$$

is cartesian, in the sense that there is defined an exact sequence of abelian groups with involution

$$0 \longrightarrow A \longrightarrow S^{-1}A \oplus B \longrightarrow T^{-1}B \longrightarrow 0 \quad .$$

Cartesian morphisms were introduced by Karoubi [1] (Appendix 5 of Part I), who proved that a cartesian morphism $f:(A,S) \longrightarrow (B,T)$ induces an isomorphism of exact categories

$$f : (\text{h.d. 1 } S\text{-torsion } A\text{-modules}) \longrightarrow (\text{h.d. 1 } T\text{-torsion } B\text{-modules}) \ ;$$

$$M \longmapsto B \otimes_A M \ (= M \text{ as an } A\text{-module}) \ .$$

As an immediate consequence of this and of the localization exact sequence of Proposition 2.4 we have:

<u>Proposition 3.1</u> A cartesian morphism $f:(A,S) \longrightarrow (B,T)$ induces excision isomorphisms of relative L-groups

$$f : L_n(A,S,\varepsilon) \longrightarrow L_n(B,T,\varepsilon) \qquad (n(\text{mod } 4)) \ ,$$

and there is defined a Mayer-Vietoris exact sequence of absolute L-groups

$$\ldots \longrightarrow L_n(A,\varepsilon) \longrightarrow L_n^S(S^{-1}A,\varepsilon) \oplus L_n(B,\varepsilon) \longrightarrow L_n^T(T^{-1}B,\varepsilon) \longrightarrow L_{n-1}(A,\varepsilon) \longrightarrow \ldots \ .$$

[]

A Mayer-Vietoris exact sequence of the above type was first obtained by Wall [6] for a cartesian square of arithmetic type (cf. Proposition 3.2 below), by a direct proof which avoided relative L-theory at the expense of invoking the strong approximation theorem. In fact, it is possible to obtain both the Mayer-Vietoris sequence and the excision isomorphisms avoiding the localization sequence, by directly constructing appropriate morphisms

$$\hat{\delta} : L_n^T(B \longrightarrow T^{-1}B,\varepsilon) \longrightarrow L_{n-1}(A,\varepsilon) \qquad (n(\text{mod } 4))$$

(generalizing the method of Wall [6]), using the characterization of the relative L-groups in terms of relative forms and formations of §1. The idea of combining a localization exact sequence with the above isomorphism of categories is due to Karoubi [1], who obtained excision isomorphisms and a

Mayer-Vietoris sequence in hermitian K-theory (with the qualifications regarding the L-groups expressed at the end of §2). Bak [2] has obtained similar results in the context of the KU-theory of Bass [2].

In §7 below we shall generalize the excision isomorphisms and the Mayer-Vietoris sequence of Proposition 3.1 to the intermediate L-groups.

Given a multiplicative subset $S \subset A$ of a ring with involution A define the <u>S-adic completion of A</u> to be the inverse limit

$$\hat{A} = \varprojlim_{s \in S} A/sA$$

of the inverse system of rings $\{A/sA \mid s \in S\}$ with structure maps the natural projections

$$A/stA \longrightarrow A/sA \quad (s,t \in S) .$$

Then $\hat{A}$ is a ring, with involution by

$$^- : \hat{A} \longrightarrow \hat{A} \ ; \ \{a_s \in A/sA \mid s \in S\} \longmapsto \{\overline{a_s} \in A/sA \mid s \in S\} .$$

The inclusion

$$f : A \longrightarrow \hat{A} \ ; \ a \longmapsto \{a \in A/sA \mid s \in S\}$$

is a morphism of rings with involution, such that the image of S is a multiplicative subset $\hat{S} = f(S) \subset \hat{A}$.

<u>Proposition 3.2</u> The inclusion $f:(A,S) \longrightarrow (\hat{A},\hat{S})$ is a cartesian morphism, so that there are induced excision isomorphisms

$$f : L_n(A,S,\varepsilon) \longrightarrow L_n(\hat{A},\hat{S},\varepsilon) \quad (n \pmod 4)$$

and there is defined a Mayer-Vietoris exact sequence

$$\dots \longrightarrow L_n(A,\varepsilon) \longrightarrow L_n^S(S^{-1}A,\varepsilon) \oplus L_n(\hat{A},\varepsilon) \longrightarrow L_n^{\hat{S}}(\hat{S}^{-1}\hat{A},\varepsilon) \longrightarrow L_{n-1}(A,\varepsilon) \longrightarrow \dots .$$

[]

In particular, we have a cartesian morphism $f:(\mathbb{Z},\mathbb{Z}-\{0\}) \longrightarrow (\hat{\mathbb{Z}},\mathbb{Z}-\{0\})$, with $\hat{\mathbb{Z}} = \varprojlim_m \mathbb{Z}/m\mathbb{Z}$ the profinite completion of $\mathbb{Z}$. The associated cartesian square

$$\begin{array}{ccc} \mathbb{Z} & \longrightarrow & \mathbb{Q} \\ \downarrow & & \downarrow \\ \hat{\mathbb{Z}} & \longrightarrow & \hat{\mathbb{Q}} \end{array}$$

is the 'arithmetic square', with $\hat{\mathbb{Q}}$ the finite adéle ring of $\mathbb{Q}$. In Wall [6] there was obtained an L-theoretic Mayer-Vietoris exact sequence for the cartesian square

$$\begin{array}{ccc} \mathbb{Z}\otimes_{\mathbb{Z}}A & \longrightarrow & \mathbb{Q}\otimes_{\mathbb{Z}}A \\ (=A) & & \\ \downarrow & & \downarrow \\ \hat{\mathbb{Z}}\otimes_{\mathbb{Z}}A & \longrightarrow & \hat{\mathbb{Q}}\otimes_{\mathbb{Z}}A \end{array}$$

for any ring with involution A such that the additive group of A is finitely generated. For torsion-free A (e.g. $A = \mathbb{Z}[\pi]$, with π a finite group) this is just the cartesian square considered in Proposition 3.2 with $S = \mathbb{Z}-\{0\}\subset A$.

Given a ring with involution A we shall say that multiplicative subsets $S,T\subset A$ are <u>coprime</u> if for any $s\in S$, $t\in T$ the ideals $sA, tA \triangleleft A$ are coprime, that is if there exist $a,b\in A$ such that

$$as + bt = 1 \in A \ .$$

The multiplicative subsets

$$ST = \{st \mid s\in S, t\in T\}\subset A \quad , \quad T^{-1}S = \{\tfrac{s}{1} \mid s\in S\}\subset T^{-1}A$$

are such that there is a natural identification

$$(T^{-1}S)^{-1}(T^{-1}A) = (ST)^{-1}A \ .$$

<u>Proposition 3.3</u> If $S,T\subset A$ are coprime multiplicative subsets then the inclusion $f:(A,S)\longrightarrow(T^{-1}A,T^{-1}S)$ is a cartesian morphism, inducing excision isomorphisms

$$f : L_n(A,S,\varepsilon)\longrightarrow L_n(T^{-1}A,T^{-1}S,\varepsilon) \quad (n(\mathrm{mod}\ 4)) \ ,$$

and there is defined a Mayer-Vietoris exact sequence

$$\ldots\longrightarrow L_n(A,\varepsilon)\longrightarrow L_n^S(S^{-1}A,\varepsilon)\oplus L_n^T(T^{-1}A,\varepsilon)\longrightarrow L_n^{ST}((ST)^{-1}A,\varepsilon)\longrightarrow L_{n-1}(A,\varepsilon)\longrightarrow\ldots \ .$$

[]

For example, if $S = \{p_1^{k_1}p_2^{k_2}\ldots p_r^{k_r} \mid k_1,k_2,\ldots,k_r\geq 0\}$ and $T = \{q_1^{j_1}\ldots q_s^{j_s} \mid j_1\ldots j_s\geq 0\}$ for some disjoint collections of primes $P = \{p_1,p_2,\ldots\}$, $Q = \{q_1,q_2,\ldots\}$ such that $P\cup Q = \{\text{all primes in } \mathbb{Z}\}$ then $S^{-1}\mathbb{Z} = \mathbb{Z}[\tfrac{1}{P}] = \mathbb{Z}_{(Q)}$ (= localization away from P = localization at Q) and $S,T\subset\mathbb{Z}$ are coprime multiplicative subsets with $(ST)^{-1}\mathbb{Z} = \mathbb{Q}$.

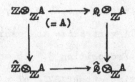

§4. Products

We shall now show that the localization sequence

$$\ldots \longrightarrow L_n(A,\varepsilon) \longrightarrow L_n^S(S^{-1}A,\varepsilon) \longrightarrow L_n(A,S,\varepsilon) \longrightarrow L_{n-1}(A,\varepsilon) \longrightarrow \ldots$$

is an exact sequence of $L^0(R)$-modules if A is an R-module for some ring with involution R. As in §1 $L^0(R)$ denotes the symmetric Witt group of R. We shall use this $L^0(R)$-action to prove that the natural maps

$$L_n(A,\varepsilon) \longrightarrow L_n^S(\mathbb{Q}\otimes_{\mathbb{Z}}A,\varepsilon) \qquad (n(\text{mod } 4)\ ,\ S = \mathbb{Z}-\{0\}\subset A)$$

are isomorphisms modulo 8-torsion for any torsion-free ring with involution A, along with other results of this nature.

A ring with involution A is an <u>R-module</u> for some ring with involution R if there is given a morphism of rings with involution

$$R\otimes_{\mathbb{Z}}A \longrightarrow A\ ;\ r\otimes a \longmapsto ra\ ,$$

with the involution on $R\otimes_{\mathbb{Z}}A$ defined by

$$\bar{\ }: R\otimes_{\mathbb{Z}}A \longrightarrow R\otimes_{\mathbb{Z}}A\ ;\ r\otimes a \longmapsto \bar{r}\otimes\bar{a}\ .$$

Note that each $r1_A \in A$ $(r\in R)$ is central in A, so that given an R-module M and an A-module N there is defined an A-module

$$M\otimes_R N = M\otimes_{\mathbb{Z}}N/\{rx\otimes y - x\otimes(r1_A)y \mid x\in M, y\in N, r\in R\}$$

with A acting by

$$A\times M\otimes_R N \longrightarrow M\otimes_R N\ ;\ (a,x\otimes y) \longmapsto x\otimes ay\ .$$

In particular, we have a pairing

(f.g. projective R-modules) $\times$ (f.g. projective A-modules)

$\longrightarrow$ (f.g. projective A-modules) ; $(M,N) \longmapsto M\otimes_R N$,

with natural identifications

$$(M\otimes_R N)^* = M^*\otimes_R N^*\ .$$

Given a multiplicative subset $S\subset A$ we have that $S^{-1}A$ is an R-module by

$$R\otimes_{\mathbb{Z}}S^{-1}A \longrightarrow S^{-1}A\ ;\ r\otimes\frac{a}{s} \longmapsto \frac{ra}{s}\ ,$$

and that there is defined a pairing

(f.g. projective R-modules) $\times$ (h.d. 1 S-torsion A-modules)

$\longrightarrow$ (h.d. 1 S-torsion A-modules) ; $(M,N) \longmapsto M\otimes_R N$, with natural identifications

$$(M\otimes_R N)^\wedge = M^*\otimes_R N^\wedge .$$

Define $L^0(R)$-actions on quadratic L-theory by

$$L^0(R)\otimes_{\mathbb{Z}} L_n(A,\varepsilon) \longrightarrow L_n(A,\varepsilon) \;;$$

$$\begin{cases} (M,\varphi)\otimes(N,\psi) \longmapsto (M\otimes_R N,\varphi\otimes\psi) \\ (M,\varphi)\otimes(N,\psi;F,G) \longmapsto (M\otimes_R N,\varphi\otimes\psi;M\otimes_R F,M\otimes_R G) \end{cases} \text{if } n = \begin{cases} 2i \\ 2i+1 \end{cases} .$$

(In terms of the products defined in Part I of Ranicki [2] these are just the composites

$$L^0(R)\otimes_{\mathbb{Z}} L_n(A,\varepsilon) \xrightarrow{\otimes} L_n(R\otimes_{\mathbb{Z}} A,1\otimes\varepsilon) \longrightarrow L_n(A,\varepsilon) \quad (n(\mathrm{mod}\ 4))).$$

Define also $L^0(R)$-actions

$$L^0(R)\otimes_{\mathbb{Z}} L_n^S(S^{-1}A,\varepsilon) \longrightarrow L_n^S(S^{-1}A,\varepsilon) \;;$$

$$\begin{cases} (M,\varphi)\otimes S^{-1}(N,\psi) \longmapsto S^{-1}(M\otimes_R N,\varphi\otimes\psi) \\ (M,\varphi)\otimes S^{-1}(N,\psi;F,G) \longmapsto S^{-1}(M\otimes_R N,\varphi\otimes\psi;M\otimes_R F,M\otimes_R G) \end{cases} \text{if } n = \begin{cases} 2i \\ 2i+1 \end{cases} ,$$

$$L^0(R)\otimes_{\mathbb{Z}} L_n(A,S,\varepsilon) \longrightarrow L_n(A,S,\varepsilon) \;;$$

$$\begin{cases} (M,\varphi)\otimes(N,\lambda,\nu) \longmapsto (M\otimes_R N,\varphi\otimes\lambda,\varphi\otimes\nu : x\otimes y \longmapsto \varphi(x)(x)\nu(y)) \\ (M,\varphi)\otimes\left(F,\left(\begin{pmatrix}\gamma\\ \mu\end{pmatrix}\right),\theta)G\right) \longmapsto \left(M\otimes_R F,\left(\begin{pmatrix}1\otimes\gamma\\ \varphi\otimes\mu\end{pmatrix}\right),\varphi\otimes\theta)M\otimes_R G\right) \end{cases} \text{if } n = \begin{cases} 2i \\ 2i+1 \end{cases} .$$

In each case the element

$$(R,1:R\longrightarrow R^*;q\longmapsto(r\longmapsto r\bar{q}))\in L^0(R)$$

acts by the identity. (In general R is not itself an R-module. However, if R is commutative then it is an R-module in the usual fashion, and the symmetric Witt group $L^0(R)$ is a commutative ring with 1).

<u>Proposition 4.1</u> Let A,R be rings with involution such that A is an R-module, and let $S\subset A$ be a multiplicative subset. The localization sequence

$$\ldots \longrightarrow L_n(A,\varepsilon) \longrightarrow L_n^S(S^{-1}A,\varepsilon) \longrightarrow L_n(A,S,\varepsilon) \longrightarrow L_{n-1}(A,\varepsilon) \longrightarrow \ldots$$

is an exact sequence of $L^0(R)$-modules.

[]

(More generally, if $f:A\longrightarrow B$ is a morphism of rings with involution which is a morphism of R-modules then the symmetric Witt group $L^0(R)$ acts on the exact sequence of Proposition 1.3

$$\ldots \longrightarrow L_n(A,\varepsilon) \xrightarrow{f} L_n(B,\varepsilon) \longrightarrow L_n(f,\varepsilon) \longrightarrow L_{n-1}(A,\varepsilon) \longrightarrow \ldots).$$

In our applications of Proposition 4.1 we shall need to know the symmetric Witt groups $L^0(\mathbb{Z}_m)$ of the finite cyclic rings $\mathbb{Z}_m = \mathbb{Z}/m\mathbb{Z}$.

Let $m = p_1^{k_1} p_2^{k_2} \dots p_r^{k_r}$ be the factorization of m into prime powers, so that

$$\mathbb{Z}_m = \bigoplus_{i=1}^{m} \mathbb{Z}_{p_i^{k_i}} \quad , \quad L^0(\mathbb{Z}_m) = \bigoplus_{i=1}^{m} L^0(\mathbb{Z}_{p_i^{k_i}}) \ .$$

Lemma 5 of Wall [4] and Theorem 3.3 of Bak [1] on reduction modulo a complete ideal (alias Hensel's lemma) apply to show that the projections

$$\begin{cases} \mathbb{Z}_{2^k} \longrightarrow \mathbb{Z}_8 \ , \ k \geqslant 3 \\ \mathbb{Z}_{p^k} \longrightarrow \mathbb{Z}_p \ , \ p \text{ odd} \ , \ k \geqslant 1 \end{cases} \quad \text{induce isomorphisms}$$

$$\begin{cases} L^0(\mathbb{Z}_{2^k}) \longrightarrow L^0(\mathbb{Z}_8) = \mathbb{Z}_8 \oplus \mathbb{Z}_2 \\ L^0(\mathbb{Z}_{p^k}) \longrightarrow L^0(\mathbb{Z}_p) = \begin{cases} \mathbb{Z}_2 \oplus \mathbb{Z}_2 & \text{if } p \equiv 1 (\text{mod } 4) \\ \mathbb{Z}_4 & \text{if } p \equiv 3 (\text{mod } 4) \ . \end{cases} \end{cases}$$

Moreover,

$$L^0(\mathbb{Z}_4) = \mathbb{Z}_4 \oplus \mathbb{Z}_2 \quad , \quad L^0(\mathbb{Z}_2) = \mathbb{Z}_2 \ .$$

For each integer $m \geqslant 2$ define the number

$$\psi(m) = \text{exponent of } L^0(\mathbb{Z}_m) = \begin{cases} 2 & \text{if } m = d \text{ or } 2d \\ 4 & \text{if } m = 4d, e, 2e \text{ or } 4e \\ 8 & \text{otherwise} \end{cases} ,$$

with

d = a product of odd primes $p \equiv 1 (\text{mod } 4)$

e = a product of odd primes, including at least one $p \equiv 3 (\text{mod } 4)$.

A ring with involution A is of characteristic m if m is the least integer $\geqslant 2$ such that $m1 = 0 \in A$, in which case $ma = 0$ for all $a \in A$ and A is a $\mathbb{Z}_m$-module.

Proposition 4.2 If the ring with involution A is of characteristic m then the localization sequence

$$\dots \longrightarrow L_n(A,\varepsilon) \longrightarrow L_n^S(S^{-1}A,\varepsilon) \longrightarrow L_n(A,S,\varepsilon) \longrightarrow L_{n-1}(A,\varepsilon) \longrightarrow \dots$$

is an exact sequence of $L^0(\mathbb{Z}_m)$-modules, so that all the L-groups involved are of exponent $\psi(m)$.

[]

The symmetric Witt groups $L^0(\hat{\mathbb{Z}}_m)$ of the rings of m-adic integers $\hat{\mathbb{Z}}_m = \varprojlim_k \mathbb{Z}/m^k\mathbb{Z}$ are computed as follows. Again, let $m = p_1^{k_1} p_2^{k_2} \dots p_r^{k_r}$ so that

$$\hat{\mathbb{Z}}_m = \bigoplus_{i=1}^{r} \hat{\mathbb{Z}}_{p_i} \quad , \quad L^0(\hat{\mathbb{Z}}_m) = \bigoplus_{i=1}^{r} L^0(\hat{\mathbb{Z}}_{p_i})$$

and

$$L^0(\hat{\mathbb{Z}}_p) = \begin{cases} L^0(\mathbb{Z}_8) = \mathbb{Z}_8 \oplus \mathbb{Z}_2 & \text{if } p = 2 \\ L^0(\mathbb{Z}_p) = \begin{cases} \mathbb{Z}_2 \oplus \mathbb{Z}_2 & \text{if } p \equiv 1 (\bmod 4) \\ \mathbb{Z}_4 & \text{if } p \equiv 3 (\bmod 4) \end{cases} \end{cases}$$

For each integer $m \geq 2$ define the number

$$\hat{\Psi}(m) = \text{exponent of } L^0(\hat{\mathbb{Z}}_m) = \begin{cases} 2 \text{ if } m \text{ is a product of odd primes } p \equiv 1 (\bmod 4) \\ 4 \text{ if } m \text{ is a product of odd primes at least one} \\ \qquad \text{of which is } p \equiv 3 (\bmod 4) \\ 8 \text{ if } m \text{ is even .} \end{cases}$$

The method of Wall [5] applies to show that the symmetric Witt group of the profinite completion $\hat{\mathbb{Z}} = \varprojlim_m \mathbb{Z}/m\mathbb{Z} = \prod_p \hat{\mathbb{Z}}_p$ is the infinite product

$$L^0(\hat{\mathbb{Z}}) = \prod_p L^0(\hat{\mathbb{Z}}_p) \quad .$$

A ring with involution A is <u>m-torsion-free</u> if $S = \{m^k | k \geq 0\} \subset A$ is a multiplicative subset, so that the localization away from m $S^{-1}A = A[\frac{1}{m}]$ is defined. The m-adic completion $\hat{A} = \varprojlim_k A/m^kA$ is a $\hat{\mathbb{Z}}_m$-module.

A ring with involution A is <u>torsion-free</u> if $S = \mathbb{Z} - \{0\} \subset A$ is a multiplicative subset, so that the localization $S^{-1}A = \mathbb{Q} \otimes_{\mathbb{Z}} A$ is defined. The profinite completion $\hat{A} = \varprojlim_m A/mA$ is a $\hat{\mathbb{Z}}$-module .

<u>Proposition 4.3</u> Let A be a ring with involution which is m-torsion-free (resp. torsion-free) and let $S = \{m^k | k \geq 0\} \subset A$ (resp. $S = \mathbb{Z} - \{0\} \subset A$). The localization sequence of the S-adic completion $\hat{A} = \varprojlim_{s \in S} A/sA$

$$\dots \longrightarrow L_n(\hat{A},\varepsilon) \longrightarrow L_n^{\hat{S}}(\hat{S}^{-1}\hat{A},\varepsilon) \longrightarrow L_n(\hat{A},\hat{S},\varepsilon) \longrightarrow L_{n-1}(\hat{A},\varepsilon) \longrightarrow \dots$$

is an exact sequence of $L^0(\hat{\mathbb{Z}}_m)$ (resp. $L^0(\hat{\mathbb{Z}})$)- modules, so that all the L-groups are of exponent $\hat{\Psi}(m)$ (resp. 8). Thus the L-groups $L_n(A,S,\varepsilon) = L_n(\hat{A},\hat{S},\varepsilon)$ are of exponent $\hat{\Psi}(m)$ (resp. 8) and the natural maps

$$L_n(A,\varepsilon) \longrightarrow L_n^S(S^{-1}A,\varepsilon) \qquad (n (\bmod 4))$$

are isomorphisms modulo $\hat{\Psi}(m)$ (resp. 8)-torsion.

[]

The integral group ring $\mathbb{Z}[\pi]$ of a group π is torsion-free, with localization $S^{-1}\mathbb{Z}[\pi] = \mathbb{Q}[\pi]$ ($S = \mathbb{Z}-\{0\}$) the rational group ring, so that as a particular case of Proposition 4.3 we have:

<u>Proposition 4.4</u> The natural maps

$$L_n(\mathbb{Z}[\pi]) \longrightarrow L_n^S(\mathbb{Q}[\pi]) \quad (n(\mathrm{mod}\ 4))$$

are isomorphisms modulo 8-torsion, for any group π.

[]

Results of this type were first obtained for finite groups π. If we take for granted the result that the natural maps $L_{2i}(\mathbb{Q}[\pi]) \longrightarrow L_{2i}(\mathbb{R}[\pi])$ are isomorphisms modulo 2-primary torsion (π finite, $i(\mathrm{mod}\ 2)$) then Theorems 13A.3, 13A.4 i) of Wall [3] can be interpreted as stating that the natural maps $L_{2i}(\mathbb{Z}[\pi]) \longrightarrow L_{2i}^S(\mathbb{Q}[\pi])$ are isomorphisms modulo 2-primary torsion. The results of Passman and Petrie [1] and Connolly [1] can be interpreted as stating that the natural maps $L_{2i+1}(\mathbb{Z}[\pi]) \longrightarrow L_{2i+1}^S(\mathbb{Q}[\pi])$ are isomorphisms modulo 8-torsion (π finite, $i(\mathrm{mod}\ 2)$).

Results similar to those of Propositions 4.3, 4.4 were first obtained by Karoubi [1], for hermitian K-theory.

§5. Dedekind algebra

We shall now investigate the general properties of the L-groups $L_n(A,S,\varepsilon)$ (n(mod 4)) in the case when the ring with involution A is an algebra over a Dedekind ring R and $S = R-\{0\}$. An S-torsion A-module has a canonical direct sum decomposition as a direct sum of $\mathcal{P}$-primary S-torsion A-modules, with $\mathcal{P}$ ranging over all the (non-zero) prime ideals of R, and there is a corresponding decomposition for $L_n(A,S,\varepsilon)$.

Given a multiplicative subset $S \subset A$ in a ring with involution A we shall say that the pair (A,S) is a Dedekind algebra if $R = S \cup \{0\}$ is a Dedekind ring with respect to the ring operations inherited from A. The localization $S^{-1}A = F \otimes_R A$ is the induced algebra over the quotient field $F = S^{-1}R$. For example, a torsion-free ring with involution A is the same as a Dedekind algebra $(A, \mathbb{Z}-\{0\})$. A Dedekind ring with involution R is the same as a Dedekind algebra $(R, R-\{0\})$. In dealing with Dedekind algebras (A,S) and the prime ideals $\mathcal{P}$ of R we shall always exclude the case $\mathcal{P} = \{0\}$.

Let (A,S) be a Dedekind algebra.

The annihilator of an S-torsion A-module M is the ideal of R defined by

$$\mathrm{ann}(M) = \{s \in R \mid sM = 0\} \triangleleft R .$$

Like all ideals of R this has a unique expression as a product of powers of distinct prime ideals $\mathcal{P}_1, \mathcal{P}_2, \ldots, \mathcal{P}_r$

$$\mathrm{ann}(M) = \mathcal{P}_1^{k_1}\mathcal{P}_2^{k_2}\ldots\mathcal{P}_r^{k_r} \quad (k_i \geqslant 1) .$$

If M is such that the natural map $M \longrightarrow M\hat{}\hat{}$ is an isomorphism (e.g. if M is h.d. 1) then

$$\mathrm{ann}(M\hat{}) = \overline{\mathrm{ann}(M)} \triangleleft R .$$

An S-torsion A-module M is $\mathcal{P}$-primary for some prime ideal $\mathcal{P}$ of R if

$$\mathrm{ann}(M) = \mathcal{P}^k$$

for some $k \geqslant 1$.

Define the <u>localization of A at $\mathcal{P}$</u> for some prime ideal $\mathcal{P}$ of R to be the ring

$$A_{\mathcal{P}} = (R-\mathcal{P})^{-1}A .$$

If $\overline{\mathcal{P}} = \mathcal{P}$ there is defined an involution

$$- : A_{\mathcal{P}} \longrightarrow A_{\mathcal{P}} ; \frac{a}{r} \longmapsto \frac{\bar{a}}{\bar{r}} \qquad (a \in A,\ r \in R-\mathcal{P}) .$$

(If $\overline{\mathcal{P}} \neq \mathcal{P}$ there is defined an involution $- : A_{\mathcal{P}} \times A_{\overline{\mathcal{P}}} \longrightarrow A_{\mathcal{P}} \times A_{\overline{\mathcal{P}}} ; (x,y) \longmapsto (\bar{y},\bar{x})$).

Given an h.d. 1 S-torsion A-module M define an h.d. 1 $\mathcal{P}$-primary S-torsion A-module

$$M_{\mathcal{P}} = A_{\mathcal{P}} \otimes_A M .$$

If $\mathrm{ann}(M) = \mathcal{P}_1^{k_1}\mathcal{P}_2^{k_2}\ldots\mathcal{P}_r^{k_r}$ it is possible to identify

$$M_{\mathcal{P}} = \begin{cases} \mathcal{P}_1^{k_1}\mathcal{P}_2^{k_2}\ldots\mathcal{P}_{i-1}^{k_{i-1}}\mathcal{P}_{i+1}^{k_{i+1}}\ldots\mathcal{P}_r^{k_r}M & \text{if } \mathcal{P}=\mathcal{P}_i \text{ for some } i,\ 1 \leqslant i \leqslant r \\ 0 & \text{if } \mathcal{P} \notin \{\mathcal{P}_1,\mathcal{P}_2,\ldots,\mathcal{P}_r\} \end{cases}$$

so that

$$M = \bigoplus_{i=1}^{r} M_{\mathcal{P}_i} \quad , \quad (M^{\wedge})_{\mathcal{P}} = (M_{\overline{\mathcal{P}}})^{\wedge} \quad , \quad \mathrm{Hom}_A(M,M') = \bigoplus_{\mathcal{P}} \mathrm{Hom}_A(M_{\mathcal{P}},M'_{\mathcal{P}}) .$$

We thus have a canonical identification of exact categories

$$(\text{h.d. 1 S-torsion A-modules}) = \bigoplus_{\mathcal{P}} (\text{h.d. 1 } \mathcal{P}\text{-primary S-torsion A-modules}) ,$$

with $\mathcal{P}$ ranging over all the prime ideals of R. The S-duality functor $M \longmapsto M^{\wedge}$ sends the $\mathcal{P}$-primary component to the $\overline{\mathcal{P}}$-primary component.

Express the spectrum of prime ideals of R as a disjoint union

$$\mathrm{spec}(R) = \{\mathcal{P}\} \cup \{\mathcal{Q}\} \cup \{\overline{\mathcal{Q}}\}$$

with $\mathcal{P}$ ranging over all the prime ideals such that $\overline{\mathcal{P}} = \mathcal{P}$.

A non-singular split ε-quadratic linking $\begin{cases}\text{form}\\ \text{formation}\end{cases}$ over (A,S) $\begin{cases}(M,\lambda,\nu)\\ (F,G)\end{cases}$ has a canonical direct sum decomposition

$$\begin{cases} (M,\lambda,\nu) = \bigoplus_{\mathcal{P}} (M_{\mathcal{P}},\lambda_{\mathcal{P}},\nu_{\mathcal{P}}) \oplus \bigoplus_{\mathcal{Q}} (M_{\mathcal{Q}} \oplus M_{\overline{\mathcal{Q}}},\lambda_{\mathcal{Q}},\nu_{\mathcal{Q}}) \\ (F,G) = \bigoplus_{\mathcal{P}} (F_{\mathcal{P}},G_{\mathcal{P}}) \oplus \bigoplus_{\mathcal{Q}} (F_{\mathcal{Q}} \oplus F_{\overline{\mathcal{Q}}},G_{\mathcal{Q}} \oplus G_{\overline{\mathcal{Q}}}) \end{cases} ,$$

such that for each $\mathcal{Q}$

$$\begin{cases} (M_Q \oplus M_{\bar{Q}}, \lambda_Q, \nu_Q) = 0 \in L_0(A,S,\varepsilon) \\ (F_Q \oplus F_{\bar{Q}}, G_Q \oplus G_{\bar{Q}}) = 0 \in L_1(A,S,\varepsilon) \quad . \end{cases}$$

For each prime ideal $\mathcal{P}$ of R such that $\bar{\mathcal{P}} = \mathcal{P}$ define the L-groups $L_n(A,\mathcal{P}^\infty,\varepsilon)$ (n(mod 4)) in the same way as $L_n(A,S,\varepsilon)$ but using only $\mathcal{P}$-primary h.d. 1 S-torsion A-modules. There is a natural identification

(h.d. 1 $\mathcal{P}$-primary S-torsion A-modules)

$$= (\text{h.d. } 1\ S_{\mathcal{P}}\text{-torsion } A_{\mathcal{P}}\text{-modules})$$

where $S_{\mathcal{P}} = \{\frac{s}{1} \in A_{\mathcal{P}} | s \in S\} \subset A_{\mathcal{P}}$, so that we can also identify

$$L_n(A,\mathcal{P}^\infty,\varepsilon) = L_n(A_{\mathcal{P}},S_{\mathcal{P}},\varepsilon) \quad (n(\text{mod } 4)) \ .$$

If $\mathcal{P} = \pi R$ is a prime ideal of R which is principal, with generator $\pi \in \mathcal{P}$, then $\bar{\pi} = \pi u \in \mathcal{P}$ for some unit $u \in R$ such that $u\bar{u} = 1 \in R$ and there is defined a multiplicative subset $S_\pi = \{\pi^j u^k | j \geqslant 0, k \in \mathbb{Z}\} \subset A$ such that

(h.d. 1 $\mathcal{P}$-primary S-torsion A-modules)

$$= (\text{h.d. } 1\ S_\pi\text{-torsion A-modules})$$

$$L_n(A,\mathcal{P}^\infty,\varepsilon) = L_n(A,S_\pi,\varepsilon) \quad (n(\text{mod } 4)) \ .$$

<u>Proposition 5.1</u> The L-groups of a Dedekind algebra (A,S) have a canonical direct sum decomposition

$$L_n(A,S,\varepsilon) = \bigoplus_{\mathcal{P}} L_n(A,\mathcal{P}^\infty,\varepsilon) \quad (n(\text{mod } 4))$$

with $\mathcal{P}$ ranging over all the prime ideals of R such that $\bar{\mathcal{P}} = \mathcal{P}$.
The localization exact sequence of (A,S) can thus be expressed as

$$\ldots \longrightarrow L_n(A,\varepsilon) \longrightarrow L_n^S(S^{-1}A,\varepsilon) \longrightarrow \bigoplus_{\mathcal{P}} L_n(A,\mathcal{P}^\infty,\varepsilon) \longrightarrow L_{n-1}(A,\varepsilon) \longrightarrow \ldots \ .$$

[]

The localization sequence in the case $(A,S) = (R,R-\{0\})$

$$\ldots \longrightarrow L_n(R,\varepsilon) \longrightarrow L_n(F,\varepsilon) \longrightarrow \bigoplus_{\mathcal{P}} L_n(R,\mathcal{P}^\infty,\varepsilon) \longrightarrow L_{n-1}(R,\varepsilon) \longrightarrow \ldots$$

is closely related to the original localization exact sequence of Milnor (Corollary IV.3.3 of Milnor and Husemoller [1]) for the symmetric Witt group of a Dedekind ring R

$$0 \longrightarrow L^0(R) \longrightarrow L^0(F) \longrightarrow \bigoplus_{\mathcal{P}} L^0(R/\mathcal{P}) \ .$$

(In the part of Ranicki [2] devoted to localization we shall extend this to an exact sequence

$$0 \longrightarrow L^0(R,\varepsilon) \longrightarrow L^0(F,\varepsilon) \longrightarrow \bigoplus_{\mathcal{P}} L^0(R/\mathcal{P},\varepsilon) \longrightarrow L^1(R,-\varepsilon) \longrightarrow 0$$

with $L^1(R,\varepsilon)$ the cobordism group of non-singular ε-symmetric formations over R). Now $L_1(F,\varepsilon) = 0$, so that the above sequence of quadratic L-groups breaks up into two sequences of the type

$$0 \longrightarrow \bigoplus_{\mathcal{P}} L_1(R,\mathcal{P}^\infty,\varepsilon) \longrightarrow L_0(R,\varepsilon) \longrightarrow L_0(F,\varepsilon) \longrightarrow \bigoplus_{\mathcal{P}} L_0(R,\mathcal{P}^\infty,\varepsilon) \longrightarrow L_1(R,-\varepsilon) \longrightarrow 0 .$$

A standard devissage argument shows that the forgetful functors

(f.d. vector spaces over the residue class field $R/\mathcal{P}$)

$\longrightarrow$ (h.d. 1 $\mathcal{P}$-primary S-torsion R-modules) ; $V \longmapsto V$

induce isomorphisms in algebraic K-theory and symmetric L-theory. There are induced morphisms in quadratic L-theory

$$L_n(R/\mathcal{P},\varepsilon) \longrightarrow L_n(R,\mathcal{P}^\infty,\varepsilon) \qquad (n(\mathrm{mod}\ 4),\ \overline{\mathcal{P}} = \mathcal{P})$$

but these may not be isomorphisms (particularly if $R/\mathcal{P}$ is a field of characteristic 2, cf. Appendix 1 of Part II of Karoubi [1]). For example, neither of the morphisms

$$L_0(\mathbb{Z}_2,1) = \mathbb{Z}_2 \longrightarrow L_0(\mathbb{Z},(2\mathbb{Z})^\infty,1) = \mathbb{Z}_8 \oplus \mathbb{Z}_2 \ ;\ 1 \longmapsto (0,1)$$

$$L_1(\mathbb{Z}_2,-1) = 0 \longrightarrow L_1(\mathbb{Z},(2\mathbb{Z})^\infty,-1) = \mathbb{Z}_2$$

is an isomorphism.

Next, we shall describe the Mayer-Vietoris exact sequence of the L-groups of a localization-completion square of a Dedekind algebra (A,S)

$$\begin{array}{ccc} A & \longrightarrow & S^{-1}A \\ \downarrow & & \downarrow \\ \hat{A} & \longrightarrow & \hat{S}^{-1}\hat{A} \end{array}$$

(Proposition 3.2) in terms of the prime ideal structure of the Dedekind ring $R = S \cup \{0\}$. We shall confine the discussion to the case when $\overline{\mathcal{P}} = \mathcal{P}$ for every prime ideal $\mathcal{P}$ of R, leaving the general case for the reader.

The <u>$\mathcal{P}$-adic completion of A</u> for some prime ideal $\mathcal{P}$ of R is the ring

$$\hat{A}_{\mathcal{P}} = \varprojlim_k A/\mathcal{P}^k A ,$$

with involution

$$- : \hat{A}_{\mathcal{P}} \longrightarrow \hat{A}_{\mathcal{P}} \ ;\ \{a_k \in A/\mathcal{P}^k A \mid k \geq 1\} \longrightarrow \{\bar{a}_k \in A/\mathcal{P}^k A \mid k \geq 1\} .$$

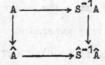

The $\mathcal{P}$-adic completion $\hat{A}_{\mathcal{P}}$ of A can be identified with the $S_{\mathcal{P}}$-adic completion of the localization $A_{\mathcal{P}}$ of A at $\mathcal{P}$

$$\hat{A}_{\mathcal{P}} = \varprojlim_{s\in S_{\mathcal{P}}} A_{\mathcal{P}}/sA_{\mathcal{P}} \; (= \varprojlim_{k} A/\pi^k A \text{ if } \mathcal{P} = \pi R,\ \pi\in\mathcal{P}).$$

Given $s\in S = R-\{0\}$ let $sR = \mathcal{P}_1^{k_1}\mathcal{P}_2^{k_2}\ldots\mathcal{P}_r^{k_r} \triangleleft R$, so that

$$A/sA = A/\mathcal{P}_1^{k_1}A \oplus A/\mathcal{P}_2^{k_2}A \oplus \ldots \oplus A/\mathcal{P}_r^{k_r}A$$

$$\frac{1}{s}\in \hat{A}_{\mathcal{P}} \subset \hat{S}_{\mathcal{P}}^{-1}\hat{A}_{\mathcal{P}} \text{ if } \mathcal{P}\notin\{\mathcal{P}_1,\mathcal{P}_2,\ldots,\mathcal{P}_r\}.$$

It is thus possible to define morphisms of rings with involution

$$\hat{A} = \varprojlim_{s\in S} A/sA \longrightarrow \prod_{\mathcal{P}} \hat{A}_{\mathcal{P}}$$

$$\hat{S}^{-1}\hat{A} \longrightarrow \coprod_{\mathcal{P}} (\hat{S}_{\mathcal{P}}^{-1}\hat{A}_{\mathcal{P}}, \hat{A}_{\mathcal{P}})$$

and hence also abelian group morphisms

$$L_n(\hat{A},\varepsilon) \longrightarrow \prod_{\mathcal{P}} L_n(\hat{A}_{\mathcal{P}},\varepsilon)$$

$$L_n^{\hat{S}}(\hat{S}^{-1}\hat{A},\varepsilon) \longrightarrow \coprod_{\mathcal{P}} (L_n^{\hat{S}_{\mathcal{P}}}(\hat{S}_{\mathcal{P}}^{-1}\hat{A}_{\mathcal{P}},\varepsilon), L_n(\hat{A}_{\mathcal{P}},\varepsilon)) \qquad (n(\bmod 4)) .$$

(The restricted product $\coprod_{\mathcal{P}}(G_{\mathcal{P}},H_{\mathcal{P}})$ of a collection of pairs of objects $(G_{\mathcal{P}},H_{\mathcal{P}})$ indexed by $\{\mathcal{P}\}$ and equipped with morphisms $H_{\mathcal{P}} \longrightarrow G_{\mathcal{P}}$ is defined to be the direct limit

$$\coprod_{\mathcal{P}} (G_{\mathcal{P}},H_{\mathcal{P}}) = \varinjlim_{I} \left(\prod_{\mathcal{P}\in I} G_{\mathcal{P}} \times \prod_{\mathcal{P}\notin I} H_{\mathcal{P}}\right)$$

taken over all the finite subsets I of $\{\mathcal{P}\}$). Wall [5] and Bak [2] have studied some of the circumstances under which the above morphisms are isomorphisms, roughly speaking when A is finitely generated as an R-module and $S^{-1}A = F\otimes_R A$ is a semi-simple F-algebra (e.g. if $(A,S) = (\mathbb{Z}[\pi],\mathbb{Z}-\{0\})$ for a finite group π, with $R = \mathbb{Z}$). At any rate, it is possible to obtain a Mayer-Vietoris exact sequence relating the L-groups of $A, S^{-1}A$ to those of all the $\mathcal{P}$-adic completions $\hat{A}_{\mathcal{P}}$, $\hat{S}_{\mathcal{P}}^{-1}\hat{A}_{\mathcal{P}}$. Propositions 3.2, 5.1 give morphisms of exact sequences

$$\begin{array}{ccccccccc}
\cdots \longrightarrow & L_n(A,\varepsilon) & \longrightarrow & L_n^S(S^{-1}A,\varepsilon) & \longrightarrow & \bigoplus\limits_{\mathcal{P}} L_n(A,\mathcal{P}^\infty,\varepsilon) & \longrightarrow & L_{n-1}(A,\varepsilon) & \longrightarrow \cdots \\
& \downarrow & & \downarrow & & \downarrow & & \downarrow & \\
\cdots \longrightarrow & L_n(\hat{A},\varepsilon) & \longrightarrow & L_n^{\hat{S}}(\hat{S}^{-1}\hat{A},\varepsilon) & \longrightarrow & \bigoplus\limits_{\mathcal{P}} L_n(\hat{A}_{\mathcal{P}},\hat{S}_{\mathcal{P}},\varepsilon) & \longrightarrow & L_{n-1}(\hat{A},\varepsilon) & \longrightarrow \cdots \\
& \downarrow & & \downarrow & & \downarrow 1 & & \downarrow & \\
\cdots \longrightarrow & \prod\limits_{\mathcal{P}} L_n(\hat{A}_{\mathcal{P}},\varepsilon) & \longrightarrow & \prod\limits_{\mathcal{P}} (L_n^{\hat{S}_{\mathcal{P}}}(\hat{S}_{\mathcal{P}}^{-1}\hat{A}_{\mathcal{P}},\varepsilon), L_n(\hat{A}_{\mathcal{P}},\varepsilon)) & \longrightarrow & \bigoplus\limits_{\mathcal{P}} L_n(\hat{A}_{\mathcal{P}},\hat{S}_{\mathcal{P}},\varepsilon) & \longrightarrow & \prod\limits_{\mathcal{P}} L_{n-1}(\hat{A}_{\mathcal{P}},\varepsilon) & \longrightarrow \cdots
\end{array}$$

involving the isomorphisms

$$L_n(A,\mathcal{P}^\infty,\varepsilon) = L_n(A_{\mathcal{P}},S_{\mathcal{P}},\varepsilon) \longrightarrow L_n(\hat{A}_{\mathcal{P}},\hat{S}_{\mathcal{P}},\varepsilon) \quad (n(\bmod 4)) \ .$$

We deduce the following exact sequence, which is valid even in the case when the Dedekind ring R has prime ideals $\mathcal{P}$ such that $\overline{\mathcal{P}} \neq \mathcal{P}$.

Proposition 5.2 **Given a Dedekind algebra (A,S) there is defined a Mayer-Vietoris exact sequence**

$$\cdots \longrightarrow L_n(A,\varepsilon) \longrightarrow L_n^S(S^{-1}A,\varepsilon) \oplus \prod_{\mathcal{P}} L_n(\hat{A}_{\mathcal{P}},\varepsilon) \longrightarrow \prod_{\mathcal{P}} (L_n^{\hat{S}_{\mathcal{P}}}(\hat{S}_{\mathcal{P}}^{-1}\hat{A}_{\mathcal{P}},\varepsilon), L_n(\hat{A}_{\mathcal{P}},\varepsilon))$$
$$\longrightarrow L_{n-1}(A,\varepsilon) \longrightarrow \cdots \ ,$$

with $\mathcal{P}$ ranging over all the prime ideals of $R = S \cup \{0\}$ such that $\overline{\mathcal{P}} = \mathcal{P}$.

[]

§6. Polynomial extensions

Given a central indeterminate x over a ring A there is defined a multiplicative subset $S = \{x^k | k \geq 0\} \subset A[x]$ with localization $S^{-1}A[x] = A[x,x^{-1}]$. An h.d. 1 S-torsion A[x]-module M is the same as a f.g. projective A-module M together with a nilpotent endomorphism $e: M \longrightarrow M; y \longmapsto xy$, and there is in fact a canonical identification of exact categories

(h.d. 1 S-torsion A[x]-modules M)

= (f.g. projective A-modules M with a nilpotent endomorphism $e: M \longrightarrow M$).

As in Chapter XII of Bass [1] it is possible to combine this identification with the localization exact sequence of algebraic K-theory

$$K_1(A[x]) \longrightarrow K_1(A[x,x^{-1}]) \longrightarrow K_1(A[x],S) \longrightarrow K_0(A[x]) \longrightarrow K_0(A[x,x^{-1}])$$

to obtain split exact sequences

$$0 \longrightarrow K_1(A[x]) \longrightarrow K_1(A[x,x^{-1}]) \longrightarrow K_1(A[x],S) \longrightarrow 0$$

$$0 \longrightarrow K_1(A) \longrightarrow K_1(A[x]) \oplus K_1(A[x^{-1}]) \longrightarrow K_1(A[x,x^{-1}]) \longrightarrow K_0(A) \longrightarrow 0 \ ,$$

i.e. the 'fundamental theorem of algebraic K-theory'.

It is likewise possible to use an L-theoretic localization exact sequence to describe the L-groups of the polynomial extensions $A[x], A[x,x^{-1}]$ of a ring with involution A, where $\bar{x} = x$. Indeed, such was the approach taken by Karoubi [1]. On the other hand, we have already shown in Part IV of Ranicki [1] that there are defined split exact sequences

$$0 \longrightarrow L_n(A[x]) \longrightarrow L_n^S(A[x,x^{-1}]) \longrightarrow L_n^K(A[x^{-1}]) \longrightarrow 0$$

$$0 \longrightarrow L_n(A) \longrightarrow L_n^K(A[x]) \oplus L_n^K(A[x^{-1}]) \longrightarrow L_n^S(A[x,x^{-1}]) \longrightarrow L_n(A) \longrightarrow 0$$

$$(n(\text{mod } 4),\ K = \text{im}(\widetilde{K}_0(A) \longrightarrow \widetilde{K}_0(A[x^{\pm 1}]))) \ ,$$

by a modification of Part II of Ranicki [1] (which concerned the L-theory of the Laurent extension $A[z,z^{-1}]$ of A, with $\bar{z} = z^{-1}$). We shall now explicitly identify

$$L_n(A[x],S,\varepsilon) = L_n^K(A[x^{-1}],\varepsilon) \quad (n(\text{mod } 4)) \ .$$

The Witt class of a non-singular split ε-quadratic linking form over (A[x],S) corresponds to the Witt class of a non-singular ε-quadratic form over $A[x^{-1}]$, whereas ε-quadratic linking forms over (A[x],S) correspond to even

ε-symmetric forms over $A[x^{-1}]$, so that the extra structure of split ε-quadratic linking forms over $(A[x],S)$ is seen to carry delicate quadratic information such as the Arf invariant.

The <u>polynomial extensions</u> of a ring with involution A are the rings $A[x],A[x^{-1}],A[x,x^{-1}]$ with involution by

$$\bar{x} = x \ .$$

Then $S = \{x^k | k \geqslant 0\} \subset A[x]$ is a multiplicative subset in the sense of §2, such that

$$S^{-1}A[x] = A[x,x^{-1}] \ , \ S^{-1}A[x]/A[x] = x^{-1}A[x^{-1}] = \sum_{j=-\infty}^{-1} x^j A \ .$$

Given an h.d. 1 S-torsion $A[x]$-module M we have a f.g. projective A-module together with a nilpotent endomorphism

$$e : M \longrightarrow M \ ; \ y \longmapsto xy \ ,$$

in which case the dual $M^* = \mathrm{Hom}_A(M,A)$ is a f.g. projective A-module with a nilpotent endomorphism

$$e^* : M^* \longrightarrow M^* \ ; \ f \longmapsto (y \longmapsto f(ey))$$

and there is defined a natural $A[x]$-module isomorphism

$$M^* \longrightarrow M^\wedge = \mathrm{Hom}_{A[x]}(M,S^{-1}A[x]/A[x]) \ ; \ f \longmapsto (y \longmapsto \sum_{j=-\infty}^{-1} x^j f(e^{-j-1}y)) \ .$$

Given h.d. 1 S-torsion $A[x]$-modules M,M' there is a natural identification

$$\mathrm{Hom}_{A[x]}(M,M') = \{ f \in \mathrm{Hom}_A(M,M') \,|\, fe = e'f \} \ .$$

An ε-symmetric linking form over $(A[x],S)$ (M,λ) is the same as a pair (M,e) (as above) together with an element $\varphi \in Q^\varepsilon(M)$ such that

$$\varphi e = e^*\varphi \in Q^\varepsilon(M) = \ker(1-T_\varepsilon : \mathrm{Hom}_A(M,M^*) \longrightarrow \mathrm{Hom}_A(M,M^*))$$

$$\lambda : M \times M \longrightarrow S^{-1}A[x]/A[x] \ ; \ (y,z) \longmapsto \sum_{j=-\infty}^{-1} x^j \varphi(y,e^{-j-1}z) \ .$$

An ε-quadratic linking form over $(A[x],S)$ (M,λ,μ) is the same as a triple (M,e,φ) (as above) such that both (M,φ) and $(M,\varphi e)$ are even ε-symmetric forms over A, that is

$$\varphi \ , \ \varphi e \in Q\langle v_0\rangle^\varepsilon(M) \equiv \mathrm{im}(1+T_\varepsilon : Q_\varepsilon(M) \longrightarrow Q^\varepsilon(M)) \ ,$$

in which case

$$\mu : M \longrightarrow Q_\varepsilon(A[x],S) = S^{-1}A[x]/\{ b + \varepsilon\bar{b} \,|\, b \in A[x]\} \ ;$$

$$y \longmapsto \sum_{j=-\infty}^{-1} x^j \varphi(y,e^{-j-1}y) \ .$$

A split ε-quadratic linking form over $(A[x],S)$ (M,λ,ν) is the same as a triple (M,e,φ) (as above) together with elements $\psi_0,\psi_1 \in Q_\varepsilon(M)$ such that

$$\varphi = \psi_0 + \varepsilon\psi_0^* \quad , \quad \varphi e = \psi_1 + \varepsilon\psi_1^* \in Q\langle v_0\rangle^\varepsilon(M) \ ,$$

in which case

$$\nu : M \longrightarrow Q_\varepsilon(S^{-1}A[x]/A[x]) = \sum_{j=-\infty}^{-1} x^j Q_\varepsilon(A) \ ;$$

$$y \longmapsto \sum_{k=-\infty}^{-1}(x^{2k+1}\psi_0(y)(e^{-2k-2}y) + x^{2k}\psi_1(y)(e^{-2k-2}y)).$$

Define an abelian group morphism

$$L_0(A[x],S,\varepsilon) \longrightarrow L_0^K(A[x^{-1}],\varepsilon) \ ; \ (M,\lambda,\nu) \longmapsto (M[x^{-1}],\psi_0 + x^{-1}\psi_1) \ ,$$

where $M[x^{-1}] = A[x^{-1}]\otimes_A M$, $K = \mathrm{im}(\widetilde{K}_0(A) \longrightarrow \widetilde{K}_0(A[x^{-1}]))$.

A split ε-quadratic linking formation over $(A[x],S)$ $(F,(\begin{pmatrix}\gamma\\ \mu\end{pmatrix},\theta)G)$ is the same as an ε-quadratic formation over A $(H_\varepsilon(F);F,\mathrm{im}(\begin{pmatrix}\gamma\\ \mu\end{pmatrix}:G \longrightarrow F\oplus F^*))$ toogether with nilpotent endomorphisms $f \in \mathrm{Hom}_A(F,F)$, $g \in \mathrm{Hom}_A(G,G)$ such that

$$\gamma g = f\gamma \in \mathrm{Hom}_A(G,F) \ , \ \mu g = f^*\mu \in \mathrm{Hom}_A(G,F^*) \ , \ \gamma^*\mu g \in Q\langle v_0\rangle^{-\varepsilon}(G) \ ,$$

in which case

$$\theta : G \longrightarrow Q_{-\varepsilon}(A[x],S) \ ; \ y \longmapsto \sum_{j=-\infty}^{-1} x^j(\gamma^*\mu g^{-j-1}y) \ .$$

Define an abelian group morphism

$$L_1(A[x],S,\varepsilon) \longrightarrow L_1^K(A[x^{-1}],\varepsilon) \ ;$$

$$(F,G) \longmapsto (H_\varepsilon(F[x^{-1}]);F[x^{-1}],\mathrm{im}(\begin{pmatrix}\gamma\\ \mu(1+x^{-1}g)\end{pmatrix}:G[x^{-1}] \longrightarrow F[x^{-1}]\oplus F[x^{-1}]^*)).$$

In this way there are defined abelian group morphisms

$$L_n(A[x],S,\varepsilon) \longrightarrow L_n^K(A[x^{-1}],\varepsilon) \quad (n(\mathrm{mod}\ 4))$$

which fit into a morphism of exact sequences

$$\begin{array}{ccccccccc}
L_{n+1}(A[x],S,\varepsilon) & \longrightarrow & L_n(A[x],\varepsilon) & \longrightarrow & L_n^S(A[x,x^{-1}],\varepsilon) & \longrightarrow & L_n(A[x],S,\varepsilon) & \rightarrow & L_{n-1}(A[x],\varepsilon) \\
\downarrow & & \downarrow 1 & & \downarrow 1 & & \downarrow & & \downarrow \\
0 & \longrightarrow & L_n(A[x],\varepsilon) & \longrightarrow & L_n^S(A[x,x^{-1}],\varepsilon) & \longrightarrow & L_n^K(A[x^{-1}],\varepsilon) & \longrightarrow & 0
\end{array}$$

The top sequence is the localization sequence given by Proposition 2.4, while the bottom sequence is one of the split exact sequences obtained in the proof of Theorem 4.1 of Part IV of Ranicki [1] (- only the case $\varepsilon = \pm 1 \in A$ was considered there, but the proof generalizes to arbitrary $\varepsilon \in A$). We deduce:

<u>Proposition 6.1</u> The abelian group morphisms

$$L_n(A[x],S,\varepsilon) \longrightarrow L^K_n(A[x^{-1}],\varepsilon) \quad (n(\mathrm{mod}\ 4))$$

are isomorphisms.

[]

Define non-singular split (-1)-quadratic linking forms over $(\mathbb{Z}[x], S=\{x^k | k\geqslant 0\})$ (M,λ,ν), (M,λ,ν') by

$$M = \mathbb{Z}\oplus\mathbb{Z} \quad , \quad xM = 0$$

$$\lambda : M\times M \longrightarrow \mathbb{Z}[x,x^{-1}]/\mathbb{Z}[x] \ ; \ ((y,z),(y',z')) \longmapsto x^{-1}(yz'-y'z)$$

$$\nu : M \longrightarrow Q_{-1}(\mathbb{Z}[x,x^{-1}]/\mathbb{Z}[x]) = \mathbb{Z}[x,x^{-1}]/(\mathbb{Z}[x]+2\mathbb{Z}[x,x^{-1}]) \ ;$$

$$(y,z) \longmapsto x^{-1}(y^2+yz+z^2) \ ,$$

$$\nu' : M \longrightarrow Q_{-1}(\mathbb{Z}[x,x^{-1}]/\mathbb{Z}[x]) \ ; \ (y,z) \longmapsto x^{-1}yz$$

with the same associated (-1)-quadratic linking form over $(\mathbb{Z}[x],S)$ (M,λ,μ)

$$\mu : M \longrightarrow Q_{-1}(\mathbb{Z}[x,x^{-1}],S) = \mathbb{Z}[x,x^{-1}] \ ; \ (y,z) \longmapsto 0 \ .$$

The isomorphism given by Proposition 6.1

$$L_0(\mathbb{Z}[x],S,-1) \longrightarrow L^K_0(\mathbb{Z}[x^{-1}],-1) = \mathbb{Z}_2\oplus ?$$

sends (M,λ,ν) to the element $(1,0)$ (= the image of the Arf invariant element $(\mathbb{Z}\oplus\mathbb{Z}, \begin{pmatrix} 1 & 1 \\ 0 & 1 \end{pmatrix} \in Q_{-1}(\mathbb{Z}\oplus\mathbb{Z})) \in L_0(\mathbb{Z},-1)$ under the map induced by the natural inclusion $\mathbb{Z} \longrightarrow \mathbb{Z}[x^{-1}]$), while (M,λ,ν') is sent to 0. Thus split ε-quadratic linking forms carry more information than ε-quadratic linking forms, in general.

§7. Change of K-theory

We shall now describe the localization exact sequence for quadratic L-theory in the case when all the algebraic K-theory around is restricted to a prescribed $*$-invariant subgroup $X \subseteq \widetilde{K}_m(A)$ ($m = 0$ or 1).

Let A, S, ε be as in §2.

An h.d. 1 S-torsion A-module M has a projective class

$$[M] = [P_0] - [P_1] \in \widetilde{K}_0(A)$$

with P_0, P_1 the f.g. projective A-modules appearing in a resolution

$$0 \longrightarrow P_1 \xrightarrow{d} P_0 \longrightarrow M \longrightarrow 0 .$$

As $d \in \mathrm{Hom}_A(P_1, P_0)$ is an S-isomorphism $[M] \in \ker(\widetilde{K}_0(A) \longrightarrow \widetilde{K}_0(S^{-1}A))$.
Given a short exact sequence of h.d. 1 S-torsion A-modules

$$\mathcal{E} : 0 \longrightarrow M \xrightarrow{i} M' \xrightarrow{j} M'' \longrightarrow 0$$

there are defined f.g. projective A-module resolutions

$$\begin{array}{ccccccccc}
0 & \longrightarrow & P_1 & \xrightarrow{d} & P_0 & \longrightarrow & M & \longrightarrow & 0 \\
 & & \downarrow{\scriptstyle i_1} & & \downarrow{\scriptstyle i_0} & & \downarrow{\scriptstyle i} & & \\
0 & \longrightarrow & P_1' & \xrightarrow{d'} & P_0' & \longrightarrow & M' & \longrightarrow & 0 \\
 & & \downarrow{\scriptstyle j_1} & & \downarrow{\scriptstyle j_0} & & \downarrow{\scriptstyle j} & & \\
0 & \longrightarrow & P_1'' & \xrightarrow{d''} & P_0'' & \longrightarrow & M'' & \longrightarrow & 0 ,
\end{array}$$

and there exists a chain homotopy $k \in \mathrm{Hom}_A(P_0, P_1'')$ such that

$$j_0 i_0 = d''k \in \mathrm{Hom}_A(P_0, P_0'') \ , \ j_1 i_1 = kd \in \mathrm{Hom}_A(P_1, P_1'') \ .$$

Thus there is defined an acyclic f.g. projective A-module chain complex

$$C(\mathcal{E}) : 0 \longrightarrow P_1 \xrightarrow{\begin{pmatrix} d \\ -i_1 \end{pmatrix}} P_0 \oplus P_1' \xrightarrow{\begin{pmatrix} i_0 & d' \\ k & j_1 \end{pmatrix}} P_0' \oplus P_1'' \xrightarrow{(-j_0 \quad d'')} P_0'' \longrightarrow 0 ,$$

giving the sum formula

$$[M] - [M'] + [M''] = 0 \in \widetilde{K}_0(A) .$$

The S-dual $M\hat{} = \mathrm{Hom}_A(M, S^{-1}A/A)$ of an h.d. 1 S-torsion A-module M has projective class

$$[M\hat{}] = [P_1^*] - [P_0^*] = -[M]^* \in \widetilde{K}_0(A) .$$

The <u>projective class</u> of a split ε-quadratic linking $\begin{cases}\text{form}\\ \text{formation}\end{cases}$ over (A,S) $\begin{cases}(M,\lambda,\nu)\\ (F,G)\end{cases}$ is defined to be

$$\begin{cases}[(M,\lambda,\nu)] = [M] \in \widetilde{K}_0(A)\\ [(F,G)] = [G] - [F^\wedge] \in \widetilde{K}_0(A)\ .\end{cases}$$

If $\begin{cases}(M,\lambda,\nu)\\ (F,G)\end{cases}$ is non-singular then

$$\begin{cases}[(M,\lambda,\nu)]^* = -[(M,\lambda,\nu)] \in \widetilde{K}_0(A)\\ [(F,G)]^* = [(F,G)] \in \widetilde{K}_0(A)\ .\end{cases}$$

Given a $*$-invariant subgroup $X \subseteq \widetilde{K}_0(A)$ let $L_n^X(A,S,\varepsilon)$ $(n \pmod 4)$ be the Witt groups of non-singular split $\pm\varepsilon$-quadratic linking forms and formations over (A,S) defined exactly as $L_n(A,S,\varepsilon)$, but using only h.d. 1 S-torsion A-modules with projective class in $X \subseteq \widetilde{K}_0(A)$. In particular,

$$L_n^{\widetilde{K}_0(A)}(A,S,\varepsilon) = L_n(A,S,\varepsilon) \quad (n \pmod 4)\ .$$

Define $*$-invariant subgroups

$$X^S = X \cap \ker(S^{-1}:\widetilde{K}_0(A) \longrightarrow \widetilde{K}_0(S^{-1}A)) \subseteq \widetilde{K}_0(A)$$

$$S^{-1}X = \{[S^{-1}P] \mid [P] \in X\} \subseteq \widetilde{K}_0(S^{-1}A)\ ,$$

so that there is defined a short exact sequence of $\mathbb{Z}[\mathbb{Z}_2]$-modules

$$0 \longrightarrow X^S \longrightarrow X \longrightarrow S^{-1}X \longrightarrow 0$$

inducing a long exact sequence of Tate $\mathbb{Z}_2$-cohomology groups

$$\ldots \longrightarrow \hat{H}^n(\mathbb{Z}_2;X^S) \longrightarrow \hat{H}^n(\mathbb{Z}_2;X) \longrightarrow \hat{H}^n(\mathbb{Z}_2;S^{-1}X) \longrightarrow \hat{H}^{n-1}(\mathbb{Z}_2;X^S) \longrightarrow \ldots\ .$$

The exact sequences of Propositions 1.2,2.4 can be generalized to the intermediate projective L-groups, as follows.

<u>Proposition 7.1</u> Given $*$-invariant subgroups $X \subseteq Y \subseteq \widetilde{K}_0(A)$ there is defined a commutative diagram of abelian groups with exact rows and columns

$$\begin{array}{ccccccccc}
 & & \vdots & & \vdots & & \vdots & & \vdots \\
 & & \downarrow & & \downarrow & & \downarrow & & \downarrow \\
\ldots \longrightarrow & & L_n^X(A,\varepsilon) & \longrightarrow & L_n^{S^{-1}X}(S^{-1}A,\varepsilon) & \longrightarrow & L_n^X(A,S,\varepsilon) & \longrightarrow & L_{n-1}^X(A,\varepsilon) & \longrightarrow \ldots \\
 & & \downarrow & & \downarrow & & \downarrow & & \downarrow \\
\ldots \longrightarrow & & L_n^Y(A,\varepsilon) & \longrightarrow & L_n^{S^{-1}Y}(S^{-1}A,\varepsilon) & \longrightarrow & L_n^Y(A,S,\varepsilon) & \longrightarrow & L_{n-1}^Y(A,\varepsilon) & \longrightarrow \ldots \\
 & & \downarrow & & \downarrow & & \downarrow & & \downarrow \\
\ldots \longrightarrow & & \hat{H}^n(\mathbb{Z}_2;Y/X) & \longrightarrow & \hat{H}^n(\mathbb{Z}_2;S^{-1}Y/S^{-1}X) & \longrightarrow & \hat{H}^{n-1}(\mathbb{Z}_2;Y^S/X^S) & \longrightarrow & \hat{H}^{n-1}(\mathbb{Z}_2;Y/X) & \longrightarrow \ldots \\
 & & \downarrow & & \downarrow & & \downarrow & & \downarrow \\
\ldots \longrightarrow & & L_{n-1}^X(A,\varepsilon) & \longrightarrow & L_{n-1}^{S^{-1}X}(S^{-1}A,\varepsilon) & \longrightarrow & L_{n-1}^X(A,S,\varepsilon) & \longrightarrow & L_{n-2}^X(A,\varepsilon) & \longrightarrow \ldots \\
 & & \downarrow & & \downarrow & & \downarrow & & \downarrow \\
 & & \vdots & & \vdots & & \vdots & & \vdots
\end{array} .$$

[]

In dealing with based A-modules we shall assume (as in §1) that f.g. free A-modules have a well-defined rank, and that $\tau(\varepsilon:A \longrightarrow A) = 0 \in \widetilde{K}_1(A)$.

An h.d. 1 S-torsion A-module M is <u>based</u> if there is given a f.g. free A-module resolution

$$0 \longrightarrow P_1 \xrightarrow{d} P_0 \longrightarrow M \longrightarrow 0$$

such that P_0 and P_1 are based, in which case there is defined a torsion

$$\tau_S(M) = \tau(S^{-1}d:S^{-1}P_1 \longrightarrow S^{-1}P_0) \in \widetilde{K}_1(S^{-1}A) \quad .$$

The S-dual $M^\wedge$ is also based, with torsion

$$\tau_S(M^\wedge) = \tau_S(M)^* \in \widetilde{K}_1(S^{-1}A) \; .$$

A short exact sequence of based h.d. 1 S-torsion A-modules

$$\mathcal{E} : 0 \longrightarrow M \longrightarrow M' \longrightarrow M'' \longrightarrow 0$$

has a torsion

$$\tau(\mathcal{E}) \equiv \tau(C(\mathcal{E})) \in \widetilde{K}_1(A)$$

such that

$$S^{-1}\tau(\mathcal{E}) = \tau_S(M) - \tau_S(M') + \tau_S(M'') \in \widetilde{K}_1(S^{-1}A) \; .$$

The <u>torsion</u> of a non-singular split ε-quadratic linking $\begin{cases}\text{form}\\ \text{formation}\end{cases}$ over (A,S) $\begin{cases}(M,\lambda,\nu)\\ (F,G)\end{cases}$ with $\begin{cases}M\\ F,G\end{cases}$ based is defined by

$$\begin{cases}\tau(M,\lambda,\nu) = (\tau(\mathcal{E}:0\longrightarrow M \xrightarrow{\lambda} M^\wedge \longrightarrow 0 \longrightarrow 0),\tau_S(M)) \\ \qquad \in \ker(\begin{pmatrix}1+T & 0\\ -S^{-1} & 1-T\end{pmatrix}: \widetilde{K}_1(A)\oplus\widetilde{K}_1(S^{-1}A)\longrightarrow\widetilde{K}_1(A)\oplus\widetilde{K}_1(S^{-1}A)) \\ \tau(F,G) = (\tau(\mathcal{E}:0\longrightarrow G \xrightarrow{\binom{\gamma}{\mu}} F\oplus F^\wedge \xrightarrow{(\varepsilon\mu^\wedge\ \gamma^\wedge)} G^\wedge \longrightarrow 0),\tau_S(G)-\tau_S(F^\wedge)) \\ \qquad \in \ker(\begin{pmatrix}1-T & 0\\ -S^{-1} & 1+T\end{pmatrix}: \widetilde{K}_1(A)\oplus\widetilde{K}_1(S^{-1}A)\longrightarrow\widetilde{K}_1(A)\oplus\widetilde{K}_1(S^{-1}A)),\end{cases}$$

with $T:x\longmapsto x^*$ the duality involution.

Given $*$-invariant subgroups $X\subseteq\widetilde{K}_1(A)$, $Y\subseteq\widetilde{K}_1(S^{-1}A)$ such that

$$S^{-1}X = \{\tau(S^{-1}f)\in\widetilde{K}_1(S^{-1}A)\,|\,\tau(f)\in X\}\subseteq Y$$

let $L_n^{X,Y}(A,S,\varepsilon)$ $(n(\mathrm{mod}\ 4))$ be the Witt groups of non-singular split $\pm\varepsilon$-quadratic linking forms and formations over (A,S) defined exactly as $L_n(A,S,\varepsilon)$, but using only based h.d. 1 S-torsion A-modules and requiring the torsions to lie in

$$\{(x,y)\in X\oplus Y\,|\,x^* = (-)^{n-1}x\ ,\ S^{-1}x = y+(-)^{n-1}y^*\}\subseteq\widetilde{K}_1(A)\oplus\widetilde{K}_1(S^{-1}A)$$

In particular,

$$L_n^{\widetilde{K}_1(A),\widetilde{K}_1(S^{-1}A)}(A,S,\varepsilon) = L_n^{\{0\}\subseteq\widetilde{K}_0(A)}(A,S,\varepsilon)\quad(n(\mathrm{mod}\ 4))\ .$$

Given a morphism of $\mathbb{Z}[\mathbb{Z}_2]$-modules

$$f : G\longrightarrow H$$

define relative Tate $\mathbb{Z}_2$-cohomology groups

$$\hat{H}^n(\mathbb{Z}_2;f:G\longrightarrow H) = \frac{\{(x,y)\in G\oplus H\,|\,x^*=(-)^{n-1}x\,,fx=y+(-)^{n-1}y^*\}}{\{(u+(-)^{n-1}u^*,fu+v+(-)^n v^*)\,|\,(u,v)\in G\oplus H\}}\quad(n(\mathrm{mod}\ 2))$$

to fit into a long exact sequence

$$\ldots\longrightarrow\hat{H}^n(\mathbb{Z}_2;G)\xrightarrow{f}\hat{H}^n(\mathbb{Z}_2;H)\longrightarrow\hat{H}^n(\mathbb{Z}_2;f)\longrightarrow\hat{H}^{n-1}(\mathbb{Z}_2;G)\longrightarrow\ldots\ .$$

The exact sequences of Propositions 1.2, 2.4, 7.1 can be generalized to the intermediate torsion L-groups, as follows.

<u>Proposition 7.2</u> Given $*$-invariant subgroups $X \subseteq X' \subseteq \widetilde{K}_1(A)$, $Y \subseteq Y' \subseteq \widetilde{K}_1(S^{-1}A)$ such that $S^{-1}X \subseteq Y$, $S^{-1}X' \subseteq Y'$ there is defined a commutative diagram of abelian groups with exact rows and columns

$$\begin{array}{ccccccccccc}
 & & \vdots & & \vdots & & \vdots & & \vdots & & \\
 & & \downarrow & & \downarrow & & \downarrow & & \downarrow & & \\
\cdots & \longrightarrow & L_n^X(A,\varepsilon) & \longrightarrow & L_n^Y(S^{-1}A,\varepsilon) & \longrightarrow & L_n^{X,Y}(A,S,\varepsilon) & \longrightarrow & L_{n-1}^X(A,\varepsilon) & \longrightarrow & \cdots \\
 & & \downarrow & & \downarrow & & \downarrow & & \downarrow & & \\
\cdots & \longrightarrow & L_n^{X'}(A,\varepsilon) & \longrightarrow & L_n^{Y'}(S^{-1}A,\varepsilon) & \longrightarrow & L_n^{X',Y'}(A,S,\varepsilon) & \longrightarrow & L_{n-1}^{X'}(A,\varepsilon) & \longrightarrow & \cdots \\
 & & \downarrow & & \downarrow & & \downarrow & & \downarrow & & \\
\cdots & \longrightarrow & \hat{H}^n(\mathbb{Z}_2;X'/X) & \longrightarrow & \hat{H}^n(\mathbb{Z}_2;Y'/Y) & \longrightarrow & \hat{H}^n(\mathbb{Z}_2;X'/X \to Y'/Y) & \longrightarrow & \hat{H}^{n-1}(\mathbb{Z}_2;X'/X) & \longrightarrow & \cdots \\
 & & \downarrow & & \downarrow & & \downarrow & & \downarrow & & \\
\cdots & \longrightarrow & L_{n-1}^X(A,\varepsilon) & \longrightarrow & L_{n-1}^Y(S^{-1}A,\varepsilon) & \longrightarrow & L_{n-1}^{X,Y}(A,S,\varepsilon) & \longrightarrow & L_{n-2}^X(A,\varepsilon) & \longrightarrow & \cdots \\
 & & \downarrow & & \downarrow & & \downarrow & & \downarrow & & \\
 & & \vdots & & \vdots & & \vdots & & \vdots & &
\end{array}$$

[]

Let

$$V_n(A,\varepsilon) = L_n^{\{0\} \subseteq \widetilde{K}_0(A)}(A,\varepsilon) = L_n^{\widetilde{K}_1(A)}(A,\varepsilon) \quad (n \pmod 4)$$

be the L-groups defined using only f.g. free A-modules, and let

$$V_n(A,S,\varepsilon) = L_n^{\{0\} \subseteq \widetilde{K}_0(A)}(A,S,\varepsilon) = L_n^{\widetilde{K}_1(A),\widetilde{K}_1(S^{-1}A)}(A,S,\varepsilon) \quad (n \pmod 4)$$

be the L-groups defined using only h.d. 1 S-torsion A-modules which admit a f.g. free A-module resolution of length 1. As a special case of either of the localization sequences of Propositions 7.1, 7.2 we have an exact sequence of V-groups

$$\cdots \longrightarrow V_n(A,\varepsilon) \longrightarrow V_n(S^{-1}A,\varepsilon) \longrightarrow V_n(A,S,\varepsilon) \longrightarrow V_{n-1}(A,\varepsilon) \longrightarrow \cdots .$$

For example, the localization exact sequence of Pardon [1] is of this type.

The excision isomorphisms and the Mayer-Vietoris exact sequence for the L-theory of the cartesian square

$$\begin{array}{ccc} A & \longrightarrow & S^{-1}A \\ f\downarrow & & \downarrow f \\ B & \longrightarrow & T^{-1}B \end{array}$$

associated to a cartesian morphism $f:(A,S)\longrightarrow(B,T)$ (Proposition 3.1) can be generalized as follows.

<u>Proposition 7.3</u> Let $f:(A,S)\longrightarrow(B,T)$ be a cartesian morphism of rings with involution and multiplicative subsets, and let $X\subseteq\widetilde{K}_m(A)$, $Y\subseteq\widetilde{K}_m(S^{-1}A)$, $Z\subseteq\widetilde{K}_m(B)$, $W\subseteq\widetilde{K}_m(T^{-1}B)$ (m = 0 or 1) be $*$-invariant subgroups such that

$$S^{-1}X\subseteq Y\ ,\ B\otimes_A X\subseteq Z\ ,\ T^{-1}Z\subseteq W\ ,\ T^{-1}B\otimes_{S^{-1}A}Y\subseteq W,\ \ker(\widetilde{K}_m(A)\to\widetilde{K}_m(S^{-1}A)\oplus\widetilde{K}_m(B))\subseteq X$$

and such that the sequence

$$0\longrightarrow X/\ker(\widetilde{K}_m(A)\longrightarrow\widetilde{K}_m(S^{-1}A)\oplus\widetilde{K}_m(B))\longrightarrow Y\oplus Z\longrightarrow W\longrightarrow 0$$

is exact. Then there are defined excision isomorphisms

$$f\ :\ L_n^{X,Y}(A,S,\varepsilon)\longrightarrow L_n^{Z,W}(B,T,\varepsilon)\quad (n(\mathrm{mod}\ 4))$$

and a Mayer-Vietoris exact sequence

$$\ldots\longrightarrow L_n^X(A,\varepsilon)\longrightarrow L_n^Y(S^{-1}A,\varepsilon)\oplus L_n^Z(B,\varepsilon)\longrightarrow L_n^W(T^{-1}B,\varepsilon)\longrightarrow L_{n-1}^X(A,\varepsilon)\longrightarrow\ldots\ .$$

[]

(In the case $m = 0$ the groups $L_n^{X,Y}(A,S,\varepsilon)$ are to be interpreted as the relative groups $L_n^{X,Y}(A\longrightarrow S^{-1}A,\varepsilon)$ appearing in the exact sequence

$$\ldots\longrightarrow L_n^X(A,\varepsilon)\longrightarrow L_n^Y(S^{-1}A,\varepsilon)\longrightarrow L_n^{X,Y}(A\longrightarrow S^{-1}A,\varepsilon)\longrightarrow L_{n-1}^X(A,\varepsilon)\longrightarrow\ldots\ .$$

For $Y = S^{-1}X$ these are the groups defined previously

$$L_n^{X,S^{-1}X}(A,S,\varepsilon) = L_n^X(A,S,\varepsilon)\quad (n(\mathrm{mod}\ 4))\ ,$$

but for general X,Y it is not possible to express these relative L-groups in terms of linking forms and formations over (A,S)).

For example, the Mayer-Vietoris sequence of Theorem 6.6 of Wall [6] is a special case of the sequence of Proposition 7.3, with $(B,T) = (\hat{A},\hat{S})$ and

$$X = \ker(\widetilde{K}_1(A)\longrightarrow\widetilde{K}_1(\hat{S}^{-1}\hat{A}))\ ,\ Y = \ker(\widetilde{K}_1(S^{-1}A)\longrightarrow\widetilde{K}_1(\hat{S}^{-1}\hat{A}))$$
$$Z = \ker(\widetilde{K}_1(\hat{A})\longrightarrow\widetilde{K}_1(\hat{S}^{-1}\hat{A}))\ ,\ W = \{0\}\subseteq\widetilde{K}_1(\hat{S}^{-1}\hat{A})\ .$$

References

A.Bak [1] K-theory of forms preprint (1978)

[2] Surgery and K-theory groups of quadratic forms over finite groups and orders preprint (1978)

and W.Scharlau

[1] Grothendieck and Witt groups of orders and finite groups Invent. math. 23, 207 - 240 (1974)

H.Bass [1] Algebraic K-theory Benjamin (1968)

[2] Unitary algebraic K-theory in the Proceedings of the Battelle Seattle Conference on Algebraic K-theory, Vol. III, Springer Lecture Notes no.343, 57 - 265 (1973)

G.Carlsson and R.J.Milgram

[1] Some exact sequences in the theory of Hermitian forms to appear in J. Pure and Applied Algebra

F.X.Connolly [1] Linking forms and surgery Topology 12, 389 - 409 (1973)

M.Karoubi [1] Localisation de formes quadratiques I. Ann. scient. Éc. Norm. Sup. (4) 7, 359 - 404 (1974) II. ibid. 8, 99 - 155 (1975)

J.Milnor and D.Husemoller

[1] Symmetric bilinear forms Springer (1973)

W.Pardon [1] The exact sequence of a localization of Witt groups in the Proceedings of the Evanston Conference on Algebraic K-theory, Springer Lecture Notes no.551, 336 - 379 (1976)

[2] Local surgery and the exact sequence of a localization for Wall groups A.M.S. Memoir no.196 (1977)

D.S.Passman and T.Petrie

[1] Surgery with coefficients in a field Ann. of Maths. 95, 385 - 405 (1972)

A.A.Ranicki [1] Algebraic L-theory

I. Foundations Proc. London Math. Soc. (3) 27, 101 - 125 (1973)

II. Laurent extensions ibid., 126 - 158 (1973)

III. Twisted Laurent extensions in the Proceedings of the Battelle Seattle Conference on algebraic K-theory, Vol. III, Springer Lecture Notes no.343, 412 - 463 (1973)

IV. Polynomial extensions Comm. Math. Helv. 49, 137 - 167 (1974)

[2] The algebraic theory of surgery

I. Foundations , II. Applications to topology to appear in Proc. London Math. Soc. other parts in preparation

R.W.Sharpe [1] Surgery on compact manifolds: the bounded even-dimensional case Ann. of Maths. 98, 187 - 209 (1973)

J.Smith [1] Complements of codimension-two submanifolds III. Cobordism theory preprint (1977)

C.T.C.Wall [1] Quadratic forms on finite groups and related topics Topology 2, 281 - 298 (1963)

[2] Surgery on non-simply-connected manifolds Ann. of Maths. 84, 217 - 276 (1966)

[3] Surgery on compact manifolds Academic Press (1970)

[4] Classification of Hermitian forms III. Complete semilocal rings Invent. math. 19, 59 - 71 (1973)

[5] IV. Adele rings ibid. 23, 241 - 260 (1974)

[6] V. Global rings ibid., 261 - 288 (1974)

[7] Periodicity in algebraic L-theory in the Proceedings of the Tokyo Conference on Manifolds, 57 - 68 (1974).

$K_2(\mathbb{Z}[\mathbb{Z}/5])$ IS GENERATED BY RELATIONS AMONG 2×2 MATRICES

by R.W. Sharpe*

§1. Introduction.

One line of effort toward computing $K_2(\Lambda)$ is to first try to show that all elements arise as relations among d×d matrices, where one attempts to make d small. The next step would be to try to explicitly compute the relations among the d×d matrices. So far as we are aware, there is no general procedure for performing this second step, even if d=2. The first step was obtained for Λ a Euclidean ring, and d=2, by Dunwoody [3], and for Λ a ring of integers in a number field with a real imbedding by Van der Kallen [4]. We give here an extension of Dunwoody's method to obtain:

Theorem 1.1 $K_2(\mathbb{Z}[\mathbb{Z}/5])$ is generated by relations among 2×2 matrices.

We note here that $K_2(\mathbb{Z}[\mathbb{Z}/n])$ has been computed for $n<5$ (cf. [6], §10; [2], [8]). Theorem 1.1 depends on the following rather delicate property of $\mathbb{Q}(\zeta_5)$ (ζ_5=primitive fifth root of unity):

Theorem 1.2 If $\alpha\in\mathbb{Q}(\zeta_5)$ has norm $N(\alpha)$ satisfying $1<N(\alpha)\leq 5$, then there is an integral unit $u\in\mathbb{Z}(\zeta_5)$ such that either

i) $N(\alpha-u)<1$

or ii) $\alpha=u(\zeta_5-1)$

Theorem 1.2 implies that $\mathbb{Z}[\zeta_5]$ is Euclidean with respect to the norm (cf. §3), but is, in fact, a much stronger condition.

*This work was supported in part by NRC grant #A4621.

Finally, we remark that it can be shown that the hypothesis that $K_2(\mathbb{Z}[\mathbb{Z}/5])$ is generated by Steinberg symbols yields the conclusion that $K_2(\mathbb{Z}[\mathbb{Z}/5]) \approx \mathbb{Z}/2$, generated by $\{-1,-1\}$. We hope to address this point more fully in a subsequent paper. The hypothesis, however, remains unverified to date. In the final section we give some calculation due to Mike Stein which eliminate a conjectured counter example to the hypothesis.

§2. Approximation by Units in $\mathbb{Q}(\zeta_5)$.

In this section we sketch a proof of 1.2. Let $\zeta=\exp(2\pi i/5)$, $K=\mathbb{Q}(\zeta)$ and $K_{\mathbb{R}}=K\otimes_{\mathbb{Q}}\mathbb{R}$. Then $K\subset K_{\mathbb{R}}$ is a dense subset, and we note that the norm $N:K\to\mathbb{Q}$ extends to $N:K_{\mathbb{R}}\to\mathbb{R}$ by sending $\alpha\to\prod_{\sigma}\sigma(\alpha)$, where σ runs over the 4 distinct $\mathbb{R}$ algebra homomorphisms $K_{\mathbb{R}}\to\mathbb{R}$.

The integral units U of K act by multiplication on K and $K_{\mathbb{R}}$ preserving the norm, and so induce an action on $R=\{\alpha\in K_{\mathbb{R}} \mid 1\le N(\alpha)\le 5\}$. Suppose we find a fundamental domain $D\subset R$ for this action such that $N(\alpha-1)\le 1$ for $\alpha\in D$, with equality only when $\alpha=u(\zeta-1)\in K$, for u a unit. Then we will be done, for if $\alpha\in R\cap K$ then $u^{-1}\alpha\in D$ for some unit u, so that $N(\alpha-u)=N(u^{-1}\alpha-1)\le 1$ with equality only when $\alpha=uv(\zeta-1)$ for some $v\in U$.

We proceed to describe such a D. We have $K_{\mathbb{R}}\approx\mathbb{C}\oplus\mathbb{C}$ so points $z\in K_{\mathbb{R}}$ can be described by pairs $(r_1e^{i\theta_1},r_2e^{i\theta_2})$, and R is then parametrized by $\{(r_1,r_2,\theta_1,\theta_2)\in\mathbb{R}^4 \mid 1\le r_1r_2\le\sqrt{5},\ -\pi<\theta_k\le\pi,\ r_k>0\ (k=1,2)\}$. In terms of these parameters, D is given by: $D=R^o\times P^o\cup(R^+\cup R^-)\times P$ where the constituent regions are described in figures 1,2, and 3. A bit of advanced calculus shows that $N(\alpha-1)$ assumes its maximum on the 2 dimensional subcomplex $L=\partial R^{\circ}\times\partial P^{\circ}\cup(\partial R^{+}\cup\partial R^{-})\times\partial P$. From this point, it took my programable pocket calculator about a week to find that $N(\alpha-1)\leq 1$ on L with equality only when $\alpha=\zeta(\zeta+1)(\zeta-1)$, and its Galois translates. Since $\zeta(\zeta+1)$ is a unit, and the Galois group fixes the ideal generated by $\zeta-1$ we are done. (We note that $\zeta(\zeta+1)(\zeta-1)$ is plotted as u in figures 1 and 2, and that a generator of the Galois group acts by reflection in the line $r_1=r_2$, in figure 1 and by a quarter turn in figure 2.)

§3. The Strong Euclidean Algorithm

We begin by isolating a property of rings which is the cornerstone of Dunwoody's paper [2]. Let Λ be a ring and $\pi\in\Lambda$.

Definition 3.1. (Λ,π) is a Euclidean pair (or, has a strong Euclidean algorithm) if there is a multiplicative function $N:\Lambda\to\{0,1,2,...\}$ (the norm) such that for all $a,b\in\Lambda$ with $a\neq 0$ and $N(b)\geq N(a)$ there is an element $q\in\Lambda$ such that

either i) $N(b-qa) < N(b)$ and $q\in(\pi)$

or ii) $b=qa$ and $q\in\Lambda^{\cdot}$ (=the units of Λ).

Note that if $\pi=1$ this is equivalent to the usual definition of a Euclidean ring, except that we insist that N be multiplicative. The next lemma describes an even stronger condition for a pair (Λ,π) which we have verified for the pair $(Z[\zeta_5],\zeta_5-1)$ in §2.

Lemma 3.2. Let D be a division ring

$\Lambda\subset D$ a subring

$\pi\in\Lambda$ an element

$N:\Lambda\to\{0,1,2...\}$ a multiplicative function.

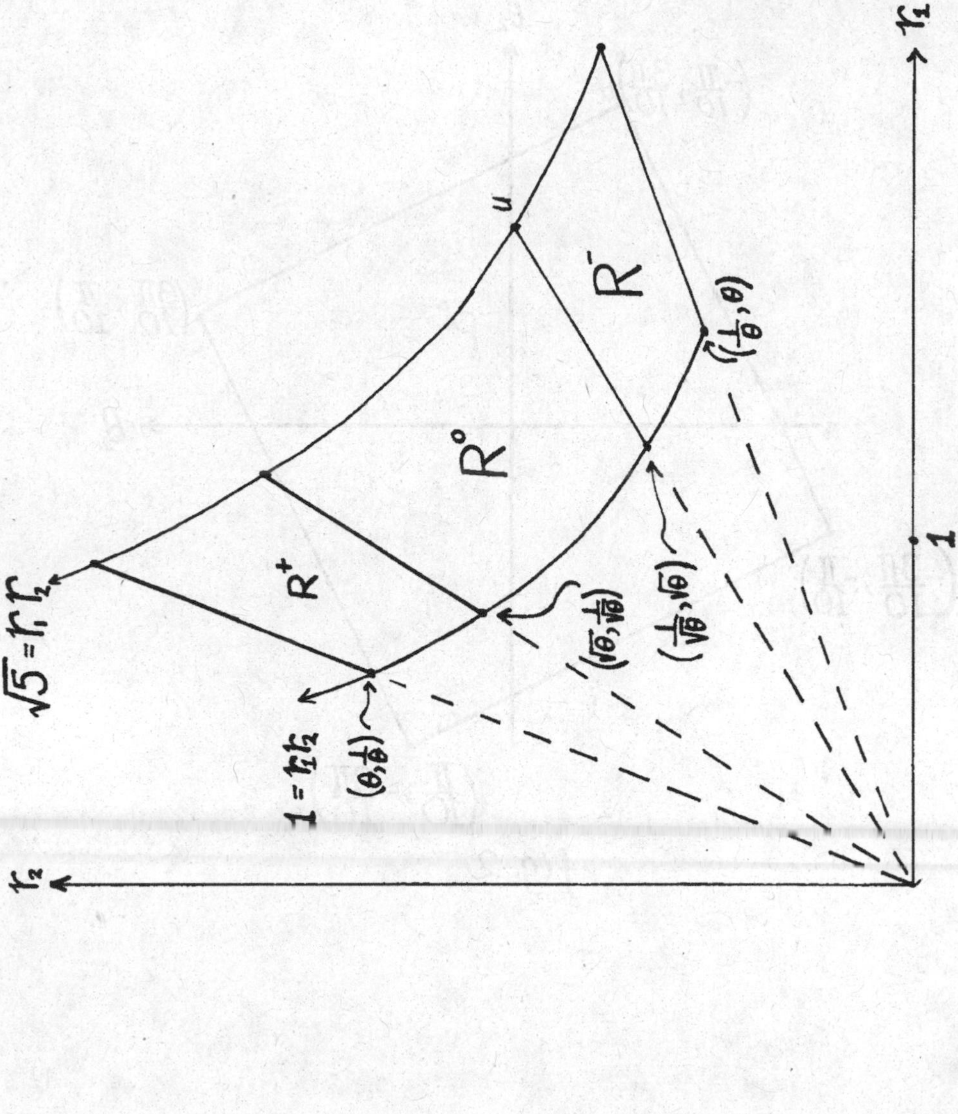

r_2
r_1
$\sqrt{5} = r_1 r_2$
$1 = r_1 r_2$
R^+
R^0
R^-
u
$(\theta, \frac{1}{\theta})$
$(\sqrt{\theta}, \frac{1}{\sqrt{\theta}})$
$(\frac{1}{\sqrt{\theta}}, \sqrt{\theta})$
$(\frac{1}{\theta}, \theta)$
1

fig. 1

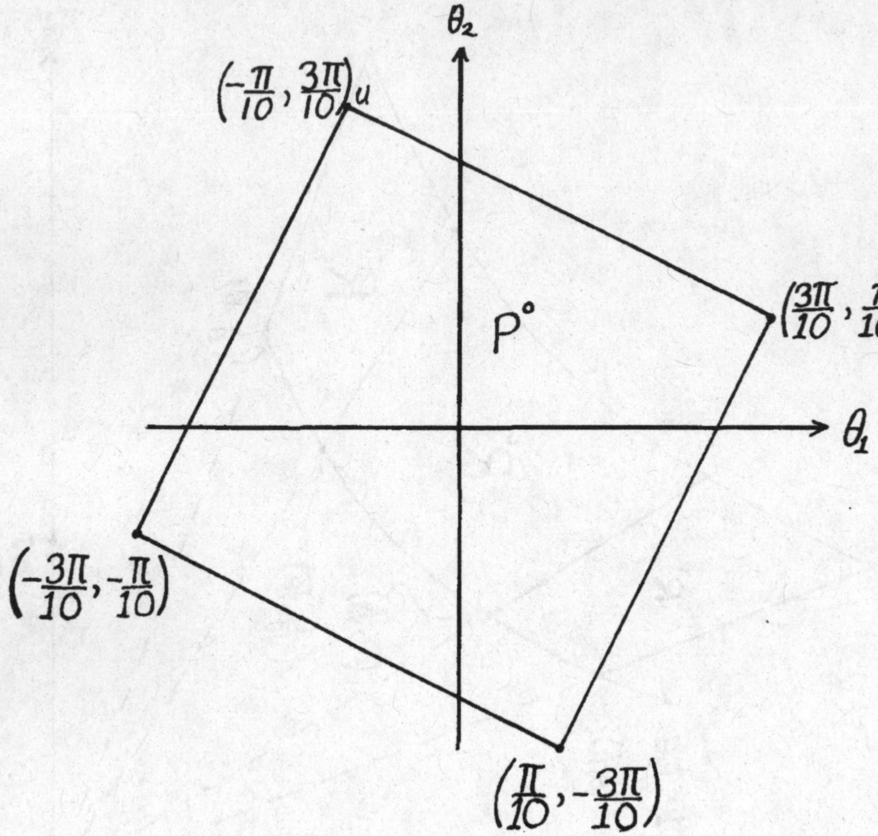
θ_2
$\left(-\frac{\pi}{10}, \frac{3\pi}{10}\right)$
u
$\left(\frac{3\pi}{10}, \frac{\pi}{10}\right)$
P°
θ_1
$\left(-\frac{3\pi}{10}, -\frac{\pi}{10}\right)$
$\left(\frac{\pi}{10}, -\frac{3\pi}{10}\right)$

fig. 2

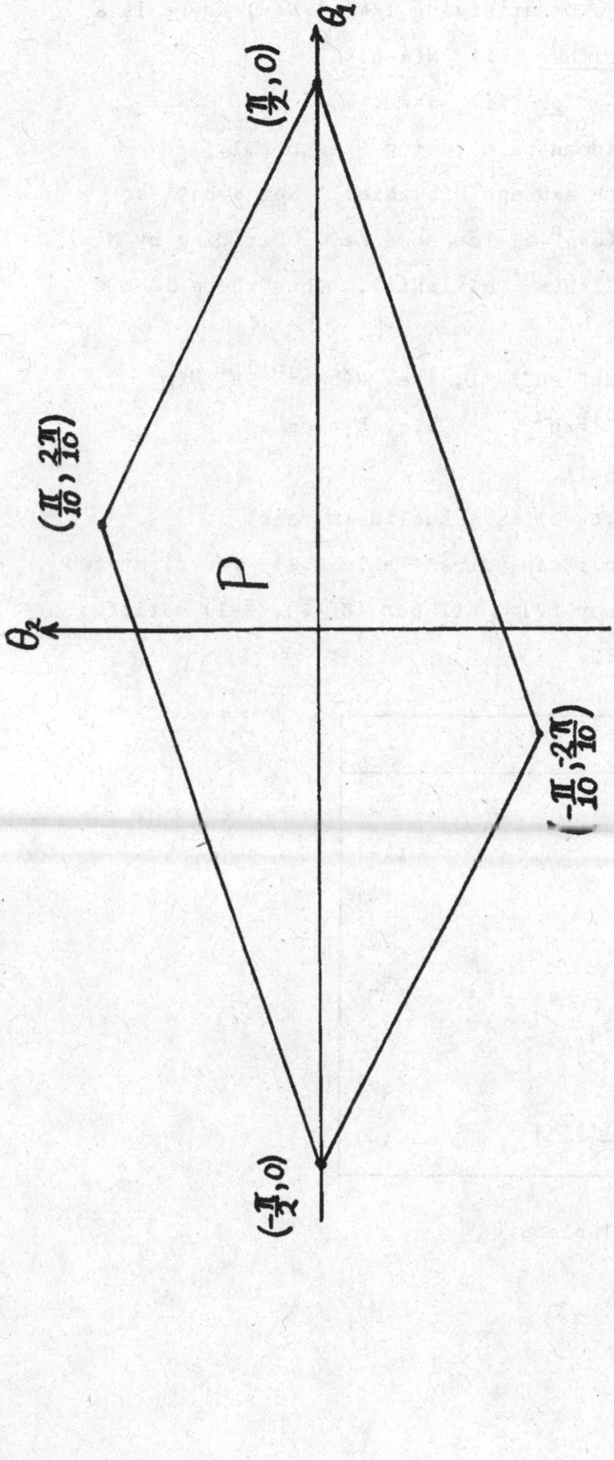
θ_1
θ_2
$(\frac{\pi}{2},0)$
$(-\frac{\pi}{2},0)$
$(\frac{\pi}{10},\frac{2\pi}{10})$
$(-\frac{\pi}{10},\frac{-2\pi}{10})$
P

fig. 3

Suppose that for all $\alpha \in D$ satisfying $1 < N(\alpha) \leq N(\pi)$ there is a unit $u \in \Lambda^{\cdot}$ such that either i) $N(\alpha - u) < 1$

or ii) $\alpha = \pi u$.

Then (Λ, π) is a Euclidean pair (using N as norm).

Proof: Let $a, b \in \Lambda$ with $a \neq 0$ and $N(b) \geq N(a)$. Set $\alpha = ba^{-1}$ so $N(\alpha) \geq 1$. Thus $N(\pi) > N(\alpha\pi^{-n}) \geq 1$ for some $n \geq 0$. Dividing by $N(\pi)$ and inverting yields $1 < N(\pi^{n+1}\alpha^{-1}) \leq N(\pi)$. Thus there is a unit $u^{-1} \in \Lambda$ such that

$$\begin{cases} \text{either } N(\pi^{n+1}ab^{-1} - u^{-1}) < 1, & \text{i.e. } N(b - u\pi^{n+1}a) < N(b) \\ \text{or } \quad \pi^{n+1}ab^{-1} = \pi u^{-1} & \text{i.e. } b = u\pi^{n}a \end{cases}$$

which implies the result.

Corollary 3.3 $(\mathbb{Z}[\zeta_5], \zeta_5 - 1)$ is a Euclidean pair.

Here is a list of some Euclidean pairs (Table 3.4) The first two cases were known to Dunwoody [2]. All but $(\mathbb{Z}[\sqrt{3}], \sqrt{3}-1)$ satisfy the hypotheses of Lemma 3.2.

Λ	π
$\mathbb{Z}$	2
$\mathbb{Z}[\zeta_3]$	$\zeta_3 - 1$
$\mathbb{Z}[i]$	$i-1$
$\mathbb{Z}[\sqrt{2}]$	$\sqrt{2}$
$\mathbb{Z}[\sqrt{3}]$	$\sqrt{3}-1$
$\mathbb{Z}[\zeta_5 + \zeta_5^{-1}]$	2
$\mathbb{Z}[\zeta_5]$	$\zeta_5 - 1$
$\mathbb{Z}[i,j,k,\frac{1+i+j+k}{2}]$	$i-1$

Table 3.4

4. Extension of the Methods of Sylvester and Dunwoody.

The aim of this section is to show:

Theorem 4.1: If (A,π) is a Euclidean pair, A has no zero divisors, and

$$\begin{array}{ccc} \mathbb{Z}[G] & \xrightarrow{\varepsilon} & \mathbb{Z} \\ \downarrow & & \downarrow \\ A & \longrightarrow & A/(\pi) \end{array}$$

is a cartesian square, with ε the augmentation, then $K_2(\mathbb{Z}G)$ is generated by relations among 2×2 matrices.

Note that 1.1 then follows in virtue of the well known cartesian square:

$$\begin{array}{ccc} \mathbb{Z}[\mathbb{Z}/5] & \to & \mathbb{Z} \\ \downarrow & & \downarrow \\ \mathbb{Z}[\zeta] & \to & \mathbb{Z}/5(\cong\mathbb{Z}[\zeta]/(\zeta-1)) \end{array}$$

and the fact (3.3) that $(\mathbb{Z}[\zeta],\zeta-1)$ is a Euclidean pair. The remainder of this section is devoted to proving 4.1. In fact, the method of Dunwoody extends with very little change, except at one point, and we outline his results in this context.

Lemma 4.3 (Dunwoody [2]). The map $St(n,\mathbb{Z}G)\to St(n,\mathbb{Z})$ induced by ε has kernel generated by $h_{ij}(g)$ $(g\in G)$ and x_{ij}^{ρ} $(\rho\in\ker\varepsilon)$ for $1\le i\ne j\le n$.

This allows us to write elements of K in the form $g_1g_2\ldots g_rh$ where g_i is of the form x_{ij}^{ρ} and $h\in H$ = the subgroup generated by $h_{ij}(g)$ $(g\in G)$.

Next, we allow K to act on A^n on the right via $K\subset St(n,\mathbb{Z}[G])\to E(n,\mathbb{Z}[G])\to E(n,A)$. Also, we extend the norm $N:A^n\to\mathbb{Z}$ by $N(a_1,a_2,\ldots a_n) = \sum_{i=1}^{n} N(a_i)$. We define $\underline{a}=(a_1,\ldots a_n)\in A^n$ to be admissible if $a_i-1\in(\pi)$ for some i and $a_j\in(\pi)$ for all $j\ne i$. The action of K on A^n preserves admissibility. We say that $g\in K$ acts smoothly on $\underline{a}$ if we can write $g=g_1g_2\ldots g_rh$ with g_i as above, and $h\in H$.

$$N(\underline{\underline{a}})>N(\underline{\underline{a}}g_1)>\ldots>N(\underline{\underline{a}}g_1\ldots g_i)\leq N(\underline{\underline{a}}g_1\ldots g_{i+1})\leq\ldots\leq N(\underline{\underline{a}}g_1\ldots g_r).$$

The hard part is to show:

Lemma 4.4 (cf. [2] lemma 3). Let $\mathbb{Z}[G]$ be as in 3.1 above, and let $\underline{\underline{a}}\in A^n$ be admissible. Then every $g\in K$ has an $\underline{\underline{a}}$ smooth representation.

Proof: The proof is virtually identical to Dunwoody's, except for "Case 5" which shows that $\chi_{12}^{\gamma}\chi_{21}^{\delta}$ has a smooth $\underline{\underline{a}}$ representation. We present this now.

We may assume that $\underline{\underline{a}}\in\Lambda^2$, not necessarily admissible. Using the fact that (Λ,π) is a Euclidean pair, we can find sequences $e_1,e_2,\ldots e_r$ and $f_1,f_2,\ldots f_s$ with terms of the form χ_{12}^{ρ} and χ_{21}^{ρ} ($\rho\in\ker\varepsilon$) such that:

$$N(\underline{\underline{a}})>N(\underline{\underline{a}}e_1)>N(\underline{\underline{a}}e_1e_2)\ldots>N(\underline{\underline{a}}e_1e_2\ldots e_r)$$

$$\text{and}\quad N(\underline{\underline{a}}g)>N(\underline{\underline{a}}gf_1)\ \ldots\ >N(\underline{\underline{a}}gf_1\ldots f_s) \qquad (*)$$

where $\underline{\underline{b}}=\underline{\underline{a}}e_1\ldots e_r$ and $\underline{\underline{c}}=\underline{\underline{a}}gf_1\ldots f_s$ are of the forms $(z,0)$, $(0,z)$ or (z,zu), where $z\in\Lambda$, $u\in\Lambda^{\cdot}$. We show next that $\underline{\underline{b}}$ and $\underline{\underline{c}}$ have the <u>same</u> form.

Let B be the matrix corresponding to the product $e_r^{-1}e_{r-1}^{-1}\ldots e_1^{-1}gf_1\ldots f_s$, so that $B\equiv I \bmod \pi$. Since B is invertible, and $\underline{\underline{b}}B=\underline{\underline{c}}$ we see that the entries in $\underline{\underline{b}}$ generate the same ideal as those in $\underline{\underline{c}}$. Factoring out the highest factor of π from $\underline{\underline{b}}$ and $\underline{\underline{c}}$ gives $\underline{\underline{b}}=\pi^k\underline{\underline{b}}'$ and $\underline{\underline{c}}=\pi^k\underline{\underline{c}}'$. Hence $\underline{\underline{b}}'B=\underline{\underline{c}}'$ and so $\underline{\underline{b}}'\equiv\underline{\underline{c}}' \bmod \pi$, which shows that $\underline{\underline{b}}$ and $\underline{\underline{c}}$ have the same form. Here is a list of possibilities:

	I	II	III	
$\underline{\underline{b}}$	$(z,0)$	$(0,z)$	(z,zu)	where $u,v,w\in\Lambda^{\cdot}$.
$\underline{\underline{c}}$	$(zv,0)$	$(0,zv)$	(zv,zw)	

Moreover $\underline{\underline{b}}'=\underline{\underline{c}}'$ modπ implies that $v\equiv 1$ and $u\equiv w$ modπ. Hence the equations:

$$\underline{\underline{c}}\, h_{21}(v) = \underline{\underline{b}} \qquad \text{(in case I)}$$

$$\underline{\underline{c}}\, h_{12}(v) = \underline{\underline{b}} \qquad \text{(in case II)}$$

$$\underline{\underline{c}}\, x_{21}^{w^{-1}(1-v)} x_{12}^{u-w} = \underline{\underline{b}} \qquad \text{(in case III)}$$

allow us to lengthen the sequence of f's to obtain $\underline{\underline{b}}=\underline{\underline{c}}$, at the expense of possibly ending (*) with one or two equalities, and allowing some $h_{ij}(v)$'s ($\{i,j\}=\{1,2\}$) in the list of f's. Now we study B using the equation $\underline{\underline{b}}B=\underline{\underline{b}}$.

If $\underline{\underline{b}}=(z,0)$, then $B=\begin{pmatrix}1 & 0\\ \beta & 1\end{pmatrix}$, $(\beta\in\pi)$ and we set $\hat{B}=x_{21}^{\beta}$.

If $\underline{\underline{b}}=(0,z)$, then $B=\begin{pmatrix}1 & \beta\\ 0 & 1\end{pmatrix}$, $(\beta\in\pi)$ and we set $\hat{B}=x_{12}^{\beta}$.

If $\underline{\underline{b}}=(z,zu)$, then $B=\begin{pmatrix}1 & -u\\ 0 & 1\end{pmatrix}\begin{pmatrix}1 & 0\\ \beta & 1\end{pmatrix}\begin{pmatrix}1 & u\\ 0 & 1\end{pmatrix}$, $(\beta\in\pi)$ and we set

$$\hat{B}=x_{12}^{-u}x_{21}^{\beta}x_{12}^{u}$$

Thus $\theta=\hat{B}^{-1}e_r^{-1}\dots e_1^{-1}gf_1\dots f_s\in K_2(2,\mathbb{Z}G)$, and so $g=e_1e_2\dots\hat{B}\theta f_1\dots f_s$ $=e_1e_2\dots e_r\hat{B}f_1\dots f_s\theta$ in $K_2(\mathbb{Z}G)$, and thus g acts smoothly on $\underline{\underline{a}}$.

It then follows as in [2] that $K_2[\mathbb{Z}G]$ is generated by relations among 2×2 matrices.

§5. Stein's Calculation.

Recall the Dennis-Stein symbol $\langle a,b\rangle \in K_2(R)$, defined whenever $u=1+ab\in R^*$. This symbol corresponds to the identity:

$$\begin{pmatrix}1 & -bu^{-1}\\ & 1\end{pmatrix}\begin{pmatrix}1 & \\ a & 1\end{pmatrix}\begin{pmatrix}1 & b\\ & 1\end{pmatrix}\begin{pmatrix}1 & \\ -u^{-1}a & 1\end{pmatrix} = \begin{pmatrix}u & \\ & u^{-1}\end{pmatrix} \quad \text{in } GL_2(R)$$

These symbols satisfy many relations, among which are:

1) $\langle a,b+c\rangle = \langle a,b\rangle, \langle a,\frac{c}{1+ab}\rangle$

2) $\langle a,b\rangle = \begin{cases}\{-a,1+ab\} & \text{if } a\in R^*\\ \{1+ab,b\} & \text{if } b\in R^*\end{cases}$

3) $\langle a,bc\rangle = \langle ab,c\rangle\langle ac,b\rangle$

(cf. [1] for more detail about these symbols).

Let t be a generator of $\mathbb{Z}/5$. Since $1+(t-t^{-1})^2=t^2+t^{-2}-1$ is a unit (with inverse $t+t^{-1}-1$), the symbol $\theta=\langle t-t^{-1},t-t^{-1}\rangle\in K_2(\mathbb{Z}[\mathbb{Z}/5])$ is defined. In an earlier version of this paper we suggested that this element might lie outside the subgroup generated by the Steinberg symbols. We give now Mike Stein's proof that $\theta=0$.

(5.1) $\theta=\langle t-t^{-1},t\rangle\langle t-t^{-1},-t^2\rangle=\{t^2,t\}\{t^3(t^2+t^{-2}-1),-t^2\}=$
$\{t^2+t^{-2}-1,-t^2\}$.

(5.2) $\theta^2=\langle 1,(t-t^{-1})^2\rangle=\{-1,1+(t-t^{-1})^2\}=\{-1,t^2+t^{-2}-1\}$; so $\theta^4=1$.

Thus $\theta=\theta^5$ (by 5.2) $=\{t^2+t^{-2}-1,-1\}$ (by 5.1) $=\theta^{-2}$ (by 5.2), so $\theta^3=1$ and hence $\theta=\theta^4\theta^{-3}=1$.

References

[1] Dennis, R.K. and Stein, M.R. K_2 of radical ideals and semi-local rings revisited, in "Lecture Notes in Mathematics" Vol. 342, pp. 281-303, Springer Verlag, Berlin 1973.

[2] Dunwoody, M. "$K_2(\mathbb{Z}\pi)$ for π a group of order two or three" J. London Math. Soc. (2) 11 (1975) 481-490.

[3] Dunwoody, M. "K_2 of a Euclidean ring". J. of Pure & Applied Algebra 7 (1976) 53-58.

[4] Van der Kallen. To appear.

[5] Lenstra, H.W. "Euclid's algorithm in cyclotomic fields" J. London Math. Soc. (2), 10 (1975) 457-465.

[6] Milnor, J. Introduction to Algebraic K-Theory, Ann. of Math. Studies 72, Princeton 1971.

[7] Ouspenski, J. "Note sur les nombres entiers dépendent d'une racine cinquieme de l'unité" Math. Ann. 66 (1909) 109-112.

[8] Snaith, V. These proceedings.

Surgery Spaces: Formulae and Structure

Laurence Taylor and Bruce Williams*

In the Fall of 1977 it became clear to us that, in our work on local surgery, we could get better theorems if we had formulae for computing surgery obstructions of problems over closed manifolds. Wall's paper [9] had come out but there was rumored to be an error and the corrigendum [10] had not yet appeared.

It seemed philosophically clear that such formulae must be contained in Ranicki's work [5] and we set out to find them from this point of view. We were of course helped by the fact that Morgan and Sullivan had worked out the answer in the important special case of the trivial group [2], and by the belief that Wall [9] could not be too far off. (He wasn't.)

The formulae herein will contain no surprises for the experts but we hope that having them explicitly written out in the literature may prove useful to others. More surprising perhaps is that we completely determine the homotopy type of Ranicki's spaces and spectra modulo our ignorance of their homotopy groups. There are geometric problems whose solution involves Ranicki's spaces, not just their homotopy groups, so the above analysis should be useful.

As an example of this last statement, Quinn [4] has shown that the obstruction to deforming a map between manifolds to a block-bundle projection has a piece involving his spaces, which are

* Both authors were partially supported by NSF Grant MCS76-07158.

homotopy equivalent to Ranicki's.

The guiding principle behind this work is that all the deep mathematics should be pushed onto others: notably Ranicki [5], Morgan and Sullivan [2], and Brumfiel and Morgan [1]. What remains is, we hope, a pleasant, if energetic romp through the stable category which is nevertheless not devoid of interest.

We conclude this section with our thanks to both John Morgan and Andrew Ranicki for conversations, correspondence, and preprints.

§1. Statement of results.

A summary of the relevant part of Ranicki's work is our first order of business. In §15 of [5], Ranicki constructs semi-simplical monoids $\mathbb{L}^m(A,\varepsilon)$, $\mathbb{L}_m(A,\varepsilon)$, and $\hat{\mathbb{L}}^m(A,\varepsilon)$ and related spectra $\mathbb{L}^\circ(A,\varepsilon)$, $\mathbb{L}_\circ(A,\varepsilon)$ and $\hat{\mathbb{L}}(A,\varepsilon)$. The k-simplices of the monoids are just

$$(m+k)\text{-dimensional}\begin{cases}\varepsilon\text{-symmetric}\\ \varepsilon\text{-quadratic}\\ \varepsilon\text{-hyperquadratic}\end{cases}\text{Poincaré }(k+2)\text{-ads over }A\text{ with}$$

some additional restrictions.

To simplify the notation, let $\mathbb{L}^\circ$ denote $\mathbb{L}^\circ(Z,1)$; $\mathbb{L}_\circ$ denote $\mathbb{L}_\circ(Z,1)$; and $\hat{\mathbb{L}}$ denote $\hat{\mathbb{L}}(Z,1)$. Further let $\mathbb{L}^\circ(\pi)$ denote $\mathbb{L}^\circ(Z\pi,1)$; $\mathbb{L}_\circ(\pi)$ denote $\mathbb{L}_\circ(Z\pi,1)$; etc.

Tensor product of chain complexes induces numerous pairings. The spectrum $\mathbb{L}^\circ$ becomes a commutative ring spectrum, and every other spectrum $\mathbb{L}^\circ(A,\varepsilon)$, $\mathbb{L}_\circ(A,\varepsilon)$, or $\hat{\mathbb{L}}(A,\varepsilon)$ becomes a module spectrum over $\mathbb{L}^\circ$. The spectrum $\hat{\mathbb{L}}$ is a commutative ring spectrum and $\hat{\mathbb{L}}(A,\varepsilon)$ is a module spectrum over it.

There are also maps between these spectra. There is a symmeterization map $(1+T)\colon \mathbb{L}_\circ(A,\varepsilon) \to \mathbb{L}^\circ(A,\varepsilon)$ whose cofibre is $\hat{\mathbb{L}}(A,\varepsilon)$. The resulting long cofibration sequence is a sequence of $\mathbb{L}^\circ$ module spectra. The map $\mathbb{L}^\circ \to \hat{\mathbb{L}}$ is even a map of ring spectra. There is also a map of $\mathbb{L}^\circ$ module spectra $e_8\colon \mathbb{L}^\circ(A,\varepsilon) \to \mathbb{L}_\circ(A,\varepsilon)$ which is given by tensor product with the even form of index 8 often denoted E_8.

Ranicki also defines two geometric maps:

$$\sigma^*\colon MSTOP \to \mathbb{L}^\circ \quad \text{and} \quad A_\pi\colon K(\pi,1)^+ \to \mathbb{L}^\circ(\pi).$$

The map σ^* is a map of ring spectra and the map A_π gives the assembly maps:

i) $MSTOP \wedge (K(\pi,1)^+) \xrightarrow{\sigma^* \wedge A_\pi} \mathbb{L}^\circ \wedge \mathbb{L}^\circ(\pi) \to \mathbb{L}^\circ(\pi)$

induces the symmetric signature map, while

ii) $MSTOP \wedge \mathbb{L}_\circ \wedge (K(\pi,1)^+) \xrightarrow{\sigma^* \wedge 1 \wedge A_\pi} \mathbb{L}^\circ \wedge \mathbb{L}_\circ \wedge \mathbb{L}^\circ(\pi) \to \mathbb{L}_\circ(\pi)$

induces the surgery obstruction.

In particular, $\pi_i(\mathbb{L}_\circ(\pi))$ is the ith Wall surgery group for oriented problems with fundamental group π (for $i \geq 0$); $\mathbb{L}_\circ$ is essentially $Z \times G/TOP$; and the above map is the old Sullivan-Wall map $\Omega_*(G/TOP \times K(\pi,1)) \to L_*(\pi)$ ([8] p.176, 13B.3).

With these definitions fixed we can state our first theorem.

<u>Theorem A</u>: The spectra $\mathbb{L}^\circ(A,\varepsilon)$ and $\mathbb{L}_\circ(A,\varepsilon)$ are generalized Eilenberg-MacLane spectra when localized at 2, and, when localized away from 2, are both $bo\Lambda_0 \vee \Sigma^1 bo\Lambda_1 \vee \Sigma^2 bo\Lambda_2 \vee \Sigma^3 bo\Lambda_3$ where $bo\Lambda_i$ denotes connective KO theory with coefficients in the group Λ_i. In

our case, Λ_i is $\pi_i(\mathbb{L}^\circ(A,\varepsilon))\otimes Z[\frac{1}{2}]$.

The spectrum $\hat{\mathbb{L}}(A,\varepsilon)$ is a generalized Eilenberg-MacLane spectrum.

While Theorem A is nice one should not read too much into it. It is true that the map A_π: $K(\pi,1)^+ \to \mathbb{L}^\circ(\pi)$ is determined by some cohomology classes and some elements in KO-theory with coefficients, but this is not much use in understanding the assembly maps unless one knows the $\mathbb{L}^\circ$ module structure of $\mathbb{L}^\circ(\pi)$ and $\mathbb{L}_\circ(\pi)$.

Since the assembly maps are basically unkown, we agree to write A_* for any of the following maps:

(1.1) $\bigoplus_i H_{*-4i}(\pi;Z_{(2)}) \oplus H_{*-(4i+1)}(\pi;Z/2) \to \pi_*(\mathbb{L}^\circ(\pi))_{(2)}$

(1.2) $KO_* \quad (K(\pi,1)) \to \pi_*(\mathbb{L}^\circ(\pi))_{(odd)}$

(1.3) $\bigoplus_i H_{*-4i}(\pi;Z_{(2)}) \oplus H_{*-(4i+2)}(\pi;Z/2) \to \pi_*(\mathbb{L}_\circ(\pi))_{(2)}$

(1.4) $KO_* \quad (K(\pi,1)) \to \pi_*(\mathbb{L}_\circ(\pi))_{(odd)}$

where 1.1 and 1.2 are induced from A_π and the pairing $\mathbb{L}^\circ \wedge \mathbb{L}^\circ(\pi) \to \mathbb{L}^\circ(\pi)$ and 1.3 and 1.4 are induced from A_π and the pairing $\mathbb{L}_\circ \wedge \mathbb{L}^\circ(\pi) \to \mathbb{L}_\circ(\pi)$.

Given an oriented topological manifold M with $\pi_1(M)=\pi$, we wish to give a formula for the symmetric signature of M, $\sigma^*(M)$. We first fix some notation.

We let $\mathcal{L}\in H^{4i}(BSTOP;Z_{(2)})$ denote the class defined by Morgan and

Sullivan [2] §7. Let $V \in H^{2i}(BSTOP;Z/2)$ denote the total Wu class. Here and below we only list a typical group in which a graded cohomology class like $\mathcal{L}$ or V lies. We have

Theorem B. Let $g: M \to K(\pi,1)$ classify the universal cover, and let $\nu: M \to BSTOP$ classify the normal bundle. Then, at 2, we have

$$(1.5)\quad \sigma^*(M)_{(2)} = A_* \, g_*(\nu^*(\mathcal{L} + V\, Sq^1 V) \cap [M]) .$$

Away from 2 M has a bo-orientation and hence a fundamental class $[M]_K$. We have

$$(1.6)\quad \sigma^*(M)_{(odd)} = A_* \, g_* \, [M]_K .$$

For the surgery obstruction we have the following formulae due to Wall [9].

Theorem C. Let g and ν be as above, and let $f: M \to \mathbb{L}_0$ classify some surgery problem. Then

$$(1.7)\quad \sigma_*(f)_{(2)} = A_* \, g_* \, ((\nu^*(\mathcal{L}) \cup f^*(\ell) + \nu^*(\mathcal{L}) \cup f^*(k) + \delta^*(\nu^*(V\, Sq^1 V) \cup f^*(k))) \cap [M]),$$

where δ^* denotes the integral bockstein. Furthermore we have

$$(1.8)\quad \sigma_*(f)_{(odd)} = A_* \, g_* \, (f^*(\Delta) \cap [M]_K)$$

The classes $\ell \in H^{4i}(\mathbb{L}_\circ;Z_{(2)})$ and $k \in H^{4i+2}(\mathbb{L}_\circ;Z/2)$ are defined below and $\Delta \in KO^0(\mathbb{L}_\circ;Z[\frac{1}{2}])$ is the equivalence from Theorem A.

Remark: Theorems B and C follow easily from 1.9 and 1.13 below. To actually carry out the proof of 1.5 and 1.7, one needs to remember that slant product gives the equivalence between $\pi_i(K(G,n) \wedge X^+)$ and $H_{n-i}(X;G)$. The formulae follow from well-known properties of the cap, cup, and slant products. Formula 1.6 is a tautology and 1.8 follows from the naturality of the cap product.

To prove our results we need to analyze $\mathbb{L}^\circ$, $\mathbb{L}_\circ$ and the various maps and pairings between them. In §3 we shall construct cohomology classes $L \in H^{4i}(\mathbb{L}^\circ;Z_{(2)})$; $r \in H^{4i+1}(\mathbb{L}^\circ;Z/2)$; $\ell \in H^{4i}(\mathbb{L}_\circ;Z_{(2)})$; and $k \in H^{4i+2}(\mathbb{L}_\circ;Z/2)$ such that these classes exhibit $\mathbb{L}^\circ$ and $\mathbb{L}_\circ$ as generalized Eilenberg-MacLane spectra at 2.

Moreover, the map σ^*: MSTOP $\to \mathbb{L}^\circ$ is determined at 2 by the formulae

(1.9) $\sigma^*(L) = \mathcal{L} \cdot U$ and $\sigma^*(r) = (V\, Sq^1 V) \cdot U$

where U is the Thom class.

The map $(1+T)$: $\mathbb{L}_\circ \to \mathbb{L}^\circ$ is determined at 2 by the formulae

(1.10) $(1+T)^*(L) = 8\ell$ and $(1+T)^*(r) = 0$.

The map e_8: $\mathbb{L}^\circ \to \mathbb{L}_\circ$ is determined at 2 by the formulae

(1.11) $e_8^*(\ell) = L$ and $e_8^*(k) = 0$.

The ring map $\rho\colon \mathbb{L}^\circ \wedge \mathbb{L}^\circ \to \mathbb{L}^\circ$ is determined at 2 by the formulae

(1.12) $\rho^*(L) = L \wedge L$ and $\rho^*(r) = r \wedge L + L \wedge r$.

The module map $m\colon \mathbb{L}^\circ \wedge \mathbb{L}_\circ \to \mathbb{L}_\circ$ is determined at 2 by

(1.13) $m^*(\ell) = L \wedge \ell + \delta^*(r \wedge k)$ and $m^*(k) = L \wedge k$.

There remains the spectrum $\hat{\mathbb{L}}$. We have a map $\hat{\mathbb{L}} \to \Sigma\mathbb{L}_\circ$ and so we get cohomology classes $\Sigma\ell \in H^{4i+1}(\hat{\mathbb{L}};Z_{(2)})$ and $\Sigma k \in H^{4i+3}(\hat{\mathbb{L}};Z/2)$. There are classes $\hat{L} \in H^{4i}(\hat{\mathbb{L}};Z/8)$ and $\hat{r} \in H^{4i+1}(\hat{\mathbb{L}};Z/2)$. The classes $\hat{L}$, $\hat{r}$, and Σk exhibit $\hat{\mathbb{L}}$ as a generalized Eilenberg-MacLane spectrum. Moreover we have the following formulae.

The classes $\hat{L}$ and $\Sigma\ell$ are related by β_8, the mod 8 bockstein, via

$$\beta_8{}^*(\hat{L}) = \Sigma\ell .$$

The map $\mathbb{L}^\circ \to \hat{\mathbb{L}}$ is determined by the formulae

(1.14) $\hat{r} \to r$; $\hat{L} \to$ mod 8 reduction of L; and $\Sigma k \to 0$.

The map $\hat{\mathbb{L}} \to \Sigma\mathbb{L}_\circ$ has been discussed.

The pairing $\hat{\rho}\colon \hat{\mathbb{L}} \wedge \hat{\mathbb{L}} \to \hat{\mathbb{L}}$ is determined by the formulae

(1.15)
$$\hat{\rho}^*(\hat{L}) = \hat{L} \wedge \hat{L} + i_*(\hat{r} \wedge \Sigma k + \Sigma k \wedge \hat{r});\quad \hat{\rho}^*(\hat{r}) = \hat{r} \wedge \hat{L} + \hat{L} \wedge \hat{r};$$
$$\hat{\rho}^*(\Sigma k) = \Sigma k \wedge \hat{L} + \hat{L} \wedge \Sigma k;\ \text{and}\ \hat{\rho}^*(\Sigma\ell) = \Sigma\ell \wedge \hat{L} + \hat{L} \wedge \Sigma\ell$$

where $i\colon Z/2 \to Z/8$ is monic.

Ranicki's hyperquadratic signature map MSG $\rightarrow \hat{\mathbb{L}}$ is a map of ring spectra. We shall take care that our class $\hat{L}$ pulls back to the Brumfiel-Morgan [1] class in $H^{4i}(BSG;Z/8)$ times the Thom class. We do not understand the pull backs of $\hat{r}, \Sigma k$, and $\Sigma \ell$ although there are obvious guesses based on Brumfiel's and Morgan's work.

The specfic formulae above yield some nice results on the surgery assembly map 1.3. It is clear that $\bigoplus_i H_{*-(4i+2)}(\pi;Z/2) \rightarrow \pi_*(\mathbb{L}_\circ(\pi))_{(2)}$ factors through the natural map $\pi_{*+1}(\hat{\mathbb{L}}(\pi)) \rightarrow \pi_*(\mathbb{L}_\circ(\pi))$ and that there are commutative diagrams

$$(1.16)\qquad \begin{array}{ccc} \oplus H_{*+1-4i}(\pi;Z/8) & \rightarrow & \pi_{*+1}(\hat{\mathbb{L}}(\pi)) \\ \downarrow{\scriptstyle(\beta_8)_*} & & \downarrow \\ \oplus H_{*-4i}(\pi;Z_{(2)}) & \rightarrow & \pi_*(\mathbb{L}_\circ(\pi))_{(2)} \end{array}$$

and

$$(1.17)\qquad \begin{array}{ccc} \oplus H_{*-4i}(\pi;Z_{(2)}) & \rightarrow & \pi_*(\mathbb{L}^\circ(\pi))_{(2)} \\ \downarrow{\scriptstyle\cong} & & \downarrow{\scriptstyle(e_8)_*} \\ \oplus H_{*-4i}(\pi;Z_{(2)}) & \rightarrow & \pi_*(\mathbb{L}_\circ(\pi))_{(2)} \end{array}$$

The first diagram follows from 1.15; the second from 1.11.

If π is a finite group whose 2 Sylow group is abelian or generalized quaternion, Stein [6] shows that the map $\oplus \tilde{H}_{*-4i}(\pi;Z_{(2)}) \rightarrow \pi_*(\mathbb{L}^\circ(\pi))_{(2)} \rightarrow \pi_*(\mathbb{L}_\circ(\pi))_{(2)}$ is trivial. Hence

<u>Corollary</u> (Stein[6]): If π is a finite group whose 2 Sylow group is abelian or generalized quaternion, then, for the surgery assembly map, we have

$$\oplus \tilde{H}_{*-4i}(\pi;Z_{(2)}) \rightarrow \pi_*(\mathbb{L}_\circ(\pi))_{(2)} \quad \text{is trivial.}$$

In particular, a surgery problem over a closed manifold M with $\pi_1(M) = \pi$ as in the corollary is solvable if its index obstruction is zero and if the associated Kervaire classes k_{4i+2} are zero. This is related to some results of Morgan and Pardon [3].

We conclude this section with a remark on notation. Henceforth we will suppress Thom classes in our formulae and write 1.9 for example as $\sigma^*(L) = \mathcal{L}$, etc.

§2. MSO-module spectra.

An MSO-module spectrum is a spectrum, E, equipped with a map $\mu\colon MSO \wedge E \to E$ such that

(2.1) the composite $S^0 \wedge E \xrightarrow{u \wedge 1} MSO \wedge E \xrightarrow{\mu} E$

is the identity, where $u\colon S^0 \to MSO$ is the unit, and

(2.2) the diagram

$$\begin{array}{ccc} MSO \wedge MSO \wedge E & \xrightarrow{1 \wedge \mu} & MSO \wedge E \\ \downarrow{\scriptstyle m \wedge 1} & & \downarrow{\scriptstyle \mu} \\ MSO \wedge E & \xrightarrow{\mu} & MSO \end{array}$$

commutes, where $m\colon MSO \wedge MSO \to MSO$ is the multiplication.

Such spectra are common. All of the Ranicki spectra discused in §1 and bordism theories like MSTOP are examples. Our first result is a proof of

<u>Theorem $A_{(2)}$</u>: Any module spectrum over MSO becomes a generalized Eilenberg-MacLane spectrum after localizing at 2.

<u>Proof</u>: Let E denote the spectrum in question, and let $K(\pi_* E_{(2)})$ be a product of Eilenberg-MacLane spectra such that

$\pi_i(K(\pi_* E_{(2)}))=\pi_i(E_{(2)})$. A map $\varphi\colon E \to K(\pi_* E_{(2)})$ determines a homomorphism $\hat{\varphi}_i\colon H_i(E;Z) \to \pi_i(E_{(2)})$ for each i, and conversely, any collection of such homomorphisms can be realized by some φ (probably several). There are projection maps $\rho_i\colon K(\pi_* E_{(2)}) \to K(\pi_i E_{(2)},i)$ and the induced map $H_i(E;Z) \xrightarrow{\varphi_*} H_i(K(\pi_* E_{(2)});Z) \xrightarrow{(\rho_i)_*} H_i(K(\pi_i E_{(2)},i);Z)$ is just $\hat{\varphi}_i$ followed by the Hurewicz map.

Now consider $\mu\colon MSO \wedge E \to E$. We can localize to get $\mu_{(2)}\colon MSO_{(2)} \wedge E \to E_{(2)}$ such that $S^0 \wedge E \to MSO_{(2)} \wedge E \to E_{(2)}$ is the localization map. But $u\colon S^0 \to MSO_{(2)}$ factors as $S^0 \to K(Z,0) \to MSO_{(2)}$ (since $MSO_{(2)}$ is a product of Eilenberg-MacLane spectra) so we get a map $\hat{\varphi}_i\colon H_i(E;Z) \to \pi_i E_{(2)}$ such that the composite $\pi_i(E) \xrightarrow{\text{Hurewicz}} H_i(E;Z) \xrightarrow{\hat{\varphi}_i} \pi_i E_{(2)}$ is localization. It is easy to see that any associated $\varphi\colon E \to K(\pi_* E_{(2)})$ is a 2-local equivalence. //

The next step in our understanding of the MSO-module spectrum E is to actually describe cohomology classes which give the equivalence between E and a generalized Eilenberg-MacLane spectrum. Morgan and Sullivan [2] have given a good way to describe 2-local cohomology classes of E: one gives homomorphisms with certain properties out of the bordism of E. Since the bordism of E is just the homotopy of MSO $\wedge$ E, one way is to give homomorphisms out of the homotopy groups of E and then use μ. This procedure gives the homomorphisms studied by Morgan and Sullivan and we wish to study it in general.

A certain amount of finiteness seems necessary, so we say

that E has finite type if $\bigoplus_{-\infty<i<r} \pi_i(E)$ is a finitely generated group for each $r \in Z$.

In what follows, R denotes either $Z_{(2)}$ or $Z/2^j$. Our basic data is a collection of homomorphisms $\Psi_R\colon \bigoplus_{-\infty<i<\infty} \pi_i(E;R) \to R$ such that

$$\text{(2.3)}\qquad \begin{array}{ccc} \oplus\,\pi_*(E;Z_{(2)}) & \to & Z_{(2)} \\ \downarrow & & \downarrow \\ \oplus\,\pi_*(E;Z/2^j) & \to & Z/2^j \end{array} \qquad \text{and} \qquad \begin{array}{ccc} \oplus\,\pi_*(E;Z/2^j) & \to & Z/2^j \\ \downarrow & & \downarrow \\ \oplus\,\pi_*(E;/2^{j+1}) & \to & Z/2^{j+1} \end{array}$$

commute.

The composites $\Psi_*\colon \oplus\,\pi_*(MSO \wedge E;R) \xrightarrow{\mu_*} \oplus\,\pi_*(E;R) \xrightarrow{\Psi_R} R$ are compatible homomorphisms in the Morgan-Sullivan sense. The requirement that they be multiplicative with respect to the index may be phrased as follows. The composite

$\pi_p(MSO;R) \otimes \pi_q(MSO \wedge E;R) \to \pi_{p+q}(MSO \wedge MSO \wedge E;R) \xrightarrow{(m\wedge 1)_*} \pi_{p+q}(MSO \wedge E;R) \to R$

is also the composite $\pi_p(MSO;R) \otimes \pi_q(MSO \wedge E;R) \xrightarrow{(\text{Index})\otimes\Psi_*} R \otimes R \to R$,

where Index : $\pi_p(MSO;R) \to R$ is just the mod R index homomorphism ([2] p. 473). The particular form of our Ψ_* permits us to get away with an apparently weaker statement. We have

Proposition 1: Suppose given an MSO-module spectrum E of finite type and a collection of homomorphisms $\Psi_R\colon \pi_*(E;R) \to R$ satisfying 2.3. Then, if

$$\text{(2.4)}\qquad \begin{array}{ccc} \pi_p(MSO;R) \otimes \pi_q(E;R) & \xrightarrow{(\text{Index})\otimes\Psi_R} & R \otimes R \\ \downarrow & & \downarrow \\ \pi_{p+q}(MSO \wedge E;R) \xrightarrow{\mu_*} \pi_{p+q}(E;R) & \xrightarrow{\Psi_R} & R \end{array}$$

commutes for all $p,q \in Z$, we have a unique cohomology class $\gamma \in H^*(E;Z_{(2)})$ such that, for any $c \in \pi_*(MSO \wedge E;R)$, the equation

$$(2.5) \qquad \Psi_R\, \mu_*(c) = \langle \mathcal{L} \wedge \gamma, c \rangle$$

holds.

The omitted proof follows easily from the work of Morgan and Sullivan [2] §4 and 2.2.

There is also a mod 2 version of Proposition 1. Given one homomorphism $\Psi: \pi_*(E;Z/2) \to Z/2$ such that 2.4 commutes, we get a unique class $\gamma \in H^*(E;Z/2)$ such that 2.5 holds. Sullivan proves this in [7]. There is even a $Z/2^r$ version in Brumfiel-Morgan [1]. We state it so as to require that the Ψ_R satisfy 2.3 and 2.4 for all $Z/2^j$ with $j \leq r$.

Addendum to Proposition 1: Suppose $\Psi_R: \pi_i(E;R) \to R$ is zero for all R under consideration. Then diagram 2.4 shows that $\Psi_R: \pi_{i+4k}(E;R) \to R$ is also zero for all negative integers k. The proof that γ is unique shows more. It shows that the components of γ in dimensions $i+4k$ must be zero for all non-postive integers k.

The classes that Morgan and Sullivan built by this method satisfy $\mu^*(\gamma) = \mathcal{L} \wedge \gamma$. Our next goal is to give a general explanation for this formula. Suppose that we have three MSO-module spectra, E_i, μ_i for $i=1,2,3$, of finite type, and three sets of homomorphisms $\Psi_i: \pi_*(E_i;R) \to R$ $i=1,2,3$ satisfying 2.3 and 2.4. We get three classes $\gamma_i \in H^*(E_i;Z_{(2)})$ (or in $H^*(E_i;Z/2^r)$).

With the above notation fixed, let us further suppose that

we have a pairing $\nu: E_1 \wedge E_2 \to E_3$ such that

$$
(2.6)\quad
\begin{array}{ccc}
MSO \wedge E_1 \wedge MSO \wedge E_2 \xrightarrow{1\wedge T\wedge 1} MSO \wedge MSO \wedge E_1 \wedge E_2 & \xrightarrow{m\wedge\nu} & MSO \wedge E_3 \\
\downarrow \mu_1\wedge\mu_2 & & \downarrow \mu_3 \\
E_1 \wedge E_2 & \xrightarrow{\quad\nu\quad} & E_3
\end{array}
$$

commutes. Then we have

<u>Theorem 1</u>: The diagram

$$
\begin{array}{ccc}
\pi_p(E_1;R) \otimes \pi_q(E_2;R) \to \pi_{p+q}(E_1 \wedge E_2;R) & \xrightarrow{\nu_*} & \pi_{p+q}(E_3;R) \\
\downarrow \Psi_1\otimes\Psi_2 & & \downarrow \Psi_3 \\
R \otimes R & \xrightarrow{\qquad\qquad} & R
\end{array}
$$

commutes for all $p,q \in Z$ and all R under consideration

$$\text{iff} \qquad \nu^*(\gamma_3) = \gamma_1 \wedge \gamma_2 \quad .$$

Remark: By applying the theorem to the map μ_i: MSO $\wedge E_i \to E_i$ one easily sees that $\mu_i^*(\gamma_i) = \mathcal{L} \wedge \gamma_i$.

This remark can be amplified slightly. Let $x \in H^*(E;Z_{(2)})$ (or $H^*(E;Z/2^r)$) be any cohomology class. Then $\langle x, \ \rangle: \pi_*(E;R) \to R$ is a set of homomorphisms satisfying 2.3. If $\mu^*(x) = \mathcal{L} \wedge x$, then they satisfy 2.4 as well, and the cohomology class determined by Proposition 1 is just x again. We have

<u>Corollary 1</u>: Given two classes $x_1, x_2 \in H^*(E;Z_{(2)})$ such that $\mu^*(x_i) = \mathcal{L} \wedge x_i$ $i=1,2$, then $x_1 = x_2$ iff $\langle x_1, \ \rangle = \langle x_2, \ \rangle: \pi_*(E;R) \to R$.

Proof of Theorem 1: See that the diagram commutes if $\nu^*(\gamma_3)=\gamma_1\wedge\gamma_2$ by using 2.1 to map $E_1\wedge E_2$ to $MSO\wedge E_1\wedge MSO\wedge E_2$ and chasing the resulting diagrams. We concentrate on the converse.

We begin by rephrasing lemma 7.1 ([2] p. 533) for spectra of finite type as

Lemma 1: Two sets of homomorphisms

$$\psi_i:\ \pi_*(\,MSO\wedge E_1\wedge E_2;R)\ \to\ R \qquad i=1,2$$

satisfying 2.3 are equal if the composites

$$\pi_p(MSO\wedge E_1;R)\otimes\pi_q(MSO\wedge E_2;R)\to\pi_{p+q}(MSO\wedge E_1\wedge MSO\wedge E_2;R)\xrightarrow{(m\wedge 1\wedge 1)_*}\xrightarrow{(1\wedge T\wedge 1)_*}\pi_{p+q}(MSO\wedge E_1\wedge E_2;R)\xrightarrow{\psi_i}R$$ for i=1,2 are equal

for all $p,q\in Z$ and all R under consideration.

Using this result we proceed to use the uniqueness part of the Morgan-Sullivan description of cohomology classes. We must show that $\langle\mathcal{L}\wedge\nu^*(\gamma_3),c\rangle=\langle\mathcal{L}\wedge\gamma_1\wedge\gamma_2,c\rangle$ for all $c\in\pi_*(MSO\wedge E_1\wedge E_2;R)$ which, by the lemma above, are of the form $c=(m\wedge 1\wedge 1)_*(1\wedge T\wedge 1)_*(c_1\wedge c_2)$. So

$$\langle\mathcal{L}\wedge\nu^*(\gamma_3),c\rangle=\ \langle\mathcal{L}\wedge\gamma_3,(1\wedge\nu)_*(c)\rangle=\Psi_3\ (\mu_3)_*\ (1\wedge\nu)_*\ (c)\quad\text{by 2.5.}$$

Now $(\mu_3)_*\ (1\wedge\nu)_*\ (c)=\nu_*\ (\mu_1\wedge\mu_2)_*\ (c_1\wedge c_2)$ as a diagram chase using 2.6 shows. But Ψ_3 of this is $\Psi_1(c_1)\cdot\Psi_2(c_2)$ since our diagram commutes. We finish by showing $\langle\mathcal{L}\wedge\gamma_1\wedge\gamma_2,c\rangle=\Psi_1(c_1)\cdot\Psi_2(c_2)$ also.

$$\begin{aligned}\langle\mathcal{L}\wedge\gamma_1\wedge\gamma_2,c\rangle&=\langle\mathcal{L}\wedge\gamma_1\wedge\gamma_2,(m\wedge 1\wedge 1)_*\ (1\wedge T\wedge 1)_*\ (c_1\wedge c_2)\rangle\\ &=\langle\mathcal{L}\wedge\gamma_1\wedge\mathcal{L}\wedge\gamma_2,c_1\wedge c_2\rangle\quad\text{since } m^*\mathcal{L}=\mathcal{L}\wedge\mathcal{L}\\ &=\langle\mathcal{L}\wedge\gamma_1,c_1\rangle\ \langle\mathcal{L}\wedge\gamma_2,c_2\rangle=\Psi_1(c_1)\cdot\Psi_2(c_2)\ \text{by 2.5. //}\end{aligned}$$

When it works, the above discussion is quite satisfactory. There are however natural homomorphisms $\Psi_R: \pi_*(E;R) \to R$ such as the deRham invariant or the surgery obstruction for which diagram 2.4 fails to commute. A more general treatment seems necessary.

We fix some class $\beta \in H^*(MSO \wedge E; Z_{(2)})$ (or $H^*(MSO \wedge E; Z/2^r)$) such that the homomorphisms $\Psi_R \mu_* + \langle \beta , \ \rangle: \pi_*(MSO \wedge E;R) \to R$ are compatible and multiplicative with respect to the index. With this data fixed, and for E of finite type, we have

Proposition 2: There exists a unique cohomology class $\ell \in H^*(E;Z_{(2)})$ (or $H^*(E;Z/2^r)$) such that, for any $c \in \pi_*(MSO \wedge E;R)$

$$\langle \mathcal{L} \wedge \ell, c \rangle = \Psi_R \, \mu_*(c) + \langle \beta, c \rangle \ . \tag{2.7}$$

Note that Proposition 1 follows from Proposition 2 using $\beta=0$, but we preferred to write out the easy, natural case first. We now need the analogue of Theorem 1, so we return to our three spectra, E_i, our pairing ν, and our commutative diagram 2.6. Our homomorphisms $\Psi_i: \pi_*(E_i;R) \to R$ satisfy 2.3 but not necessarily 2.4. We have classes $\beta_i \in H^*(MSO \wedge E_i; Z_{(2)})$ (or $H^*(MSO \wedge E_i; Z/2^r)$) as in Proposition 2.

To complicate matters still further, the diagram in Theorem 1 will not commute in cases of interest to us. Hence we fix a class $\alpha \in H^*(E_1 \wedge E_2; Z_{(2)})$ (or $H^*(E_1 \wedge E_2; Z/2^r)$) such that

$$\begin{array}{ccc} \pi_p(E_1;R) \otimes \pi_q(E_2;R) & \rightarrow & \pi_{p+q}(E_1 \wedge E_2;R) \\ \downarrow \Psi_1 \otimes \Psi_2 & & \downarrow \Psi_3\ \nu_*(\ ) + \langle\alpha,\ \rangle \\ R \otimes R & \longrightarrow & R \end{array}$$

commutes. With all of these hypotheses we have

<u>Theorem 2</u>: There exists a class $\mathfrak{s} \in H^*(E_1 \wedge E_2;Z_{(2)})$ (or $H^*(E_1 \wedge E_2;Z/2^r)$) such that

i) the mod R reductions of $(1_\wedge T_\wedge 1)^*\ (m_\wedge 1_\wedge 1)^*\ (1_\wedge \nu)^*(\beta_3)$ and $\mathfrak{L}_\wedge \ell_1{}_\wedge \beta_2 + \beta_1{}_\wedge \mathfrak{L}_\wedge \ell_2 - \beta_1{}_\wedge \beta_2 + (\mu_1{}_\wedge \mu_2)^*(\alpha) + (1_\wedge T_\wedge 1)^*(\mathfrak{L}_\wedge \mathfrak{L}_\wedge \mathfrak{s})$ evaluate the same on $\pi_*(MSO \wedge E_1 \wedge MSO \wedge E_2;R)$;

ii) $\nu^*(\ell_3) = \ell_1 \wedge \ell_2 + \mathfrak{s}$

both hold. Either condition determines $\mathfrak{s}$ uniquely.

The analogue of Corollary 1 is

<u>Corollary 2</u>: Given two cohomology classes $x_1,\ x_2 \in H^*(E;Z_{(2)})$ such that $\mu^*(x_i) = \mathfrak{L}_\wedge x_i + \beta$ i=1,2, then

$x_1 = x_2$ iff $\langle x_1,\ \rangle = \langle x_2,\ \rangle : \pi_*(E;R) \rightarrow R$.

<u>Proof of Theorem 2</u>: We begin with the uniqueness statement. Clearly ii) determines $\mathfrak{s}$ uniquely. The Morgan-Sullivan uniqueness result shows that condition i) can be satisfied by at most one $\mathfrak{s}$.

Let us define $\mathfrak{s}$ so that condition ii) holds. Then we need to verify that i) is satisfied. By Lemma 1, it suffices to do this for all c of the form $c = c_1 \wedge c_2$, where $c_i \in \pi_*(MSO \wedge E_i;R)$.

$<(1_\wedge T_\wedge 1)^* \ (m_\wedge 1_\wedge 1)^* \ (1_\wedge \nu)^*(\beta_3), c> = <\beta_3, (1_\wedge \nu)_* \ (m_\wedge 1_\wedge 1)_* \ (1_\wedge T_\wedge 1)_*(c)>$.

This, by 2.7, is just

$<\mathfrak{L}_\wedge \ell_3, (1_\wedge \nu)_* \ (m_\wedge 1_\wedge 1)_* \ (1_\wedge T_\wedge 1)_*(c)> - \Psi_3 \ (\mu_3)_*(1_\wedge \nu)_*(m_\wedge 1_\wedge 1)_*(1_\wedge T_\wedge 1)_*(c) =$

$<\mathfrak{L}_\wedge \mathfrak{L}_\wedge \nu^* \ell_3, (1_\wedge T_\wedge 1)_*(c)> - \Psi_3 \ \nu_* \ (\mu_1 \wedge \mu_2)_*(c_1 \wedge c_2)$ by 2.6. This in turn is

$<\mathfrak{L}_\wedge \ell_1 \wedge \mathfrak{L}_\wedge \ell_2, c_1 \wedge c_2> - \Psi_1(c_1) \cdot \Psi_2(c_2) + <\alpha, (\mu_1 \wedge \mu_2)_*(c_1 \wedge c_2)> +$

$$<(1_\wedge T_\wedge 1)^* \ (\mathfrak{L}_\wedge \mathfrak{L}_\wedge s), c_1 \wedge c_2>$$

by condition ii) and the definition of α.

This in its turn is equal, by 2.7, to

$<\mathfrak{L}_\wedge \ell_1 \wedge \mathfrak{L}_\wedge \ell_2, c_1 \wedge c_2> - (<\mathfrak{L}_\wedge \ell_1, c_1> - <\beta_1, c_1>) \ (<\mathfrak{L}_\wedge \ell_2, c_2> - <\beta_2, c_2>) \ +$

$$<\alpha, (\mu_1 \wedge \mu_2)_*(c_1 \wedge c_2)> + <(1_\wedge T_\wedge 1)^* \ (\mathfrak{L}_\wedge \mathfrak{L}_\wedge s), \ c_1 \wedge c_2> \ .$$

Multiplying out and simplifying we get

$$<\mathfrak{L}_\wedge \ell_1 \wedge \beta_2 + \beta_1 \wedge \mathfrak{L}_\wedge \ell_2 - \beta_1 \wedge \beta_2 + (\mu_1 \wedge \mu_2)^*(\alpha) + (1_\wedge T_\wedge 1)^* \ (\mathfrak{L}_\wedge \mathfrak{L}_\wedge s), c_1 \wedge c_2>$$

which really is what we wanted to get. //

§3. The spectra $\mathbb{L}^\circ$, $\mathbb{L}_\circ$, and $\hat{\mathbb{L}}$ at 2.

The goal of this section is to use the techniques and results of the previous section to analyze the Ranicki spectra $\mathbb{L}^\circ$, $\mathbb{L}_\circ$ and the maps $\rho\colon \mathbb{L}^\circ \wedge \mathbb{L}^\circ \to \mathbb{L}^\circ$; $m : \mathbb{L}^\circ \wedge \mathbb{L}_\circ \to \mathbb{L}_\circ$; $\sigma^*\colon \mathrm{MSTOP} \to \mathbb{L}^\circ$; $(1+T)\colon \mathbb{L}_\circ \to \mathbb{L}^\circ$; and $e_8\colon \mathbb{L}^\circ \to \mathbb{L}_\circ$.

The map ρ makes $\mathbb{L}^\circ$ into a ring spectrum, and the composite $\mathrm{MSO} \to \mathrm{MSTOP} \to \mathbb{L}^\circ$ is a map of ring spectra. The map m makes $\mathbb{L}_\circ$ into an $\mathbb{L}^\circ$ module spectrum, so $\mathbb{L}^\circ$ and $\mathbb{L}_\circ$ (and indeed $\mathbb{L}^\circ(A,\varepsilon)$ and $\mathbb{L}_\circ(A,\varepsilon)$)

are all module spectra over MSO so the theory in §2 applies. We let μ°: MSO $\wedge \mathbb{L}^\circ \to \mathbb{L}^\circ$ and $\mu_\circ$: MSO $\wedge \mathbb{L}_\circ \to \mathbb{L}_\circ$ denote the structure maps.

We consider four homomorphisms. We have the index $\pi_*(\mathbb{L}^\circ) \to Z_{(2)}$; the deRham invariant $\pi_*(\mathbb{L}^\circ) \to Z/2$; the surgery obstruction $\pi_*(\mathbb{L}_\circ) \to Z_{(2)}$; and the Kervaire invariant $\pi_*(\mathbb{L}_\circ) \to Z/2$.

By the universal coefficients theorem and a bit of luck, $\pi_{4i}(\mathbb{L}^\circ;R) = \pi_{4i}(\mathbb{L}^\circ) \otimes R$ and $\pi_{4i}(\mathbb{L}_\circ;R) = \pi_{4i}(\mathbb{L}_\circ) \otimes R$. The two homomorphisms into $Z_{(2)}$ are zero unless $* \equiv 0 \pmod 4$ and extend uniquely to homomorphisms $\pi_*(\mathbb{L}^\circ;R) \to R$ and $\pi_*(\mathbb{L}_\circ;R) \to R$ which are zero unless $* \equiv 0 \pmod 4$. Both homomorphisms satisfy 2.3.

The index homomorphism satisfies 2.4. To see this, note that $\pi_*(MSO;R) \to \pi_*(\mathbb{L}^\circ;R)$ is onto except when $* = 1$ or 2. Proposition 6.6 of Morgan-Sullivan [2] shows that diagram 2.4 commutes, at least if q is not 1 or 2. Since the map $\pi_4(\mathbb{L}^\circ;R) \otimes \pi_i(\mathbb{L}^\circ;R) \to \pi_{i+4}(\mathbb{L}^\circ \wedge \mathbb{L}^\circ;R) \xrightarrow{\rho_*} \pi_{i+4}(\mathbb{L}^\circ;R)$ is an isomorphism by Ranicki [5] §7, diagram 2.4 must commute even if $q = 1$, or 2. Hence we get a unique class $L \in H^{4i}(\mathbb{L}^\circ;Z_{(2)})$ satisfying 2.5.

The Kervaire invariant $\pi_*(\mathbb{L}_\circ;Z/2) \to Z/2$ is determined by the map $\pi_*(\mathbb{L}_\circ) \to Z/2$ and does satisfy 2.4. For this, use the map constructed by Ranicki [5] (Proposition 15.5) from $Z \times G/TOP$ to the 0th space in the Ω-spectrum for $\mathbb{L}_\circ$ which is a homotopy equivalence. One can then use Sullivan's result [7] that the Kervaire invariant is multiplicative with respect to the index to see that diagram 2.4 commutes. Let $k \in H^{4i+2}(\mathbb{L}_\circ;Z/2)$ denote the class promised by the mod 2 version of Proposition 1.

The deRham invariant $\pi_*(\mathbb{L}^\circ;Z/2) \to Z/2$ definitely does not satisfy 2.4. However, Morgan and Sullivan [2] (Proposition 6.6 and Lemma 8.2) show that "deRham" + $\langle V\,Sq^1V \wedge L, \ \rangle$ is multiplicative with respect to the index (at least on $\pi_*(MSO;Z/2)$ and we use our usual trick.) This requires that, under the map $MSO \to \mathbb{L}^\circ$, L pulls back to $\mathcal{L}$, but this is easy to see from Corollary 1. Let $r \in H^{4i+1}(\mathbb{L}^\circ;Z/2)$ denote the class promised by the mod 2 version of Proposition 2.

Finally, the surgery obstruction $\pi_*(\mathbb{L}_\circ;R) \to R$ also does not satisfy 2.4. However "surgery obstruction" + $\langle \delta(V\,Sq^1V \wedge k), \ \rangle$ is multiplicative with respect to the index, where δ denotes the bockstein associated to the exact sequence $0 \to Z_{(2)} \to Z_{(2)} \to Z/2 \to 0$. This follows by the usual trick from Morgan-Sullivan [2], Proposition 8.6. We let $\ell \in H^{4i}(\mathbb{L}_\circ;Z_{(2)})$ denote the class we get from Proposition 2.

This completes the definitions of the four classes which exhibit $\mathbb{L}^\circ$ and $\mathbb{L}_\circ$ as generalized Eilenberg-MacLane spectra at 2. Our next task is to analyze Ranicki's maps.

Proof of 1.9: We must show $\sigma^*(L) = \mathcal{L}$ and $\sigma^*(r) = V\,Sq^1V$. Since Morgan and Sullivan used the homomorphism

$$\pi_*(MSO \wedge MSTOP;R) \to \pi_*(MSTOP;R) \to \pi_*(\mathbb{L}^\circ;R) \xrightarrow{\text{Index}} R$$

to define $\mathcal{L}$ ([2] §7) the first equation is Corollary 1. The second follows by a similar argument plus Lemma 8.2 of Morgan-Sullivan [2].

Proof of 1.11: We are to show that $e_8^*(k) = 0$ and $e_8^*(\ell) = L$. Clearly the diagram

$$\begin{array}{ccc} \pi_*(\mathbb{L}^\circ;Z/2) & \xrightarrow{(e_8)_*} & \pi_*(\mathbb{L}_\circ;Z/2) \\ & {\scriptstyle 0}\searrow \quad \swarrow{\scriptstyle \text{Kervaire invariant}} & \\ & Z/2 & \end{array}$$

commutes, so Corollary 1 shows $e_8^*(k) = 0$. Equally clearly

$$\begin{array}{ccc} \pi_*(\mathbb{L}^\circ;R) & \xrightarrow{(e_8)_*} & \pi_*(\mathbb{L}_\circ;R) \\ \text{Index} \searrow & & \swarrow \text{Surgery obstruction} \\ & R & \end{array}$$

commutes. Corollary 2 shows that $e_8{}^*(\ell) = L$.

An analogous argument shows that $(1+T)^*(L) = 8\ell$; $(1+T)^*(r) = 0$, which is 1.10.

Proof of 1.12: We want to show $\rho^*(L) = L \wedge L$ and $\rho^*(r) = r \wedge L + L \wedge r$. First check that diagram 2.6 commutes with $E_1 = E_2 = E_3 = \mathbb{L}^\circ$; $\nu = \rho$; and $\mu_i = \mu^\circ$ $i=1,2,3$. If we let each Ψ_i be the index homomorphism, the trick that we used in showing diagram 2.4 commuted for Ψ_1 also shows that the diagram in Theorem 1 commutes. Hence $\rho^*(L) = L \wedge L$.

To get the second equation, we let Ψ_1 be the index homomorphism and let Ψ_2 and Ψ_3 both be the deRham homomorphism. Then $\beta_1 = 0$ and $\beta_2 = \beta_3 = (V\, Sq^1 V) \wedge L$. If we let $\alpha = r \wedge L$, it is easy to check that the necessary diagram commutes, so Theorem 2 applies. Alas we have not yet calculated $(\mu_1 \wedge \mu_2)^*(\alpha)$. The correct answer is easy to guess: $\mu_2{}^*(L) = \mathcal{L} \wedge L$ and $\mu_1{}^*(r) = \mathcal{L} \wedge r + (V\, Sq^1 V) \wedge L$. We accept this answer provisionally and proceed. If we take $\theta = r \wedge L$ it is a laborious calculation to see that i) is satisfied. Hence ii) also holds, so $\rho^*(r) = L \wedge r + r \wedge L$.

The maps μ_1 and μ_2 above are both the map μ°: $MSO \wedge \mathbb{L}^\circ \to \mathbb{L}^\circ$. To justify the above calculations we must analyze this map. Let $E_1 = MSO$, $E_2 = E_3 = \mathbb{L}^\circ$; let $\nu = \mu^\circ$; let Ψ_i be the index homomorphism. Theorem 1 applies, so $(\mu^\circ)^*(L) = \mathcal{L} \wedge L$. To get the other equation, change Ψ_2 and Ψ_3 to be the deRham homomorphism. Then $\beta_1 = 0$ and

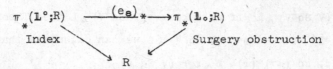

$\beta_2 = \beta_3 = (V\,Sq^1V) \wedge L$. If we take $\alpha = (V\,Sq^1V) \wedge L$ then Theorem 2 applies. If we take $\mathfrak{s} = (V\,Sq^1V) \wedge L$ we can calculate that i) is satisfied, so $(\mu^\circ)^*(r) = \mathcal{L} \wedge r + (V\,Sq^1V) \wedge L$.

This finishes the proof of 1.12.

Proof of 1.13: To analyze the map $\mathfrak{m}$ take $E_1 = \mathbb{L}^\circ$; $E_2 = E_3 = \mathbb{L}_\circ$; and $\nu = \mathfrak{m}$. The map $\mu_1 = \mu^\circ$ and $\mu_2 = \mu_3 = \mu_\circ$. Diagram 2.6 commutes.

We always take Ψ_1 to be the index homomorphism. To show $\mathfrak{m}^*(k) = L \wedge k$, which is half of 1.13, let us take $\Psi_2 = \Psi_3$ to be the Kervaire invariant. The diagram in Theorem 1 commutes, so the result follows.

To show $\mathfrak{m}^*(\ell) = L \wedge \ell + \delta(r \wedge k)$, which is the remainder of 1.13, let us take $\Psi_2 = \Psi_3$ to be surgery obstruction. Then $\beta_1 = 0$ and $\beta_2 = \beta_3 = \delta((V\,Sq^1V) \wedge k)$. If we take $\alpha = \delta(r \wedge k)$, Theorem 2 is seen to apply. We have not yet calculated $(\mu_\circ)^*(k)$ so we assume the correct answer, $\mathcal{L} \wedge k$. Then, with $\mathfrak{s} = \delta(r \wedge k)$ the reader can check that condition i) of Theorem 2 is satisfied, so our result follows.

To calculate $(\mu_\circ)^*(k)$ we apply Theorem 1 with $E_1 = MSO$; $E_2 = E_3 = \mathbb{L}_\circ$; $\nu = \mu_\circ$; $\Psi_1 =$ Index ; $\Psi_2 = \Psi_3 =$ Kervaire.

Our analysis of $\hat{\mathbb{L}}$ is less satisfactory. We have a map of $\mathbb{L}^\circ$ module spectra $\hat{\mathbb{L}} \to \Sigma\mathbb{L}_\circ$ so we have perfectly satisfactory classes $\Sigma\ell \in H^{4i+1}(\hat{\mathbb{L}}; Z_{(2)})$ and $\Sigma k \in H^{4i+3}(\hat{\mathbb{L}}; Z/2)$ obtained by pulling back the suspensions of ℓ and k respectively. If $\hat{\mu}: \mathbb{L}^\circ \wedge \hat{\mathbb{L}} \to \hat{\mathbb{L}}$ denotes the module pairing, we have $\hat{\mu}^*(\Sigma k) = L \wedge \Sigma k$. and $\hat{\mu}^*(\Sigma\ell) = L \wedge \Sigma\ell + \delta(r \wedge \Sigma k)$.

The next step in understanding $\hat{\mathbb{L}}$ is to construct the classes $\hat{L}$ and $\hat{r}$, but to do this we need to understand the pairing $\pi_p(\hat{\mathbb{L}};R) \otimes \pi_q(\hat{\mathbb{L}};R) \to \pi_{p+q}(\hat{\mathbb{L}} \wedge \hat{\mathbb{L}};R) \xrightarrow{\hat{\rho}_*} \pi_{p+q}(\hat{\mathbb{L}};R)$. This is accomplished as above except that we use the hyperquadratic signature map $MSG \to \hat{\mathbb{L}}$ and the results of Brumfiel-Morgan [1].

There is a map Index: $\pi_{4i}(\hat{\mathbb{L}};R) \to Z/8 \otimes R$ given by taking the index of the hyperquadratic form associated to the element in $\pi_{4i}(\hat{\mathbb{L}};R)$. Brumfiel and Morgan define an index homomorphism $\pi_{4i}(MSG;R) \to Z/8 \otimes R$. It is not clear that their homomorphism is the composite $\pi_{4i}(MSG;R) \to \pi_{4i}(\hat{\mathbb{L}};R) \to Z/8 \otimes R$, but it is true that we can find a homomorphism $\mathscr{I}\colon \pi_{4i}(\hat{\mathbb{L}};R) \to Z/8 \otimes R$ such that the Brumfiel-Morgan index is $\pi_{4i}(MSG;R) \to \pi_{4i}(\hat{\mathbb{L}};R) \xrightarrow{\mathscr{I}} Z/8 \otimes R$ and such that $\pi_{4i}(\mathbb{L}^\circ;R) \to \pi_{4i}(\hat{\mathbb{L}};R) \xrightarrow{\mathscr{I}} Z/8 \otimes R$ is still the index reduced mod 8.

The deRham invariant of a hyperquadratic form defines a homomorphism $\pi_{4i+1}(\hat{\mathbb{L}};Z/2) \to Z/2$.

The results of Brumfiel and Morgan [1] suffice to determine the pairing $\pi_p(\hat{\mathbb{L}};R) \otimes \pi_q(\hat{\mathbb{L}};R) \to \pi_{p+q}(\hat{\mathbb{L}};R)$. From this it is easy to understand the pairing $\pi_p(\mathbb{L}^\circ;R) \otimes \pi_q(\hat{\mathbb{L}};R) \to \pi_{p+q}(\hat{\mathbb{L}};R)$ induced by the module structure.

One sees that $\mathscr{I}$ does not cause diagram 2.4 to commute, but that $\mathscr{I} + \langle V\, Sq^1 V \wedge \Sigma k, \rangle$ is multiplicative with respect to the index (essentially [1] Theorem 8.4). Let $\hat{L}$ denote the resulting cohomology class.

Likewise the deRham homomorphism does not make diagram 2.4

commute, but by adding $\langle V\, Sq^1 V \wedge \hat{L} \,,\, \rangle$ we do get a homomorphism that is multiplicative with respect to the index and so defines a class $\hat{r}$.

We can now use Theorem 2 to study $\hat{\mu} : MSO \wedge \hat{\mathbb{L}} \to \hat{\mathbb{L}}$. The results are that $(\hat{\mu})^*(\hat{L}) = \mathcal{L} \wedge \hat{L} + \iota((V\, Sq^1 V) \wedge \Sigma k)$ where $\iota : Z/2 \to Z/8$ is the non-trival map and $(\hat{\mu})^*(\hat{r}) = \mathcal{L} \wedge \hat{r} + (V\, Sq^1 V) \wedge \hat{L}$.

The class Σk can be defined using the only non-trival homomorphism $\pi_{4i+3}(\hat{\mathbb{L}}; Z/2) \to Z/2$. Theorem 1 can be applied to show $(\hat{\mu})^*(\Sigma k) = \mathcal{L} \wedge \Sigma k$.

Now apply Theorem 2 to study the map $\hat{\rho}: \hat{\mathbb{L}} \wedge \hat{\mathbb{L}} \to \hat{\mathbb{L}}$. Equations 1.14 and 1.15 should be clear.

§4. Periodic, connective bo-module spectra.

We say that a connective spectrum E, which is a module spectrum over bo, is periodic if the maps

$$\pi_4(bo) \otimes \pi_q(E) \to \pi_{q+4}(bo \wedge E) \to \pi_{q+4}(E)$$

are isomorphisms for all non-negative q.

Since $\mathbb{L}^\circ$ becomes bo after localization away from 2, one set of examples of connective, periodic bo-module spectra are the spectra $\mathbb{L}^\circ(A,\varepsilon)$ and $\mathbb{L}_\circ(A,\varepsilon)$ after localizing away from 2.

We have

Theorem $A_{(odd)}$: Let E be a connective, periodic bo-module spectrum. Then $E_{(odd)}$ is equivalent to

$$bo\Lambda_0 \vee \Sigma^1 bo\Lambda_1 \vee \Sigma^2 bo\Lambda_2 \vee \Sigma^3 bo\Lambda_3$$

where $bo\Lambda_i$ is bo with coefficients $\Lambda_i = \pi_i(E) \otimes Z[\frac{1}{2}]$.

Proof: Since $\pi_*(bo)$ is odd torsion free, the universal coefficients theorem says that $\pi_*(bo\Lambda_i) = \pi_*(bo) \otimes \Lambda_i$. Let $M(\Lambda_i)$ be the Moore spectrum whose only non-zero homology group is Λ_i in dimension zero. Then $bo\Lambda_i$ is just $bo \wedge M(\Lambda_i)$.

We can map $M(\Lambda_0) \to E_{(odd)}$ so that, on π_0 , the map is an isomorphism. Similarly, we can map $\Sigma^1 M(\Lambda_i) \to E_{(odd)}$ so that, on π_1 , the map is an isomorphism.

Now periodicity shows that the composite

$$bo \wedge M(\Lambda_0) \to bo \wedge E_{(odd)} \to E_{(odd)}$$

induces an isomorphism on π_{4i} and the zero map on $\pi_{4i+\varepsilon}$ for $\varepsilon=1,2,3$ and all i.

There is a similar statement for $M(\Lambda_1)$, $M(\Lambda_2)$, and for $M(\Lambda_3)$. The theorem follows easily.

Note added in proof: L. Jones has had a proof of Theorem A for $\mathbb{L}_0(Z\pi,\varepsilon)$ for some years: see The non-simply connected characteristic variety theorem, Proc. Symp. Pure Math. Vol. 32 Part I, 131 - 140.

REFERENCES

[1] G. Brumfiel and J. Morgan, Homotopy-theoretic consequences of N. Levitt's obstruction theory to transversality for spherical fibrations, Pacific J. Math. 67(1976), 1-100.

[2] J. Morgan and D. Sullivan, The transversality characteristic class and linking cycles in surgery theory, Ann. of Math. 99(1974), 463-544.

[3] J. Morgan and W. Pardon, The surgery obstruction on closed manifolds, lecture.

[4] F. Quinn, A geometric formulation of surgery, in Topology of Manifolds (Proceedings of the 1969 Georgia Conference), Markham Press, 500-511.

[5] A. Ranicki, The algebraic theory of surgery, preprint, Princeton University, 1978.

[6] E. Stein, Surgery on products with finite fundamental group, Topology 16(1977), 473-493.

[7] D. Sullivan, Triangulating and smoothing homotopy equivalences and homeomorphisms, Geometric Topology seminar notes, Princeton University, 1967.

[8] C. T. C. Wall, Surgery on Compact Manifolds, Academic Press, 1970.

[9] C. T. C. Wall, Formulae for surgery obstructions, Topology 15(1976), 189-210.

[10] C. T. C. Wall, Formulae for surgery obstructions: corrigendum Topology 16(1977), 495-496.

University of Notre Dame
Notre Dame, In. 46556

BALANCED SPLITTINGS OF SEMI-FREE ACTIONS ON HOMOTOPY SPHERES

Douglas R. Anderson* and Ian Hambleton**

Let Σ^{n+k} be a homotopy (n+k)-sphere and $\rho : G \times \Sigma \rightarrow \Sigma$ a smooth semi-free action of a finite group G on Σ with fixed-point set a manifold F^n of dimension n. A decomposition of Σ into two G-invariant disks will be called a splitting of the action and the induced splitting of Σ^G denoted $F = F_1 \cup F_2$. We ask whether every such action has a splitting with $H_i(F_1) \cong H_i(F_2)$ for $i \geq 0$ (these are called balanced splittings).

One class of actions for which balanced splittings exist is obtained by the "twisted double" construction. Namely, let $\rho : G \times D^{n+k} \rightarrow D^{n+k}$ be a semi-free action of G on an (n+k)-disk. Let $\Sigma = D \cup_{\varphi} D$ where $\varphi : \partial D \rightarrow \partial D$ is an equivariant diffeomorphism. Our interest in the problem considered here arose from trying to understand the conditions under which a given semi-free action is a twisted double. An action that admits a balanced splitting resembles a twisted double at least homologically and thus exhibits some symmetry. On the other hand, an action with no balanced splitting is rather strongly asymmetrical.

In this paper we introduce a semi-characteristic invariant of the action to detect the existence of balanced splittings and construct some examples of actions whose semi-characteristic invariant is nonzero. Such actions have no balanced splitting. For most of our results, the arguments are outlined here so that the reader who is familiar with work in this area (e.g., by L. Jones [1] and R. Oliver [2]) can follow them. Full details will appear elsewhere.

Before beginning a precise description of the results, we remark that the fixed-point set F will be assumed nonempty and connected throughout to avoid

*Partially supported by National Science Foundation grant MCS 76-05997.

**Partially supported by an NSF grant at the Institute for Advanced Study, Princeton, New Jersey.

trivial cases. In addition, although the structure of the groups G is known (see [7]), since they all admit free linear representations, this classification is not used in the present situation.

The first named author would like to thank McMaster University for its hospitality during the period when the research contained in this paper was done.

1. Statement of Results

In order to provide an algebraic setting for the invariant, two categories of finitely-generated $\mathbb{Z}G$-modules will be useful. Let $\mathfrak{D}(G)$ denote the category of finite Abelian groups of order prime to $|G|$. If we regard the groups in $\mathfrak{D}(G)$ as trivial G-modules, then there is an inclusion $\mathfrak{D}(G) \to \mathcal{C}(G)$ (from a result of Rim [4]) where $\mathcal{C}(G)$ is the category of cohomologically trivial modules. The Grothendieck groups of these categories are $G_0(\mathfrak{D}(G))$ and $G_0(\mathcal{C}(G))$ and a further result of Rim [4] allows the identification,

$$\widetilde{G}_0(\mathcal{C}(G)) \cong \widetilde{K}_0(\mathbb{Z}G).$$

Finally, let $A(G) = \mathrm{Im}(G_0(\mathfrak{D}(G)) \to \widetilde{K}_0(\mathbb{Z}G))$ and note that $A(G)$ is just the image of $\partial : K_1(\mathbb{Z}/|G|) \to \widetilde{K}_0(\mathbb{Z}G)$ considered by Swan [5].

Now let F^n be the fixed-point set of a semi-free action of G on Σ^{n+k}. It follows from Smith theory that $\widetilde{H}_i(F) \in \mathfrak{D}(G)$ for $i < n$. Similarly, if the decomposition $F = F_1 \cup F_2$ is induced by a splitting of the action, then $\widetilde{H}_i(F_j) \in \mathfrak{D}(G)$ $(j = 1, 2)$ for all i, and $\widetilde{H}_i(F_0) \in \mathfrak{D}(G)$ for $i < n-1$ where $F_0 = F_1 \cap F_2$. Any decomposition of F satisfying these necessary conditions will be called a splitting of F.

Definition 1. Let X be a finite CW complex with $\widetilde{H}_i(X) \in \mathfrak{D}(G)$ for $i \geq 0$. Then

$$\chi_G(X) = \sum_{i \geq 0} (-1)^i [\widetilde{H}_i(X)] \text{ in } \widetilde{K}_0(\mathbb{Z}G).$$

Theorem A. *Let (Σ, ρ) be a smooth semi-free action of a finite group G on a homotopy $(n+k)$-sphere Σ with fixed-point set F^n. If $1 \leq n \leq k-2$, then a splitting $F = F_1 \cup F_2$ is induced by a splitting of the action if and only if $\chi_G(F_1) = 0$.*

The sufficiency part of Theorem A is proved by an equivariant handle attaching argument similar to the argument given by Jones [1]. The necessity is obtained by observing that $\chi_G(F_1) = -\sum_{i \geq 0} (-1)^i [H_i(D_1, F_1)] = \sum_{i \geq 0} (-1)^i [C_i(D_1, F_1)] = 0$ where D_1 is a G-invariant disk such that $F_1 = D_1 \cap F$. Note that $\chi_G(F_1) = \pm\chi_G(F_2)$, so that statement does not depend on the ordering of F_1 and F_2.

If we now assume that $F = F_1 \cup F_2$ is a balanced splitting (that is, $H_i(F_1) \cong H_i(F_2)$ for $i \geq 0$ in addition to the conditions above), then $\chi_G(F_1)$ determines a semi-characteristic invariant of the action.

Definition 2. *Let X be a finite CW complex of dimension n with $\widetilde{H}_i(X) \in \mathfrak{D}(G)$ for $i < n$ and $|H_m(X)| = q^2$ when $n = 2m+1$. In $G_0(\mathfrak{D}(G))$ set*

$$\widetilde{\chi}_{\frac{1}{2}}(X) = \begin{cases} \sum_{i=1}^{m-1} (-1)^i [H_i(X)] & \text{if } n = 2m \\ \sum_{i=1}^{m-1} (-1)^i [H_i(X)] + (-1)^m [\mathbb{Z}/q] & \text{if } n = 2m+1 . \end{cases}$$

Now let $x \mapsto \bar{x}$ be the involution on $\widetilde{K}_0(\mathbb{Z}G)$ induced by sending $[P]$ to $-[P^*]$. Then we wish to define $\chi_{\frac{1}{2}}(X)$ to be the cohomology class in $H^n(\mathbb{Z}/2; A(G))$ represented by the image of $\widetilde{\chi}_{\frac{1}{2}}(X)$. This will make sense if we assume that when $n = 2m+1$, $2\widetilde{\chi}_{\frac{1}{2}}(X) = 0$ in $A(G)$ (because the subgroup $A(G)$ is actually fixed by the involution $-$). However, we assert that $\chi_{\frac{1}{2}}(F^n)$ is well defined for $F = \Sigma^G$ provided $|H_m(F)|$ is a square when $n = 2m+1$. Since the resulting class is an invariant of the action, we denote it $\chi_{\frac{1}{2}}(\Sigma, \rho)$.

Theorem B. *Let (Σ, ρ) be a smooth semi-free G-action on a homotopy $(n+k)$-*

sphere with fixed-point set F^n and $1 \leq n \leq k-2$ $(n \neq 3,4)$.

i) If $n = 2m$, the action has a balanced splitting.

ii) If $n = 2m+1$, the action has a balanced splitting if and only if $|H_m(F)|$ is a square and $\chi_{\frac{1}{2}}(\Sigma, \rho) = 0$.

Among the actions obtained by the twisted double construction are those for which $\Sigma = D \cup_{\varphi} D$ with φ = identity. These are called strong doubles. If the homological conditions on the splitting of F^n are strengthened to include

$$\ker(H_{m-1}(F_0) \to H_{m-1}(F_1)) = \ker(H_{m-1}(F_0) \to H_{n-1}(F_2))$$

for $m = [\frac{n}{2}]$, we call it a strong balanced splitting. Such a condition is satisfied for all $m \geq 0$ if the splitting arises from a splitting of (Σ, ρ) as a strong double.

Theorem C. Under the same hypotheses as Theorem B, (Σ, ρ) has a strong balanced splitting if and only if $\chi_{\frac{1}{2}}(\Sigma, \rho) = 0$ and $|H_m(F^n)|$ is a square (when $n = 2m+1$).

Remark. A (strong) balanced splitting of the fixed-point set F^n exists always for $n = 2m$ and for $n = 2m+1$ if and only if $|H_m(F)|$ is a square. There are examples of actions with $n = 2m+1$ and $|H_m(F)|$ a nonsquare (e.g., a Brieskorn example of an involution on S^5 with fixed-point set a lens space).

Finally, we have some examples to show that the semi-characteristic invariant can be nonzero. Since the exponent of $A(G)$ divides the Artin exponent of G [6], if $|G|$ is odd $\chi_{\frac{1}{2}}(\Sigma, \rho)$ is always zero. Otherwise the Sylow 2-subgroup $Syl_2(G)$ is cyclic or generalized quaternion $Q2^{\ell}$ of order 2^{ℓ} [7], and our examples concern this second case. The fixed-point sets of these examples are actually strong doubles.

Theorem D. Let G be a finite group with $Syl_2(G) = Q2^{\ell}$ admitting a free linear representation of dimension d and n, k integers such that $5 \leq n \leq k-2$. Then there exists a semi-free G-action ρ on a homotopy (n+k)-sphere Σ with $\dim(\Sigma^G) = n$ and $\chi_{\frac{1}{2}}(\Sigma,\rho) \neq 0$ provided $k \equiv 0 \pmod d$ and $n \not\equiv 1 \pmod 4$ (when $\ell = 3$) or $n \not\equiv 0, 1 \pmod 4$ (when $\ell \geq 4$).

Remarks.

1) If $n \equiv 2 \pmod 4$, we have the more complete result (valid for any G admitting a free representation) that each element of $H^n(Z/2; A(G))$ can arise as the $\chi_{\frac{1}{2}}$-obstruction to the existence of a strong balanced splitting.

2) If (Σ,ρ) and (Σ,ρ') are concordant actions, then $\chi_{\frac{1}{2}}(\Sigma,\rho) = \chi_{\frac{1}{2}}(\Sigma,\rho')$. This means that the examples above are not concordant to linear actions.

2. Proof of Theorem B.

In this section we outline the proof of Theorem B.

Lemma 1. *If F^n $(n \neq 3, 4)$ is a closed orientable manifold with $\widetilde{H}_i(F) \in \mathfrak{D}(G)$ for $i < n$, then F has a (strong) balanced splitting if and only if $|H_m(F)|$ is a square when $n = 2m+1$.*

If $n = 2m$, we start with a handlebody $N_0 \subset F$ that carries all handles of F of index $\leq m-1$ in a given handle decomposition. Only enough handles of index m are added to make $H_{m-1}(N_0) \cong H_{m-1}(F)$ without creating any m-dimensional homology. If $n = 2m+1$ and $|H_m(F)|$ is a square, there exists a short exact sequence of the form

$$0 \to T \to H_m(F) \to T \to 0.$$

To $N_0 \subset F$, as before consisting of handles of index $\leq m-1$, can be attached handles of index m and m+1 to get $N_1 \subset F$ such that $H_i(N_1) \cong H_i(F)$ for $i < m-1$, $H_m(N_1) \cong T$ and $H_i(N_1) = 0$ for $i > m$.

For the necessity when $n = 2m+1$, let $F = F_1 \cup F_2$ be a balanced splitting and factor the exact sequence of the pair (F, F_1) into

$$0 \to A \to H_{m-1}(F_1) \to H_{m-1}(F) \to \dots \to H_1(F, F_1) \to 0$$

$$0 \to B \to H_m(F_1) \to H_m(F) \to H_m(F, F_1) \to A \to 0$$

$$0 \to H_{2m}(F_1) \to H_{2m}(F) \to \dots \to H_{m+1}(F, F_1) \to B \to 0$$

and use formal manipulations and Poincaré duality to show $[A] = [B]$ in $G_0(\mathfrak{D}(G))$. The middle sequence then shows

$$[H_m(F)] = 2[H_m(F_1)] - 2[A] \quad \text{in} \quad G_0(\mathfrak{D}(G)).$$

Since $G_0(\mathfrak{D}(G))$ is the free Abelian group on generators $[\mathbb{Z}/p]$ where p is a prime not dividing $|G|$, $|H_m(F)|$ is a square.

The next step is to determine the relationship between $\chi_G(F_1)$ and $\chi_{\frac{1}{2}}(\Sigma, \rho)$ when $F = \Sigma^G$ has a balanced splitting $F = F_1 \cup F_2$.

Lemma 2. Suppose $F^n = F_1 \cup F_2$ is a balanced splitting.

i) If $n = 2m+1$, $\widetilde{\chi}_{\frac{1}{2}}(F) = \chi_G(F_1)$ and $2\widetilde{\chi}_{\frac{1}{2}}(F) = 0$.

ii) If $n = 2m$,

$$\widetilde{\chi}_{\frac{1}{2}}(F) = \chi_G(F_1) - 2\left(\sum_{i=m}^{2m-1} (-1)^i [H_i(F_1)]\right) + (-1)^m [A]$$

where

$$A = \ker(H_{m-1}(F_1) \to H_{m-1}(F)).$$

From the sketch proof for Lemma 1,

$$\tilde{\chi}_{\frac{1}{2}}(F) = \sum_{i=1}^{m-1} (-1)^i [H_i(F)] + (-1)^m [\mathbb{Z}/q]$$

(where $|H_m(F)| = q^2$ and $n = 2m+1$)

$$= (-1)^m [A] + \sum_{i \neq m} (-1)^i [H_i(F_1)] + (-1)^m ([H_m(F_1)] - [A])$$

$$= \chi_G(F_1).$$

By taking the full sequence of the pair $(F\text{-pt}, F_1)$, we obtain $2\chi_G(F_1) = \chi_G(F\text{-pt}) = 0$. The argument for i) is similar.

Recall now that $\chi_{\frac{1}{2}}(\Sigma, \rho)$ lies in

$$H^n(\mathbb{Z}/2; A(G)) = \begin{cases} A(G)/2A(G) & , \quad n = 2m \\ \{x \in A(G) \mid 2x = 0\}, & n = 2m+1. \end{cases}$$

Clearly, Lemma 2 shows that $\chi_{\frac{1}{2}}(\Sigma, \rho)$ is well defined. The proof of Theorem B when $n = 2m+1$ is now an immediate consequence of Lemmas 1, 2 and Theorem A.

For $n = 2m$ there is one more ingredient. This is a simple method for changing one balanced splitting to a new one. Let $M_n(\ell, p)$ denote a regular neighborhood of the complex $S^\ell \cup_p e^{\ell+1}$ embedded in S^n (where $1 \leq \ell \leq n-4$ and $(p, |G|) = 1$). Then S^n has a splitting into two thickened Moore spaces

$$S^n(\ell, p) = M_n(\ell, p) \cup (S^n - \mathring{M}_n(\ell, p)).$$

If $F^n = F_1 \cup F_2$ is a splitting, there is another splitting $F = F_1' \cup F_2'$ obtained by connected sum along F_0 and ∂M:

$$F^n \approx F^n \# S^n(\ell, p) = (F_1 \# M_n(\ell, p)) \cup (F_2 \# (S^n - \mathring{M}_n(\ell, p))).$$

If a splitting of F is understood, $F \# S^n(\ell, p)$ means this new splitting.

<u>Lemma</u> 3. <u>Suppose</u> $F^n = F_1 \cup F_2$ <u>is a (strong) balanced splitting</u> $(p, |G|) = 1$ <u>and</u> $1 \leq \ell \leq n-4$.

i) $F^n \# S^n(\ell, p) \# S^n(n-\ell-2, p)$ <u>is a balanced splitting (a strong balanced splitting unless</u> $n = 2m$ <u>and</u> $\ell = m-1$).

ii) $F^{2m} \# S^{2m}(m-1, p)$ <u>is a balanced splitting.</u>

iii) <u>If</u> $F^n = F_1' \cup F_2'$ <u>is the new splitting in</u> i)

$$\chi_G(F_1') = \begin{cases} \chi_G(F_1) & \text{if } n = 2m+1 \\ \chi_G(F_1) + (-1)^{\ell} 2[\mathbb{Z}/p] & \text{if } n = 2m . \end{cases}$$

iv) <u>For the splitting in</u> ii),

$$\chi_G(F_1') = \chi_G(F_1) + (-1)^{m-1}[\mathbb{Z}/p].$$

If $n = 2m$, a balanced splitting of Σ^G can now be obtained with $\chi_G(\Sigma^G) = 0$ using Lemma 1 and this construction. Essentially the same argument gives the proof of Theorem C also.

3. Construction of Examples

If X is any finite CW complex with $\tilde{H}_i(X) \in \mathcal{P}(G)$ for $i \geq 0$, a G any group with a free linear representation, let N_1 be a regular neighborhood of X in $\mathbb{R}^n$ and $M^m = \mathbb{R}^n \times \mathbb{V}^k$ where $5 \leq n \leq k-2$ and $\mathbb{V}^k$ is a free representation space for G of real dimension k. By the same method as that used for Theorem A, there exists a $([\frac{m}{2}]-1)$-connected G-invariant compact manifold M_1^m (M_1 is a submanifold of M except possibly when $m = 2s$) such that $M_1^G = N_1$, $\tilde{H}_i(M_1) = 0$ for $i \neq [\frac{m}{2}]$ and $(-1)^s[H_s(M_1)] = \chi_G(N_1) = \chi_G(X)$ for $s = [\frac{m}{2}]$. If $m = 2s$, the final handle attaching to produce M_1 must be done so that the cycles representing $H_s(M_1)$ have zero equivariant self-intersection. Using this manifold M_1, we will try to construct an action on some $(n+k)$-sphere with fixed-point set F^n the double of N_1. Since $A(G)$ is also the image of the Swan homomorphism, $(-1)^{[\frac{m}{2}]} \chi_G(X) = [\mathbb{Z}/r]$ ($= \partial r$) for some integer r prime to $|G|$. If $m = 2s+1$, after a further surgery on M_1 (using the bundle map $\nu_{M_1} \to \nu_M$), we obtain M_1' with $\widetilde{H}_i(M_1') = 0$, $i \neq s$ and $H_s(M_1') = \mathbb{Z}/r$. Let W^m be the double of M_1 $(m = 2s)$ or M_1' $(m = 2s+1)$. This is a smooth, semi-free, $([\frac{m}{2}]-1)$-connected G-manifold with W^G the double of N_1 and

$$H_s(W) = \begin{cases} P \oplus P^* & , \; m = 2s \\ \mathbb{Z}/r \oplus \mathbb{Z}/r \, , & m = 2s+1. \end{cases}$$

where P is a projective $\mathbb{Z}G$-module. Moreover, the geometric self-intersection (self-linking) is trivial on P and P^* (both copies of $\mathbb{Z}/r$), and the intersection form (linking form) is hyperbolic. This means that the obstruction doing surgery on W to obtain a homotopy sphere can be formulated in terms of the "hyperbolic map" in the Ranicki-Rothenberg sequence [3]:

$$\ldots \to L^p_{m+1}(\mathbb{Z}G) \to H^m(\mathbb{Z}/2;\widetilde{K}_0(\mathbb{Z}G)) \xrightarrow{\mathbb{H}} L^h_m(\mathbb{Z}G) \to L^p_m(\mathbb{Z}G) \, .$$

<u>Proposition</u> 1. W^m <u>is equivariantly cobordant to a semi-free action</u> (Σ, ρ) <u>on a homotopy</u> m-<u>sphere with</u> $\Sigma^G = F$ <u>if</u> $\mathbb{H}(\chi_G(X)) = 0$.

To obtain the examples referred to in the Remarks following Theorem D, we note that $\mathbb{H}(x) = 0$ for any $x \in A(G)$ when $\mathbb{H} : H^0(\mathbb{Z}/2, \widetilde{K}_0(\mathbb{Z}G)) \to L^h_2(\mathbb{Z}G)$. Clearly, any element of A(G) arises as $\chi_G(X)$ for some suitable X (e.g., a Moore space). For Theorem D itself, we note that when $H \leq G$ is a subgroup, the restriction $A(G) \to A(H)$ is onto (Ullom [6]). The existence of the desired examples now follows from:

<u>Proposition</u> 2. <u>Let</u> $\mathbb{H} : H^m(\mathbb{Z}/2;K_0(\mathbb{Z}G)) \to L^h_m(\mathbb{Z}G)$ <u>be the hyperbolic map</u>.

i) <u>If</u> $G = Q8$, $\mathbb{H} = 0$ <u>when</u> $m \not\equiv 1$ (<u>mod</u> 4) <u>and</u> $\mathbb{H}$ <u>is injective when</u> $m \equiv 1$ (<u>mod</u> 4).

ii) <u>If</u> $G = Q2^{\ell}$ $(\ell \geq 4)$, $\mathbb{H} = 0$ <u>when</u> $m \equiv 2, 3$ (<u>mod</u> 4).

REFERENCES

[1] L. Jones, "A converse to the fixed-point theory of P. A. Smith", Ann. of Math., (2) 94 (1971), 52-68.

[2] R. Oliver, "Fixed-point sets of group actions on finite acyclic complexes", Comment. Math. Helv., 50 (1975), 155-177.

[3] A. Ranicki, "The algebraic theory of surgery", preprint.

[4] D. S. Rim, "Modules over finite groups", Ann. of Math., (2) 69 (1959), 700-712.

[5] R. G. Swan, "Periodic resolutions for finite groups", Ann. of Math., (2) 72 (1960), 267-291.

[6] S. Ullom, "Non-trivial lower bounds for class groups of integral group rings", Ill. J. Math., 20 (1976), 361-371.

[7] J. Wolf, Spaces of Constant Curvature, McGraw-Hill, New York, 1967.

Syracuse University
Syracuse, NY 13210

McMaster University, Hamilton, Ontario
The Institute for Advanced Study, Princeton, NJ 08540

Some Examples of Finite Group Actions

Amir H. Assadi

Introduction. This paper describes some examples of finite group actions on euclidean spaces, disks, and spheres. It is based on part of the author's thesis [1]. The organization of the paper is as follows: Section 0 states a general extension problem and some special cases. Section 1 outlines some pertinent definitions and results of [1]. Section 2 describes a general method to construct stationary point free actions on euclidean spaces, which leads to the classification of the groups which can have such actions. The method yields infinitely many non-equivalent smooth actions for each desired family of isotropy subgroups. Section 3 is concerned with low dimensional examples. Section 4 constructs examples of smooth actions on disks and spheres with certain stationary point sets. In Section 5 we show that all finite simplicial complexes can occur as the stationary point set of PL G-actions on PL spheres. There is also a counterpart for topological actions. Section 6 describes examples of actions on spheres with non-equivalent representations at different connected components of stationary point set.

I would like to express my sincere thanks to Professor William Browder for his encouragement and support during my research, and for many helpful conversations and suggestions which led to formulation and improvement of results in [1]. I would like to thank Professors Deane Montgomery, Wu Chung Hsiang, Robert Oliver, and Ted Petrie for encouraging and helpful conversations. This work has been supported partially by an NSF Grant through the Institute for Advanced Study.

§ 0. Let G be a finite group, and let $\mathcal{F}$ and $\mathcal{F}_1 \subset \mathcal{F}$ be two collections of subgroups of G. Let M be a smooth G-manifold with isotropy subgroups belonging to $\mathcal{F}$ and such that M is an embedded submanifold of the smooth manifold W. Under the above hypotheses, consider the following question:

(0.1) When is it possible to extend the G-action on M to a smooth G-action on W such that the isotropy subgroups of W - M lie in $\mathcal{F}_1$?

Some special cases of (*) are interesting in their own right:

(0.2) Let M have the trivial G-action, and let $\mathcal{F}_1 = \mathcal{F} - \{G\}$. Then (0.1) reduces to the problem of deciding whether we can find a smooth G-action on W with isotropy subgroups belonging to $\mathcal{F}$ and such that $W^G = M$.

(0.3) Let $M = \phi$, and let $\{G\} \notin \mathcal{F}$. Then the problem will be to find a G-action on W with isotropy subgroups in $\mathcal{F}$. In particular this action will have no stationary points.

In [1], we have formulated and studied the above problem under suitable hypotheses. Of course, one does not expect to have a concrete answer to (0.1) in such a general setting. Instead, the following general approach is suggested in [1]: We divide the problem into two parts. (a) Try to construct a G-action on a smooth manifold W' homotopy equivalent to W having M as an embedded submanifold and such that the isotropy subgroups of the action on W' - M lie in $\mathcal{F}_1$. (b) Try to do the above construction in such a fashion that the homotopy equivalence between W' and W can be "deformed" or "modified" to a diffeomorphism.

To be able to do (a), one should solve the following "homotopy problem". Namely, to construct a G - CW complex X homotopy equivalent to W such that $M \subset X$ as a subcomplex, and the isotropy subgroups of X - M be in $\mathcal{F}_1$. If this is achieved, then one may try to use an "equivariant thickening" technique to replace equivariant cells in X - M by "equivariant handles" to obtain W'. Part (b) is already a deep problem in differential topology, and only under severe restrictions it is possible to decide when two homotopy equivalent smooth manifolds are diffeomorphic. With these restrictions in mind, one naturally tries to carry out the construction of part (a) to obtain W' and a homotopy equivalence to W satisfying the correct hypotheses to be deformable to a diffeomorphism.

Roughly speaking, under the assumption that W is simply-connected and open, $\mathcal{F}$ and $\mathcal{F}_1$ satisfying some reasonable conditions, and that M meets certain necessary conditions depending on $\mathcal{F}$, $\mathcal{F}_1$ and W, one is able to carry out this program. However, if W is compact (still simply-connected), then the homotopy problem of part (a) does not always have a solution. It can be solved if a well-defined (algebraic) obstruction vanishes. The equivariant thickening of such a G - CW complex can be achieved if a certain K_G-theoretic condition is met.

In this note, we will sketch some applications of the above techniques to solve some problems in the theory of transformation groups. We will use the notions and results in Chapters I-III of [1]. Most of the terminology and notation is based on those in Bredon's book on transformation groups [2].

§1. Let G be a finite group in the following unless otherwise specified. A collection of subgroups of G is called a family if it is closed under conjugation. The family of prime power order subgroups of G is denoted by $\mathcal{P}(G)$. Let $\mathcal{F}$ be a family and $\mathcal{C}$ be a collection of subgroups of G. $\mathcal{F}$ inherits a natural partial ordering induced by inclusion of subgroups. With this partial ordering in mind, we say $\mathcal{F}$ has unique minimals for $\mathcal{C}$ if for each $H \in \mathcal{C}$ there exists a unique minimal element $m(H) \in \mathcal{F}$ such that $m(H) \supseteq H$. For instance, let $\mathcal{F}$ be the family of isotropy subgroups of a smooth G-action on a manifold M with M^H connected and non-empty for all $H \subseteq G$. Then $\mathcal{F}$ has unique minimals for all subgroups of G. If M is an acyclic G-manifold, then M^H is connected and nonempty for all $H \in \mathcal{P}(G)$. Consequently, in the latter example $\mathcal{F}$ has unique minimals for $\mathcal{P}(G)$.

Let G be a finite group acting smoothly and effectively on a euclidean space; then the family of isotropy subgroups $\mathcal{F}$ of this action has the following property: (1.1) $\mathcal{F}$ contains the trivial subgroup and has unique minimals for $\mathcal{P}(G)$. Let $m(\mathcal{P}) = \{H \in \mathcal{F} \mid H = m(P)$ for some $P \in \mathcal{P}(G)$ or $H = 1\}$. Denote by $\#(\mathcal{L})$ the number of elements of $\mathcal{L} \subset \mathcal{F}$. Let $\ell(\mathcal{F}) = \max\{\#(\mathcal{L}) \mid \mathcal{L} \subset m(\mathcal{P})$ is a linearly ordered subset$\}$. For instance, let q be an odd integer, and D_q be the dihedral group of order 2q. Then $\mathcal{F} = \{$all subgroups isomorphic to $\mathbb{Z}_2\} \cup \{\mathbb{Z}_q, 1\}$ has unique minimals for $\mathcal{P}(D_q)$, and $\ell(\mathcal{F}) = 2$. A corollary of Theorem II.1.4 of [1] is the following:

1.2. Corollary: Let G be a finite group not of prime power order, and let $\mathcal{F}$ be a family of subgroups of G satisfying (1.1), and such that $G \notin \mathcal{F}$. Then there exists a contractible countable G - CW complex X with isotropy subgroups in the family $\mathcal{F}$. In particular $X^G = \phi$, and X can be chosen to have $\dim X \leq \ell(\mathcal{F}) + 1$. ■

To obtain smallest dimension of such contractible G - CW complexes X with $X^G = \phi$, one simply looks for the families $\mathcal{F}$ satisfying (1.1), $G \notin \mathcal{F}$, and such that $\ell(\mathcal{F})$ is the smallest possible integer. For instance, if G is nilpotent, one can take $\mathcal{F}$ to be the family of all Sylow subgroups and the trivial group. Then $\ell(\mathcal{F}) = 2$, and $\dim X = 3$. This is the smallest possible dimension in general. We will use this corollary to prove that any finite group with order not a prime power has a smooth action on a euclidean space (of a reasonable dimension) with no stationary points. The first example of this kind was introduced by Conner and Floyd who constructed a smooth action of $\mathbb{Z}_{pq}$, the cyclic group of order pq, $(p, q) = 1$, on some euclidean space with no stationary points.

Let us briefly mention the technique which was originated by Conner and Floyd (cf. [4]) and extended by others to produce such examples.

The basic step in their construction was to define an equivariant map $f : S^3 \to S^3$ of degree zero, where $\mathbb{Z}_{pq}$ acts on S^3 with no stationary points. By taking an equivariant regular neighborhood of the mapping telescope of f, which was embedded in some linear representation space of $\mathbb{Z}_{pq}$, they obtained a contractible open subset of the euclidean space which becomes diffeomorphic to some $\mathbb{R}^n$ upon multiplying it by a copy of the real line. This $\mathbb{Z}_{pq}$-action on $\mathbb{R}^n$ is smooth and has no stationary points.

Kister ([8], [9]) using a refinement of this construction, provided stationary-point free actions of $\mathbb{Z}_{pq}$ on $\mathbb{R}^8$. P. A. Smith in [13] has shown that $\mathbb{Z}_{pq}$ cannot act without stationary points on $\mathbb{R}^6$. So the question of existance of such actions on $\mathbb{R}^7$ has been left unanswered ever since.

As far as the generalization of the above results to the action of compact Lie groups are concerned, Conner and Montgomery constructed the first example of smooth SO(3)-actions on the euclidean space $\mathbb{R}^n$, $n \geq 12$ (cf. [5]). Their method was based on the construction of an equivariant map $f : S^4 \to S^4$ of degree zero, $(S^4)^{SO(3)} = \phi$.

Hsiang and Hsiang in [7] extended the above result to any compact connected non-abelian Lie group G, by constructing an equivariant map $f : S^n \to S^n$ of degree zero with $(S^n)^G = \phi$, and then applying the above mentioned argument.

Our point of view in constructing such examples is to make use of (1.2) which provides us with contractible G-complexes with no stationary points, then apply the equivariant thickening discussed in Chapter III of [1]. We would like to mention that the following generalization of (1.2) to compact Lie groups is possible, but we will confine ourselves here to the finite group case:

(1.3) Let G be a compact Lie group which is not an extension of a torus by a prime power order finite group. Then there exists a finite dimensional contractible G-complex on which G acts with no stationary points.

An application of the equivariant thickening technique then yields:

(1.4) Let G be as in (1.3). Then G has a smooth stationary point free action on some euclidean space.

We remark that the above theorem completely classifies the compact Lie groups which can act on a euclidean space with no stationary points, since if G is an extension of a torus by a prime power order group, then P. A. Smith theory requires the action of any such G on an acyclic finite dimensional complex to have stationary points.

The proof of the following theorem can be found in Chapter III of [1]. (Cf. [1] Theorem III.1.3.)

(1.5) Let G be a compact Lie group and let M^n be a smooth G-manifold with isotropy subgroups in $\mathcal{F}$, and embedded in the G - CW complex X with the isotropy subgroups of X - M in $\mathcal{F}_1 \subset \mathcal{F}$. Suppose ξ is a G-bundle over X such that:

(i) $(\xi | M) \simeq T(M) \oplus \nu$ is a decomposition of smooth G-bundles such that the family of isotropy subgroups of the associated sphere bundle $S(\nu)$ is $\mathcal{F}_1$.

(ii) $2 \dim X^H < \dim(\xi | X^H)^H + \dim N(H)/H$ for all $H \in \mathcal{F}_1$.

(iii) For each $H \in \mathcal{F}_1$ and each connected component X_α^H of X^H the $\dim E(\xi | M_{i(\alpha)})^H$ is constant for all connected component $M_{i(\alpha)}$ of M such that $M_{i(\alpha)} \subset X_\alpha^H$.

Then there exists a smooth G-manifold N in which M is G-embedded, and there exists an equivariant map $f : N \to X$ exending the identity on M, such that f is an equivariant homotopy equivalence, and $f^*(\xi)$ is equivalent to T(M) as G-bundles. Moreover, N - M has isotropy subgroups in $\mathcal{F}_1$. ■

In fact f is an equivariant simple homotopy equivalence when X is assumed to be compact, and this leads to a uniqueness argument for such equivariant thickenings in some special cases.

We should mention that the idea of thickening a cell complex to a manifold using a bundle over it goes back to Mazur (cf. [10] e.g.) whose work was, in turn, motivated by Whitehead's theory of simple homotopy types and Smale's results on the structure of manifolds. So (1.5) is really an equivariant version of this idea which is slightly technical in the presence of a compact Lie group action.

§ 2. Let $\rho : G \to GL(V)$ be a (real) linear representation of G, and let $\mathcal{F}$ by the set of proper isotropy subgroups of the action of G on V induced by ρ. Then $\mathcal{F} \cup \{G\}$ has unique minimals for all subgroups of G (cf. [1] Chapter I). But it may happen that G is the unique minimal element containing some prime power order subgroup. So we shall make the following assumption on $\mathcal{F}$:

(2.1) $\mathcal{F}$ contains unique minimals for $\mathcal{P}(G)$.

By looking at some appropriate subrepresentation of the regular representation of G, one observes that (2.1) is satisfied for many representations ρ in general.

Since $\mathcal{F}$ satisfies (1.1), by (1.2) we may assume that we have constructed a contractible G - CW complex X with isotropy subgroups in $\mathcal{F}$, and $X^G = \phi$, $\dim X \leq \ell(\mathcal{F}) + 1$. For instance, if G is nilpotent, or the dihedral group D_q, (q odd) we can arrange to have $\dim X = 3$. (Of course, not all elements of $\mathcal{F}$ may occur as isotropy subgroups, but this does not matter in the construction of smooth actions. Moreover, we may always arrange to have all elements of $\mathcal{F}$ as isotropy subgroups, but in that case $\dim X = \ell(\mathcal{F}) + 1$ in general.)

If ρ has no trivial component so that $V^G = \{0\}$, then it is not difficult to see that $\dim V \geq \ell(\mathcal{F}) + 1$. Since $V \oplus V$ has the same family of isotropy subgroups as V does, we may replace V by $V \oplus V$,(*) and add some trivial representation (if necessary) to V or $V \oplus V$ to make it satisfy:

(2.2) For each isotropy subgroup $H \in \mathcal{F}$ of X, $\dim V^H > 2 \dim X^H$.

Suppose $H \in$ Isotropy (X) is a maximal subgroup, i.e., if $K \in$ Isotropy (X) and $K \supseteq H$, then $K = H$. By taking a smooth regular neighborhood of $X^H/N(H)/H$ in $(V^{(H)})^H/(N(H)/H)$, and passing to the appropriate smooth covering space, we get a smooth $N(H)/H$-manifold with free action of dimension equal to $\dim V^H$. By repeating this process for other conjugacy classes of H, we obtain a smooth manifold $M(H)$ with $\dim M(H) = \dim V^H$ and G/H as the only orbit type. The tangent bundle of $M(H)^H$ is equivalent to $V^H \times M(H)^H$ and $M(H)$ has the G-homotopy type of $X^{(H)}$. We can repeat this for all other maximal subgroups $K \in$ Isotropy (X). Let M be the disjoint union of all $M(K)$ for $K \in$ Isotropy (X) a maximal subgroup. So M has the equivariant homotopy type of the union of X^K for such maximal subgroups K, and we may assume that X has been gotten from M by attaching orbits of cells of type G/L, where $L \in \mathcal{F}$ is not maximal. Let ξ be the bundle over X with total space $E(\xi) = V \times X$, where G acts diagonally on $V \times X$. Then $\xi | M$ is seen to be $T(M) \oplus \nu$. Furthermore, it is easily seen that the conditions of (1.5) are satisfied. So we obtain a G-manifold W with isotropy subgroups in $\mathcal{F}$, $\dim W = \dim V$, and W is contractible. The proof of Theorem (1.5) in [1] implies that we can arrange to have W simply-connected at infinity, applying van Kampen's theorem and the fact that $\dim W$ is sufficiently high. By the Browder-Levine-Livesay Theorem [3] we can find a (simply-connected) boundary for W, and W will be diffeomorphic to the interior of a disk; i.e. a euclidean space. So we have:

(2.3) Theorem. Let G, $\mathcal{F}$, and V be as above. Then there is a smooth action of G on a euclidean space $\mathbb{R}^n$, $n = \dim V$, and such that the isotropy subgroups of the action are in the family $\mathcal{F}$. In particular this action has no stationary points. ■

(2.4) Corollary. There are infinitely many non-equivalent actions of the above kind on $\mathbb{R}^n$, distinguished by their equivariant homotopy types. ■

The proof of the corollary is based on the observation that there are infinitely many G - CW complex X as in (1.2) which are distinguished by their equivariant homotopy types and that W is equivariantly homotopy equivalent to X.

§ 3. In many cases, the knowledge of a particular group yields better result. For instance, take the case of the cyclic group of order pq, $(p, q) = 1$. It is easy to construct $\mathbb{Z}_p$-acyclic and $\mathbb{Z}_q$-acyclic 4-manifolds with free $\mathbb{Z}_q$-$\mathbb{Z}_p$-actions respectively; e.g., take S^1 with the usual free $\mathbb{Z}_p$-action, and choose p points which lie on one orbit of actions on S^1, say $a_1, a_2, \ldots, a_p$. Then attach one copy of S^1 at each point a_i to this circle, and redefine the $\mathbb{Z}_p$-action to be the same as before on the old copy of S^1 and permute these new copies. Now choose a new orbit of points $b_1, \ldots, b_p$ and do the same. In this fashion we obtain a 1-complex X_1, with $H_1(X_1)$ being the direct sum of infinite copies of $\mathbb{Z}$. So $H_1(X_1;\mathbb{Z}_q)$ is free over the semisimple ring $\mathbb{Z}_q[\mathbb{Z}_p]$, as it is seen in Theorem II.1.4. of [1]. By attaching 2-cells appropriately, we obtain a 2-complex X_2 which has free $\mathbb{Z}_p$-action and is $\mathbb{Z}_q$-acyclic. To obtain the above mentioned 4-manifold, observe that $X_1/\mathbb{Z}_p$ has the homotopy type of a smooth 4-manifold M_1 with boundary, by taking copies $S^1 \times D^3$ instead of S^1, with $\mathbb{Z}_p$-action the same as before on S^1 and trivial on D^3 and mimicking the above construction. By van Kampen's theorem, $\pi_1(\partial M_1) \cong \pi_1(M_1)$. So we can represent the attaching maps of the above-mentioned 2-cells by embedded circles in ∂M_1, which are also seen to have trivial normal bundles. So we can attach 2-handles to M_1 along ∂M_1, the same way that we attached 2-cells to $X_1/\mathbb{Z}_p$ to obtain M_2 which is homotopy equivalent to $X_2/\mathbb{Z}_p$. Taking $\widetilde{M}_2$ to be the covering space of M_2 corresponding to X_2 over $X_2/\mathbb{Z}_p$, we obtain the desired $\mathbb{Z}_q$-acyclic 4-manifold with free $\mathbb{Z}_p$-action, as the group of covering translations. Moreover, if $\rho_t : \mathbb{Z}_{pq} \to SO(2)$ denotes the representation $\rho_t(e^{\frac{2\pi i}{pq}}) =$

$$\begin{pmatrix} \operatorname{Cos}\frac{2\pi it}{pq} & \sin\frac{2\pi it}{pq} \\ -\operatorname{Sin}\frac{2\pi it}{pq} & \operatorname{Cos}\frac{2\pi it}{pq} \end{pmatrix}$$

then it is easy to verify that we can take $\widetilde{M}_2$ to have a tangent bundle equivalent to $V(2\rho_q) \times \widetilde{M}_2$. We can do a similar construction to obtain a $\mathbb{Z}_p$-acyclic 4-manifold $\widetilde{M}_2'$ with free $\mathbb{Z}_q$-action and a similar equivariant tangent bundle. As in the proof of (1.2), (cf. [1]), we attach free orbits of $\mathbb{Z}_{pq}$-cells of dimension ≤ 3 to obtain the 3-dimensional contractible $\mathbb{Z}_{pq}$-complex X with no stationary points. Take ξ over X to have $E(\xi) = X \times V(2\rho_p \oplus 2\rho_q)$ and apply (1.5) to this situation. Arguing as in the above theorem, we have

(3.1) Corollary. There is a smooth action of $\mathbb{Z}_{pq}$ on $\mathbb{R}^8$ with no stationary points. ■

The above result recovers the theorem of Kister [9]. A similar result holds for many other groups, such as the dihedral groups, abelian groups and many other nilpotent groups.

Using the theorem of P. A. Smith (cf. [13]), it follows that if such an action existed on $\mathbb{R}^7$, then $(\mathbb{R}^7)^{\mathbb{Z}_p}$ and $(\mathbb{R}^7)^{\mathbb{Z}_p}$ must be 3-dimensional manifolds satisfying a condition similar to the above conditions for 4-dimensional manifolds. So if we could construct three dimensional manifolds with properties similar to $\widetilde{M}_2$, then a modification of the above argument would go through, and we could construct a $\mathbb{Z}_{pq}$-action on $\mathbb{R}^7$ with no stationary point. Thus, this problem is reduced to a question about three manifolds: Given X_2 as above, does $X_2/\mathbb{Z}_p$ have the homotopy-type of a smooth three manifold (with trivial tangent bundle)?

§ 4. Consider the problem (0.2) above. A special case of (0.2) is when we consider W to be a disk. The following theorem ([1], Theorem II.6.2.) gives a complete solution to the part (a) of the proposed program to solve (0.1).

(4.1) Theorem. Suppose the family $\mathcal{F} - \{G\}$ has unique minimals for $\mathcal{P}(G)$, and $\{1\} \in \mathcal{F}$. Then: (i) F can be realized as the stationary point set of an action of G on a contractible finite complex X if and only if $\chi(F) \equiv 1 \bmod N(\mathcal{F})$, for some well-defined integer $N(\mathcal{F})$ depending on the family $\mathcal{F}$ only. (ii) X in (i) can be constructed to satisfy $\dim X \leq \max\{\dim F + \ell(\mathcal{F}), \ell(\mathcal{F}) + 2\}$. The order of G is assumed to be not a prime power. ■

This can be considered as a generalization of a theorem of R. Oliver (cf. [11]). When $\mathcal{F}$ is taken to be the family of all subgroups, Oliver has calculated completely the integers N (all subgroups), and he has denoted it by n_G. $N(\mathcal{F})$ is calculable from the information about the family $\mathcal{F}$, and the value of n_H for $H \in \mathcal{F}$, maximal subgroups.

The following application to G-actions on spheres is rather interesting:

(4.2) Theorem. Assume that G is a finite group not of prime power order, and let F be a smooth manifold bounding a parallelizable manifold of odd Euler characteristic, and let $\mathcal{F}$ be the family of isotropy subgroups of a linear representation such that $\mathcal{F} - \{G\}$ has unique minimals for $\mathcal{P}(G)$. Then there exists a smooth G-action on some sphere S^n with $(S^n)^G = F$, and the family of isotropy subgroups of the action being $\mathcal{F}$.

Proof: Let $F = \partial W$, with W parallelizable and $\chi(W) =$ odd. By doing trivial surgeries below the middle dimension appropriately, we may assume that $F = \partial W'$, W' parallelizable, $\chi(W') \equiv 1 \bmod N(\mathcal{F})$, (or $\chi(W') = 1$ for instance). Then (4.1) shows that we can construct a contractible finite $\mathcal{F}$ -complex X with $X^G = W'$. Take a (real) linear representation space V of G, and let ξ be the bundle over X with $E(\xi) = V \times X$. Since W' is parallelizable, we can easily choose V so that ξ satisfies the requirements of (1.5). After equivariant thickening, we obtain a contractible compact manifold $\cup^{n+1}$ with $\cup^G = W'$. By Smale's theorem, $\partial\cup^{n+1}$ is the sphere S^n. Evidently, $(S^n)^G = (\partial\cup^{n+1})^G = \partial W' = F$. ∎

§ 5. It is well-known that if a compact Lie group G acts smoothly on a closed manifold M, then M^G is a smooth closed manifold. We show, however, that if the action is only piecewise linear, then this property fails in a strong sense. By a piecewise linear action, we mean an action which is simplicial in some appropriate triangulation of the underlying manifold. Note that if p is a prime and G is a finite p-group, then by P. A. Smith theory, the stationary point set must be a $\mathbb{Z}_p$-homology manifold. Let $\Phi = \{G \mid G$ is a compact Lie group which has an action on a contractible finite complex with no stationary points.$\}$ Let K be any finite simplicial complex, and W be a smooth regular neighborhood of it. Then we can construct a smooth action on a disk D^n with $(D^n)^G = W$. Let D'^m be another disk on which G acts smoothly with no stationary points. Then $D'^m \cup c(\partial D'^m)$ is PL-homeomorphic to a sphere S^m with $(S^m)^G =$ one point. Here $c(\partial D'^m)$ denotes the cone on $\partial D'^m$ with the obvious G-action coming from an action on $\partial D'^m$. Let $\Sigma = D^n \times S^m \cup \partial D^n \times c(S^m)$. So Σ is also PL homeomorphic to a sphere and it is easily seen that $\Sigma^G = W \times (\text{point}) \cup \partial W \times [0,1] \cong W$. Since the smooth actions on D^n and D'^m are also piecewise linear, in some triangulation of Σ the action is simplicial. Since W is a regular neighborhood of K, it is simple homotopy equivalent to K. Therefore, we can collapse the simplices of W to K. From the theory of regular neighborhoods it follows that after collapsing $W \subset \Sigma$ to K, the simplicial complex Σ' obtained from Σ in this fashion is still PL homeomorphic to the standard sphere S^{n+m}. But $(\Sigma')^G = K$. So we have:

(5.1) Theorem. Let $G \in \Phi$, and let K be any finite simplicial complex. Then there is a piecewise linear G-action of G on a sphere S with $S^G = K$. ∎

It is clear from our previous discussion how to calculate the dimension of S, in terms of the family of isotropy subgroups, etc. The class Φ is determined by Oliver (cf. [12]).

For topological actions, we have the following version. The order of G is assumed to be not a prime power.

(5.2) Corollary. Let $\mathcal{F}$ be the family of proper isotropy subgroups of a linear representation of the finite group G, having unique minimals for $\mathcal{P}(G)$. Let K be any finite simplicial complex with $\chi(K) - 1$ a multiple of $N(\mathcal{F})$. Then there is a topological G-action on a sphere Σ with $\Sigma^G = K$, and the family of isotropy subgroups $\mathcal{F} \cup \{G\}$. ∎

In the above corollary, we first obtain a smooth action of G on some $\mathbb{R}^k$ with $\mathcal{F}$ as the family of isotropy subgroups. The one-point compactification of $\mathbb{R}^k$ is the sphere S^k with $(S^k)^G$ = one point. In the mean time, there is a smooth G-action on a disk D^m with $(D^m)^G = W$, where W is a regular neighborhood of K, since $\chi(K) - 1$ is a multiple of $N(\mathcal{F})$ - using (4.1) and (1.5). The remainder of the argument is similar to the preceding one.

§ 6. A. Edmonds and R. Lee in [6] constructed smooth actions of $\mathbb{Z}_{pq}$ on some euclidean space with non-equivalent representations of $\mathbb{Z}_{pq}$ at different stationary points. Similarly, we can construct smooth actions on a sphere S^n with non-equivalent representations at different connected components of the stationary point set. For instance, let F be the disjoint union of a 2-disk D^2, a 2-sphere S^2, and a 2-torus minus a disk $T^2 - D^2$. Then F is a stably parallelizable 2-manifold with ∂F the disjoint union of two circles, and $\chi(F) = 1$. Taking G to be a finite group not of prime power order, we may construct an action of G on a contractible finite complex X with stationary point set F. For example, if $G = \mathbb{Z}_{pq}$ or D_{pq}, the dihedral group of order $2pq$, $\dim X = 4$ is possible. Let ξ be a G-bundle on X such that the linear representations of G on the fibres of $\xi | D^2$ and $\xi | T^2 - D^2$ are non-equivalent. For instance, for $G = \mathbb{Z}_{pq}$, one can take Edmonds-Lee example, or if $\mathbb{Z}_{pq} \subset G$, (e.g. $G = D_{pq}$) one might take appropriate representations of G induced by the above mentioned non-equivalent representations of $\mathbb{Z}_{pq} \subset G$. Then we can use the bundle ξ to thicken up equivariantly X to obtain a smooth G-disk D^{n+1} with $(D^{n+1})^G = F$. It follows that the linear representations of G on various connected components of F are non-equivalent. In particular, this G-action on $S^n = \partial D^{n+1}$ has non-equivalent linear representations of G on the two different copies of S^1 which are left stationary by the G action on S^n.

Remarks. 1. We should mention that Ted Petrie has constructed examples of smooth action of some odd order abelian groups on homotopy spheres with non-equivalent representations at different stationary points. These groups seem to be rather restricted; e.g., they are assumed to have at least three non-cyclic Sylow subgroups.(**)

2. Equivariant thickening for the case of G finite abelian has been done by Edmonds and Lee (cf. [6]). I would like to thank Frank Quinn for bringing their paper to my attention; this resulted in a simplification of our earlier proof of the equivariant thickening theorem.

Added in proof: (*) This is true because G is assumed to be finite and the isotropy subgroups of a linear representation of a finite group forms a family which is closed under intersection. Cf. [1].

(**) I have been unable to find a reference containing the proof of this result. According to Petrie this follows from his results on G-surgery.

Amir H. Assadi
School of Mathematics
The Institute for Advanced Study
Princeton, NJ 08540

References

[1] Assadi, A. H.: "Finite Group Actions on Simply-connected CW Complexes and Manifolds", Thesis, Princeton University.

[2] Bredon, G.: "Introduction to Compact Transformation Groups", Academic Press, 1972.

[3] Browder, W.-Levine, J.-Livesay, G.: "Finding a Boundary for an Open Manifold."

[4] Conner, P. E. and Floyd, E. E.: "On the Construction of Periodic Maps Without Fixed Points." Proc. Amer. Soc. 10 (1959), 354-360.

[5] Conner, P. E. and Montgomery, D.: "An Example for SO(3)." Proc. Nat. Acad. Sci. USA 48 (1962), 1918-1922.

[6] Edmonds, A. and Lee, R.: "Fixed Point Sets of Group Actions on Euclidean Space", Topology 14 (1975), 339-345.

[7] Hsiang, W. C. and Hsiang, W. Y.: "Differentiable Actions of Compact Connected Classical Groups, I." Am. J. Math. 89 (1967), 705-786.

[8] Kister, J. M.: "Examples of Periodic Maps on Euclidean Spaces Without Fixed Points", Bull. AMS, 67 (1961), 471-474.

[9] Kister, J. M.: "Differentiable Periodic Actions on E^8 Without Fixed Points", Amer. J. Math. 85 (1963), 316-319.

[10] Mazur, B.: "Differential Topology From the Point of View of Simple Homotopy Theory". Pub. I.H.E.S.

[11] Oliver, R.: "Fixed-Point Sets of Group Actions on Finite Acyclic Complexes", Comment. Math. Helv. 50 (1975), 155-177.

[12] Oliver, R.: "Smooth Compact Lie Group Actions on Disks", Math. Z. 149 (1976), 79-96.

[13] Smith, P. A.: "New Results and Old Problems in Finite Transformation Groups." Bull. AMS 66 (1960), 401-415.

The Homotopy Structure of Finite Group Actions on Spheres

T. tom Dieck
Mathematical Institute
University of Göttingen
Fed. Republic of Germany

T. Petrie
Rutgers University
U.S.A.

1. Introduction

Our aim is to develop a theory of actions of finite groups on homotopy spheres in analogy with complex representation theory (linear theory). Much of the study of transformation groups concentrates on emphasizing the diversity of actions on spheres. In contrast we seek to organize the common features. The mechanism for this is the homotopy representation group V(G) (section 2) associated to the finite group G. This is the Grothendieck group of equivalence classes of actions of G on spheres with addition defined by join.

In order to describe V(G) we require "characters" i. e. apriori defined invariants, "orthogonality relations" i. e. relations between characters and "irreducible" objects i. e. additive generators of V(G). The dimension function (2.4 - 2.6) and the degree function are examples of characters. Remark 2.6 deals with orthogonality relations for the dimension function. The exact sequence 2.26 expresses an orthogonality relation for the degree function. This is algebraic.

On the geometric side we seek to identify the linear theory in V(G).

(Associate to a complex representation W the class of its unit sphere S(W) in V(G).) We also seek to identify the classical exotic examples of actions on spheres in V(G). For example in section 3 we identify the exotic free actions of the metacyclic group $Z_{p,q}$ on homotopy spheres in $V(Z_{p,q})$ and show that they together with the linear examples generate.

The flexibility of our viewpoint allows the study of actions on spheres with additional specified properties provided they are compatible with join. Here we emphasize two extremes: (i) Require the fixed set of each subgroup to be a homotopy sphere (section 2) and (ii) No additional requirement i. e. arbitrary actions on spheres(§4). Of course each requirement leads to a different group but the treatment is conceptually the same. The group associated to the requirement that G should act freely is important and interesting but only implicitly treated here.

Among the many invariants of actions on spheres, equivariant homotopy type is the one which plays a distinguished role. The invariant is interesting in its own right, it is compatible with join and carries much of the algebra associated to the study of smooth actions.

This invariant is built into the definition of V(G) and appears explicitly in the treatment of arbitrary actions in section 4. In particular section 4 treats the general problem of realizing equivariant homotopy types of actions on spheres. The main result of this section (4.8) generalizes in a geometric way work of Swan on free actions on homotopy spheres [7] .

It is worth noting here these items which appear in the development of our theory: representation theory, mapping degrees of equivariant maps (2.13), invertible modules over the Burnside ring (2.11), surgery in the homotopy category (section 4), the projective class group (2.15 and section 4), actions on Brieskorn varieties (2.28) and Smith theory (2.31 and 4.8). The latter singles out the class $\mathcal{T}$ of groups of prime power order for a special role in the theory.

The style of this paper is informal. The main structure theorems are interspersed with many examples which illustrate them.

Acknowledgement: Rothenberg has introduced a group S(G) of semi-linear actions on homotopy spheres analogous to our Z(G) of Section 1. The two points of view should be compared. See Rothenberg Torsion invariant and finite transformation groups, Proc. of Symposia in Pure and Applied Math 32 (1978).

2. Semi-linear spheres.

The closest approximation to the complex representation ring, which is almost entirely expressed in terms of homotopy theory, is the group $V(G)$ of semi-linear spheres, to be defined below. We repeat that G is always a finite group.

Definition 2.1. A semi-linear sphere is a finite G-CW-space X such that for each subgroup H of G the fixed point set X^H is an n(H)-dimensional space which is homotopy-equivalent to the sphere $S^{n(H)}$. Moreover we assume that n(H) is odd and that the NH (= normalizer) action on $H_{n(H)}(X^H;Z)$ is trivial. We orient X by choosing a generator for each group $H_{n(H)}(X^H;Z)$.

Remark 2.2. The condition on the dimension of the spaces involved excludes examples of the following type. Let $G = Z/pZ$ be the cyclic group of prime order p. Then there exists a finite G-CW-complex X such that X and X^G are homotopy-equivalent to the same sphere S^n and the degree of the inclusion $X^G \longrightarrow X$ is different from one. This phenomenon can not occur for a closed manifold X, but X could be $S^n \times D^k$. The other assumptions in 2.1 are made in order to imitate complex linear spheres.

The join $X_1 * X_2$ of semi-linear spheres X_1 and X_2 is again a semi-linear sphere. If X_1 and X_2 are oriented, there is a canonical way to orient $X_1 * X_2$ such that forming the "oriented join" is associative.

Definition 2.3. Let X_1 and X_2 be oriented semi-linear spheres. They are called oriented homotopy-equivalent, in symbols

$$X_1 \sim X_2 ,$$

if there exists a G-map $f : X_1 \longrightarrow X_2$ such that the degree of f^H with respect to the given orientations is one for all subgroups H of G .

Let $V^+(G)$ be the semi-group of oriented G-homotopy types of semi-linear spheres with join as composition law, and let V(G) be the associated Grothendieck group. We write the group law in V(G) additively.

Let $\phi(G)$ be the set of conjugacy classes of subgroups of G and let

$$C := C(G) := C(\phi(G), Z)$$

be the ring of all functions $\phi(G) \longrightarrow Z$.

Definition 2.4. Let X be a semi-linear sphere. Its dimension function Dim(X) is given by

$$\mathrm{Dim}(X)(H) = \frac{1}{2}(\dim X^H + 1) .$$

The assignment $X \longmapsto \mathrm{Dim}(X)$ induces a homomorphism $\mathrm{Dim}: V(G) \longrightarrow C(G)$. The kernel will be denoted v(G). Hence by definition we have an exact sequence

(2.5) $$0 \longrightarrow v(G) \longrightarrow V(G) \xrightarrow[\mathrm{Dim}]{} C(G)$$

We are therefore facing the problems: Determine the image of Dim. Compute v(G).

Remark 2.6. The function Dim should be interpreted as a "character" function for V(G). The fact that Dim is in general not onto is expressed by "orthogonality relations" for this "character". We need more "characters" (= a priori invariants for spheres) to detect elements in v(G).

We should point out that working with V(G) we only consider spheres from a "stable" point of view. In this connection the following result is useful.

Proposition 2.7. $X_1 \sim X_2 \iff X_1 * X_3 \sim X_2 * X_3$.

Example 2.8. (Linear Spheres) If V is a complex representation then SV is a semi-linear G-sphere. Using the complex structure of V we define a canonical orientation (in the sense of 2.1) for SV. Let $R_h(G)$ be the subgroup of the complex representation ring R(G) consisting of elements V-W such that SV $\sim$ SW. Put $J(G) = R(G)/R_h(G)$. Then we have an exact sequence

$$(2.9) \qquad 0 \longrightarrow j(G) \longrightarrow J(G) \xrightarrow[\text{Dim}]{} C(G)$$

The group j(G) was computed in [2] for $G \in \mathcal{P}$. See [4] for the general case. We have injective homomorphisms

$$(2.10) \qquad j(G) \longrightarrow v(G) \; , \; J(G) \longrightarrow V(G) .$$

One of our aims is to show that these maps are, in general, not surjective, i. e. we show that non-linear homotopy types occur. Again this problem splits into two: Non-linear dimension functions and non-linear degree functions (see below for the latter).

Let Inv(G) be the group of invertible modules over the Burnside ring A(G), introduced in [4]. This is essentially the Picard group of A(G), but taking into account orientations. Inv(G) is a finite group of order

$$\left| \prod_{(H)} (Z/|WH| Z)^* \right| ,$$

where $WH = NH/H$ for $H < G$, and the product is taken over $(H) \in \phi(G)$. (R^* denotes units of the ring R, and $|G|$ denotes the order of G.)

Proposition 2.11. Let $X_1 - X_2 \in v(G)$. Then the group $\omega(X_1,X_2)$ of G-equivariant stable homotopy classes from X_1 to X_2 is an invertible module over $A(G)$. The assignment $X_1 - X_2 \longmapsto \omega(X_1,X_2)$ induces an injective homomorphism

$$(2.12) \qquad u : v(G) \longrightarrow Inv(G).$$

Proof. Analogous to [4], Theorem 2.

Definition 2.13. Let $X_1 - X_2 \in v(G)$. Let $f : X_1 \longrightarrow X_2$ be a G-map. The degree function $Deg(f)$ of f is the map

$$Deg(f) : \phi(G) \longrightarrow Z : (H) \longmapsto \text{degree } f^H .$$

A function $z : \phi(G) \longrightarrow Z$ is called invertible if $z(H)$ is prime to $|G|$ for all $H < G$.

Proposition 2.11 and the results of [4] give the following

Corollaries 2.14. (i) $v(G)$ is a finite group.
(ii) Given $X_1 - X_2 \in v(G)$ there exists a G-map $f : X_1 \longrightarrow X_2$ with invertible degree function.
(iii) The module $\omega(X_1,X_2) \in Inv(G)$ is determined by any invertible degree function $Deg(f)$, $[f] \in \omega(X_1,X_2)$.

Our next aim is to determine the image of u (see 2.12). This uses a certain obstruction map from Inv(G) into projective class groups expressing conditions for an invertible degree function to come from a map between spheres.

Let ZG be the integral group ring of G and let $\tilde{K}_o(ZG)$ be the projective class group of ZG, i. e. the Grothendieck group of finitely generated projective ZG-modules modulo free modules. (Equivalently one can use modules of finite homological dimension.) In the theory of transformation groups the so called Swan homomorphism (see e. g. [6][7])

(2.15) $$s_G : K_1(Z/|G|Z) \cong (Z/|G|Z)^* \longrightarrow \tilde{K}_o(ZG)$$

is important in various contexts. Our obstruction invariant is a generalization of this homomorphism, namely if we put

(2.16) $$\tilde{\mathcal{K}}_o(G) := \prod_{(H)\in\phi(G)} \tilde{K}_o(ZWH)$$

then our invariant will be a certain homomorphism

(2.17) $$\sigma : \mathrm{Inv}(G) \longrightarrow \tilde{\mathcal{K}}_o(G)$$

which in special cases (e. g. for G abelian) reduces essentially to the product of the Swan homomorphisms s_{WH}, $(H)\in\phi(G)$.

Let now $f : X \longrightarrow Y$ be a G-map with invertible degree function, $X - Y \in v(G)$. Let $Z = C(f)$ be the mapping cone of f and put $Z_s = \bigcup_{H\neq 1} Z^H$. Then

(2.18) G acts freely on Z/Z_s away from the base point.

(2.19) $$\tilde{H}_*(Z/Z_s\ ;\ Z/|G|Z) = 0\ .$$

If 2.19) holds for a space B instead of Z/Z_s then $\tilde{H}_i(B;Z)$ is a ZG-module of homological dimension at most one. Hence we can define

$$\chi(B) := \sum (-1)^i \tilde{H}_i(B;Z) \in \tilde{K}_o(ZG) . \tag{2.20}$$

Then 2.18 yields $\chi(Z/Z_s) = 0$, hence

$$\chi(Z) - \chi(Z_s) = \chi(Z/Z_s) = 0. \tag{2.21}$$

Moreover $\chi(Z) = s_G(\text{degree } f)$.

Now it turns out that $\chi(Z_s)$ can be computed from the $s_{NH}(\deg f^H)$, $H \neq 1$; i. e. there exists a relation of the type

$$\chi(Z_s) = \sum_{H \neq 1} a_H \operatorname{Ind}_{NH}^G s_{NH}(\deg f^H) \tag{2.22}$$

where $\operatorname{Ind}_{NH}^G : \tilde{K}_o(ZNH) \longrightarrow \tilde{K}_o(ZG)$ is the induction homomorphism and where the a_H are integers depending on H (and G) but not on f. In view of 2.21 and 2.22 we therefore put for an invertible function $z: \phi(G) \to Z$

$$\sigma(z)(1) = s_G(z(1)) - \sum_{H \neq 1} a_H \operatorname{Ind}_{NH}^G s_{NH}(z(H)) \in \tilde{K}_o(ZG) \tag{2.23}$$

and more generally

$$\sigma(z)(H) := \sigma(z^H)(1) \in \tilde{K}_o(ZWH) \tag{2.24}$$

where $z^H : \phi(WH) \longrightarrow Z$ is the induced function $z^H(K/H) = z(K)$, $NH > K > H$.

<u>Proposition 2.25.</u> <u>The</u> <u>assignment</u> $z \longmapsto \sigma(z) \in \tilde{\mathfrak{K}}_o(G)$ <u>induces a homomorphism</u>

$$\sigma : \operatorname{Inv}(G) \longrightarrow \tilde{\mathfrak{K}}_o(G)$$

with image of u *contained in its kernel*.

We should point out that the conditions 2.23, namely $\sigma(\text{Deg } f)(1) = 0$, for a degree function Deg f for a map f between spheres, is analogous to the congruences for degree functions obtained in [4]. In [4] we used equivariant K-theory where as here we use algebraic K-theory.

The proof follows from 2.21 - 2.24 using the presentation of Inv(G) given in [4], (32). Actually the sequence

$$(2.26) \qquad 0 \longrightarrow v(G) \xrightarrow[u]{} \text{Inv}(G) \xrightarrow[\sigma]{} \tilde{\mathcal{R}}_0(G)$$

should be exact. We have verified this in some cases and the general result is within reach. In order to explain the next result we need one more tool.

If X is a G-CW-space then the stable equivariant homotopy group $\omega(X,S)$ of mappings of X into a semi-linear sphere S is a module over the Burnside ring. If all the X^H are oriented, i. e. have a preferred generator of the top-dimensional homology group, then it might happen that elements in $\omega(X,S)$ are determined by the degrees of the fixed point mappings $X^H \longrightarrow S^H$, $H < G$, and that $\omega(X,S)$ defines an invertible module. Such modules are called *geometric modules*.

Theorem 2.27. If Inv(G) *is generated by geometric modules then* 2.26 *is exact. This holds in particular for abelian groups* G *and groups of prime power order*.

The proof of 2.27 uses the fact that the obstruction map σ also measures the obstruction for converting a suitable approximation of a sphere into a sphere (via attaching cells). It is a surprising fact that the latter obstruction is detected by a degree function.

It turns out that in this context the Brieskorn varieties are of fundamental importance. They can be used to construct geometric modules. Let $V^d(A)$ be the Brieskorn variety consisting of points $(z_o,\dots,z_n) \in C^{n+1}$ such that

$$z_o^d + z_1^2 + \dots + z_n^2 = 0$$

$$\sum_{i=o}^{n} |z_i|^2 = 1$$

and with G-action induced by the representation $A\ ;\ G \longrightarrow O(n) \subset U(n)$ acting on the last n coordinates $z_1,\dots,z_n$. The map

$$\varphi : V^d(A) \longrightarrow S(A)$$

$$(z_o,\dots,z_n) \longmapsto \left(\sum_{i=1}^{n} |z_i|^2\right)^{-1/2} (z_1,\dots,z_n)$$

is equivariant and has degree d. It induces an analogous map between H-fixed point sets. The module $\omega(V^d(A),\ S(A))$ is geometric if $(d,|G|) = 1$.

<u>Theorem 2.28.</u> <u>Let</u> G <u>be</u> <u>abelian</u> <u>or</u> <u>of</u> <u>prime</u> <u>power</u> <u>order</u>. <u>Then</u> Inv(G) <u>is</u> <u>generated</u> <u>by</u> <u>modules</u> $\omega(V^d(A),S(A))$.

We now turn our attention to dimension functions.

<u>Remark 2.29.</u> Let V be a complex representation of G. Denote its character with the same letter. Then the orthogonality relation

$$|G| \dim V^G = \sum_{g\in G} V(g)$$

can be rewritten as

$$(2.30) \qquad |G| \dim V^G = \Sigma_C (\Sigma_{D<C} \mu(|C/D|) \; |D| \dim V^D)$$

where C runs through the cyclic subgroups of G and μ is the Möbius function. The equality 2.30 shows in particular that dim V^G is determined by the dim V^C, C< G cyclic. Moreover it shows that the dim V^C cannot be arbitrary integers because the left hand side of 2.30 is zero mod $|G|$.

The equality 2.30 is still valid for the dimension function of a semi-linear sphere provided G has prime power order. In fact Borel [1] has shown that for G elementary abelian 2.30 is true for an arbitrary (finitistic) Z/pZ -homology sphere. This can be used to show

<u>Theorem 2.31. Let</u> X <u>be a semi-linear sphere. Let</u> G <u>be of prime power order. Then</u> Dim(X) = Dim(SV) <u>for a suitable complex representation</u> V.

Theorems 2.27 and 2.31 together yield a computation of V(G) in case G has prime power order. The general case is difficult. In particular if G is not of prime power order (and not cyclic) then there always exist non-linear dimension functions. We shall describe a construction of such functions in section 4 (but not for semi-linear spheres).

3. Examples.

We illustrate the preceding results by describing in detail two examples, the abelian group G = Z/pZ x Z/pZ and the meta-cyclic group G = $Z_{p,q}$. But to begin with we briefly mention cyclic groups.

3a. G = Z/nZ.

In this case the map 2.10

$$J(G) \longrightarrow V(G)$$

is an isomorphism. This is due to the fact the map $j(G) \longrightarrow \mathrm{Inv}(G)$ is an isomorphism and that $\mathrm{Dim} : J(G) \longrightarrow C(G)$ is surjective. But the reader should be warned: This does not mean that the image of $V^+(G)$ in $V(G)$ is generated by linear spheres; or in other words: a semi-linear sphere may have stably the homotopy type of a linear sphere without being of linear homotopy type itself.

3b. $G = Z/pZ \times Z/pZ$, p an odd prime.

There are p+1 subgroups of order p, denoted $H_o,\dots,H_p$. Therefore $\phi(G) = \{G,H_o,\dots,H_p,1\}$. The Burnside ring $A(G) \subset C(G)$ consists of all functions z such that $z(G) \equiv z(H_i) \bmod p$, $z(1) \equiv (1-p)\sum_{i=o}^{p} z(H_i) \bmod p^2$. The group Inv(G) has in general the following presentation ([4] , 32)

$$0 \longrightarrow (A(G)/nC)^* \longrightarrow C/nC^* \longrightarrow \mathrm{Inv}(G) \longrightarrow 0$$

where $n = |G|$, $C = C(G)$. Let $d(H) : C(G) \longrightarrow Z$ be the evaluation at (H). Introduce new coordinates u for $(C/nC)^*$ by

(3.1) $$u(G) = d(G)$$
$$u(H_i) = d(H_i)d(G)^{-1}$$
$$u(1) = d(1)\cdot \prod_{i=o}^{p} d(H_i)^{-1}\cdot d(G)^p.$$

Then

$$u := (u(H)) : (C/nC)^* \longrightarrow \prod_{(H)} (Z/|G/H|Z)^*$$

has kernel $(A(G)/nC)^*$ (see [3] , Theorem 5) so that we obtain a canonical isomorphism

(3.2) $$\tilde{u} : \mathrm{Inv}(G) \cong \prod_{(H)} (Z/|G/H|Z)^* .$$

Using 3.2 the invariant σ of 2.25 is converted into the product of the Swan homomorphisms $s_{G/H}$. It is known that s_H is zero for H cyclic and that s_G has kernel precisely the (p-1)-torsion. Theorem 2.27 tells us that

$$v(G) \longrightarrow \mathrm{Inv}(G) \cong (Z/p^2Z)^* \times \prod_{i=o}^{p} (Z/pZ)^*$$

has as image the subgroup of elements whose first component is a p-th power. The image of the linear spheres $j(G) \longrightarrow \mathrm{Inv}(G)$ is the subgroup $\{1\} \times \prod_{i=o}^{p} (Z/pZ)^*$. So there exist exotic homotopy types for this group G.

The geometric meaning of this additional factor in v(G) can be explained as follows. Let X be a semi-linear sphere with $X^G = \emptyset$, for simplicity. Then one can find representations A_i of G with kernel H_i such that there exists a G-map

$$f : X \longrightarrow S(A_o \oplus A_1 \oplus \ldots \oplus A_p)$$

with the property: $\deg f^{H_i} = 1$ for $0 \leq i \leq p$. The fact that X is exotic is then reflected in the fact that the degree of f is not 1 mod p^2. This degree has to be a p-th power mod p^2 hence can be detected mod p^2 by reduction mod p. A non-standard linking property of the fixed point sets X^{H_i} is responsible for this phenomenon. In order to explain this, note that in the equivariant cohomology theory $H^*(EGx_G -; Z/pZ)$ the space X has an Euler-class $e(X) \in H^nBG$, $n = \dim X + 1$. We must have

$$e(X) = d\, e(X^{H_0})\, e(X^{H_1}) \cdot \ldots \cdot e(X^{H_p})$$

with $d \equiv$ degree f mod p.

The reader should observe that the results described so far allow the characterization of the linear homotopy types purely in homotopical terms.

Remark. The considerations above can be generalized to arbitrary p-groups.

3c. The meta-cyclic group $Z_{p,q}$.

Let p and q be odd prime such that $q|p-1$. The group $Z_{p,q}$ has generators x,y with relations

$$x^p = 1,\ y^q = 1,\ y^{-1}\, x\, y = x^a$$

with a of order q in $(Z/pZ)^*$. We have an exact sequence

$$1 \longrightarrow H \longrightarrow Z_{p,q} \longrightarrow K \longrightarrow 1$$

with H generated by x and K generated by y. The group $1, H, Z_{p,q}$ are normal, and NK = K. Hence $\Phi(G) = \{1, H, K, G\}$.

The group G has q one-dimensional irreducible representations, lifted from K; and $(p-1)/q$ q-dimensional irreducible representations, induced from H. The Galois group Γ of pq-th roots of unity acts on these irreducible representations with three orbits. Representatives 1,V, and W of these orbits have the dimension functions

	1	H	K	G
Dim S(1)	1	1	1	1
Dim SV	1	1	0	0
Dim SW	q	0	1	0

and the image of Dim : J(G) $\longrightarrow$ C(G) is generated by these functions.

Now there exists a sphere X of dimension 2q-1 with free G-action. (See Theorem 4.9 for a homotopy-theoretic construction.) Hence

	1	H	K	G
Dim X	q	0	0	0

.

The functions Dim S(1), Dim SV, Dim SW, Dim X generate a subgroup of C(G) of index q. On the other hand, using equivariant cohomology, it is easy to show that for any semi-linear sphere Y

(3.3) $$\dim Y \equiv \dim Y^H \mod q.$$

Hence 3.3 describes the image of Dim : V(G) $\longrightarrow$ C(G), and we have exhibited generators.

The Burnside ring A(G) $\subset$ C(G) consists of functions z such that

$$z(G) \equiv z(H) \mod q$$
$$z(H) \equiv z(1) \mod p$$
$$z(K) \equiv z(1) \mod q .$$

From this it is seen that the map

$$(C/pqC)^* \longrightarrow Z/qZ^* \times Z/pZ^* \times Z/qZ^*$$

$$z \longmapsto (z(H)z(G)^{-1}, z(1)z(H)^{-1}, z(1)z(K)^{-1})$$

induces an isomorphism

(3.4) $\tilde{u} : \mathrm{Inv}(G) \cong Z/qZ^* \times Z/pZ^* \times Z/qZ^*$.

In generals terms: The first factor is $Z/|WH|Z^*$ the last two factors give $Z/|W1|Z^*$. Using this isomorphism the map

$$j(G) \longmapsto \mathrm{Inv}(G)$$

is given as follows:

$$SV - S(\psi^k V) \longmapsto (k,1,k)$$

$$SW - S(\psi^b W) \longmapsto (1,b^q,1)$$

(ψ^a is the Adams operation). In order to compute v(G) we need to know the map σ of 2.25., which in terms of 3.4 is the product of the Swan homomorphisms $s_{WH} = 0$ and s_G. It can be shown that the kernel of s_G consists of the q-th powers mod p. If X_1, X_2 are two spheres with free action of dimension 2q-1 then $\omega(X_1,X_2) \in \mathrm{Inv}(G)$ is via 3.4 an element of the form (1,c,c) and any q-th power mod p is realizable by a suitable c. We might think of X_2 as being a "Galois conjugate" of X_1. Hence we obtain the result: Dim V(G) is freely generated by the image of the "irreducibles" S(1), SV, SW, X. These elements together with their "Galois conjugates" generate V(G). The congruence 3.3 and the description of the kernel of s_G may be considered as "Orthogonality relations".

4. Arbitrary actions on homotopy spheres

In this section we discuss actions on spheres without any restriction on fixed sets of subgroups. The set $\mathcal{P}$ of groups of prime power order play a special role as is known from Smith Theory because Σ^H must be a mod p homology sphere whenever H acts on the sphere Σ, moreover for p-groups there are important connections between representations and actions on homotopy spheres. Here are two indications: Let $P \in \mathcal{P}$ and P act on Σ. Then

(4.1) $\mathrm{Dim}(\Sigma) = \mathrm{Dim}(SV)$ for some complex representation V of P. (See 2.31)

(4.2) If P acts freely on Σ, then Σ has the same P-homotopy type as S(V) for some complex representation V of P.

These considerations impose conditions on an action of an arbitrary group on a homotopy sphere Σ. Conversely we want to use these restrictions imposed by the groups in $\mathcal{P}$ and their representations to construct actions of general groups on homotopy spheres and describe their equivariant homotopy types.

Let $\mathcal{T}$ denote the set Sylow subgroups of G up to conjugacy. Let $P \in \mathcal{T}$ be a Sylow subgroup and V_P a complex representation of P. We require $\dim_C V_P$ to be independent of $P \in \mathcal{T}$ and set

(4.3) $$\mathcal{V} = \{ V_P \mid P \in \mathcal{T} \}$$

If Σ is a G-homotopy sphere, $\mathrm{Res}_P \Sigma$ denotes the P-homotopy sphere obtained by restricting the action to P. Suppose for each $P \in \mathcal{P}$ there is a P-map $f_P : \mathrm{Res}_P \Sigma \longrightarrow S(V_P)$. This we abbreviate by writing

$f : \Sigma \longrightarrow S(\mathcal{V})$.

(4.4) Realization problem: Given $\mathcal{V}$ when does there exist a G-homotopy sphere Σ and $f : \Sigma \longrightarrow S(\mathcal{V})$ such that each f_P is a homotopy equivalence?

If such a Σ exists we say that $\mathcal{V}$ is realized by Σ.

Consider for example the case $\mathcal{V}$ is free i. e. P acts freely on $S(V_P)$ for $P \in \mathcal{T}$. If $\mathcal{V}$ is realized, then G acts freely on Σ and this implies that G has periodic cohomology and dim $\mathcal{V}$ is a multiple of $\frac{1}{2}$ period G. Here dim $\mathcal{V}$ = dim V_P for any $P \in \mathcal{T}$.

Definition 4.5. $\mathcal{V}$ is G invariant up to homotopy if for each $P \in \mathcal{T}$ and $H \subset P \cap gPg^{-1}$ we have $\mathrm{Res}_H V_P \sim \mathrm{Res}_{g^{-1}Hg} V_P$ (see 2.3 for notation).

Remark. $\mathrm{Res}_H V_P$ and $\mathrm{Res}_{g^{-1}Hg} V_P$ are regarded as representations of H and 4.5 asserts that these representations are oriented homotopy-equivalent.

If X is a G space, $\mathrm{Iso}_G(X)$ denotes the set of isotropy groups of points of X. $\mathrm{Iso}(\mathcal{V})$ denotes the set of subgroups of G which are conjugate to a group in $\mathrm{Iso}_P(S(V))$ for some $P \in \mathcal{T}$. Let $B(\mathcal{V})$ denote the subgroup of $\tilde{K}_o(Z(G))$ generated by elements z having this property:

(4.6) $z = H_k(X)$ where X is a pointed G-complex with base point $p \in X^G$, $\mathrm{Iso}_G(X-p) \subset \mathrm{Iso}(\mathcal{V})$, $\tilde{H}_i(X) = o$ $i \neq k$ and $H_k(X)$ is a projective $Z(G)$-module.

Then $B(\mathcal{V}) \supset B(\mathcal{W})$ whenever $\mathrm{Iso}(\mathcal{V}) \supset \mathrm{Iso}(\mathcal{W})$ and

(4.7) $$B(\mathcal{V}) = 0 \text{ if } \mathcal{V} \text{ is free.}$$

The answer to 4.4 is given by

<u>Theorem 4.8.</u> $\mathcal{V}$ <u>is realized by</u> Σ <u>iff</u> $\mathcal{V}$ <u>is</u> G <u>invariant up to homotopy and an invariant</u> $\chi(\mathcal{V}) \in \tilde{K}_o(Z(G))/B(\mathcal{V})$ <u>vanishes; moreover</u> $\chi(\mathcal{V} \oplus \mathcal{W}) = \chi(\mathcal{V}) + \chi(\mathcal{W})$ <u>whenever</u> $Iso(\mathcal{V}) = Iso(\mathcal{W})$. <u>There is an integer</u> $n = n(G)$ <u>such that</u> $n\mathcal{V}$ <u>is always realized.</u>

To give a feeling for this theorem let $G = Z_{p,q}$ the meta-cyclic group introduced in section 3c. Then if $\mathcal{V}$ is free, it is invariant up to homotopy iff $\dim \mathcal{V} \equiv 0(q)$. Note $2q$ = period $Z_{p,q}$. Let $n = \dim V_p$ and $\lambda^n(V_p) \in Z/|P|Z^*$ be the integer mod $|P|$ obtained by noting the n-th exterior power of V_p is a one dimensional representation of P. This is defined by an integer mod $|P|$ which is a unit mod $|P|$ because P acts freely on V. This integer mod $|P|$ is denoted $\lambda^n(V_p)$.

<u>Theorem 4.9.</u> ([5] , [8]) <u>Let</u> $G = Z_{p,q}$, $\dim \mathcal{V} \equiv 0(q)$ and $\mathcal{V}$ <u>free</u>. <u>Then</u> $\chi(\mathcal{V}) = 0$ <u>iff</u> $\lambda^n(V_p)$ <u>is a</u> q-th <u>power</u> mod p.

At the other extreme from $\mathcal{V}$ free we have the case where $Iso(\mathcal{V})$ is as large as possible i. e. $Iso(\mathcal{V})$ is the full set of subgroups of G of prime power order. Then $B(\mathcal{V})$ is the kernel of ($\rho : \tilde{K}_o(Z(G)) \longrightarrow \tilde{K}_o(\mathfrak{M})$) where $\mathfrak{M}$ is a maximal order for $Z(G)$. This is due to Oliver. In this case $\chi(\mathcal{V})$ is zero iff $\rho\chi(\mathcal{V}) = 0$ in $\tilde{K}_o(\mathfrak{M})$.

<u>Question 4.10.</u> Is $\rho\chi(\mathcal{V}) = 0$ always? The answer is yes for $G = Z_{p,q}$. An answer in either direction would be interesting.

<u>Remark 4.11.</u> Theorem 4.8 is a generalization of the results of [7] where the case $\mathcal{V}$ is free was treated. The terminology and viewpoint of [7] is algebraic.

It is possible only to briefly mention the ideas involved with Theorem 4.8. Choose integers $\underline{a} = \{ a_P | P \in \mathcal{T} \}$ such that $\sum a_P |G|/|P| = 1$. Let $X(\underline{a},\mathcal{V}) = \coprod a_P G x_P S(V_P)$ where $S(V_P)$ is oriented by the complex structure of V_P and the sign of a_P is incorporated into the orientation of $X(\underline{a},\mathcal{V})$. The hypothesis that $\mathcal{V}$ is G invariant up to homotopy has two geometric consequences:

i) There exists a map $f : X(\underline{a},\mathcal{V}) \longrightarrow S(\mathcal{V})$ such that degree $f_P = 1$ for all $P \in \mathcal{T}$.

ii) By zero dimensional G surgery and by attaching handles of type $G/H \times D^i$ to $X(\underline{a},\mathcal{V})$, we produce a finite G complex X and a map $f : X \longrightarrow S(\mathcal{V})$ where $H_i(X) = o$ unless $i = o$, $m = 2 \dim \mathcal{V} - 1$ and $i = m-1$, $H_o(X) = H_m(X) = Z$, $H_{m-1}(X)$ is a projective $Z(G)$-module and degree $f_P = 1$ for all $P \in \mathcal{T}$. Then $\chi(\mathcal{V})$ is the class of $H_{m-1}(X)$ in $\tilde{K}_o(Z(G))/B(\mathcal{V})$.

If $\chi(\mathcal{V}) = 0$, there exists a homotopy sphere $\Sigma \supset X$ and an extension $f' : \Sigma \longrightarrow S(\mathcal{V})$ so that Σ realize $\mathcal{V}$. Moreover, $Iso_G(\Sigma) = Iso(\mathcal{V}) \subset \mathcal{P}$ and $\dim \Sigma^H = \dim S(V_P)^H$ whenever $H \subset P$.

This short discussion gives a hint of how the representation theory of the Sylow subgroups of G is used to construct actions of G on homotopy spheres. The constructions involved in Theorem 4.8 are applicable to the case of semi-linear actions. In this case, however, we encounter obstructions in $\tilde{K}_o(Z(W(H)))$ for all $H < G$ and not just $H = 1$.

There is an analog of Theorem 4.8 in the smooth category. This requires additional hypothesis on $\mathcal{V}$ and G. For example if $G = D_q$ is the dihedral group of order 2q with q odd, there are free actions of this group on a CW complex Σ but not on a smooth Σ. For a finite group of odd order, Theorem 4.8 remains true in the smooth category with slightly stronger hypothesis on V .

References

1 Borel, A.: Fixed point theorems for elementary commutative groups. In: Seminar on transformation groups. Princeton University Press, Princeton 1960.

2 tom Dieck, T.: Homotopy-equivalent group representations. J. reine angew. Math. 298, 182 - 195 (1978).

3 tom Dieck, T.: Homotopy equivalent group representations and Picard groups of the Burnside ring and the character ring. Manuscripta math. To appear.

4 tom Dieck, T., and T. Petrie: Geometric modules over the Burnside ring. Invent. math. 47, 273 - 287 (1978).

5 Petrie, T.: Representation theory, surgery and free actions of finite groups on varieties and homotopy spheres, Springer Verlag Lecture Series 168 (1970).

6 Petrie, T.: G maps and the projective class group, Comm. Math. Helv. 39 (51) 611 - 626 (1977).

7 Swan, R.: Periodic resolutions for finite groups, Ann. of Math. 72 (1960) 267 - 291.

8 Wall, C. T. C.: Periodic Projective Resolutions, Preprint.

ADDITION OF EQUIVARIANT SURGERY OBSTRUCTIONS

Karl Heinz Dovermann

§1. Notation and statement of results

Let G be a finite group. All manifolds are smooth compact oriented G manifolds and all maps are equivariant; if not otherwise stated they are also smooth. Let X be a smooth G manifold, then $\pi(X)$ is a tabulation of the fixed point sets for the subgroups H of G, the dimensions of the fixed point sets, slice representations etc. If $f: X \to Y$ is an equivariant map of G manifolds then $\lambda(f) = (\pi(X), \pi(Y), \hat{f}, \mu)$ describes the combinatorical structure of f. Here $\hat{f}: \pi(X) \to \pi(Y)$ is the map induced by f and μ is a listing of degrees. An h map (X,f,b,C), $f: X \to Y$, is a G map of smooth G manifolds together with bundle data b and C (for details see §2,4). h maps are the maps for which we can handle the

G Surgery Problem: When can we do G surgery [6] on (X,f,b,C) to obtain a new h map (X',f',b',C') such that $f': X' \to Y$ is a homotopy equivalence?

Then we say the surgery problem (X,f,b,C) is solvable and f' is an h Equivalence. The obstruction to converting an h map with combinatorical structure λ by G surgery into an h Equivalence is an element in the set $I(G,\lambda)$ [6]. The definition of $I(G,\lambda)$ is a generalization of the geometric definition of Wall groups in §9 [9]. The purpose of this paper is to introduce the notion of addition for equivariant surgery obstructions and to give some applications. $I(G,\lambda)$ is only defined in geometric terms, and we can only consider objects corresponding to the restricted objects defined by Wall. Thus we also have to defined the addition in these terms. First we give the notion of G Poset pairs describing the combinatorical structure of a G map. To carry out the addition we need a notion of transversality as we want to do surgery in the source and target space on 0 and 1 dimensional spheres. There are non vanishing obstructions to equivariant transversality and thus we define the weaker notion of quasitransversality, which is strong enough for our purpose (chapter 3). In chapter 4 we define h maps and $I^h(G,\lambda)$, the superscript h will be suppressed as we give the details only in the case of the category h [6]. Remarks about other categories can be found in chapter 7. In chapter 5 we give the construction for the addition in $I(G,\lambda)$. In chapter 6 we state and prove the main theorems. Namely under some additional assumptions we have

(i) $I(G,\lambda)$ is an abelian group (compare theorem 6.3)

(ii) Let $Y = S(V)$ be the unitsphere in a G representation V and $\dim Y^G \geq 3$. Then the addition is already defined in $N_G(Y,\lambda)$ and the obstruction map $\sigma: N_G(Y,\lambda) \to I(G,\lambda)$ is a homomorphism (compare theorem 6.6).

Elements in $N_G(Y,\lambda)$ are represented by h maps with target Y. It is worth mentioning that we only need $\dim Y^G \geq 3$, and we do not have to assume anything about X (considering $f: X \to Y$). In particular X^G may be empty.

Theorem 6.7. (Assumptions as in 6.6.) If $N_G(Y,\lambda) \neq \phi$, then there exists an h Equivalence with target Y and the combinatorical structure λ ($hS_G(Y,\lambda) \neq \phi$).

It should be pointed out that often $id: Y \to Y$ is no appropriate candidate for an element in $N_G(Y,\lambda)$.

In chapter 7 we give an application of theorem 6.7 due to Ted Petrie [2]. Here we use it to present one idea how to use the group structure of $I(G,\lambda)$ in concrete geometric problems. There $X^G = \phi$, $Y^G \neq \phi$ (see remark to (ii)) and $\pi(X) \neq \pi(Y)$ (see remark to 6.7.).

I want to thank my adviser Ted Petrie. He suggested this problem to me and gave me much valuable mathematical and moral support. This paper contains a major part of my Ph.D. thesis written at Rutgers University, N.J.

§2. G Poset pairs

For the convenience of the reader we give a brief outline of G Poset pairs. The general reference is §1 [6] where this material was introduced.

Let π be a partially ordered finite G set, in short G poset. Maximal components are not interchanged by the G action.

Examples:

2.1) $\mathcal{S}(G)$ is the set of subgroups of G, where $H \le K$ if $H \supseteq K$.

2.2) Let X be a G manifold. Then $\Pi(X) = \coprod_{H \in \mathcal{S}(G)} \pi_0(X^H)$ with the obvious map $\rho: \Pi(X) \to \mathcal{S}(G)$. If $\alpha, \beta \in \Pi(X)$ then $\alpha \le \beta$ if $|\alpha| \subseteq |\beta|$ and $\rho(\alpha) \le \rho(\beta)$. $|\,|$ denotes the underlying space.

2.3) $\pi(X) = \{\alpha \in \Pi(X) \mid \rho(\alpha) = G^{\alpha} = \bigcap_{x \in |\alpha|} G_x\}$.

There is a retraction map $r: \Pi(X) \to \pi(X)$ defined by $|\alpha| = |r\alpha|$. $\pi(X)$ determines $\Pi(X)$ if X is a smooth G manifold.

2.4 Let $H \subseteq \mathcal{S}(G)$. Then $\Pi(X,H) = \rho^{-1}(H) \subseteq \Pi(X)$ and $\pi(X,H) = r\Pi(X,H)$.

Let $P \subseteq \mathcal{S}(G)$ be the set of p-subgroups of G, then $\pi(X,P)$ will be of particular interest. X is an oriented G manifold if $|\alpha|$ is oriented for all $\alpha \in \pi(X)$. Then $|\alpha|$ is oriented as well for $\alpha \in \Pi(X)$.

Definition 2.5. For $\alpha \in \pi(X)$ we define $W(\alpha) = G_{\alpha}/G^{\alpha}$. Here G_{α} is the group leaving α fixed as an element in $\Pi(X)$. $W(\alpha)$ is a group.

Definition 2.6. $A_{\alpha} = Z_{(P)}$. P is the set of primes p such that there exists a p group H and $\alpha \in \pi(X,H)$. $A_m = Z$ if m is a maximal component.

Definition 2.7. (with a slight abuse of notation). $\pi = (\pi, d, s, w)$ is a G Poset if:

a) $d: \pi \to Z^{+}$

b) for $\alpha \in \pi$ $s(\alpha) \in R(G^{\alpha})$

c) $w(\alpha): W(\alpha) \to Z_2$.

Here R denotes the real representation ring. Furthermore we assume that G^{α} is given for $\alpha \in \pi$.

Example: If $\pi = \pi(X)$ then $d(\alpha)$ = dimension $(|\alpha|)$, $s(\alpha)$ is the slice representation of $|\alpha|$ and $w(\alpha)$ is the orientation homomorphism given by the action of $W(\alpha)$ on $|\alpha|$.

A G Poset map $\tau\colon \pi_1 \to \pi_2$ of two G Posets π_1 and π_2 is a map of the underlying G posets, which means it is equivariant and order preserving. τ is an isomorphism of G Posets if the induced map for the G posets is a homeomorphism and $G_\alpha = G_{\tau(\alpha)}$, $G^\alpha = G^{\tau(\alpha)}$, $s(\alpha) = s(\tau(\alpha))$, and $w(\alpha) = w(\tau(\alpha))$. τ is an equivalence of G Posets if τ is an isomorphism of G Posets and $d(\alpha) = d(\tau(\alpha))$.

Definition 2.8. $\lambda = (\pi_1,\pi_2,\tau,\mu,\gamma,\delta)$ is a G Poset pair if

a) π_1 and π_2 are G Posets

b) $\tau\colon \pi_1 \to \pi_2$ is a G Poset map

c) $\mu\colon \bar{\pi}_2{}^* \to Z$ where $\bar{\pi}_2{}^* = \{\beta \in \bar{\pi}_2 \mid dr(\alpha) = dr(\beta) \text{ for } \alpha \in \bar{\tau}^{-1}(\beta)\}$.

Here $\bar{\pi}_i$ denotes the completion of π_i. If $\pi_i = \pi(X)$ then $\bar{\pi}_i = \Pi(X)$. $\bar{\tau}\colon \bar{\pi}_1 \to \bar{\pi}_2$ is the map induced by τ and r is the retraction $\bar{\pi}_i \to \pi_i$.

d) for $\beta \in \pi_2$ we have $\gamma(\beta) \in R(G^\beta)$ and $\delta(\beta) \in R(G^\beta)$.

d) is needed for our addition and is not contained in the definition of G Poset pairs in [6]. Let $\lambda^i = (\pi_1{}^i,\pi_2{}^i,\tau^i,\mu^i,\gamma^i,\delta^i)$ be G Poset pairs, $i = \{1,2\}$. Then $j\colon \lambda^1 \to \lambda^2$ is an isomorphism of G Poset pairs if it consists of a pair of G Poset isomorphisms (j_1,j_2), $j_i\colon \pi_i{}^1 \to \pi_i{}^2$, such that

$$j_2\tau^1 = \tau^2 j_1$$

$$\mu^1(\alpha) = \mu^2(j_2(\alpha)) \quad \text{for all} \quad \alpha \in (\bar{\pi}_2{}^1)^*$$

j is an equivalence of G Poset pairs if j_i is an equivalence of G Posets, $i = \{1,2\}$ and for all $\beta \in \pi_2{}^1$ we have $\gamma^1(\beta) = \gamma^2(j_2(\beta))$ and $\delta^1(\beta) = \delta^2(j_2(\beta))$.

2.9 Example. For $f\colon X \to Y$ we have $\lambda(f) = (\pi(X),\pi(Y),\hat{f},\mu,\gamma,\delta)$. f induces maps $\tilde{f}\colon \Pi(X) \to \Pi(Y)$ and $\hat{f}\colon \pi(X) \to \pi(Y)$. Let $\beta \in \bar{\pi}_2{}^*$ and

$$\{\alpha_1,\ldots,\alpha_r\} = \{\alpha \in \tilde{f}^{-1}(\beta) \mid dr(\alpha) = dr(\beta)\}.$$

Then degree $(f_{\alpha_i} : |\alpha_i| \to |\beta|)$ is defined and $\mu(\beta) = \sum_{i=1}^{r} \deg f_{\alpha_i}$. Here f_{α_i} denotes the map $f|_{|\alpha_i|}: |\alpha_i| \to |f(\alpha_i)|$. γ and δ will be given later for h maps.

§3. The concept of Quasitransversality

In our later construction for the addition in $I(G,\lambda)$ we would like to have to concept of G transversality. But there exist obstructions to making a G map $f: X \to Y$ transverse to a submanifold $B \subset Y$, even if B is just a point. Thus we introduce the weaker notion of quasi transversality. In our context there does not occure any obstructions to quasitransversality.

Definition 3.1. Let $f: X \to Y$ be a G map and $B \subset Y$ a G invariant closed submanifold. Then f is quasitransverse to B $(f \tilde{\pitchfork} B)$ if

3.2) $A = f^{-1}(B)$ is a closed G invariant submanifold of X

3.3) there exists a closed G tubular neighbouhood T_1 of A and T_2 of B such that f induces a map of triads:

$$(X, X-\mathring{T}_1, T_1) \to (Y, Y-\mathring{T}_2, T_2)$$

and $f|_{T_1}: T_1 \to T_2$ is fiber and norm preserving.

3.4) $f|_{X-A}: X-A \to Y$ is smooth.

Remark: T_i is the closed unit disk bundle of the normal bundle with respect to some equivariant Riemannian metric. $\mathring{T}_i = \{x \in T_i \mid ||x|| < 1\}$.

Lemma 3.5. Let $f: X \to Y$ and B a G invariant closed submanifold of Y. Let $f \pitchfork B$. Then $f \tilde{\pitchfork} B$.

This is trivial and expresses that quasitransversality is a weaker notion than transversality.

Technical preparation:

Lemma 3.6. Let V and W be H modules and for $K \subseteq H$ $\dim V^k \leq \dim W^k$. Then there exists a norm preserving H map $c: V \to W$ which is smooth off zero.

Proof: It is enough to give a smooth map $c'\colon S(V) \to S(W)$. Then c is given as the radial extension of c'. The existence of c' is a consequence of standard obstruction theory [8] and the fact that $\dim V^k \le \dim W^k$.

Definition 3.7. Let Y be a G manifold. Then we define

$$b_\#\colon\ Y \to \Pi(Y) \quad \text{as follows:}$$

Let $q \in Y$, $\beta \in \Pi(Y,G_q)$ and $q \in |\beta|$. Then $b_\#(q) = r\beta$. (r is the retraction: $\Pi(Y) \to \pi(Y)$ in chapter 2.)

Definition 3.8. Let Y be a G manifold and $B \subset Y$. Then B is locally homogenous if $b_\#|_B\colon B \to \pi(Y)$ is continuous.

Lemma 3.9. Let $q \in Y$ then $\rho(b_\#(q)) = G_q$.

Proof: Let $\beta \in \Pi(Y,G_q)$ and $q \in |\beta|$. Then

$$G_q = \rho(\beta) \subseteq \rho(b_\#(q)) = \bigcap_{x\in|\beta|} G_x \subseteq G_q .$$

Assumptions for quasitransversality:

Q : X and Y are smooth G manifolds and $\dim X = \dim Y$. $f\colon (X,\partial X) \to (Y,\partial Y)$ is an equivariant map. For all $\alpha \in \pi(X)$ we assume that $d(\alpha) \le d(\hat{f}(\alpha))$.

Q_μ: If $\alpha,\alpha' \in \pi(X)$, $\tilde{f}(\alpha) = \tilde{f}(\alpha')$ and $d(\alpha) = d(\alpha') = d(r\tilde{f}(\alpha))$ then $\alpha = \alpha'$.

We make the following choice:

C_α: For all $\alpha \in \pi(X)$ such that $d(\alpha) = d(f(\alpha))$ we choose a norm preserving $\rho(\alpha)$ map $c_\alpha\colon s(\alpha) \to s(\hat{f}(\alpha))$ such that

3.10) c_α is smooth off zero (compare 3.6)

3.11) $c_{g\alpha} = g c_\alpha g^{-1}$

ad Q_μ: Define $\mu_1\colon \pi(X) \to Z$ by

$$\text{3.12)}\quad \mu_1(\alpha) = \begin{cases} \mu(\tilde{f}(\alpha)) & \text{if } d(\alpha) = d(\hat{f}(\alpha)) \\ 0 & \text{otherwise} \end{cases}$$

Then $\mu_1(\alpha)$ = degree $(f_\alpha\colon |\alpha| \to |\tilde{f}(\alpha)|)$.

The choice C_α is possible because of

Lemma 3.13. Assume Q and α such that $d(\alpha) = d(\hat{f}(\alpha))$. Restrict the $\rho(\hat{f}(\alpha))$ action on $s(\hat{f}(\alpha))$ to a $\rho(\alpha)$ action. Then $V = s(\alpha)$, $W = s(\hat{f}(\alpha))$, and $H = \rho(\alpha)$ satisfy the assumptions of lemma 3.6.

Proof: For all $K \subseteq H$ we have to show that $\dim V^K \leq \dim W^K$. Let $K \subseteq H$ and $\gamma \in \Pi(X,K)$, $|\gamma| \supseteq |\alpha|$. Then $\dim V^K = d(r\gamma) - d(\alpha)$ and $\dim W^K = d(r\tilde{f}(\gamma)) - \dim(\hat{f}(\alpha))$. As $d(r\gamma) \leq d(\tau\tilde{f}(\gamma))$ we obtain $\dim V^K \leq \dim W^K$.

The technique to make a map quasitransverse

Let $f\colon (X,\partial X) \to (Y,\partial Y)$ be as in Q and $B \subset \operatorname{int} Y$ a locally homogeneous submanifold. Let $\Sigma \subset \pi(X)$, Σ closed (i.e. Σ is G invariant and if $\beta \in \Sigma$, $\alpha \in \pi(X)$ and $\alpha \leq \beta$, then $\alpha \in \Sigma$), and α minimal in $\pi(X)-\Sigma$. Let $U(\Sigma)$ be a closed neighbourhood of $|\Sigma|$ such that $f(\partial U(\Sigma)) \cap B = \phi$ and $f|_{U(\Sigma)} \pitchfork B$.

3.14) Then $f \simeq {}^1f$ rel $U(\Sigma) \cup \partial X$ (as a G map) such that ${}^1f_\alpha|_{|\alpha|-U(\Sigma)}\colon |\alpha|-U(\Sigma) \to |\tilde{f}(\alpha)|$ is transverse to B. The construction of ${}^1f_\alpha$ is a standard application of transversality [4], [10] as $W(\alpha)$ operates freely on $|\alpha|-U(\Sigma)$. A homotopy extension defines ${}^1f$.

3.15) Let $A_\alpha = ({}^1f_\alpha|_{|\alpha|-U(\Sigma)})^{-1}(B)$ and assume that $\nu(|\alpha|,X)|_{A_\alpha} = A_\alpha \times s(\alpha)$ and $\nu(|\hat{f}(\alpha)|,Y)|_B = B \times s(\hat{f}(\alpha))$. $\nu(A_\alpha,|\alpha|) \times D(s(\alpha))$ and $\nu(B,|\hat{f}(\alpha)|) \times D(s(\hat{f}(\alpha)))$ are closed neighborhoods of A_α and B. ${}^1f|_{\nu(A,|\alpha|)\times D(s(\alpha))}$ and ${}^1f_\alpha|_{D(\nu(A_\alpha,|\alpha|))} \times c_\alpha\colon D(\nu(A_\alpha,|\alpha|)) \times D(s(\alpha)) \to D(\nu(B,|\hat{f}(\alpha)|)) \times D(s(\hat{f}(\alpha)))$ are both homotopic to ${}^1f_\alpha| \times 0$ (0 the zero map). Thus ${}^1f \simeq {}^2f$ rel $U(\Sigma) \cup \partial X$, where ${}^2f$ is an extension of ${}^1f_\alpha \times c_\alpha$, such that there exists a closed neighborhood U of $|\alpha|-U(\Sigma)$ with the properties:

$$({}^2f|_U)^{-1}(B) = A_\alpha, \quad {}^2f(\partial U) \cap B = \phi$$

$GU \cup U(\Sigma)$ is a closed neighborhood of $|\Sigma \cup G\alpha|$ and

${}^2f|_{GU\cup U(\Sigma)} \tilde{\pitchfork} B$.

3.16) To obtain a quasitransverse map we do induction over the partial order of $\pi(X)$ starting with $\Sigma = \phi$. The induction step is given by 3.14 and 15.

These three steps 3.14-16 describe a technique how to obtain a quasitransverse map. Relative versions will be used later and they are based on relative versions of 3.14.

Results for quasitranversality:

Theorem 3.17. Let $f\colon (X,\partial X) \to (Y,\partial Y)$ satisfy Q, Q_μ and make a choice C_α. Let $B \subset \operatorname{int} Y$ be a finite G set and let $b_\#\colon B \to \pi(Y)$ the map defined as in 3.7. Then

a) $f \simeq {}^1f$ rel ∂X such that ${}^1f \tilde{\pitchfork} B$ (existence).

b) Let $A = ({}^1f)^{-1}(B)$, and let $a_\#\colon A \to \pi(X)$ be defined as in 3.7. Then A can be chosen to be a finite G set, and $(A,a_\#)$ can be chosen uniquely depending only on $(B,b_\#,C_\alpha)$ and $(\pi(Y),\pi(Y),\hat{f},\mu)$ (geometric realization of the algebraic degree).

c) A has a unique decomposition $A = A_+ \amalg A_-$ where

$$A_\pm = \{p \in A \mid \text{sign}\det d({}^1f_{a_\#(p)})_p = \pm 1\}.$$

Proof: a) As $d(\alpha) \leq d(\hat{f}(\alpha))$ it follows that $\dim A_\alpha \leq \dim B$ for all $\alpha \in \pi(X)$ in step 3.14. Thus A_α is a finite G set and $\nu(|\alpha|,X)|_{A_\alpha}$ and $\nu(|\hat{f}(\alpha)|,Y)|_B$ are product bundles and 3.15 applies. Then 3.16 implies 3.17a). b-c) We want to show that we can construct A together with its map $a_\#\colon A \to \pi(X)$ in a unique way. Choose an order preserving surjective map $\phi\colon \pi(X) \to \{1,\ldots,r\}$ where $\phi(m) = r$ and $\phi(\alpha) = \phi(\alpha')$ iff $\alpha = g\alpha'$ for some $g \in G$. Then we define $\Sigma_i = \{\alpha \in \pi(X) \mid \phi(\alpha) \leq i\}$ and $\Sigma_0 = \phi$. Now we give a proof by induction. We have to show:

3.18) for each point $q \in B$ the set $({}^1f)^{-1}(q) \cap |\Sigma_i\uparrow|$ is uniquely given.

3.19) for $p \in ({}^1f)^{-1}(q) \cap |\Sigma_i|$ with $a_\#(p) = \gamma$ sign det $d({}^1f_\gamma)_p$ is given.

3.18-19) is trivial for Σ_0. Thus assume it is true for Σ_k. Then we have to show it for Σ_{k+1}. Let $G\alpha = \Sigma_{k+1} - \Sigma_k$ and $H = \rho(\alpha)$. Then we have to give $({}^1f)^{-1}(q) \cap (|\alpha| - |\Sigma_k|)$ in a unique way. To do this we compute d = degree $f_\alpha \big| |\alpha| - U(\Sigma_k)$ rel $\partial U(\Sigma_k)$, where $U(\Sigma_k)$ is a small neighborhood of $|\Sigma_k|$ in X for which $f^{-1}(q) \cap |\Sigma_k| = f^{-1}(q) \cap U(\Sigma_k)$. Here f denotes the map which is quasitransverse restricted to $U(\Sigma_k)$ and ${}^1f$ the map constructed by applying 3.14-15 with $\Sigma = \Sigma_k$. ${}^1f$ will be quasitransverse if we restrict it to $U(\Sigma_{k+1})$, $\Sigma_{k+1} = \Sigma_k \cup G\alpha$.

${}^1f$ can be chosen such that the number of points $p \in ({}^1f)^{-1}(q) \cap (|\alpha| - |\Sigma_k|)$ is uniquely given by $|d|$, and sign det $d({}^1f_\alpha)_p$ = sign d. This is the standard argument that the algebraic degree can be realized geometrically.

<u>Computation of degree</u> $f_\alpha \big| |\alpha| - U\Sigma_k$ <u>rel $\partial U\Sigma_k$</u> $= d$.

First let us compute $\deg f_{\alpha|U(\Sigma_k)}$. Denote $\{p(1),\dots,p(n)\} = f_\alpha^{-1}(q) \cap \Sigma_k$ and $D_{p(i)}$ a small disk in $|\alpha|$ with center $p(i)$. Let $\gamma = a_\#(p(j))$. Then

3.20) $f\big|_{D_{p(j)}} = ((df_\gamma)_{p(j)}, (c_\gamma)^H) : D_{p(j)} \to D_q$

where D_q is a disk in $|f(\alpha)|$ with center q. Define

3.21) $\nu(j) = (\text{sign det } (df_\gamma)_{p(j)}) \cdot \deg(c_\gamma)^H$.

Sign $\det(df_\gamma)_{p(j)}$ is given by assumption of the induction and $\deg(c_\gamma)^H$ is given by the choice C_α. Then

3.22) $\deg f_{\alpha|U(\Sigma_k)}$ rel $\partial U(\Sigma_k) = \sum_{j=1}^{n} \nu(j)$.

Now:

3.23) $d = \begin{cases} \deg f_\alpha - \deg f_{\alpha|U(\Sigma_k)} \text{ rel } \partial U(\Sigma_k) & \text{if } d(\alpha) = d(\hat{f}(\alpha)) \\ 0 \text{ otherwise} & . \end{cases}$

$\deg f_\alpha$ is defined because of the assumption C_μ. Furthermore it is an easy observation that the computation of $(A, a_\#)$ does not depend on the choice of ϕ. This completes the proof of 3.17.

§4. h maps and $I(G,\lambda)$

Let $f: X \to Y$ be a pseudo equivalence of G manifolds (i.e. f is a homotopy equivalence and a G map). Let $H \subset G$ be a p group and $f^H: X^H \to Y^H$ the restriction of f to the H fixed point sets. Smith theory tells us that f^H is a mod p homology equivalence. Thus to convert f by G surgery into a pseudo equivalence we have to do surgery on components in $\pi(X,P)$. Then we do induction over the partial order of $\pi(X,P)$. To do surgery on $|\alpha|$, $\alpha \in \pi(X,P)$, we need bundle data for $|\alpha|$ and $|\tilde{f}(\alpha)|$. To do this inside of X (ambient surgery) we need normal data. This gives the motivation for the bundle data below. This surgery procedure gives a stepwise obstruction theory [7]; for $\alpha \in \pi(X,P)$ we obtain the obstructions

$\chi_\alpha(f)$ a projective obstruction

$\sigma_\alpha(f)$ a Wall obstruction.

To be able to handle the projective obstruction we need the condition

(i) $[M_f] \equiv 0 \; (\Omega(G,\Pi(Y),\tilde{f}\pi(X))+\Delta(G,\Pi(Y),\tilde{f}\pi(X) \cup \pi(Y)))$.

M_f is the mapping cone of f and $[M_f]$ its class in the generalized Burnside ring $\Omega(G,\Pi(Y),\tilde{f}\pi(X) \cup \pi(Y))$ [6]. In short we write $[M_f] \equiv 0 \; (\theta(\lambda))$. $\theta(\lambda)$ only depends on $\pi(X),\pi(Y)$, and $\hat{f}$. We do not need these definitions. It is enough to point out that $[M_f]$ is computed in terms of Euler characteristics. Then we'll show the necessary results for the Euler characteristics.

4.1 Let $\hat{f}: \pi_1 \to \pi_2$. In the analysis of the projective obstruction there occurs integer $n(\lambda)$ (actually $n(\pi_2 \cup \tilde{f}\pi_1)$). $n(\lambda)$ is introduced in [5].

Theorem 4.2. [5], [6]. Let $f: X \to Y$. Either assume that G is abelian or that $\rho(\pi(Y,P)) \subset P$. Then $n(\lambda) \leq 1$.

The definition of a π vectorbundle $\eta_\bullet$ and a π vectorbundle isomorphism $b: \eta_\bullet \to \eta'_\bullet$ is given in [6]. In short: Let Y be a G manifold. Then a $\Pi(Y)$ vectorbundle $\eta_\bullet$ over Y is a collection of G_α vectorbundles over $|\alpha|$, one for each element α in $\Pi(Y)$. These bundles satisfy some compatibility conditions. In particular $\{\nu(|\beta|,Y)\}_{\beta\in\Pi(Y)}$ is a $\Pi(Y)$ vector bundle denoted by $\nu(\cdot,Y)$. A $\Pi(Y)$ vectorbundle $\eta_\cdot$ can be restricted to a $\pi(Y)$ bundle, again denoted by $\eta_\cdot$.

<u>Definition 4.3</u>. An <u>ambient G map</u> $\phi = (X,f,b,C)$ consists of a G map $f: X \to Y$ between smooth G manifolds together with

(ii) a G vectorbundle ξ over Y and a stable vector bundle isomorphism $C: TX \to f^*\xi$ ($\cong_s$)

(iii) a $\Pi(Y)$ vectorbundle $\eta_\cdot$ over Y and a $\pi(X)$ vector bundle isomorphism $b: \nu(\cdot,X) \to f^*\eta_\cdot$.

<u>Definition 4.4</u>. An <u>h normal map</u> is an ambient G map $\phi = (X,f,b,C)$, $f: X \to Y$ is a map of oriented G manifolds and for each $\alpha \in \pi(X,H)$ with $H \in P$ we have

(iv) $d(\alpha) = d(\hat{f}(\alpha))$

(v) $\hat{f}$ induces a surjection of $\pi(X,H)$ onto $\pi(Y,H)$.

<u>Definition 4.5</u>. An <u>h map</u> $\phi = (X,f,b,C)$ is an h normal map such that

(vi) $\hat{f}$ and $\partial\hat{f}$ induce homeomorphisms on $\pi(\cdot,H)$ for all $H \in P$

(i) $[M_f] \equiv 0(\theta(\lambda))$

(vii) for each $\alpha \in \pi(X,P)$ we have that $\deg f_\alpha: |\alpha| \to |\hat{f}(\alpha)|$ is a unit in A_α

(viii) $\pi_i(|\beta|) = 0$ **for** $i \le n(\lambda)$ **and** $\beta \in \pi_2(P)$.

<u>Definition 4.6</u>. An h Equivalence $\phi = (X,f,b,C)$ is an h map such that $f: X \to Y$ is a pseudo equivalence.

Let $\beta \in \pi(Y)$, $q \in Y$, and $b_\#(q) = |\beta|$. Then $(\eta_\beta)|_q$ and $\xi|_q$ are G^β representations. So in continuation of 2.9 we define

4.7 $$\gamma(\beta) = (\eta_\beta)|_q \quad \text{and} \quad \delta(\beta) = \zeta|_q .$$

<u>Definition 4.8</u>. Let $\lambda = (\pi_1,\pi_2,\tau,\mu,\gamma,\delta)$ be a G Poset pair and $\phi = (X,f,b,C)$ an h map. Then $\phi \in Ɫ(G,\lambda)$ if

(ix) $\lambda(f) = \lambda$

(x) ∂f is an h Equivalence.

Definition 4.9. A G triad (W,X_0,X_1) consists of a G manifold W with $\partial W = X_0 \cup X_1$ and $X_0 \cap X_1 = \partial X_0 = \partial X_1$. The submanifolds X_0 and X_1 of ∂W are G invariant. A G map of triads preserves this structure.

Definition 4.10. Let $f_i\colon X_i \to Y$ be two G maps $i = 0,1$. If there is a G manifold pair (W,P) and a G map $F\colon (W,P) \to (Y\times I, \partial Y\times I)$ such that $\partial W = X_0 \cup X_1 \cup P$, $\partial X_0 \cup \partial X_1 = \partial P = P \cap (X_0 \cup X_1)$, and $F_i|_{X_i} = f_i$ with $F|_{X_i}\colon X_i \to Y\times i$, we say that (W,P,F) is a G cobordism between (X_0,f_0) and (X_1,f_1). If $\partial X_0 = \partial X_1$ and $P = \partial X_i\times I$ and $F(x,t) = (f_0(x),t) = (f_1(x),t)$ we say that (W,P,F) is a cobordism relative to the boundary (rel ∂).

Definition 4.11. An h normal cobordism between two h normal maps $\phi_i = (X_i,f_i,b_i,C_i)$, $f_i\colon X_i \to Y$, $i = 0,1$, is an h normal map $\Phi = (W,F,\tilde{b},\tilde{C})$ with $\partial W = X_1 \cup X_2 \cup P$, $\tilde{b}$ and $\tilde{C}$ restrict to b_i and C_i and (W,P,F) is a G cobordism between (X_0,f_0) and (X_1,f_1).

Definition 4.12. An h cobordism between h maps $\phi_i = (X_i,f_i,b_i,C_i)$, $i = 0,1$, is an h normal cobordism $\Phi = (W,F,b,C)$ between them, where Φ is an h map and the inclusions $X_i \to W$ together with the inclusion $Y\times i \to Y\times I$ induce isomorphisms of G Poset pairs $\lambda(f_i) \to \lambda(F)$.

Let $\phi = (X,f,b,C)$ be an h map, then we denote by $-\phi$ the h map $(-X,f,b,C)$ $f\colon -X \to -Y$. $-$ denotes the G manifold where the orientations of all components of the fixed point sets are reversed (for all $H \subset G$). Then λ is unchanged.

Define an equivalence relation $\sim$ on $\mathfrak{F}(G,\lambda)$ as follows: Let $\phi_i = (X_i,f_i,b_i,C_i)$, $i = 0,1$. Then $\phi_0 \sim \phi_1$ if there is an h map of triads $\Phi = (W,F,\tilde{b},\tilde{C})$ with

4.13. a) $F\colon (W,X_0 \cup -X_1,X_2) \to (Z,Y_0 \cup -Y_1,Y_2)$

b) $(X_2,f_2,\tilde{b}|_{X_2},\tilde{C}|_{X_2})$ is an h Equivalence

c) The inclusion of X_i in W and Y_i in Z induce a G Poset pair isomorphism $\lambda = \lambda(f_i) \to \lambda(F)$.

<u>Definition 4.14</u>. $I(G,\lambda) = \mathcal{I}(G,\lambda)/\sim$.

0 is distinguished in the set $I(G,\lambda)$. It is represented by any h map $\phi_0 = (X_0,f_0,b_0,C_0)$ which occurs in an h map of triads $\Phi = (W,F,\tilde{b},\tilde{C})$ with

a) $F = (W,X_0,X_1) \to (Z,Y_0,Y_1)$

b) (X_1,f_1,b_1,C_1) is an h Equivalence

c) The inclusions $X_0 \to W$ and $Y_0 \to Z$ induce an isomorphism $\lambda(f_0) \to \lambda(F)$.

Let $N_G(Y,\lambda)$ denote the subset of $\mathcal{I}(G,\lambda)$ consisting of those h maps whose target space is Y. An equivalence relation $\sim$ on $N_G(Y,\lambda)$ is defined by saying $\phi_i = (X_i,f_i,b_i,C_i)$, $i = 0,1$, are equivalent if ϕ_0 and ϕ_1 are h cobordant rel ∂.

<u>Definition 4.15</u>. $N_G(Y,\lambda) = N_G(Y,\lambda)/\sim$.

Let <u>$hS_G(Y,\lambda)$</u> denote the subset of $N_G(Y,\lambda)$ consisting of h Equivalences modulo an equivalence relation which is of no interest here.

Then we obtain a sequence

4.16 $$hS_G(Y,\lambda) \to N_G(Y,\lambda) \to I(G,\lambda).$$

Let $\lambda = (\pi_1,\pi_2,\tau,\mu,\gamma,\delta)$ be a G Poset pair satisfying the

4.17 <u>Gaphypothesis</u>: If $\alpha \in \pi_1(P)$ and $\alpha' \in \pi_1$, $\alpha' < \alpha$, then $d(\alpha') \le \frac{1}{2} d(\alpha)-1$.

<u>Main zero theorem 4.18</u>. [6] If $d(\alpha) \ge 3$ for $\alpha \in \pi_1$ and $d(\alpha) \ge 6$ for $\alpha \in \pi_1(P)$, then the sequence 4.16 is exact.

<u>Proof</u>: The proof is given in [6]. The only observation we have to make is: Let $\phi_i \in \mathcal{I}(G,\lambda)$, $i = 0,1$. Then $\phi_0 \sim \phi_1$ if and only if they are equivalent with respect to the equivalence relation given in [6]. Furthermore $\phi_0 \sim 0$ if and only if ϕ_0 is equivalent to zero in the sense of [6].

§5 The construction defining the addition

Now we want to give the details of the construction by which we add elements in $I(G,\lambda)$.

Let $\phi^i = (X^i, f^i, b^i, C^i)$ represent $x^i \in I(G,\lambda)$ with $f^i: X^i \to Y^i$, $\lambda = (\pi_1, \pi_2, \tau, \mu, \gamma, \delta)$ with the bundles ξ^i and the $\bar{\pi}_2$ bundles $\eta^i_{\bullet}$, $i \in \{', "\}$. Then we want to construct an h map $\phi = (X,f,b,C)$, $f: X \to Y$, together with the bundles ξ and the $\bar{\pi}_2$ bundle $\eta_{\bullet}$, such that $\lambda(f) = \lambda$. So $\phi \in \Phi(G,\lambda)$ and represents $X = X'+X"$. For convenience we also denote ϕ by $\phi'+\phi"$.

Make a choice C_α, and make the assumptions Q_μ (§3) $n(\lambda) \leq 1$ (§4) and the dimension assumptions:

D(i) $d(\alpha) \geq 3$ for $\alpha \in \pi_i$

(ii) $d(\alpha) \geq 6$ for $\alpha \in \pi_i(\mathcal{P})$ $i = 1,2$

(iii) $d(\alpha) \leq d(\tau(\alpha))$ for $\alpha \in \pi_1$.

The addition is a procedure in several steps. Each step will start with an ambient G map $f: X \to Y$. Then we construct a new ambient G map $f_N: X_N \to Y_N$ together with an ambient G map of triads $F: (W,X,X_N) \to (Z,Y,Y_N)$. Here $W = X\times I \cup$ handles and $Z = Y\times I \cup$ handles. We say that f_N arises from surgery in source and target. Occasionally the notation will be reduced to the data we are just working with. The basic ideas are used already in chapter 9 [9] but we extend them to solve our problem here.

Let us describe what we want to achieve in the single steps.

Step 1: Construct $f_N: X_N \to Y_N$ with $\pi(Y_N) = \pi_2$.

Step 2: Construct f_N such that $\pi_1(|\beta|) = 0$ for $\beta \in \pi_2(\mathcal{P})$.

Step 3: Do surgery to obtain $\pi(X_N) = \pi_1$

Step 4: Give surgery steps to satisfy $[M_{f_N}] \equiv 0 \quad (\theta(\lambda))$.

From this construction it will be obvious that $\lambda(f) = \lambda$. A representative of an element in $I(G,\lambda)$ had to satisfy (i)-(x) in Chapter 4. Step 4 takes care for (i). (ii) is maintained throughout the construction. (iv) and (vi) are conditions depending only on λ and there they are obvious. (viii) is taken care for in step 2.

(ix) follows as we obtain $\lambda(f) = \lambda$. (x) will be satisfied as we leave boundaries untouched.

We proceed as follows. To obtain W and Z we have to tell how we attach handles (5.1). Then we tell when we can construct $F: (W,X,X_N) \to (Z,Y,Y_N)$ (5.2.). To obtain an ambient G map we have to extend the bundle data (5.3.). This describes the technique by which we attach handles in source and target and we say that $f_N: X_N \to Y_N$ arises from surgery in source and target on $f: X \to Y$. By this technique we'll carry out step 1.-4.

We will need two standard results in K theory. Here H and G will always be finite groups.

5.1 a) If H acts trivially on X and R denotes the representation ring:

$$K_H(X) \cong K(X) \otimes R(H)$$

b) If $G \supset H$

$$K_G(G\times_H X) \cong K_H(X).$$

5.2 <u>Assumptions to attach handles</u>

Let $B = G\times_H S^k$ be imbedded in the interior of Y and T_2 be a closed disk bundle in $\nu(B,Y)$. <u>Assume that H operates trivially on S^k</u>. Then $B = G/H \times S^k$ and identify $H/H \times S^k$ with S^k. Let $\beta = b_\#(S^k)$ and $d(\beta) = d$. If $k = 0$ it is possible that $\beta = \beta' \cup \beta''$. In this case assume that $s(\beta') = s(\beta'')$. Then $\rho(\beta) = H$. Futhermore, <u>assume that $\nu(S^k,Y)$ is trivial</u> after forgetting the group action. Then $T_2 = G\times_H S^k \times D^{n-k} \times D(s(\beta))$. Let $\bar{T}_2 = G\times_H D^{k+1} \times D^{n-k} \times D(s(\beta))$, then we can form $Z = Y \times I \cup_{T_2} \bar{T}_2$. $\partial Z = Y \cup_{\partial Y} \partial Y \times I \cup_{\partial Y} Y_N$. Furthermore we can apply the same procedure if $B = \coprod B_i$ and each B_i satisfies the above assumptions.

<u>Proof</u>: As $\nu(S^k,Y)$ is a trivial 5.1.a) implies that $\nu(S^k,Y)$ is an H product-bundle and it splits as $\nu(S^k,|\beta|) \oplus \nu(|\beta|,Y)|_{S^k} = S^k \times R^{d-k} \times s(\beta)$. By 5.1.b) $\nu(B,Y) = G\times_H \nu(S^k,Y)$ and thus $T_2 = G\times_H S^k \times D^{n-k} \times D(s(\beta))$. $T_2 \subset \partial\bar{T}_2$ and Z arises by identifying T_2 with its image in $Y \times 1$.

5.3 Assumptions to construct $F\colon (W,X,X_N) \to (Z,Y,Y_N)$.

Let $B = B_i$ as in 5.2. Assume that $f \simeq {}^1f$ and ${}^1f \tilde{\pitchfork} B$ with ${}^1f^{-1}(B) = A$. Assume that $A = \amalg A_i$ and $A_i = G \times_K S^k$ such that A and T_1, a closed diskbundle in $\nu(A,X)$, satisfy the assumptions in 5.2. (K depends on i). Then we construct $W = X \times I \cup_{T_1} \bar{T}_1$. Let $h_t\colon f \simeq {}^1f$ be the homotopy between f and ${}^1f$ and $H = X \times I \to Y \times I$ be defined by $H(x,s) = (h_s(x),s)$. Let $S_A{}^k \subset A$, $S_B{}^k \subset B$ and ${}^1f(S_A{}^k) = S_B{}^k$. $a_\#(S_A{}^k) = \alpha$, $\tau(\alpha) = \delta$ and $d = d(\alpha) = d(\delta)$. Thus $b_\#(S_B{}^k) = \beta \le \delta$. Assume ${}^1f$ arises from the technique given in 3.14-15). With the decomposition $\nu(S_A{}^k,X) \cong \nu(S_A{}^k,|\alpha|) + \nu(|\alpha|,X)\Big|_{S^k_A}$ and $\nu(S_B{}^k,Y) = \nu(S_B{}^k,|\delta|) + \nu(|\delta|,Y)\Big|_{S^k_B}$ assume that ${}^1f\Big|_{\nu(S^k_A,X)}$ decomposes as ${}^1f_\alpha\Big|_{\nu(S^k_A,|\alpha|)} \times c_\alpha$, and with the trivilizations ϕ_A and ϕ_B the following diagram commutes:

5.4
$$\begin{array}{ccc}
\nu(S^k_A,|\alpha|) & \xrightarrow{\ {}^1f_\alpha\ } & \nu(S^k_B,|\delta|) \\
\phi_A \downarrow & & \downarrow \phi_B \\
S^k_A \times R^{d-k} & \xrightarrow{\ Id\ } & S^k_B \times R^{d-k}
\end{array}$$

Then H_t extends to a map $F'\colon (W,\partial W) \to (z,\partial Z)$.

Proof: Obviously Id extends to $Id\colon D_A{}^{k+1} \times R^{d-k} \to D_B{}^{k+1} \times R^{d-k}$. By 5.1 and the fact that the restriction of ${}^1f$ to $\nu(|\alpha|,X)\Big|_{S^k_A}$ is given as $Id \times c_\alpha$ it follows that ${}^1f$ extends to a map $(\bar{T}_1,T_1,\partial\bar{T}_1-\mathring{T}_1) \to (\bar{T}_2,T_2,\partial\bar{T}_2-\mathring{T}_2)$. Glueing H_t and the extension of ${}^1f$ together gives a map

$$F'\colon (W,X,\partial X \times I,X_N) \to (Z,Y,\partial Y \times I,Y_N).$$

As ${}^1f$ was smooth away from A it follows that $F'\Big|_{\partial W}$ is smooth. Approximating F' rel ∂W by a smooth map gives

$$F\colon (W,X,X_N) \to (Z,Y,Y_N).$$

Lemma 5.5. Let (X,f,b,C), $f\colon X \to Y$ be an ambient G map and assume 5.2.-3). Let $k \le 2$ and if $k = 0$ and $\{\pm 1\} = S^0 \subset B$ assume that $\xi\Big|_{+1} = \xi\Big|_{-1}$ and $\eta.\Big|_{+1} = \eta.\Big|_{-1}$. Then $F\colon (W,\partial W) \to (Z,\partial Z)$ gives rise to an ambient G map $(W,F,\tilde{b},\tilde{C})$.

<u>Proof</u>: It follows from 5.2 that $(TX \oplus \varepsilon)\big|_{S^k}$ and $\nu(\cdot,X)\big|_{S^k}$ are trivial bundles (without group action). ε is the trivial 1-dimensional bundle. As f was an ambient G map it follows from 5.4 and (ii) and (iii) chapter 4 that $(\xi \oplus \varepsilon)\big|_{S^k}$ and $\eta_{\bullet}\big|_{S^k}$ are trivial and by 5.1 they extend over Z. Let $\bar{\delta}$ be the image of δ under the map $\pi(Y) \to \pi(Z)$, induced by the inclusion. Let $|\alpha|^* = \{x \in |\alpha| \mid a_{\#}(x) = \alpha\}$, for $\alpha \in \pi_1$, $|\bar{\delta}|^*$ similiar. Then we have a commutative diagram

$$\begin{array}{ccc} G_\alpha \times_K S^k \times D^{d-k} & \longrightarrow & |\alpha|^* \\ \downarrow & & \downarrow \\ G_\alpha \times_K D^{k+1} \times D^{d-k} & \longrightarrow & |\bar{\delta}|^* \end{array}$$

and it follows from §5 [6] that $b\colon \nu(\cdot,X) \xrightarrow{\cong} f^*\eta_{\bullet}$ extends to $\tilde{b}\colon \nu(\cdot,W) \xrightarrow{\cong} F^*\tilde{\eta}_{\bullet}$ and the stable isomorphism $C\colon TX \to f^*\xi$ to $\tilde{C}\colon TW \to F^*\tilde{\xi}$.

5.6 <u>Description of $\lambda(F)$ and $\lambda(F_N)$</u>.

We assumed that $d(\alpha) \geq 3$ for $\alpha \in \pi_i$ and we only do surgery on k dimensional spheres with $k \leq 2$ and $k = 2$ only if $d(\alpha) \geq 4$. Thus we do not decompose any $|\alpha|$. As we did surgery on $G \times_H S^k$ where H operated trivially on S^k we do not introduce new fixed point components. Thus

a) $\lambda(f_N)$ is isomorphic to $\lambda(F)$.

If $1 \leq k$ it follows that:

b) $\lambda(f)$ is equivalent to $\lambda(f_N)$.

Now let $k = 0$; $S^0 = \{\pm 1\}$ imbedded in X respectively Y, with $\alpha_{\#}(+1) = \alpha'$, $\alpha_{\#}(-1) = \alpha''$, $b_{\#}(+1) = \beta'$, and $b_{\#}(-1) = \beta''$, $\alpha^i \in \pi(X)$ and $\beta^i \in \pi(Y)$. Then α' and α'' respectively β' and β'' will be identified in $\pi(X_N)$ respectively in $\pi(Y_N)$. Furthermore, let $\gamma' \geq \alpha'$ and $\gamma'' \geq \alpha''$ and $\rho(\gamma') = \rho(\gamma'')$. Then also γ' and γ'' are identified in $\pi(X_N)$. The equivalent statement holds for $\pi(Y_N)$. Denote these identifications by $\sim$. Then

c) $\lambda(f_N)$ is equivalent to $(\pi(X)/\sim, \pi(Y)/\sim, \tau_\sim, \mu_\sim, \gamma, \delta)$. In general $\tau_\sim$ and $\mu_\sim$ are difficult to describe but in our application they will take care for themselves and are given in the obvious way.

After these preparations we are ready to come back to the concrete construction of our addition. As stated in the beginning of this chapter we want to add x' and x'', elements in $I(G,\lambda)$ which are represented by $\phi^i = (X^i, f^i, b^i, C^i)$ with $f^i: X^i \to Y^i$ and $\lambda(f') = \lambda(f'') = \lambda = (\pi_1, \pi_2, \tau, \mu, \gamma, \delta)$, $i \in \{', ''\}$.

Step 1 is surgery on 0-dimensional spheres in source and target. Choose a finite G set ${}^oB$ with imbeddings $b^i: {}^oB \to Y^i$ such that each minimal component $\beta \in \pi_2$ is in $\mathrm{im}(b^i_\#({}^oB))$. Then assume that we have a commutative diagram

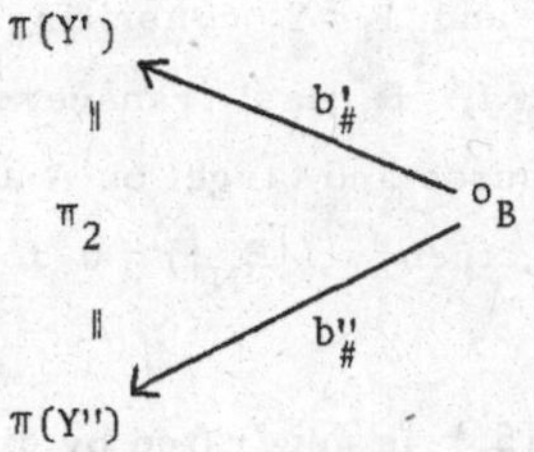

By 3.17 $f^i \simeq {}^1f^i$ and ${}^1f^i \widetilde{\pitchfork}$ $\mathrm{im}(b^i({}^oB))$. This gives rise to a unique G set ${}^oA$ with imbeddings $a^i: {}^oA \to X^i$ such that $a^i({}^oA) = ({}^1f^i)^{-1}({}^oB)$ and the following diagram commutes:

5.7

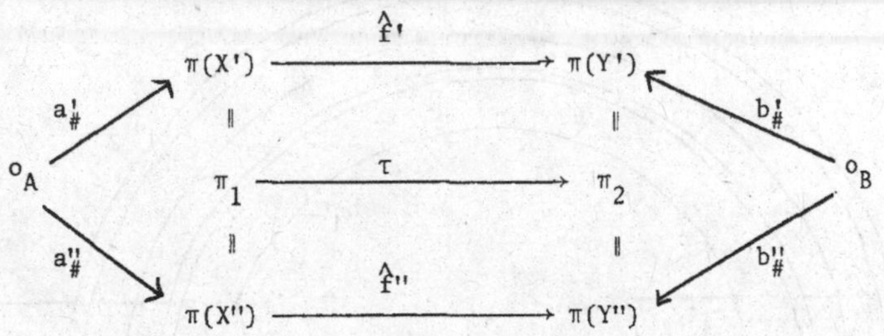

We assume that we constructed ${}^1f^i$ in the way outlined in 3.14-15). Now let $A = \mathrm{im}(a'({}^oA)) \amalg \mathrm{im}(a''({}^oA))$ and $B = \mathrm{im}(b'({}^oB)) \amalg \mathrm{im}(b''({}^oB))$.

<u>Lemma 5.8</u>. We can do 0-dimensional surgery on A and B. Let $f_N: X_N \to Y_N$ be the result of these surgery steps, then $\pi(Y_N) = \pi_2$. This is step 1.

<u>Proof</u>: Let $q \in {}^oA$ and $S^o = \{a'(q), a''(q)\}$. Then 5.7 implies that 5.2 holds for A and $X = X' \amalg X''$, as well as for B and $Y = Y' \amalg Y''$. The assumption that we

constructed ${}^{1}f^{i}$ in the way given in 3.14-15 implies that we can apply 5.3. Again by 5.7 it follows that we can apply lemma 5.5, thus $f_N: X_N \to Y_N$ is an ambient G map. As we assumed that each minimal component in π_2 was hit, it follows from 5.6 that $\pi(Y_N) = \pi_2$.

Let T_1 and T_2 be closed disk bundles in $\nu(A, X' \amalg X'')$ and $\nu(B, Y' \amalg Y'')$ given by quasitransversality and 5.2. Then we can carry out step 1 with this T_1 and T_2.

Denote the result after step 1 again by $f: X \to Y$. Now let us discuss step 2. Property (viii) chapter 4 of an h map is that $\pi_i(Y) = 0$ if $i \leq n(\lambda)$ and we assumed that $n(\lambda) \leq 1$.

Lemma 5.9. There exist $A \subset X$ and $B \subset Y$ consisting of 1-dimensional spheres and satisfying 5.2, ${}^{1}f: X \to Y$, ${}^{1}f \sim f$, ${}^{1}f$ quasi transverse to B and $({}^{1}f)^{-1}(B) = A$ such that we can do surgery in source and target on A and B. Let $f_N: X_N \to Y_N$ be the result of these surgery steps, then $\pi_1(|\beta_N|) = 0$ for all $\beta_N \in \pi_2(\varphi)$. This is step 2.

Proof. After the first step, $\pi_1(|\beta|)$ is generated by differences of handles attached in step 1 along points $b^i(q) \in |\beta^i|$, $q \in B_0$. Expressed in 1-dimensional spheres:

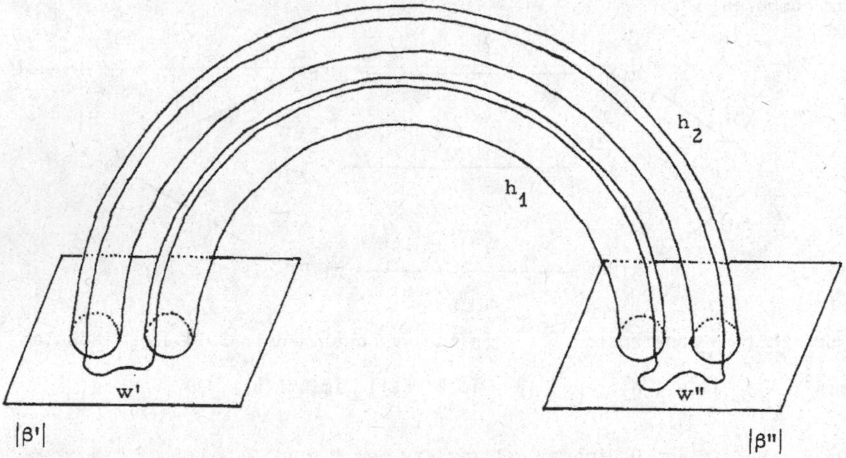

Start at a point in $|\beta'|$, run along the handle h_1, along a path w" in $|\beta''|$ to the handle h_2, come back along h_2 to $|\beta'|$ and in $|\beta'|$ back to the point we started out with.

Let $Q_\beta = \{q \in B_0 \mid b_\#(q) \le \beta\} = \{q_1, \ldots, q_r\}$. Let $q_j^i = b^i(q_j)$ and $D(q_j^i)$ the disk $T_2\big|_{q_j} \cap |\beta^i|$. We have given homeomorphisms $\phi_j : D(q_j') \approx D(q_j'')$. For a pair of points q_j and q_k choose a path w_{jk}^i imbedded in $|\beta^i|^*$ such that $w_{jk}^i(0) \in \partial(D(q_j^i))$, $w_{jk}^i(1) \in \partial(D(q_k^i))$, $\phi_j(w_{jk}'(0)) = w_{jk}''(0)$ and $\phi_k(w_{jk}'(1)) = w_{jk}''(1)$. $|\beta^i|^* = \{y \in Y^i \mid b_\#(y) = \beta^i\}$. Furthermore, w_{jk}^i misses $\overset{o}{D}(q_m^i)$ for all $m \in \{1, \ldots, r\}$ and hits $\partial(D(q_j^i))$ and $\partial(D(q_k^i))$ at its endpoints.

Choose enough paths to connect all the disks $D(q_j^i)$. Choose them equivariantly as a G-set for all components in $G\beta^i$ and do this for all $\beta \in \pi_2(\mathcal{P})$.

Denote the starting points by $Q^i(-) = \coprod_{j,k} w_{jk}^i(0)$ and the endpoints by $Q^i(+) = \coprod_{j,k} w_{jk}^i(1)$. By changing the chosen c_α in C_α by a homotopy, we can assume that $f^i \overset{\sim}{\pitchfork} Q^i(\pm)$ and let $P^i(\pm) = (f^i)^{-1}(Q^i(\pm))$. We have a homeomorphism $P'(\pm) \approx P''(\pm)$, and furthermore, $P^i(+) \approx P^i(-)$ by quasi transversality. Define a map $Q^i(-) \to Q^i(+)$ by $\omega_{jk}^i(0) \to \omega_{jk}^i(1)$. Then we obtain the commutative diagram

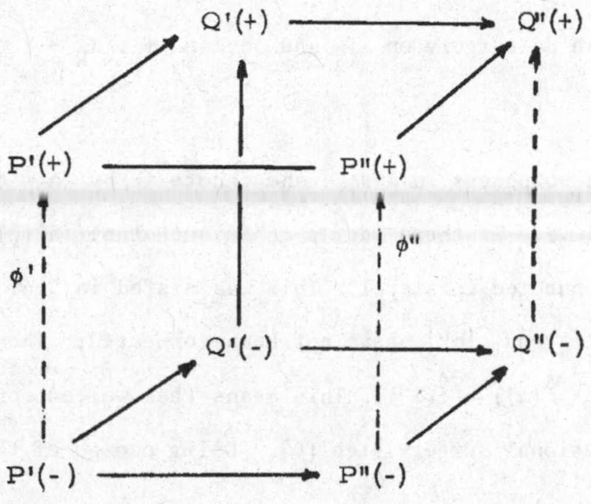

The maps $P^i(\pm) \to Q^i(\pm)$ are induced by f, and ϕ^i can be filled in such that the triangle

$$\begin{array}{ccc} P^i(+) & & \\ & \searrow & \\ \phi^i \big\uparrow & & \pi \\ & \nearrow & \\ P^i(-) & & \end{array}$$

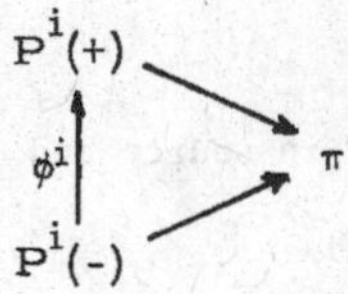

commutes.

For $p \in P^i(-)$ choose a path $v_p^i: I \to X^i - \overset{o}{T}_1$ such that $v_p^i(0) = p$ and $v_p^i(1) = \phi^i(v_p^i(0))$. Complete the paths in the obvious way to imbedded circles, and denote their image by A in X and B in Y. As $|\alpha'|$ and $|\alpha''|$ were 1-connected, $\alpha \in \pi_1$, we can assume that f(A) = B; for a single circle we can even assume that the map is identity and th the map on a tubular neighborhood is a product. Thus f restricted to a tubular neighborhood of A is quasitransverse. Now make $f \tilde{\pitchfork} B$ rel A, then $f^{-1}(B) = A \cup S^1$'s. These S^1's are locally homogeneous submanifolds of X such that $a_\#(S^1) = \alpha \in \pi(X;\mathcal{P})$. As the map was quasitransverse to B on the handles, we can assume furthermore that each S^1 either lies in $X' - T_1$ or in $X'' - T_1$. As components in $(X')^P$ were 1-connected, P a p-group, we can cancel the additional S^1. Hence we can assume that $f^{-1}(B) = A$, $f \tilde{\pitchfork} B$, and by having the map f on a neighborhood of A chosen as a product it follows that the assumptions for 5.2, 5.3, and 5.5 are satisfied. Doing surgery on A in B completes the proof of Lemma 5.9.

Denote the result after step 2 again by $f: X \to Y$.

Lemma 5.10. We can do surgery on f and obtain $f_N: X_N \to Y$ such that $\pi(X_N) = \pi_1$. This is step 3.

Proof: Consider a component $\alpha \in \pi_1$, and denote it by α^i considered as an element in $\pi(X^i)$, $i \in \{','' \}$. If there was $p \in {}^oA$ such that $a_\#^i(p) \leq \alpha^i$, then $|\alpha'|$ and $|\alpha''|$ have been connected in step 1. This was stated in 5.6. Consider α, α', and α'' such that $|\alpha'|$ and $|\alpha''|$ have not been connected. Then $d(\alpha') = d(\alpha'')$, $s(\alpha') = s(\alpha'')$ and $\hat{f}(\alpha') = \hat{f}(\alpha'')$. This means that we can connect $|\alpha'|$ and $|\alpha''|$ by a usual 0-dimensional surgery step [6]. Doing enough of these steps completes the proof.

Step 4. Denote the result of step 3 again by $f: X \to Y$. An h map is supposed to satisfy the congruence $[M_f] \equiv 0\ (\theta(\lambda))$. (Compare (i) in chapter 4).

Lemma 5.11. We can do surgery (in source and target) on $f: X \to Y$ to obtain $f_N: X_N \to Y_N$ and $[M_{f_N}] \equiv 0(\theta(\lambda))$.

5.12 Remark: It should be pointed out that we are free not to apply step 4 in case the congruence is satisfied already after step 3.

Proof of the lemma: Let $\theta(\lambda) = \Omega+\Delta$ as in (i) chapter 4. Surgery steps in the source change $[M_f]$ by an element in 2Ω, thus the equivalence class of $[M_f]$ mod 2Ω is unchanged. Let β' and β'' be again the components in $\pi(Y')$ and $\pi(Y'')$ belonging to $\beta \in \pi_2$, and $\tilde{\beta} \in \pi(Y)$ the component we constructed from these during step 1 and 2. After step 4 we denote this component by β_N. Then it is sufficient to show $\chi(|\beta_N|) = \chi(|\beta'|)+\chi(|\beta''|)$. Then $[M_{f_N}] = [M_{f'}]+[M_{f''}]+\varepsilon$, where $\varepsilon \in 2\Omega$. The effect on Euler characteristics by doing surgery on a k-dimensional sphere is again annihilated if we do surgery on a further (k+1)-dimensional sphere. So we have to do surgery on 1- and 2-dimensional spheres in the target (thus also in the source).

Imbed the spheres in Y in small disks (surely as a locally homogeneous submanifold). Then f can be made quasitransverse to the centers of these disks, and it is an easy check that we can do surgery on these spheres and their inverse images. For β π_2, $|\beta| \neq Y$, we have to do surgery only on spheres of dimension 1. Thus the assumptions in 5.6 are satisfied.

The result after step 4 is the result of our addition. It is denoted by $\phi = (X,f,b,C)$, $f: X \to Y$ with bundles η and ξ.

5.13 Remark: We should still make a remark about the ambient G map of triads we constructed during the addition (compare 5.5)

$$F: (W, X' \amalg X'', X) \to (Z, Y' \amalg Y'', Y).$$

Clearly $\lambda(F)$ is isomorphic to $\lambda(f)$. It is a standard argument that $\pi_1(Z) = 0$. $[M_F] = [M_{f'}]+[M_{f''}]+ \frac{\varepsilon}{2}$, where $\varepsilon \in 2\Omega$ is as in the proof of lemma 5.11. Thus $[M_F] \equiv 0 \quad (\theta(\lambda))$ (compare chapter 4 (i)) and F is an h map of triads.

§6 <u>The group $I(G,\lambda)$</u>

Having all notation available, let us state the main theorems and prove them. Let $\lambda = (\pi_1,\pi_2,\tau,\mu,\gamma,\delta)$ be a G Poset pair. Then we assume

6.1) $n(\lambda) \leq 1$ (4.1)

$d(\alpha) \geq 3$ for $\alpha \in \pi_i$

$d(\alpha) \geq 6$ for $\alpha \in \pi_i(P)$ $i = 1,2$

$d(\alpha) \leq d(\tau(\alpha))$ for $\alpha \in \pi_1$

the gaphypothesis (4.17)

Q_μ (chapter 3).

Let $\phi = (X,f,b,C)$, $f\colon X \to Y$ represent an element x in $I(G,\lambda)$ and let $\alpha \in \pi(X,P)$. Then we define

6.2) $\operatorname{sign}_\alpha(x) = \operatorname{sign}(W(\alpha),|\tilde{f}(\alpha)|)-\operatorname{sign}(W(\alpha),|\alpha|) \in R(W(\alpha))$

$|\alpha|$ and $|\tilde{f}(\alpha)|$ are $W(\alpha)$ manifolds and R is the representation ring. $\operatorname{sign}_\alpha(x)$ is well defined.

<u>Theorem 6.3</u>. Make the assumptions 6.1 and let the addition be defined as in chapter 5. Then $I(G,\lambda)$ is an abelian group and $\operatorname{sign}\colon I(G,\lambda) \to \prod_{\alpha\in\pi_1(P)} R(W(\alpha))$ is a homomorphism. Sign denotes the collection of the maps $\operatorname{sign}_\alpha$ in 6.2.

<u>Proof</u>: First we show that the addition is well defined. Let ϕ^i_j represent elements in $I(G,\lambda)$ where $i \in \{','' \}$ and $j = 1,2$ and ϕ^i_1 and ϕ^i_2 represent the same element in $I(G,\lambda)$. $\phi^i_j = (X^i_j,f^i_j,b^i_j,C^i_j)$ and $f^i_j\colon X^i_j \to Y^i_j$. Then we want to show that $\phi_j = \phi'_j+\phi''_j$ represent the same element in $I(G,\lambda)$. Let $\Phi^i = (W^i,F^i,\tilde{b}^i,\tilde{C}^i)$ be h maps of triads representing the equivalence between ϕ^i_1 and ϕ^i_2, $F^i\colon W^i \to Z^i$. Furthermore $\Phi_j = (W_j,F_j,\tilde{b}_j,\tilde{C}_j)$ the map of triads given during the construction of the addition in chapter 5. Stack them on top of eachother as indicated in 6.4.

6.4)

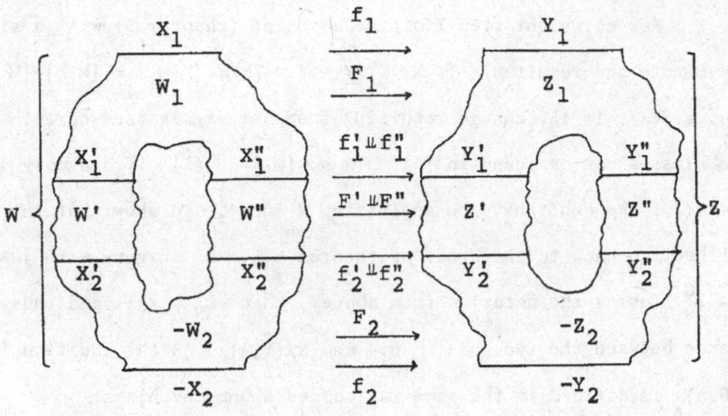

Denote the result by $\Phi = (W,F,\tilde{b},\tilde{C})$, $F: W \to Z$. To show that ϕ_1 and ϕ_2 represent the same class in $I(G,\lambda)$ we have to show that Φ is an h map. Φ is an h normal map of triads. To change Φ to an h map of triads we still have to satisfy 4 (i) and (viii), i.e. $[M_F] \equiv 0\ (\theta(\lambda))$ and components of $Z^{\mathcal{P}}$ a 1-connected, $\mathcal{P}$ a p-group. We apply a similar step as step 2 in Ch. 5 to achieve this. By assumption we had that $[M_{F_1}]$, $[M_{F_2}]$, $[M_{F'}]$ and $[M_{F''}]$ are in $\theta(\lambda)$.

Thus by a Mayer-Vietoris argument $[M_F] \in \theta(\lambda)$. We changed this by making Z 1-connected. But further surgery as in step 4 chapter 5 makes sure that $[M_{F_N}] \in \theta(\lambda)$, F_N being the result after the surgeries just described. This shows that the addition is well defined. The rest of the proof are standard arguments.

<u>Corollary 6.5</u>. Let λ be a G Poset pair satisfying 6.1 and $I(G,\lambda) \neq \phi$. Then there exists a G manifold Y and $hS_G(Y,\lambda) \neq \phi$.

This corollary is still very unsatisfactory as we can't say anything about Y. Thus let $Y = S(V)$ be the unit sphere in a G representation and $Y^G \neq \phi$. Y^G is the fixed point set of Y. Let λ again satisfy 6.1.

<u>Theorem 6.6</u>. If $[Y \amalg pt] \equiv 0(\theta(\lambda))$ the addition is defined in $N_G(Y,\lambda)$ and $\sigma: N_G(Y,\lambda) \to I(G,\lambda)$ is a homomorphism (of semigroups).

<u>Proof</u>: Let ϕ' and ϕ'' represent elements in $N_G(Y,\lambda)$, $\phi^i = (X^i,f^i,b^i,C^i)$ with

$f^i: X^i \to Y$. Now carry out step 1 of the addition (chapter 5) with a single point $q \in Y^G$. Denote the result by $f: X \to Y\#Y = Y$. Then $[M_f] = [M_{f'}]+[M_{f''}]+\alpha_0-[Y_{\amalg}pt]$ where $\alpha_0 \in \theta(\lambda)$ is the change resulting from surgery in the source. The analysis of these classes can be found in [6]. Thus $[M_f] \in \theta(\lambda)$ if and only if $[Y_{\amalg}pt] \in \theta(\lambda)$. We don't have to apply step 2 and 4. To show that the addition is well defined, go back to the proof of theorem 6.5. Do surgery with just a single $S^1 = B \subset Z^G$ (using the notation from above), then $Z_N = Y\times I$ and thus we obtain an equivalence between the two ends in 6.4 in $N_G(Y,\lambda)$. As the addition in $N_G(Y,\lambda)$ and $I(G,\lambda)$ is defined in the same way σ is a homomorphisms.

Theorem 6.7. Let Y be as in 6.6 and assume $N_G(Y,\lambda) \neq \phi$. Then there exists an h-Equivalence $\phi = (X,f,b,C)$, $f: X \to Y$ and ϕ represents an element in $N_G(Y,\lambda)$ (i.e. $hS_G(Y,\lambda) \neq \phi$).

Proof: This is an easy consequence of the facts: Given a representant ϕ_0 of an element in $N_G(Y,\lambda)$. Then $\phi_o+(-\phi_o) = \phi$ represents zero in $I(G,\lambda)$. Using the addition as in 6.6, observing that $Y\#(-Y) = Y$ and the main zero theorem (chapter 4) gives us the result.

Corollary 6.8. Let $Y = S(V)$ be the unit sphere in some G representation and the dimension of Y^H is odd for all $H \subset G$. Then theorem 6.6 and 6.7 apply.

Proof: The only observation we have to make is that $\chi((Y_{\amalg}pt)^H) = \chi(pt)$ for all $H \subset G$ and $[pt] \in \theta(\lambda)$.

§7 Applications and Generalizations

To give at least one application let us state the following problem.

Artin relation: Given a group G and a ring A. Does there exist a family $\mathcal{H} = \mathcal{H}(G,A)$ of subgroups of G such that $G \notin \mathcal{H}$ and a function ψ, such that for all smooth closed homology A spheres

$$\dim Y^G = \psi\{\dim Y^H \mid H \in \mathcal{H}\}?$$

There are some positive results:

Theorem 7.1. (Artin [3]) If $Y = S(V)$ is the unit sphere in some G representation $\mathcal{H}$, then $\mathcal{H}$ can be chosen to be the set of cyclic subgroups of G.

The proof is a consequence of the induction theorem for characters.

Example 7.2. If $G = D_q$, the dihedral group with $2q$ elements then

$$2 \dim Y^{Z_2} + \dim Y^{Z_q} = \dim Y^1 + 2 \dim Y^{D_q}.$$

Theorem 7.3. (Borel [1]) If Y is a Z_p cohomology sphere and G an elementary abelian p group, then

$$\dim Y - \dim Y^G = \sum_{H \in \mathcal{H}} (\dim Y^H - \dim Y^G)$$

where $\mathcal{H}$ is the set of subgroups H of G of index p.

There is no similiar result if we only assume that Y is a closed homotopy sphere and G is a finite group. This is expressed as follows:

Theorem 7.4. (Dovermann and Petrie [2]): Let G be a finite group and $\mathcal{H}$ be a family of subgroups of G, $G \notin \mathcal{H}$. Assume there exists a function ψ such that for all closed homotopy spheres Y

$$\dim Y^G = \psi\{\dim Y^H \mid H \in \mathcal{H}\}.$$

Then G is a p group.

The idea of the proof is the following. Let G be not a p group. Then we construct an h map $f: X \to Y$ such that

$$\begin{aligned} \pi(Y) &\approx S(G) && \text{(chapter 2)} \\ \pi(X) &\approx S(G)-G && \text{(i.e. } X^G = \phi) \\ d(\alpha) &= d(\hat{f}(\alpha)) && \text{for all } \alpha \in \pi(X) \end{aligned}$$

and $\lambda(f)$ satisfies all assumptions in 6.1 and 6.8. Thus 6.7 applies and we obtain an h Equivalence $f_N: X_N \to Y$, such that $\dim X_N^H = \dim Y^H$ for all $H \neq G$, but $\dim X_N^G \neq \dim Y^G$.

We want to make some remarks how our results extend to other categories defined in [6]

a) h_p denotes the category in which we want to obtain a homology equivalence with coefficients in Z_p. All results we proved for $I(G,\lambda)$ are valid in this category and the proofs are the same.

b) s denotes the category in which we want to obtain a G homotopy equivalence. Here the choice C_α (chapter 3) will mean $c_\alpha = Id$ for all $\alpha \in \pi_1$. The proof is even simpler because we can still apply usual transversality theory.

c) wh denotes a category in which we require a weaker condition for the tangent bundle data (compare ii chapter 4), namely for $f: X \to Y$ there exists a bundle ξ over Y such that $(TX)_G \cong_s f_G^*(\xi_G)$. Here E is a contractiable space with free G action and $Y_G = E\times_G Y$. $f_G: X_G \to Y_G$ is the induced map. Furthermore, assume that ξ is orientiable.

Theorem 7.5. Assume 6.1 for λ. Then $I^{wh}(G,\lambda)$ is an abelian group.

Proof: Almost all the proof for the category h still holds for wh. We only have to consider the extension of the tangent bundle data. In the construction of the addition we do surgery on $B \subset Y$, $B = \amalg B_i$ and $B_i = G \times_H S^k$, where H operates trivially on S^k. If $k = 0$ the extension of ξ is trivial. The case of step 4 is reduced to the situation of a point. In step 2 we do surgery on copies of $G \times S^1$. As ξ is orientiable it extends to a bundle over the handles we attach.

The consequences 6.5-6.8 are proved for the categories $c = h_p, s, wh$ as in chapter 6 once we know that $I^c(G,\lambda)$ is a group.

REFERENCES

[1] Borel, A., et al., Seminar on transformation groups, Ann. of Math. Studies 46, Princeton University Press, (1960).

[2] Dovermann, K.H. and Petrie, T., Dihedral actions on homotopy spheres and a theorem of Artin, to appear.

[3] Feit, W., Characters of finite groups, Benjamin, N.Y. (1967).

[4] Milnor, J., Differential topology, notes, Princeton University (1957).

[5] Oliver, R. and Petrie, T., G surgery in the homotopy category and $K_0(Z(G))$, to appear.

[6] Petrie, T., G surgery II, to appear.

[7] __________, Pseudo equivalences of G manifolds, Proc. Sym. Pure Math., Vol. 32 (1978), 119-163.

[8] Steenrod, N.E., The topology of fiber bundles, Princeton University Press, Princeton, N.J. (1951).

[9] Wall, C.T.C., Surgery on compact manifolds, Academic Press, N.Y., (1970).

[10] Wassermann, A., Equivariant differential topology, Topology 8 (1967), 127-150.

Obstructions to Realizing Equivariant Witt Classes

by

John Ewing *

In many ways this is a mathematical apology for some often brutal calculations performed in [3]. Rather than an act of contrition, it is an effort to indicate the precise significance of the results of these calculations, and to show that the final result of [3] was wholly to be expected. (This is, however, hindsight rather than foresight.)

The work in [3] concerned the problem of realizing equivaraint Witt classes by smooth cyclic group actions on manifolds. Suppose we fix an odd prime p and are given a smooth, oriented, closed manifold M^{2n}, together with a smooth self-map $T: M^{2n} \to M^{2n}$ such that $T^p = 1$. From the pair (M^{2n}, T) we can manufacture an interesting algebraic invariant. There are three ingredients:

1. $V = H^n(M^{2n};\mathbb{Z})/\text{Torsion}$; a free $\mathbb{Z}$-module.
2. $\beta: V \times V \to \mathbb{Z}$; a non-singular, $(-1)^n$-symmetric bilinear form (induced by the cup product pairing.)
3. $T = T^*: V \to V$; a β-isometry of period p.

In general, for a fixed prime p, one would like to know which triples (V, β, T) can be realized in this way from a smooth periodic map on some manifold. At the moment such a question is hopelessly difficult.

We can gain a good deal of information, however, by putting an equivalence relation on the set of such triples which is natural both

---

* Partially supported by NSF Grant #MCS 76-05973.

algebraically and geometrically.

Definitions

1. We say (V,β,T) is metabolic if there exists a T-invariant submodule $U \subseteq V$ such that

$$0 \to U \hookrightarrow V \xrightarrow{ad} \mathrm{Hom}_{\mathbb{Z}}(U,\mathbb{Z}) \to 0$$

is exact, where $ad(x) = \beta(x,\cdot)$.

2. We say (V_1,β_1,T_1) is Witt equivalent to (V_2,β_2,T_2) if they are isomorphic after adding metabolic triples to each. (Addition, of course, means orthogonal sum.) The Witt class of a triple will be denoted by $[V,\beta,T]$.

3. Let W_0 (resp., W_2) denote the group of equivalence classes of symmetric (resp., skew-symmetric) triples (V,β,T). Addition is induced by orthogonal sum; the zero element is the metabolic class. We will generically write W_* for either group.

We can now ask a more reasonable question: up to Witt equivalence, which triples (V,β,T) are realizable by a smooth action on some smooth manifold? In other words, if we let $R_* \subseteq W_*$ denote the subgroup of Witt classes which are realizable, we want to compute R_* . Since W_* is well-known, (it's just a free abelian group on $(p+1)/2$ or $(p-1)/2$ generators, resp.), it is sufficient to calculate W_*/R_* . That's precisely the task we set about in [3].

The determination of R_* is, in principle, straight-forward.

One first notes that the Witt class is determined by the G-signature [1]. From the G-signature Theorem one knows that the G-signature is determined by the fixed point data. Armed with this, one simply considers all possible sets of fixed point data and determines all possible G-signatures. There are, however, a number of steps to fill in.

We shall return to the results of this calculation in a moment, but first, it is interesting to ask why, in general, one would expect that some Witt classes are not realizable. Indeed, on the basis of calculations for small primes p, one might conjecture the following.

Conjecture A: $R_* = W_*$ for all p.

In fact it is a relatively old result that conjecture A is false, and we shall see why in a moment. First, we need to briefly digress to "recall" some facts from number theory.

Digression. As before, p will always denote an odd prime. We let $\lambda = e^{2\pi i/p}$ and $Q(\lambda)$ = cyclotomic field. Inside $Q(\lambda)$ there is the ring of (algebraic) integers $\mathbb{Z}[\lambda]$, which consists of all (rational) integer combinations of powers of λ. For our purposes, however, it will be convenient to work with the slightly larger ring $P = \mathbb{Z}[\lambda, 1/p]$, which we get by inverting p in $\mathbb{Z}[\lambda]$. The ring P is also a Dedekind domain.

Recall that the ideal class group of P consists of equivalence classes of fractional P-ideals $\mathfrak{A} \subseteq Q(\lambda)$. Two ideals $\mathfrak{A}_1$ and $\mathfrak{A}_2$ are equivalent if there are numbers $\alpha_1, \alpha_2 \in Q(\lambda)$ such that $\alpha_1 \mathfrak{A}_1 = \alpha_2 \mathfrak{A}_2$. The group operation is induced by multiplication of ideals; the class of principal ideals serves as identity. We shall denote the ideal class group by $C = C(P)$. (One can show that $C(P) = C(\mathbb{Z}[\lambda])$.)

The order of C is usually denoted by h and can be factored into two factors, $h = h_1 h_2$, which are called the first and second factors.

The first factor, which is the more tractable of the two, can be described as the order of the subgroup

$$C^- = \{I \in C \mid I\bar{I} = 1\}.$$

(Here the bar denotes complex conjugation.) While h_1 is in principle computable for any p, it tends to grow rather rapdily: $h_1 = 1$ for $p \leq 19$; $h_1 = 3$ for $p = 23$; $h_1 \simeq 10^{27}$ for $p = 163$.

The second factor can be described in terms of P^*, the group of units of P. In general it's quite difficult to determine these units. Of course, there are certain obvious units, for example, $\pm\lambda^i$. Moreover, since

$$p = \prod_{k=1}^{p-1} (\lambda^k - \lambda^{-k}),$$

and p is invertible, we see that $\lambda^i - \lambda^{-i}$ are also "obvious" units. We let $E \subseteq P^*$ denote the group generated by these obvious units; in general it's pretty tough to find any unit <u>not</u> in E. The group E is often referred to as the group of "cyclotomic" units, but it might better be called the group of obvious units. The second factor h_2 is the index of E in

P^* , and from this description it ought to be clear that h_2 is extraodinarily difficult to compute.

Finally we can mention the "number theory" associated to P from a more modern point of view. Since P is a Dedekind domain, it follows that any finitely generated, torsion free P-module X is projective and moreover,

$$X \approx \mathfrak{a}_1 \oplus \dots \oplus \mathfrak{a}_n \ ,$$

where the $\mathfrak{a}_i$ are fractional P-ideals. The isomorphism class of X is determined by the rank n and the class $[\mathfrak{a}_1 \dots \mathfrak{a}_n] \in C$. From this point of view we recognize C as $\tilde{K}_0(P)$.

Returning from our digression, we must show that in general R_* is a proper subgroup of W_* . We begin by defining the ideal class invariant.

Given a triple (V,β,T) we temporarily forget β . Let $\tilde{V} = V \otimes \mathbb{Z}[1/p]$ and let $\tilde{V}_0 \subseteq \tilde{V}$ denote the submodule of elements fixed by T . Having inverted p , $\tilde{V}_0$ is a direct summand of $\tilde{V}$. In fact, if we let $S = 1+T+T^2+\dots+T^{p-1}$ then

$$0 \to \operatorname{Ker} S \to \tilde{V} \xrightarrow{S} \tilde{V}_0 \to 0$$

is split exact. Now we can think of Ker S as a projective P-module by letting $\lambda x = Tx$. From our comments above it follows that

$$\operatorname{Ker} S \cong \mathfrak{a}_1 \oplus \dots \oplus \mathfrak{a}_n \ ,$$

where the $\mathfrak{a}_i$ are fractional P-ideals, and we know that Ker S is determined by n and $[\mathfrak{a} = \mathfrak{a}_1 \mathfrak{a}_2 \cdots \mathfrak{a}_n] \in C$.

Definition: The ideal class invariant of (V,β,T) is

$$\iota(V,\beta,T) = [\sigma\tau] \in C .$$

But what happened to β ? Notice that β induces an isomorphism $V \xrightarrow{\simeq} \mathrm{Hom}_{\mathbb{Z}}(V,\mathbb{Z})$ by $x \longmapsto \beta(x,\cdot)$. If we let $Tf(x) = f(T^{-1}x)$ we see that this isomorphism is equivariant. Now carefully following our construction through we find that the equivariant isomorphism induced by β induces a P-module isomorphism

$$\sigma\tau \xrightarrow{\simeq} \mathrm{Hom}_P(\overline{\sigma\tau},P) = \overline{\sigma\tau}^{-1}$$

Hence $\sigma\tau\,\overline{\sigma\tau} \simeq P$ and $\iota(V,\beta,T) = [\sigma\tau] \in C^-$.

To summarize: the existence of β restricts the ideal class invariant to that special subgroup of the ideal class group which is associated to the first factor.

What if (V,β,T) is metabolic? Then the ideal class invariant is even more restricted. Indeed, suppose $U \subseteq V$ is T-invariant and and

$$0 \to U \to V \to \mathrm{Hom}_{\mathbb{Z}}(U,\mathbb{Z}) \to 0$$

is exact. Following the construction through once more we see that

$$\iota(V,\beta,T) = \iota(U)\overline{\iota(U)}^{-1} \in C .$$

We conclude that the ideal class invariant induces a homomorphism

$$\hat{\iota}: W_* \to \frac{C^-}{\{I\bar{I}^{-1} \mid I \in C\}} = H^1(\mathbb{Z}_2;C)$$

By explicitly providing a sufficient number of Witt classes it can be shown that $\hat{\iota}$ is onto. (See [2] or [3].)

Of course the range of $\hat{\imath}$ is an elementary abelian 2-group. Moreover, it is apparent that $H^1(\mathbb{Z}_2;C)$ is trivial whenever h_1 is odd. It is definitely <u>not</u> trivial for some values of p ; the first three are 29, 113, and 197.

Now here's the key result, proved by Swan in [5].

<u>Theorem (Swan)</u>: Given (M^{2n},T) as before,

$$\hat{\imath}[H^n(M^{2n};\mathbb{Z})/\mathrm{Tor},\beta,T] = 0 \quad ;$$

that is, R_* is contained in the kernel of $\hat{\imath}$ and so W_*/R_* is mapped epimorphically onto $H^1(\mathbb{Z}_2;C)$.

Since $H^1(\mathbb{Z}_2;C)$ is not always trivial, it is clear that conjecture A is false. The ideal class invariant is an obstruction to realizing a Witt class. Is it the only one?

<u>Conjecture B</u>: $W_*/R_* \approx H^1(\mathbb{Z}_2;C)$; that is, the ideal class invariant is the only obstruction.

The evidence looks good! We can mention first of all two results of the calculations done in [3].

<u>Theorem 1</u>: W_*/R_* is a 2-group.

<u>Theorem 2</u>: If h_1 is odd then $R_* = W_*$

(As we mentioned before, if Conjecture B is true both of these must be.)

But perhaps the most convincing evidence comes from the PL-case. We can go back to the beginning and ask only that our manifolds and maps be PL rather than smooth. We then have $R_*^{PL} \subseteq W_*$, the group of Witt classes which are realized by PL-actions.

Theorem 3: $W_*/R_*^{PL} \approx H^1(\mathbb{Z}_2; C)$.

This follows quite easily from work of Petrie [4], together with some elementary facts from surgery theory. Very briefly, the argument is as follows. We know that the question of which Witt classes can be realized is equivalent to the question of which G-signatures can be. By the realization theorem of surgery theory we can realize any surgery obstruction in $L^h(\mathbb{Z}_p)$ by a manifold with two boundary components, one of which is a standard lens space. Taking the universal cover, Petrie shows how to identify the G-signature with the multi-signature of the surgery obstruction. We can then cap one end of the manifold with a disk (orthogonal action) and cone-off the other end; the G-signature is undisturbed. Finally, the possible multi-signatures are well-known.

That's all very convincing; there is, however, a major difficulty arising from one more result obtained in [3].

Theorem 4: If $R_* = W_*$ then h_1 is odd.

This is, of course, a converse to Theorem 2. What's wrong? Suppose h_1 is even for some prime but $H^1(\mathbb{Z}_2; C) = 0$. Then our obstruction group is trivial, yet Theorem 4 insists that there is some obstruction. Can this ever happen? Yes; the first case is $p = 163$. (It can only happen, however, when h_2 is even.) So we must abandon Conjecture B; there is another obstruction.

To see exactly what it is we ought to concentrate on those Witt classes for which the first obstruction vanishes. Suppose

$[V,\beta,T] \in \operatorname{Ker} \hat{\imath}$. We can assume that for some representative $\imath(V,\beta,T)$ is trivial in C. That means that $\imath(V,\beta,T) = [\alpha P]$ for some $\alpha \in Q(\lambda)$. Let's go through our construction of the ideal class invariant once more. Recall that β induces an equivariant isomorphism $V \xrightarrow{\cong} \operatorname{Hom}_{\mathbb{Z}}(V,\mathbb{Z})$ which in turn induces a P-module isomorphism αP to $\bar{\alpha}^{-1}P$. Now such an isomorphism is given by multiplication by $u/\alpha\bar{\alpha}$, where $u \in P^*$, the units of P . Of course the unit u is only determined up to complex norms, but nonetheless, it is an important piece of information which we threw away before. Letting $N \subseteq P^*$ denote the group of complex norms, we have shown if $\imath(V,\beta,T)$ is trivial then we can define an invariant $\delta(V,\beta,T) \in P^*/N$. The invariant δ should properly be called the determinant of (V,β,T) .

Again we must compute the δ-invariant for metabolic triples, and it is quite easy to check that if (V,β,T) is metabolic then $\delta(V,\beta,T)$ is plus or minus a complex norm. Hence δ induces a homomorphism:

$$\hat{\delta}: \operatorname{Ker} \hat{\imath} \to P^*/\pm N$$

As before, we can show that this is onto by providing sufficient examples.

Now where is the analogue of Swan's Theorem for this invariant? It is contained in the following result which, once again, is a consequence of the calculation of [3, sec. 5].

<u>Theorem 5</u>: Given (M^{2n},T) as before,

$$\hat{\delta}[H^n(M^{2n};\mathbb{Z})/\mathrm{Tor},\beta,T] = \pm \prod_{k=1}^{(p-1)/2} (\lambda^k - \lambda^{-k})^{\varepsilon_k}$$

where $\varepsilon_k = 0$ or 1 depending on the fixed point data. In other words, if $[V,\beta,T] \in R_*$ then

$$\hat{\delta}[V,\beta,T] \in E/\pm N \subseteq P^*/\pm N$$

We might briefly say that for a Witt class to be realizable, not only must the ideal class invariant be trivial, but the determinant must be "cyclotomic".

We note that if h_2 is odd the groups $E/\pm N$ and $P^*/\pm N$ are in fact identical. In general, however, they are not. Unpublished results of the author show that, of all primes less than 1000, they differ for nine primes: 163, 277, 349, 397, 547, 607, 709, 853, and 937. Of course it follows that for each of these, h_2 is even.

Now it seems natural to try one more conjecture.

<u>Conjecture C</u>: $W_*/R_* \approx H^1(\mathbb{Z}_2;C) \oplus \frac{P^*/\pm N}{E/\pm N}$;

that is, these are the only obstructions.

The evidence is strong. In fact the final result of [3] which we mention is a direct consequence of Corollary 5.3 of that paper.

<u>Theorem 6</u>: $W_*/R_* \otimes \mathbb{Z}_2 \approx H^1(\mathbb{Z}_2;C) \oplus \frac{P^*/\pm N}{E/\pm N}$

The only question remaining, therefore, is whether or not W_*/R_* is an <u>elementary</u> 2-group; that is, is twice every Witt class realizable? We note that this is indeed the case for all primes less than 1000. Moreover, number theoretic considerations strongly suggest it is true in general. Nonetheless, it is still unproved, which is at least unsettling!

Finally, we note that while we have attempted here to carefully separate the two obstructions for the sake of exposition, P. Conner has very elegantly combined the two into one; namely, the discriminant. A thorough treatment can be found in [2]. We would like to acknowledge our great indebtedness to all this work.

References

[1] J.P. Alexander, P.E. Conner, G.C. Hamrick and J.W. Vick, Witt classes of integral representations of an abelian p-group, Bull. of AMS, 80(1974), 1179-1181.

[2] J.P. Alexander, P.E. Conner and G.C. Hamrick, Odd order group actions and Witt classification of innerproducts, Lec. Notes No. 625, Springer-Verlag, Berlin 1977.

[3] J.H. Ewing, The image of the Atiyah-Bott map, to appear in Math. Z.

[4] T. Petrie, The Atiyah-Singer invariant, the Wall groups $L_n(\pi,1)$ and the function te^x+1/te^x-1, Ann. of Math., 92(1970), 174-187.

[5] R.G. Swan, Invariant rational functions and a problem of Steenrod, Inventiones, 7(1969), 148-158.

STABLE G-SMOOTHING

Richard Lashof

Introduction

A theory of smoothing finite groups actions was given in [L2]. In this paper we investigate stable G-smoothing for G an arbitrary compact Lie group. A stable G-smoothing of a topological G-manifold M is a G-smoothing of the product of M with an orthogonal G-space $R(\rho)$. The main result is a bijective correspondence between isotopy classes of stable G-smoothings and isotopy classes of stable G-vector bundle reductions of the tangent microbundle τM; i.e., G-vector bundle reductions of $\tau M \oplus \varepsilon_\sigma$, where ε_σ is the product bundle $M \times R(\sigma)$.

In §1 we give the elementary theory of stable G-smoothing and in particular show that for manifolds with a single orbit type, stable G-smoothing implies G-smoothing. Section 1 only requires a knowledge of G-vector bundles. To make further progress, one needs the theory of G-microbundles as developed in [L2] and independently by Le Dimet [L3]. In [L2] we also gave a theory of normal G-microbundles following Milnor's arguement in the G-trivial case. In §2 we take the opportunity of developing a more complete theory of normal G-microbundles following Hirsch's argument in the G-trivial case. (Ibisch also developed the theory of normal G-microbundles much earlier [I1], but only an announcement has appeared so far as I can discover. He was of course also interested in developing a G-smoothing theory [L3]). The proof of the main result is given in §3.

Stable G-smoothing Theory I

In this section we give a direct argument to show that if the tangent microbundle of a G-manifold M with finitely many orbit types reduces to a G-vector bundle then $M \times R(\sigma)$ is G-smoothable, $R(\sigma)$ some orthogonal G-space. Neither the notion of reduction of the tangent microbundle nor the proof of stable G-smoothability actually requires the theory of G-microbundles (cf. [Cl]).

To formulate reducibility we introduce the category $S_G(X)$ of G-spaces over a G-space X with cross-section. I.e., an object in $S_G(X)$ is a triple (E,p,s) where E is a G-space, $p: E \to X$ is a G-map called <u>projection</u> and the <u>cross-section</u> $s: X \to E$ is an equivariant map such that $ps = 1_X$. A morphism $\varphi: (E_1,p_1,s_1) \to (E_2,p_2,s_2)$ is an equivariant map $\varphi: E_1 \to E_2$ such that $p_1 = p_2\varphi$ and $\varphi s_1 = s_2$.

<u>Example 1</u>: A G-vector bundle $\xi = (E(\xi),p_\xi,s_\xi)$, where p_ξ is projection and s_ξ is the 0-section.

<u>Example 2</u>: $\tau M = (M \times M, pr_1, d)$, where $pr_1: M \times M \to M$ is projection onto the first factor and $d: M \to M \times M$ is the diagonal.

<u>Definition</u>: Let M^n be a topological G-manifold. We will say that the "tangent microbundle of M reduces to a G-vector bundle" if there exists an n-dimensional G-vector bundle ξ and an $S_G(X)$ embedding $\varphi: \xi \to \tau M$. I.e., $\varphi: E(\xi) \to M \times M$ is an $S_G(X)$ morphism and a topological embedding.

We also need the notion of inverse-bundle for G-vector bundles. First note that the Whitney sum $\xi \oplus \eta$ of two G-vector bundles over X is again a G-vector bundle. Also recall that a G-trivial bundle $\varepsilon_\sigma = \varepsilon_\sigma(X,G)$ is a G-vector bundle equivalent to a G-product bundle $X \times R(\sigma)$, for some orthogonal G-space $R(\sigma)$. More generally, if S is an H-space we will write $\varepsilon_\sigma(S,H)$ for the G-vector bundle $G \times_H (S \times R(\sigma))$ over $G \times_H S$.

<u>Definition</u>: Let ξ be a G-vector bundle over X. A G-vector bundle η over X is called an <u>inverse</u> to ξ if for some orthogonal representation σ of G, $\xi \oplus \eta \simeq \varepsilon_\sigma$.

To construct inverse bundles we need a condition on X:

Definition: A space X is called finitistic if every open cover has a finite dimensional open refinement. A G-space X is called G-finitistic if X/G is finitistic. This means that every open G-cover has a finite dimensional open G-refinement. A compact space or a space with finite covering dimension is finitistic.

Proposition 1.1: Let ξ be a G-vector bundle over a paracompact G-space X such that X is G-finitistic. If $E(\xi)$ has only finitely many orbit types, then ξ has an inverse bundle.

The proof of Proposition 1.1 will appear elsewhere. For a slightly weaker result see [W1].

Corollary 1.2: A G-vector bundle ξ over G-space X has an inverse if

a) X is compact, or

b) X is a finite dimensional separable metric space and $E(\xi)$ has finitely many orbit types.

Theorem 1.3: Suppose M^n is a topological G-manifold with a finite number of orbit types such that its tangent microbundle τM reduces to a G-vector bundle ξ. Then there is an orthogonal G-space $R(\sigma)$ such that $M \times R(\sigma)$ has a G-smoothing with tangent G-vector bundle $pr_1^*\xi \oplus \varepsilon_\sigma$.

First we note:

Lemma 1.4: Let M be a topological G-manifold such that τM reduces to a G-vector bundle, then for each $x \in M$ there is a smoothing of a G_x invariant neighborhood U_x such that G_x acts smoothly on U_x.

Proof: φ enbeds $E(\xi)_x$ into $(x) \times M$ such that $\varphi_x(0) = (x,x)$. Thus $pr_2\varphi$ is a G_x homeomorphism of $E(\xi)_x$ with an open G_x invariant neighborhood U_x of x in M. Since G_x acts linearly on $E(\xi)_x$ we get a G_x smoothing of U_x,

Corollary 1.5: For M as in (1.4), if H is a closed subgroup of G, M^H is a locally flat submanifold.

Proof: U_x^H is a smooth submanifold and hence topologically locally flat. But $U_x^H = U_x \cap M^H$. Hence M^H is locally flat.

Proof of Theorem 1.3: First consider the case where $\xi \simeq \varepsilon_\rho$ and $E(\xi) \simeq M \times R(\rho)$. Let $\varphi: M \times R(\rho) \to M \times M$ be the $S_G(M)$ embedding. Now M embeds equivariantly in some orthogonal G-space $R(\alpha)$. Since M^H is a manifold and hence an ANR, Jaworowski's extension theorem [J1] implies there is an invariant neighborhood N of M in $R(\alpha)$, and an equivariant refraction $r: N \to M$. Define $\theta: M \times R(\alpha) \to R(\alpha)$ by $\theta(x,y) = x + y$. Then θ is equivariant. Further there is an invariant map $\delta: M \to R_+$ such that $\theta(x,y) \in N$ for $|y| < \delta(x)$. Define $\theta': M \times R(\alpha) \to N$ by $\theta'(x,y) = \theta(x,\delta(x)y)$. Then θ' is equivariant and $\theta'(x,0) = x$. For δ sufficiently small we also have $(r\theta'(x,y),x) \subset \varphi(M \times R(\rho)) \subset M \times M$. Define $\psi: M \times R(\alpha) \to R(\alpha) \times R(\rho)$ by $\psi(x,y) = (\theta'(x,y), pr_2\varphi^{-1}(r\theta'(x,y),x))$, $pr_2: M \times R(\rho) \to R(\rho)$. Then ψ is equivariant and

(1) $\psi(x,0) = (x, pr_2\varphi^{-1}(x,x)) = (x,0)$

(2) ψ is injective. I.e.,

$$\psi(x_1,y_1) = \psi(x_2,y_2) \Longrightarrow pr_2\varphi^{-1}(r\theta'(x_1,y_1),x_1)$$

$$= pr_2\varphi^{-1}(r\theta'(x_2,y_2),x_2)$$

and $\theta'(x_1,y_1) = \theta'(x_2,y_2)$. Hence $r\theta'(x_1,y_1) = r\theta'(x_2,y_2)$

and since the map $x \to pr_2\varphi^{-1}(x_0,x)$ is injective, we must have $x_1 = x_2$. But $\theta'(x,y) = x + \delta(x)y$ and hence $x_1 = x_2$ implies $y_1 = y_2$.

By invariance of domain, ψ is open. Thus ψ pulls back a G-smoothing with tangent bundle $\varepsilon_{\alpha\oplus\rho} = \varepsilon_\alpha \oplus \varepsilon_\rho$, proving the special case of $\xi \simeq \varepsilon_\rho$.

In the general case, let η be an inverse of ξ, say $\xi \oplus \eta \simeq \varepsilon_\rho$. Let $\varphi\colon E(\xi) \to M \times M$ be the $S_G(M)$ embedding. Then the lemma below gives a G-vector bundle reduction $\hat{\varphi}\colon E(\eta) \times R(\rho) \simeq E(p_\eta^*(\xi \oplus \eta)) \to E(\eta) \times E(\eta)$. Thus by the special case, for some α, $E(\eta) \times R(\alpha)$ is G-smoothable with tangent bundle $\varepsilon_{\rho\oplus\alpha}$. Now $M \times R(\rho) \times R(\alpha) = E(\xi \oplus \eta) \times R(\alpha) = E(p_\eta^*\xi) \times R(\alpha)$ is a G-vector bundle over $E(\eta) \times R(\alpha)$; and hence can be given the structure of a smooth G-vector bundle over the smoothing of $E(\eta) \times R(\alpha)$. Thus $M \times R(\rho \oplus \alpha)$ admits a G-smoothing with tangent bundle $pr_1^*\xi \oplus \varepsilon_{\rho\oplus\alpha}$.

<u>Lemma 1.6</u>: Let η be a G-vector bundle over a G-manifold M. A G-vector bundle reduction $\varphi\colon E(\xi) \to M \times M$ defines a G-vector bundle reduction $\hat{\varphi}\colon E(p_\eta^*(\varepsilon \oplus \eta)) \to E(\eta) \times E(\eta)$.

<u>Proof</u>: By definition, $pr_1 \circ \varphi = p_\xi$ and $\varphi s_\varepsilon = d$. Since the 0-section of $E(\xi)$ is a G-deformation retract, $s_\xi p_\xi \sim 1_\xi$. Hence $p_\xi \sim pr_1 \circ \varphi s_\xi p_\xi = pr_1 \circ d \circ p_\xi = pr_2 \circ d \circ p_\xi = pr_2 \circ \varphi \circ s_\xi p_\xi \sim pr_2 \circ \varphi$. Thus there is a G-vector bundle map $\tilde{\varphi}_2 E(p_\xi^*\eta) \to E(pr_2^*\eta)$ covering φ. Since $pr_2^*\eta$: $M \times E(\eta) \xrightarrow{1\times p_\eta} M \times M$, $E(\xi \oplus \eta) = E(p_\xi^*\eta)$ and $p_{\xi\otimes\eta}\colon E(p_\xi^*\eta) \to E(\xi) \xrightarrow{p_\xi} M$; we have $pr_1\tilde{\varphi}_2 = p_{\xi\otimes\eta}$ and $\tilde{\varphi}_2 s_{\xi\otimes\eta} = (1 \times s_\eta)\circ d$. Thus $\tilde{\varphi}_2\colon E(\xi \oplus \eta) \to M \times E(\eta)$ is an $S_G(M)$ embedding ($\tilde{\varphi}_2$ is an ambedding since it covers the embedding φ). Pulling these G-spaces over M back over $E(\eta)$ by p_η, $\tilde{\varphi}_2$ defines an $S_G(E(\eta))$ embedding $\hat{\varphi}\colon E(p_\eta^*(\xi \oplus \eta)) \to E(\eta) \times E(\eta)$.

Remark: The above argument will prove the corresponding result for a G-R^n bundle η over M and a G-R^n bundle reduction $\varphi: \xi \to \tau M$, provided the 0-section of $E(\xi)$ is a G-deformation retract.

Next we give an example of a stably G-smoothable but not G-smoothable G-manifold, and on the other hand some conditions under which stable G-smoothability implies G-smoothability.

Example: Let S be a non-locally Euclidean topological space such that $S \times R = E^n$. Then for any compact Lie group G, $G \times S$ is a manifold and by taking G acting on the first factor by left translation it is a G-manifold. Then if $M = G \times S$ were G-smoothable, $S = M/G$ would be a smooth manifold. On the other hand, $M \times R = G \times S \times R = G \times R^n$ is G-smoothable.

Proposition 1.7: Let M be a locally smoothable G-manifold with just one orbit type (H). If $M \times R(\sigma)$ is G-smoothable for some orthogonal G-space $R(\sigma)$, then M is G-smoothable, provided $\dim M/G \geq 5$.

Proof: First consider the case where the action of G on M is free. Then $M \times_G R(\sigma)$ has a smooth structure. (Here we consider M to be a right G-space under the action $xg = g^{-1}x$.) On the other hand, $\pi: M \times_G R(\sigma) \to M/G$ is a vector bundel over M/G. Let η be an inverse bundle over M/G. Then $E(\pi^*\eta) \to M \times_G R(\sigma)$ admits a smooth vector bundle structure. But $E(\pi^*\eta) = M/G \times R^n$. But this implies M/G is smoothable and hence that M is G-smoothable.

Now suppose M has orbit type H, then $M = G/H \times_{\overline{N}(H)} M^H$, where $\overline{N}(H) = N(H)/H$ acts freely on M^H. Suppose $M \times R(\sigma)$ is G-smoothable. Then $(M \times R(\sigma))^H = M^H \times R(\sigma)^H$ is $\overline{N}(H)$ smoothable. $R(\sigma)^H$ is of course an orthogonal $\overline{N}(H)$ space. By the special case. M^H is $\overline{N}(H)$ smoothable. (Note that $\dim M^H/\overline{N}(H) = \dim M/G \geq 5$.) Hence M is G-smoothable.

Finally we note a relation between local G_x-smoothability (see (1.4)) and local G-smoothability.

Proposition 3.8: Suppose M^n is a topological G-manifold such that each $x \in M$ has a G_x smoothable G_x invariant neighborhood. Then there is an orthogonal G-space $R(\sigma)$ such that $M \times R(\sigma)$ is locally G-smoothable.

Proof: Let L be the Lie algebra of G. With respect to a biinvariant metric, the adjoint action of G makes L into an orthogonal G-space $R(\sigma)$. We claim $M \times R(\sigma)$ is locally G-smoothable.

Let S_x be a slice through $x \in M$. If D is an open normal disc to G_x in G through $1 \in G$; then $D \times S_x \subset G \times_{G_x} S_x$ is a G_x invariant neighborhood of $[1,x]$, and $DS_x \subset GS_x$ is a G_x invariant neighborhood of x in M. Now $GS_x \times R(\sigma) \simeq (G \times_{G_x} S_x) \times R(\sigma)$ is a G-invariant neighborhood of $(x,0)$ in $M \times R(\sigma)$. Let $\sigma' = \sigma|G_x$. Then $\varphi: G \times_{G_x} (S_x \times R(\sigma')) \to (G \times_{G_x} S_x) \times R(\sigma)$, $\varphi[g,(y,z)] = ([g,y], \sigma(g)z)$ is a G-equivalence. But $R(\sigma') = R(\sigma_1) \times R(\sigma_2)$, where $R(\sigma_2)$ is the Lie algebra of G_x under the adjoint action and $R(\sigma_1) = R(\sigma_2)^{\perp}$. Also $R(\sigma_1)$ may be identified with D under the exponential map. Thus $G \times_{G_x} (S_x \times R(\sigma') \simeq G \times_{G_x} (S_x \times D \times R(\sigma_2)) \simeq G \times_{G_x} (DS_x \times R(\sigma_2))$. By assumption DS_x contains a G_x smoothable neighborhood U_x. Thus $\varphi(G \times_{G_x} (U_x \times R(\sigma_2)) \subset GS_x \times R(\sigma) \subset M \times R(\sigma)$ is a G-smoothable neighborhood of $(x,0)$ in $M \times R(\sigma)$. The union of these smoothable neighborhoods is a G-invariant neighborhood N of $M \times 0$ in $M \times R(\sigma)$. But there is an $S_G(M)$ embedding of $M \times R(\sigma)$ in N. Hence $M \times R(\sigma)$ is locally G-smoothable.

Normal G-microbundles

Definition: A G-R^n bundle ξ is an R^n bundle $p: E \to X$ with 0-section $s: X \to E$, such that E and X are G-spaces and p and s are G-maps. A G-R^n bundle is called locally linear if each

$x \in X$ has a G-invariant neighborhood U_x such that $\xi|U_x$ is equivariantly equivalent to a G-vector bundle.

Definition: A G-microbundle $\mu: X \overset{s}{\to} E \overset{p}{\to} X$ is a microbundle such that E and X are G-spaces and p and s are G-maps. If X is paracompact, we say two G-microbundles $\mu_i: X \xrightarrow{s_i} E_i \xrightarrow{p_i} X$, $i = 1,2$, are equivalent if there exist invariant neighborhoods V_i of $s_i(X)$ in E_i, and a G-equivalence $\Phi: V_1 \to V_2$ such that $p_2\varphi = p_1$ and $s_2 = \varphi s_1$.

A G-microbundle μ is locally linear if each $x \in X$ has a G-invariant neighborhood U_x such that $\mu|U_x$ is equivalent (as a G-microbundle) to a G-vector bundle.

A locally linear $G\text{-}R^n$ bundle is a locally linear G-microbundle. Conversely we proved in [L2] (see also [L3]):

Theorem A: Let μ be an n-dimensional locally linear G-microbundle over a paracompact G-space X. Let U_0 be an invariant neighborhood of the closed invariant subspace $A \subset X$. Then

a) If ξ_0 is a locally linear $G\text{-}R^n$ bundle over U_0 contained in $\mu|U_0$, there is a locally linear $G\text{-}R^n$ bundle ξ in μ such that $\xi|A = \xi_0|A$.

b) If ξ_1, ξ_2 are locally linear $G\text{-}R^n$ bundles contained in μ with $\xi_1|U_0 = \xi_2|U_0$, there is a homotopy of the inclusion $E(\xi_1)$ in $E(\mu)$ through equivariant microbundle embeddings, rel A, to a $G\text{-}R^n$ bundle equivalence of $E(\xi_1)$ onto $E(\xi_2)$.

Corollary 1: There is a bijective correspondence between equivalence classes of locally linear G-microbundles and $G\text{-}R^n$ bundles over paracompact G-spaces X.

Corollary 2: The 0-section of a locally linear $G\text{-}R^n$ bundle over a paracompact G-space X is an equivariant fibrewise deformation retract.

If X is a G-space, a <u>G-chart</u> (S,H) on X is an H-invariant subspace S such that $G(S)$ is open and the action $G \times S \to G(S)$ induces a G-homeomorphism of $G \times_H S$ onto $G(S)$. I.e., S is a generalized slice in the sense of Palais [B2].

In this section we assume that all G-microbundles are locally linear and have paracompact base spaces.

If $\mu\colon X \overset{s}{\to} E \overset{p}{\to} X$ is a G-microbundle and $f\colon X^1 \to X$ is an equivariant map of G-spaces, then the <u>induced</u> G-microbundle $f^*\mu$: $X^1 \xrightarrow{s'} E' \xrightarrow{p'} X'$ is defined in the usual way: $E' = \{(x',z) \subset X' \times E \mid f(x') = p(z)\}$, $p'(x',z) = z'$, $s'(x') = (x',f(x'))$. Also for $pr_2\colon X' \times E \to E$ induces the natural map $\pi_2\colon E' \to E$.

If $\mu_i\colon X_i \xrightarrow{s_i} E_i \xrightarrow{p_i} X_i$, $i = 1,2$, are G-microbundles their <u>product</u> $\mu_1 \times \mu_2\colon X_1 \times X_2 \xrightarrow{s_1 \times s_2} E_1 \times E_2 \xrightarrow{p_1 \times p_2} X_1 \times X_2$ is a G-microbundle with $\dim \mu_1 \times \mu_2 = \dim \mu_1 + \dim \mu_2$. The <u>Whitney sum</u> $\mu_1 \oplus \mu_2$ of two G-microbundles over X is defined to be $d^*(\mu_1 \times \mu_2)$, where $d\colon X \to X \times X$ is the diagonal map. Note that $\mu_1 \oplus \mu_2 \simeq \mu_2 \oplus \mu_1$.

Given two G-microbundles $\mu\colon X \overset{s}{\to} E \overset{p}{\to} X$. $\nu\colon E \overset{t}{\to} D \overset{q}{\to} E$ the <u>composition</u> $\mu\circ\nu$ is $\mu\circ\nu\colon X \xrightarrow{t\circ s} D \xrightarrow{p\circ q} X$. Again $\dim \mu\circ\nu = \dim \mu + \dim \nu$. It is easy to check that $\mu_1 \oplus \mu_2 \simeq \mu_1 \circ p_1^* \mu_2 \simeq \mu_2 \circ p_2^* \mu_1$.

<u>Definition</u>: Let μ be a G-microbundle over X. A G-microbundle ν over X is called an <u>inverse</u> if $\mu \oplus \nu \simeq \varepsilon_\sigma$, for some orthogonal representation σ of G.

Further in this section we will assume all G-manifolds are locally smoothable or at least that each $x \in M$ has a G_x smoothable G_x neighborhood. This last is equivalent to requiring the tangent microbundle τM to be a locally linear G-microbundle (see 1.4). We also assume all G-manifolds have a finite number of orbit types.

Let M be a G-manifold and let V be a G-manifold containing M as an invariant subspace. Also let K be a closed invariant subset of M. Throughout this section we will assume K, M, V are related in this fashion.

Definition: A normal microbundle on M in V is a G-microbundle $\nu: M \xrightarrow{i} E \xrightarrow{p} M$, where E is a neighborhood of M in V. Let ν_j: $M \xrightarrow{i_j} E_j \xrightarrow{p_j} M$, $j = 0,1$ be normal microbundles on M in V. A rel K isotopy from ν_0 to ν_1 is a G-isotopy $f_t: E \to V$ such that:

1) $E \subset V$ is a neighborhood of M in E_0

2) f_t is fixed on M and on $E_0 \cap p_0^{-1}K$

3) f_1 is a microbundle equivalence of ν_0 with ν_1.

If such an isotopy exists we will write $\nu_0 \cong \nu_1$ rel K. If $\nu_0 \cong \nu_1$ rel a neighborhood of K we write $\nu_0 \cong \nu_1$ rel $[K]$.

Consider pairs (U,ν) where U is an invariant neighborhood of K in M and ν is a normal microbundle on U in V. (U,ν) and (U',ν') are in the same K-germ if there exists a neighborhood W of K in $E(\nu) \cap E(\nu')$ such that $p|W = p'|W$. The class of (U,ν) is called the K-germ of ν and is denoted $[\nu]_K$. We say $[\nu]_K$ extends over M if there exists a normal microbundle ν on M such that $[U]_K = [\nu]_K$.

If M is a G-manifold write $\tau_1(M) = \tau(M)$: $M \xrightarrow{d} M \times M \xrightarrow{p_r} M$, and $\tau_2(M)$: $M \xrightarrow{d} M \times M \xrightarrow{pr_2} M$. $\tau_2(M)$ is called the 2nd tangent microbundle. Note that $\tau_1(M)$ and $\tau_2(M)$ are equivalent as G-microbundles. We may also consider $\tau_i(M)$, $i = 1,2$, as normal microbundle on the diagonal in $M \times M$.

Definition: A G-manifold is symmetric if $\tau_1(M) \cong \tau_2(M)$ as normal microbundles on $d(M)$ in $M \times M$.

Lemma 2.1:

a) Every G-smoothable G-manifold is symmetric.

b) Every invariant open subset of a symmetric G-manifold is a symmetric G-manifold.

<u>Proof</u>:

a) Take an invariant Riemannian metric with respect to the smooth structure and let TM be the tangent G-vector bundle. Let φ_i: $TM \to M \times M$, $i = 1,2$, $\varphi_1(x,X) = (x, \exp_x X)$, $\varphi_2(x,X) = (\exp_x X, x)$. Then φ_i embeds a neighborhood of the 0-section of TM as a neighborhood of the diagonal in $M \times M$. Further φ_i is a G-microbundle equivalence of TM with $\tau_i(M)$, $i = 1,2$. By the uniqueness of smooth G-normal tubes [B3], $\tau_1(M) \cong \tau_2(M)$.

b) This follows trivially from the definition.

Write d again for the composition $M \xrightarrow{d} M \times M \xrightarrow{1\times i} \mu \times V$. Let ν be a normal microbundle on M in V and consider the two normal microbundles on D(M) in $M \times V$:

$$\tau_V|M\colon M \xrightarrow{d} M \times V \xrightarrow{pr_1} M, \text{ and}$$

$$\nu^*\colon M \xrightarrow{d} M \times E(\nu) \xrightarrow{p\circ pr_2} M.$$

Observe that ν^* is the composition $\nu^* = \tau_2(M)\circ(M \times \nu)$, where $M \times \nu\colon M \times M \xrightarrow{1\times s} M \times E(\nu) \xrightarrow{1\times p} M \times M$. On the other hand, $\tau_V|M = \tau_1(M)\circ(M \times \nu)$. (One can assume $V = E(\nu)$.) If M is symmetric, the isotopy of $\tau_1(M)$ to $\tau_2(M)$ induces an isotopy of $\tau_V|M$ to ν^* in $M \times V$. Thus we have:

<u>Lemma 2.2</u>. If M is symmetric $\tau_V|M \cong \nu^*$.

<u>Corollary 2.3</u>: If ν_0 and ν_1 are normal microbundles on M in V then $\nu_0^* \cong \nu_1^*$.

<u>Corollary 2.4</u>: If ν_0 and ν_1 are normal microbundles on M in V and $[\nu_0]_K = [\nu_1]_K$, then $\nu_0^* \cong \nu_1^*$ rel [K].

<u>Proof</u>: Since locally linear G-microbundles contain locally linear

$G\text{-}R^n$ bundles, we can consider $M \times \nu_i$ to be locally linear $G\text{-}R^n$ bundles over $M \times M$. The isotopy of ν_i^* to $\tau_V|M$ may be viewed as a G-bundle covering homotopy of the isotopy of $\tau_2(M)$ to $\tau_1(M)$. Since we may assume $\nu_0|U = \nu_1|U$, U a neighborhood of K, we may assume the isotopies of ν_0^* and ν_1^* to $\tau_V|M$ are the same over a neighborhood of K. (This can be seen by viewing the covering homotopy property as a question of extending cross-sections in an associated bundle.) But then the isotopy of ν_0^* to ν_1^* obtained by composing the first isotopy with the inverse of the second is constant over a neighborhood of K.

<u>Definition</u>: We will say that M^n is <u>G-parallelizable</u> if $\tau(M) \simeq G \times_H (S \times R(\sigma))$, where $M = G \times_H S$ and $\sigma: H \to O(n)$.

<u>Lemma 2.5</u>: Assume M is G-parallelizable and that $r: V \to M$ is a G-retraction. There is a G-embedding $h: G \times_H (r^{-1}(S) \times R(\sigma)) \to M \times V$ making the diagram below commutative. (Here M and $\tau(M)$ are as above.):

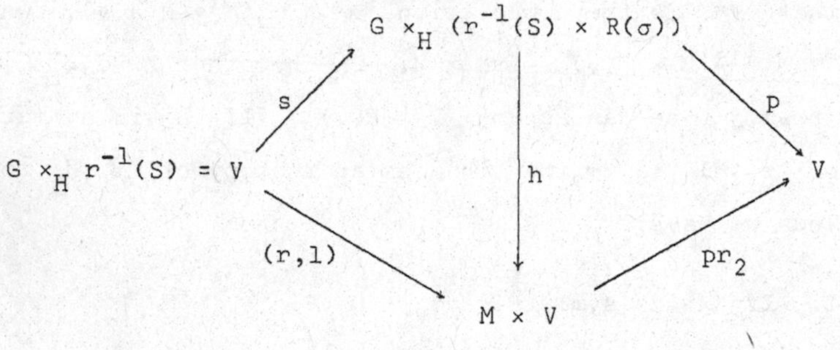

where s is the 0-section and p is the projection onto $G \times_H r^{-1}(S) = V$.

<u>Proof</u>. The top edge of the diagram represents $r^*\tau_1 M$, while the bottom edge represents $r^*\tau_2 M$. But these are equivalent. Hence h exists.

<u>Proposition 2.6</u>: Let M be G-parallelizable, say $M \simeq G \times_H S$, $\tau M \simeq G \times_H (S \times R(\sigma))$. Let $\bar{\sigma}$ be an orthogonal representation of G containing σ as a subrepresentation. Then

a) M has a normal microbundle in $V \times R(\bar{\sigma})$.

b) If also M is symmetric and ν_0, ν_1 are normal microbundles on M in V having the same K-germ,

$$\nu_0 \oplus \varepsilon_{\bar{\sigma}} \cong \nu_1 \oplus \varepsilon_{\bar{\sigma}} \text{ rel } [K].$$

<u>Proof</u>: By [J1] we can replace V be a neighborhood which retracts to M.

a) From the commutativity of the diagram in (2.5), $hs|M = d: M \to M \times V$. Since $M \times V$ is the total space of $\tau_V|M$, and h is a G-homeomorphism onto a neighborhood of $d(M)$ in $M \times V$, $s(M) = h^{-1}(dM)$ has a normal microbundle in $G \times_H (r^{-1}(S) \times R(\sigma))$.

Now $\bar{\sigma}|H \simeq \sigma \oplus \sigma'$, and $V \times R(\bar{\sigma})$ is G-equivalent to $G \times_H (r^{-1}(S) \times R(\sigma) \times R(\sigma'))$, which is a G-vector bundle over $G \times_H (r^{-1}(S) \times R(\sigma))$. Thus M has a normal microbundle in $V \times R(\bar{\sigma})$.

b) We can assume M is a G-deformation retract of $E(\nu_i)$ (Corollary 2 of Theorem A). Thus $r|E(\nu_i)$ is homotopic through retractions to $p_i: E(\nu_i) \to M$. Note that if h_i is the equivalence in (2.5) corresponding to the retraction p_i, then h_i is an isomorphism of $\nu_i \oplus \varepsilon_\sigma(S,H)$ with ν_i^*. If we take $r = p_0$, the homotopy of p_0 with p_1 can be assumed fixed over a neighborhood of K. Since $\nu_0^* \cong \nu_1^*$ rel $[K]$, it follows that $\nu_0 \oplus \varepsilon_\sigma(S,H) \cong \nu_1 \oplus \varepsilon_\sigma(S,H)$ rel $[K]$ in $E(r^*\varepsilon_\sigma(S,H))$ and hence $\nu_0 \oplus \varepsilon_\sigma \cong \nu_1 \oplus \varepsilon_{\bar{\sigma}}$ rel $[K]$ in $V \times R(\bar{\sigma})$, $\bar{\sigma}|H \simeq \sigma \oplus \sigma'$.

Proposition 2.7 (Isotopy Extension Theorem): Let E be an invariant neighborhood of K in V. Suppose $f_t: E \to V$ is a G-isotopy fixed on $M \cap E$. There exists an invariant neighborhood E' of M in $V \times R$ (trivial action on R) and a G-isotopy $H_t: E' \to V \times R$ which is fixed on M and has the same K-germ as $f_t \times 1$.

Proof. Choose invariant neighborhoods E_0, E_1 of K in V with $\bar{E}_0 \subset E_1$ and $\bar{E}_1 \subset E$. Let $\varphi: E \to R$ be an invariant function such that $\varphi = 0$ on $\bar{E}_0$ and $\varphi = -3$ outside E_1. Let $T: E \times R \to E \times R$ be the G-equivalence $T(x,s) = (x, s + \varphi(x))$. Now let $F_t: E \times R \to V \times R$ be the isotopy such that $F_t | E \times [-1,1] = f_t \times \mathrm{id}$ and $F_t =$ identity outside $E \times [-2,2]$; i.e., slow the isotopy down to zero as we go from -1 to -2 and from 1 to 2. Let $G_t = T^{-1}F_tT$: $E \times [-1,1] \to V \times R$. If x is outside E_1, $G_t(x,s) = T^{-1}F_t(x,s-3) = (x,s)$, and if $x \in M$, $G_t(x,s) = T^{-1}F_t(x,s-\varphi(x)) = T^{-1}(x,s-\varphi(x)) = (x,s)$. Then G_t extends by the identity to H_t on a neighborhood E' of M in $V \times (-1,1)$. Since $G_t | E_0 \times [-1,1] = f_t \times 1$, H_t agrees with $f_t \times 1$ on a neighborhood of K.

Corollary 2.8: Let $\nu: M \overset{i}{\to} V \overset{p}{\to} M$ be a normal microbundle on M in V and let U be a neighborhood of K. Let E be a neighborhood of U in $p^{-1}(U)$ and $f_t: E \to p^{-1}(U)$ a G-isotopy of $\nu|U$ to a normal microbundle η on U in V. Then there is a G-isotopy h_t of $\nu \oplus \varepsilon_0$, ε_0 the trivial line bundle $M \times R$ over M, to a normal microbundle ξ on M in $V \times R$ such that

a) h_t has the same K-germ as $f_t \times 1: E \times R \to V \times R$.

b) ξ has the same K-germ as $\eta \oplus \varepsilon_0$.

c) The K-germ of $\xi \oplus \varepsilon_0$ extends over M in $V \times R$.

Proposition 2.9: Assume that $M - K$ is G-parallelizable and symmetric. If ν is a normal bundle on U in V, U a neighborhood of K in M; the K-germ of $\nu \oplus \varepsilon_\sigma$, some orthogonal G-representation

σ, extends over M in $V \times R(\sigma) \times R$.

Proof: Let K' be a closed invariant neighborhood of K in U. By replacing M by $M - K$, U by $U - K$ and K by $K' - K$, we may assume M is G-parallelizable and symmetric.

Now $\tau_V|U$ and ν^* are isotopic normal microbundles on $d(U)$ in $M \times V$. By Corollary 2.8, $\nu^* \oplus \varepsilon_0$ extends to a normal microbundle η on $d(M)$ in $M \times V \times R$; i.e., η has the same K-germ. It follows from (2.5) (see proof of (2.6)) that there is a G-vector bundle ξ such that $\eta \oplus \xi$ is a G-normal microbundle on M in $V \times R(\sigma) \times R$ and that $\nu^* \oplus \xi|U \oplus \varepsilon_0 \simeq \nu \oplus \varepsilon_\sigma \oplus \varepsilon_0$. Thus $\eta \oplus \xi$ extends $\nu \oplus \varepsilon_\sigma \oplus \varepsilon_0$.

Theorem 2.10: For some orthogonal G-representation σ (σ not necessarily the same in a), b), c)):

a) M has a normal microbundle in $V \times R(\sigma)$.

b) If ν is a normal microbundle on U in V, U a neighborhood of K in M, the K-germ of $\nu \oplus \varepsilon_\sigma$ extends over M in $V \times R(\sigma)$.

c) If ν_0 and ν_1 are normal microbundles on M in V having the same K-germ, then $\nu_0 \oplus \varepsilon_\sigma \cong \nu_1 \oplus \varepsilon_\sigma$ rel [K].

Proof: By replacing M by $M \times R(\alpha)$ and V by $V \times R(\alpha)$, α an orthogonal G-representation, we can assume M is locally smoothable (1.7). Since M has a finite number of orbit types, it follows that we can cover M by orthogonal tubes having only finitely many slice types. Since M is finite dimensional it may be covered by finitely many smoothable charts $U_i = (S_i, H_i)$ such that $\tau(U_i) \simeq G \times_{H_i} (S \times R(\sigma_i))$. Thus the theorem follows from Proposition 2.6, Corollary 2.8, and Proposition 2.9.

Corollary 2.11: Let μ be a G-microbundle over a G-ENR such that $E(\mu)$ (or some neighborhood of the 0-section in $E(\mu)$) has only finitely many orbit types. Then μ has an inverse.

Proof: Embed X in some orthogonal G-space $R(\sigma)$, and let N be an invariant neighborhood which equivariantly retracts to X, say $r: N \to X$. Then $E(r^*\mu)$ also has only finitely many orbit types. We can take $E(r^*\mu)$ to be a locally linear $G\text{-}R^n$ bundle over N, and hence $E(r^*\mu)$ is a locally smoothable G-manifold. Embed $E(r^*\mu)$ in same orthogonal G-space $R(\alpha)$. By the theorem we can assume $E(r^*\mu)$ has a normal microbundle ν. Now on the one hand (see Lemma 1.6 and Remark following) $(\tau E(\nu))|N \simeq \tau N \oplus r^*\mu \otimes \nu|N = \varepsilon_\sigma \oplus r^*\mu \oplus \nu|N$, and on the other hand $(\tau E(\nu))|N \simeq \varepsilon_\alpha$. Thus $r^*\mu$ has the inverse $\nu|N \oplus \varepsilon_\sigma$. Hence μ has the inverse $\nu|X \oplus \varepsilon_\sigma$.

Stable G-smoothing theory II

We first review the definitions of equivalent, isotopic and (sliced) concordant smoothings for G-manifolds, and the corresponding stable notions.

Definition: A G-smoothing of a topological G-manifold M is a G-homeomorphism $\alpha: V \to M$, where V is a smooth G-manifold. Two G-smoothings $\alpha_i: V_i \to M$, $i = 0,1$, are called

1) Equivalent, if there exists a G-diffeomorphism $d: V_0 \to V_1$ such that $\alpha_1 d = \alpha_0$.
2) Isotopic, if α_0 is homotopic through G-homeomorphisms to α_0' where $\alpha_1^{-1}\alpha_0'$ is a G-diffeomorphism.
3) Concordant, if there exists a smooth G-manifold with boundary V, a G-homeomorphism $\alpha: V \to M \times I$ and G-diffeomorphisms $d_i: V_i \to \partial_i V$ with $(\alpha|\partial_i V)\circ d_i = \alpha_i$, $i = 0,1$.

4) <u>Sliced Concordant</u>, if there is a concordance $\alpha\colon V \to M \times I$ such that $pr_2 \circ \alpha\colon V \to I$ is a smooth submersion.

<u>Remark</u>: Given a G-smoothing $\alpha\colon V \to M$ we frequently use α to identify the underlying topological G-manifold of V with M, and thus consider the G-smoothing α to be a smooth G-structure M_α on M. With this identification our definitions translate as follows:

1') α_0, α_1 are equivalent if $id_M\colon M_{\alpha_0} \to M_{\alpha_1}$ is a G-diffeomorphism.

2') α_0, α_1 are isotopic if $id_M\colon M_{\alpha_0} \to M_{\alpha_1}$ is homotopic through G-homeomorphisms to a G-diffeomorphism.

3') α_0, α_1 are concordant if there is a smooth G-structure $(M \times I)_\alpha$ on $M \times I$ such that $\alpha|M \times 0 = \alpha_0$ and $\alpha|M \times 1 = \alpha_1$.

4') α_0, α_1 are sliced concordant if there exists a concordance $(M \times I)_\alpha$ between them with $pr_2\colon (M \times I)_\alpha \to I$ a smooth submersion.

For G-smoothings α_0, α_1 of M we have:

Equivalent $\Longrightarrow$ Isotopic $\Longrightarrow$ Sliced Concordant $\Longrightarrow$ Concordant

<u>Definition</u>: A <u>stable</u> G-smoothing of M is a G-smoothing of $M \times R(\rho)$, some orthogonal G-space $R(\rho)$. Two stable G-smoothings $(M \times R(\rho_0))_\alpha$ and $(M \times R(\rho_1))_{\alpha_1}$ are <u>stably</u> equivalent (isotopic, sliced concordant, concordant) if there are orthogonal G-spaces $R(\sigma_0)$ and $R(\sigma_1)$ such that $\rho_0 \oplus \sigma_0 = \rho_1 \oplus \sigma_1 = \theta$ and the smooth G-structures $(M \times R(\rho_0))_{\alpha_0} \times R(\sigma_0)$ and $(M \times R(\rho_1))_{\alpha_1} \times R(\sigma_1)$ define equivalent (isotopic, sliced concordant, concordant) G-smoothings of $M \times R(\theta)$.

Next we review the definitions of equivalent and isotopic G-vector bundle reductions of a G-microbundle, and the corresponding stable notions.

<u>Definition</u>: A G-vector bundle <u>reduction</u> of a G-microbundle μ over a paracompact G-space X is a G-microbundle equivalence $\varphi: E(\xi) \to E(\mu)$, where ξ is a G-vector bundle over X. Two G-vector bundle reductions $\varphi_i: E(\xi_i) \to E(\mu)$, $i = 0,1$, are

1) <u>Equivalent</u> if there exists a G-vector bundle equivalence $\psi: E(\xi_0) \to E(\xi_1)$ such that $\varphi_1\psi = \varphi_0$.

2) <u>Isotopic</u> if φ_0 is homotopic through G-microbundle equivalences to φ_0' where $\varphi_1^{-1}\varphi_0'$ is a G-vector bundle equivalence.

For G-vector bundle reductions φ_0, φ_1 of μ we have:

$$\text{Equivalent} \Longrightarrow \text{Isotopic}$$

<u>Definition</u>: A <u>stable</u> G-vector bundle reduction of μ is a G-vector bundle reduction of $\mu \oplus \varepsilon_\rho$, $\varepsilon_\rho = \varepsilon_\rho(X)$, for some orthogonal G-representation ρ. Two stable G-vector reductions $\varphi_i: E(\xi_i) \to E(\mu \oplus \varepsilon_{\rho_i})$, $i = 0,1$, are <u>stably</u> equivalent (isotopic) if there exist orthogonal representations σ_0, σ_1 of G such that $\rho_0 \oplus \sigma_0 = \rho_1 \oplus \sigma_1 \simeq \theta$ and the G-vector bundle reductions $\varphi_i \oplus \mathrm{id}: E(\xi_i \oplus \varepsilon_{\sigma_i}) \to E(\mu \oplus \varepsilon_{\rho_i} \oplus \varepsilon_{\sigma_i})$ give equivalent (isotopic) reductions of $E(\mu \oplus \varepsilon_\theta)$.

Throughout this section we will assume that all G-manifolds have a finite number of orbit types. The main result of this section will be a bijective correspondence between stable isotopy classes of G-smoothings of M and stable isotopy classes of G-vector bundle reductions of τM.

We begin by defining functions:

E: Sliced concordance classes of smoothing of $\tau M \Longrightarrow$ Isotopy classes of reductions of τM,

F: Isotopy classes of reductions of $\tau M \Longrightarrow$ <u>stable</u> sliced concordance classes of smoothings of M.

Definition of E: If M_α is a smoothing of the G-manifold M, it defines a G-vector bundle reduction of τM as follows. Take a G-invariant Riemannian metric on M_α and let TM_α be the tangent G-vector bundle; then $\varphi: TM_\alpha \to M_\alpha \times M_\alpha$, $\varphi(x,X) = (x, \exp_x X)$ will define an isotopy class of G-vector bundle reductions. In fact, φ is actually an embedding only on a neighborhood of the 0-section. But we can take a smooth G-microbundle embedding $\delta: TM_\alpha \to TM_\alpha$ such that $\varphi\circ\delta$ is a smooth G-microbundle embedding in $M_\alpha \times M_\alpha$. By the usual proof for the uniqueness of smooth G-normal bundles [B3], any two smooth G-vector bundle reductions are isotopic. We denote this isotopy class of reductions by $E(\alpha)$.

Lemma 3.1: $E(\alpha)$ depends only on the sliced concordance class $[\alpha]$ of α.

Proof: Let α_i, $i = 0,1$, be G-smoothings of M and $(M\times I)_\alpha$ a sliced concordance between them. Since pr_2 is a submersion, each slice $pr_2^{-1}(t)$ is a smoothing M_{α_t} of M. The tangent bundle along the fibres (i.e., slices) is G-vector bundle equivalent to $TM_{\alpha_0} \times I$ and we get a family of G-vector bundle equivalences $\lambda_t: TM_{\alpha_0} \to TM_{\alpha_t}$. Take a G-invariant Riemannian metric on $(M\times I)_\alpha$. It defines a G-invariant metric on each slice. Thus we get a family $\varphi_t\lambda_t: TM_{\alpha_0} \to M_{\alpha_t} \times M_{\alpha_t}$ of G-vector bundle reductions depending continuously on $t \in I$. But $\varphi_t\lambda_t$ is an isotopy between $\varphi_0: TM_{\alpha_0} \to M_{\alpha_0} \times M_{\alpha_0}$ and $\varphi_1: TM_{\alpha_1} \times M_{\alpha_1} \times M_{\alpha_1}$. (We are assuming that φ_t has been uniformly shrunk as above so as to be a smooth G-microbundle embedding.) Since $\varphi_0 \in E(\alpha_0)$ and $\varphi_1 \in E(\alpha_1)$ we have that $E(\alpha_0) = E(\alpha_1)$.

Remark: It is not difficult to show that concordant G-smoothings of M define stably isotopic G-vector bundle reductions of τM.

Definition of F: If $\varphi: \xi \to \tau M$ is a G-vector bundle reduction it defines a stable G-smoothing as follows: Embed M in some orthogonal G-space $R(\rho)$ with a normal microbundle ν (see §2), which we can take to be a locally linear $G\text{-}R^k$ bundle. Let μ be a locally linear $G\text{-}R^n$ bundle contained in τM. By (1.6) we have a G-microbundle embedding of $E(p_\nu^*(\mu \oplus \nu))$ in $E(\nu) \times E(\nu)$. On the other hand, $E(\nu)$ inherits a smooth structure $E(\nu)_\rho$ as an open subset of $R(\rho)$, and hence $TE(\nu)_\rho = \varepsilon_\rho$. A smooth reduction $TE(\nu)_\rho \to E(\nu)_\rho \times E(\nu)_\rho$ is isotopic to a $G\text{-}R^{n+k}$ bundle equivalence $\hat{\theta}: TE(\nu)_\rho \to Ep_\nu^*(\mu \oplus \nu) \subset E(\nu) \times E(\nu)$. Hence by restriction over M one get a $G\text{-}R^{n+k}$ bundle equivalence $\theta: \varepsilon_\rho(M) \to \mu \oplus \nu$. This gives a G-equivalence $E(p_\nu^*\mu) = E(\mu \oplus \nu) \xrightarrow{\theta^{-1}} E(\varepsilon_\rho(M)) = M \times R(\rho)$.

Now we can assume φ is a $G\text{-}R^n$ bundle equivalence with $\mu \subset \tau M$, and hence $p_\nu^*\varphi: E(p_\nu^*\xi) \to E(p_\nu^*\mu)$ is a $G\text{-}R^{n+k}$ bundle equivalence. But $E(p_\nu^*\xi)$ has a smooth G-structure $E(p_\nu^*\xi)_\rho$ as a G-vector bundle over the smooth G-manifold $E(\nu)_\rho$. Thus $\theta^{-1} \circ p^*\varphi: E(p_\nu^*\xi)_\rho \to M \times R(\rho)$ is a stable G-smoothing of M.

Fixing the embedding of M and its normal bundle in $R(\rho)$ we get immediately that an isotopy class of reductions of τM defines an isotopy class of G-smoothings of $M \times R(\rho)$. However, it will be necessary to know what happens when we also allow the embedding and normal bundle to vary.

Lemma 3.2: An isotopy class of G-vector bundle reductions of τM gives a well defined stable sliced concordance class of smoothings of M, independent of all choices in the above construction.

Proof: First if we stabilize the embedding $M \subset R(\rho) \subset R(\rho \oplus \sigma)$, we stabilize ν to $\nu \oplus \varepsilon_\sigma$. Let $\rho' = \rho \oplus \sigma$, $\nu' = \nu \oplus \varepsilon_\sigma$. Then $\theta': \varepsilon_{\rho'}(M) \to \mu \oplus \nu'$ is $\theta \oplus 1: \varepsilon_\rho \oplus \varepsilon_\sigma \to \mu \oplus \nu \oplus \varepsilon_\sigma$ and $p_{\nu'}^*\varphi = p_\nu^*\varphi \times 1: E(p_\nu^*\xi) \times R(\sigma) \to E(p_\nu^*\mu) \times R(\sigma)$, up to isotopy. Thus the smoothing of $M \times R(\rho) \times R(\sigma)$ defined by $\theta'^{-1} \circ p_{\nu'}^*\varphi$ is the product with $R(\sigma)$ of that defined on $M \times R(\rho)$ by $\theta^{-1} \circ p_\nu^* \varphi$, up to isotopy.

Since any two embeddings of M in orthogonal G-spaces and any two normal bundles are stably isotopic, it only remains to see how the smoothing changes under a G-isotopy $f_t \colon E(\nu) \to R(\rho)$: Let $E(\nu)_{f_0^{-1}}$ and $E(\nu)_{f_1^{-1}}$ be the smoothings pulled back by f_0, f_1; then the smoothing $(E(\nu) \times I)_{f^{-1}}$ pulled back from $f \colon E(\nu) \times I \to R(\rho) \times I$, is a sliced concordance between them. Also a G-invariant metric in $(E(\nu) \times I)_{f^{-1}}$ induces a G-invariant metric on the slice $E(\nu)_{f_t^{-1}}$ and reduction $TE(\nu)_{f_t^{-1}} \to E(\nu)_{f_t^{-1}} \times E(\nu)_{f_t^{-1}}$, which depends continuously on $t \in I$. Thus we get an isotopy $\theta_t \colon \varepsilon_\rho(M) \to \mu \oplus \nu$. Further, if we pull ξ back over $E(\nu) \times I$ by $p_\nu \circ pr_1 \colon E(\nu) \times I \to E(\nu) \to M$, the smoothing $(E(p_\nu^* \xi) \times I)_{f^{-1}}$ is a sliced concordance between $E(p_\nu^* \xi)_{f_0^{-1}}$ and $E(p_\nu^* \xi)_{f_1^{-1}}$ since a smooth G-vector bundle projection is a submersion. Thus $\theta^{-1} \circ pr^* p_\nu^* \varphi \colon (E(p_\nu^* \xi) \times I)_{f^{-1}} \to M \times R(\rho) \times I$ is a sliced concordance of the smoothings induced by f_0, f_1 from φ.

Thus the isotopy class of φ give a well defined stable sliced concordance class of smoothings.

If α is a smoothing of M, let $[\alpha]_s$ be the stable sliced concordance class of α.

<u>Lemma 3.4</u>: $FE[\alpha] = [\alpha]_S$.

<u>Proof</u>: Let $\varphi \colon TM_\alpha \to M_\alpha \times M_\alpha$ be a smooth reduction. Then $E[\alpha] = [\varphi]$. Choose a smooth equivariant embedding of M_α in $R(\rho)$ and let ν be the normal G-vector bundle. Let μ be a locally linear G-R^n bundle in τM, and let $\varphi' \colon TM_\alpha \to \mu$ be isotopic to φ. Then since $TE(\nu)_\rho = p_\nu^*(TM_\alpha \oplus \nu)$, $\hat{\theta} = p_\nu^* \varphi' \oplus 1_\nu$ and $\theta = (\varphi' \oplus 1) \circ \gamma$, $\gamma \colon \varepsilon_\rho \to TM_\alpha \oplus \nu$ a smooth G-vector bundle equivalence. Then identifying $E(p_\nu^* \varepsilon)_\rho = E(TM_\alpha \oplus \nu)$, $E(p_\nu^* \mu) = E(\mu \oplus \nu)$, $p_\nu^* \varphi' = \varphi' \oplus 1$ and $\theta^{-1} \circ p_\nu^* \varphi \colon$ $E(TM_\alpha \oplus \nu) \to E(\varepsilon_\rho) = M \times R(\rho)$ is just γ^{-1}, a smooth G-vector

bundle equivalence onto $M_\alpha \times R(\rho)$. Thus $F[\varphi] = [\alpha]_S$, and $FE[\alpha] = [\alpha]_S$.

If ψ is a G-vector bundle reduction of $\tau(M \times R(\rho))$, $\psi|M$ is a G-vector bundle reduction of $\tau M \oplus \varepsilon_\rho$; i.e., a stable G-vector bundle reduction of τM. This gives a function $[\psi] \to [\psi]|M = [\psi|M]$ on isotopy classes. Also if φ is any G-vector bundle reduction of τM, we denote by $[\varphi]_S$ its stable isotopy class; i.e., if $\varphi: \xi \to \tau M$, $\varphi \oplus 1: \xi \oplus \varepsilon_\rho \to \tau M \oplus \varepsilon_\rho$.

<u>Lemma 3.5</u>: $(EF[\varphi])|M = [\varphi]_S$.

<u>Proof</u>: Given $\varphi: \xi \to \mu \subset \tau M$, take an embedding $M \subset R(\rho)$ with normal G-R^n bundle ν. Let $\hat{\theta}: TE(\nu)_\rho \to E(p_\nu^*(\mu \oplus \nu))$ and $\theta: \varepsilon_\rho(M) \to \mu \oplus \nu$, $\theta = \hat{\theta}|M$, be as in the definition of the function F. Then $F[\varphi] = [\theta^{-1} \circ p_\nu^* \varphi]_S$, $\theta^{-1} \circ p_\nu^* \varphi: E(p_\nu^* \xi)_\rho \to M \times R(\rho)$.

Let $\hat{\psi}: TE(p_\nu^*\xi)_\rho \to \tau E(p_\nu^*\xi)_\rho$ be a smooth reduction. Then $E[\theta^{-1} \circ p_\nu^* \varphi] = [(\tau(\theta^{-1} \circ p_\nu^* \varphi) \circ \hat{\psi})|M]$.

Now $\hat{\psi}$ is determined up to isotopy by $\hat{\psi}|M$. First consider $\hat{\psi}|E(\nu)$: take an embedding of $p_\nu^*(\mu \oplus \nu) \oplus p_\nu^* \mu$ in $\tau E(p_\nu^*\xi)|E(\nu)$ so as to make the diagram below commutative:

$$\begin{array}{ccc}
TE(p_\nu^*\xi)_\rho|E(\nu) & \xrightarrow{\hat{\psi}|E(\nu)} & \tau E(p^*\xi)|E(\nu) \\
\Big\downarrow = & & \cup \\
TE(\nu)_\rho \oplus p_\nu^*\xi & \xrightarrow{\hat{\theta} \oplus \hat{\phi}} & p_\nu^*(\mu\oplus\nu)\oplus(p_\nu^*\mu),
\end{array}$$

$\hat{\phi} = p_\nu^*\varphi$. Then $\hat{\psi}|M: TE(p_\nu^*\xi)_\rho|M \to \tau E(p_\nu^*\xi)|M$ has image $\mu \oplus \nu \oplus \mu$. Further we have the communitative diagram:

$$\begin{array}{ccccccccc}
\tau E(p_\nu^*\xi)|M & \leftarrow & \tau E(p_\nu^*\mu)|M & \leftarrow & \tau E(\nu\oplus\mu)|M & \leftarrow & \tau E(\mu\oplus\nu)|M & \leftarrow & \tau(M\times R(\rho)|M \\
\uparrow & & \uparrow & & \uparrow & & \uparrow & & \uparrow \\
\tau E(\nu)|M\oplus\xi & \leftarrow & \tau E(\nu)|M\oplus\mu & \leftarrow & \tau M\oplus\nu\oplus\mu & \leftarrow & \tau M\oplus\mu\oplus\nu & \leftarrow & \tau M\oplus\varepsilon_\rho \\
\uparrow & & \uparrow & & \uparrow & & \uparrow & & \uparrow \\
\mu\oplus\nu\oplus\mu & \leftarrow & \mu\oplus\nu\oplus\mu & \leftarrow & \mu\oplus\nu\oplus\mu & \xleftarrow{1\oplus S} & \mu\oplus\mu\oplus\nu & \xleftarrow{1\oplus\theta} & \mu\oplus\varepsilon_\rho
\end{array}$$

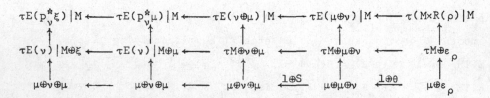

where all the maps are G-microbundle embeddings and the top and bottom rows are all G-equivalences. (S just switches factors.) Thus $(\tau(\theta^{-1} \circ p_\nu^* \varphi) \circ \hat{\psi})|M$ is

$$\begin{array}{ccccccc}
TE(p_\nu^*\xi)_\rho|M & \xrightarrow{\hat{\psi}|M} & \mu\oplus\nu\oplus\mu & \xrightarrow{1_\mu \oplus S} & \mu\oplus\mu\oplus\nu & \xrightarrow{1\oplus\theta^{-1}} & \mu\oplus\varepsilon_\rho \subset \tau(M\times R(\rho))|M \\
\| & \nearrow_{\theta\oplus\varphi} & & & & & \\
\varepsilon_\rho \oplus \xi & & & & & &
\end{array}$$

If $T\colon \xi \oplus \varepsilon_\rho \to \varepsilon_\rho \oplus \xi$ is $T_x(y,u) = (u,-y)$, then since T is linear, $\chi = (1 \oplus \theta^{-1}S)\circ(\theta \oplus \varphi)\circ T$ is an equivalent reduction of $\tau(M \times R(\rho))|M$.

Also, since $1_{\xi\oplus\xi}$ is isotopic to $R\colon \xi \oplus \xi \to \xi \oplus \xi$, $R_x(y_1,y_2) = (-y_2,y_1)$, by the rotation $(y_1,y_2) \to (\cos\theta y_1+\sin\theta y_2, -\sin\theta y_1+\cos\theta y_2)$, $0 \geq \theta \geq -\pi/2$; $1_{\mu\oplus\mu}$ is isotopic to $R' = \varphi \oplus \varphi\circ R\circ\varphi^{-1} \oplus \varphi^{-1}$, $R'_x(z,\varphi(-y)) = (\varphi(y),z)$. Thus χ is isotopic to $\chi' = 1 \oplus \theta^{-1}\circ R' \oplus 1 \circ 1 \oplus S\circ\theta \oplus \varphi\circ T$. Further, a computation gives $\chi'(y,u) = (\varphi(y),u)$. Thus $\chi' \in [\varphi]_S$ and $(EF[\varphi])|M = [\varphi]_S$.

From (3.4) and (3.5) it follows trivially that there is a bijective correspondence between stable sliced concordance classes of G-smoothings of M and stable isotopy classes of G-vector bundle reductions of τM. Our main result will then follow from:

<u>Proposition 3.6</u>: Sliced concordant G-smoothings are isotopic, provided no stratum in M/G has dimension 4.

<u>Proof</u>: The proof is exactly the same as in the G-finite case (or indeed the G-trivial case), see §6 of [L2]. The only change is in the condition for the G-engulfing theorem. The proof given in [L2] works if G is compact under the above hypothesis on M/G. (If G is finite this condition translates into the condition $\dim M^H \neq 4$ for any closed $H \subset G$).

Corollary 6.7: Sliced concordant G-smoothing on stably isotopic.

Thus we have proved:

Theorem 6.8: Let M be a G-manifold with only finitely many orbit types. Then the e is a bijective correspondence between stable isotopy classes of G-smoothings of M and stable isotopy classes of G-vector bundle reductions of τM.

Remark: One may now give an obstruction theory for stable G-smoothing as in [L2], provided one starts with a locally G_x smoothable manifold.

REFERENCES

[B1] Bierstone, E., The equivariant covering homotopy property for differentiable G-fibre bundles, J. Diff. Geom. 8 (1973), 615-622.

[B2] Borel, A., Seminar on Transformation Groups, Ann. of Math. Studies 46 (1960).

[B3] Bredon, G., Intro. to Compact Transformation Groups, Academic Press, N.Y., 1972.

[C1] Curtis, M. and Lashof, R., On product and bundle neighborhoods, Proc. AMS 15 (1972), 934-37.

[H1] Hirsch, M., On normal microbundles, Topology 5 (1966), 229-240.

[I1] Ibisch, H., Microfibrés normaux équivariants, C.R.A.S. Paris 279 (1974), 155-156.

[J1] Jaworowski, J., Extensions of G-maps and Euclidean G-retracts, Math. Z. 146 (1976), 145-148.

[K1] Kirby, R. and Sichenmann, L., Essays on Topological Manifolds, Smoothings and Triangulations, Ann. of Math. Studies 88 (1977).

[L1] Lashof, R., The immersion approach to triangulation and smoothing, Proc. Symp. on Pure Math. XXII (1971), AMS, Providence.

[L2] Lashof, R. and Rothenberg, M., G-smoothing theory, Proc. Sump. on Pure Math (Stanford Conference on Topology 1976).

[L3] Le Dimet, G-Varietés topologiques et G-microfibrés, Cahiers Topologie Géom. Différentielle XVII (1976), pp. 1-39.

[W1] Wasserman, A.G., Equivariant Differential Topology, Topology 9 (1969), 127-150.

LINEAR ACTIONS ON FRIENDLY SPACES

Arunas Liulevicius*

1. Statement of results. Let G be a topological group, X a topological space. Problem: consider the totality of G-actions on X and classify them under i) G-isomorphism, ii) G-homotopy equivalence, iii) G-pseudoequivalence. The third notion has been introduced and studied in depth by T.Petrie. A G-map $f: (X,\alpha) \longrightarrow (X,\beta)$ is said to be a pseudoequivalence if $f: X \longrightarrow X$ is a homotopy equivalence (when we forget about the action of G). In general pseudoequivalence is neither symmetric nor transitive.

Even if G is a reasonable group and X a reasonable space the original problem is too difficult to handle: there are too many G-isomorphism classes of actions, they coalesce in subtle and devious ways into G-homotopy classes, and these in turn are lumped in dark and mysterious fashion into families under pseudoequivalence. In this paper we shall show that if we impose certain reasonable topological conditions on X and require the condition of linearity on the actions of G, then the situation improves dramatically.

Let C^n be the standard n-dimensional vector space over the complex numbers C. Let $U = U(n)$ be the unitary group associated with the standard Hermitian scalar product on C^n. Consider an action of U on a topological space X, $\mu: U \times X \longrightarrow X$. If $\alpha: G \longrightarrow U$ is a representation (continuous homomorphism) of a compact topological group G into U, then $\mu(\alpha \times 1): G \times X \longrightarrow X$ is called a linear action of G on X. We denote this G-structure by (X,α), suppressing the underlying fixed U-action μ in the notation.

We will study a class of U-spaces X which have all orbits of the same type U/H, and the orbits are not too big (H is a closed connected subgroup of U of maximal rank) and not too small (there is a line L in

* Partially supported by NSF Grant # MCS 77-01623.

C^n which is fixed by H U = U(n)). We shall also avoid complications by assuming that $X = U/H \times B$ as U-space, where U/H has the standard left action of U and B has the trivial action of U. In particular B is the orbit space X/U. We shall also assume that B is non-empty, simply connected, and far from CP^{n-1} (complex projective n-1 space) with respect to cohomology: if $b \in H^2(B;Z)$ and $b^n = 0$, then $b^{n-1} = 0$ as well. A U-space X satisfying all of these properties will be called a <u>friendly</u> space.

Our main theorem says that linear G-actions on friendly spaces enjoy a surprising homotopy rigidity property:

<u>Theorem</u> 1. If X is a friendly U-space, $\alpha, \beta : G \longrightarrow U$ representations of a compact group G, then a pseudo-equivalence $f:(X,\alpha) \longrightarrow (X,\beta)$ exists if and only if there exists a linear character $\chi : G \longrightarrow S^1$ such that β or its complex conjugate $\bar{\beta}$ is similar to $\chi\alpha$ as representations of G.

<u>Corollary</u> 2. If X is a friendly U-space, then on the set of linear G-actions on X the following relations coincide: G-equivalence = G-homotopy equivalence = G-pseudoequivalence (in particular, G-pseudo-equivalence is an equivalence relation). Moreover (X,α) is equivalent to (X,β) if and only if β or $\bar{\beta}$ is similar to α as projective representations of G.

Let us point out that it is no surprise that projective representations come in. Let $C = S^1$ be the center of U = U(n), then $C \subset H$ since H is of maximal rank in U, so C acts trivially on U/H. Thus if $\chi : G \longrightarrow C = S^1$ is a homomorphism (i.e. a linear character of G) then $(X,\chi\alpha) = (X,\alpha)$.

Denote complex conjugation of matrices by $c: U \longrightarrow U$ and choose H in its conjugacy class so that $c(H) = H$. We then have induced maps $\underline{c} : U/H \longrightarrow U/H$ and $\hat{c} = \underline{c} \times 1 : U/H \times B \longrightarrow U/H \times B$ such that

$\hat{c} : (X,\beta) \longrightarrow (X,\bar{\beta})$ is a G-equivalence.

These two comments taken together show one half of Theorem 1: if β or $\bar{\beta}$ is similar to $\chi\alpha$ where $\chi: G \longrightarrow S^1$ is a linear character, then there exists a G-pseudoequivalence $f: (X,\alpha) \longrightarrow (X,\beta)$. In fact we can take f to be the identity map or the map $\hat{c}$ constructed above.

A word or two about our hypotheses. If we just assume that X is a completely regular U-space with all orbits of the type U/H, then it follows [5] that $X = U/H \times_W E$, where $W = N/H$, $N = \{ g \in U \mid gHg^{-1} = H \}$ $E = X^H = \{ x \in X \mid hx=x \text{ for all } h \text{ in } H \}$, and the orbit map $p: E \longrightarrow B=X/U$ is a principal W-bundle. Our simplifying hypothesis is that p is the trivial principal W-bundle. It does help keep the argument simple, but is unnecessary (see [10]). Regarding the conditions on H: it is not clear that connectedness is essential (although we badly need it for our argument); the condition of maximal rank probably cannot be weakened - for example, there are all sorts of interesting pseudoequivalences of linear actions on spheres, studied by Atiyah and Tall [2], Lee and Wasserman [8], Meyerhoff and Petrie [12]. We exploit the condition that H fixes a line in C^n to obtain a very useful map $U/H \longrightarrow CP^{n-1}$ which then remains at the center of our argument. Our method certainly requires this condition, however recent work of A.Back on homotopy rigidity of linear actions on complex Grassmann manifolds [3] indicates that this condition can probably be weakened. We should point out that the condition implies that $H \neq U(n)$ if $n \neq 1$, which we certainly need as a hypothesis for Theorem 1.

2. <u>Method of proof</u>. The proof can be summed up in the slogan: "Follow that line bundle!". Let $h: S^{2n-1} \longrightarrow CP^{n-1}$ be the Hopf bundle. It is a U-line bundle over the U-space $CP^{n-1} = U(n)/U(1)xU(n-1)$. If $\gamma: G \longrightarrow U$ is a representation of G, then h is a G-line bundle over (CP^{n-1}, γ). Now let L be a line in C^n fixed under H, and consider the

map $q : U/H \longrightarrow CP^{n-1}$ defined by $q(uH) = uL$, as well as the map $r: X \longrightarrow CP^{n-1}$ which is the composition $r = q\pi_1$, where π_1 is the projection on the first factor $X = U/H \times B \longrightarrow U/H$.

We notice that $W = N/H$ is the group of U-equivalences of U/H (where N is the normalizer of H in U) and we use it to normalize the homotopy equivalence $f: X \longrightarrow X$ given to us.

Theorem 3. If X is a friendly space and $f: X \longrightarrow X$ is a homotopy equivalence then there exists a $w \in W=N/H$ such that $(w \times 1)^*f^*r^*h = r^*h$, or $= r^*h^{-1}$, where $r: X \longrightarrow CP^{n-1}$ is the map described above, and h is the Hopf bundle on CP^{n-1}.

This is a key step in the argument and we shall come back to it later. Since $w \times 1 : (X,\alpha) \longrightarrow (X,\alpha)$ is a G-equivalence, we replace f by f preceeded by $w \times 1$. If now $f^*r^*h = r^*h^{-1}$, we replace β by $\bar{\beta}$ and f by $\hat{c}f$. For our new $f: (X,\alpha) \longrightarrow (X,\beta)$ we have $f^*r^*h = r^*h$.

Notice that $r_\alpha : (X,\alpha) \longrightarrow (CP^{n-1},\alpha)$ is a G-map. We inspect to see what happens when we apply K_G, the functor of equivariant K-theory.

Proposition 4. $r_\alpha^! : K_G(CP^{n-1},\alpha) \longrightarrow K_G(X,\alpha)$ is a monomorphism.

Since $r = q\pi_1$, and $\pi_1^!$ is a monomorphism (because B is non-empty), it is sufficient to show that $q: U/H \longrightarrow CP^{n-1}$ induces a monomorphism on K_G. Let T be a maximal torus of U in H, then the natural map $k: U/T \longrightarrow CP^{n-1}$ induces a monomorphism in K_G (see Atiyah and Segal [1] and Segal [15]), but k factors through q, so q induces a monomorphism in K_G.

We cannot resist the temptation to present an alternative argument shown to us by Vic Snaith - it is very useful in the more general setting where $p: E \longrightarrow B$ is not assumed to be a trivial principal W-bundle.

Let $K = \{u \in U \mid uL = L\}$, where L is the line through 0 in C^n fixed by H which we have been using in the construction of the map q. We have the inclusions $H \subset K \subset U$, and q is the canonical quotient map $q : U/H \longrightarrow U/K$. A geometric argument of J.McLeod [11] shows that $K_G(U/H, \alpha) = R(G) \otimes_{R(U)} R(H)$, where $\alpha^*: R(U) \longrightarrow R(G)$ and $i^*: R(U) \longrightarrow R(H)$ give the R(U)-module structures, $i: H \longrightarrow U$ being the inclusion. The same is true of course for K, and the map $q^!$ is precisely $1 \otimes j^* : R(G) \otimes_{R(U)} R(K) \longrightarrow R(G) \otimes_{R(U)} R(H)$, where $j: H \longrightarrow K$ is the inclusion. Now R(H) is a free R(K)-module under j^* (see Pittie [14] and Steinberg [17]), so not only is $1 \otimes j^*$ a monomorphism, but it makes $R(G) \otimes_{R(U)} R(H)$ into a free module over $R(G) \otimes_{R(U)} R(K)$. Steinberg [17] has constructed a natural R(K)-free basis of R(H) which is very useful both for theoretical considerations and for calculations (see [6]).

Let us continue with the proof of Theorem 1. If $\gamma : G \longrightarrow U$ is a representation, we let $h = h(\gamma) : (S^{2n-1}, \gamma) \longrightarrow (CP^{n-1}, \gamma)$ be the Hopf bundle. Then $K_G(CP^{n-1}, \gamma) = R(G)[h] / (c(\gamma))$, where $c(\gamma) = h^n - \gamma h^{n-1} + (\Lambda^2\gamma)h^{n-2} - \ldots + (-1)^n \Lambda^n\gamma \cdot 1$, and $\Lambda^i\gamma$ denotes the i-th exterior power of the representation γ (see [1] and [15]). In particular $K_G(CP^{n-1}, \gamma)$ is a free R(G)-module on $1, \ldots, h^{n-1}$.

Let us inspect the diagram of maps in K_G-cohomology:

$$\begin{array}{ccc} K_G(X, \beta) & \xrightarrow{f^!} & K_G(X, \alpha) \\ \uparrow r^!_\beta & & \uparrow r^!_\alpha \\ K_G(CP^{n-1}, \beta) & & K_G(CP^{n-1}, \alpha) \end{array}.$$

Remember that we have normalized f so that $f^* r^* h = r^* h$, where we are using * to indicate the pullback of non-equivariant line bundles. Let $s = h(\alpha)$, $t = h(\beta)$, both of these are G-equivariant line bundles having h as the underlying line bundle. Our normalization of f makes sure that $f^! r^! t$ has the same underlying nonequivariant line bundle as $r^! s$, namely $r^* h$. We now claim: there exists a linear character $\chi: G \longrightarrow S^1$ such that $f^! r_\beta^! t = \chi r_\alpha^! s = r_\alpha^! (\chi s)$. This is an immediate consequence of

Theorem 5. If X is a nonempty connected G-space and $H^1(X;Z) = 0$, a and b G-equivariant line bundles over X with the same underlying nonequivariant line bundle, then there exists a linear character $\chi: G \longrightarrow S^1$ such that $a = \chi b$.

For the proof of Theorem 5 please refer to [9]. The main ingredient is Segal's cohomology of groups [16].

We now let $\psi = (r_\alpha^!)^{-1} f^! r_\beta^! : K_G(CP^{n-1}, \beta) \longrightarrow K_G(CP^{n-1}, \alpha)$, and we notice that ψ is a map of R(G)-algebras defined by $\psi(t) = \chi s$. Applying ψ to the basic relation between $t^n, t^{n-1}, \ldots, t, 1$ we get

$$\chi^n s^n - \beta \chi^{n-1} s^{n-1} + \ldots + (-1)^i \Lambda^i \beta \chi^{n-i} s^{n-i} + \ldots + (-1)^n \Lambda^n \beta = 0,$$

and if we multiply by χ^{-n}, we obtain

$$s^n - \beta\chi^{-1} s^{n-1} + \ldots + (-1)^i \Lambda^i (\beta\chi^{-1}) s^{n-i} + \ldots + (-1)^n \Lambda^n (\beta\chi^{-1}) = 0.$$

Comparing this to the basic relation between $s^n, s^{n-1}, \ldots, s, 1$ we see $\alpha = \beta\chi^{-1}$, or $\beta = \chi\alpha$, as claimed in Theorem 1. The reader of course remembers that β may have been replaced by $\overline{\beta}$ during the course of the normalization of the pseudoequivalence f.

3. Proof of Theorem 3. This section is really an advertisement for a very pleasant argument due to John Ewing. Before proving Theorem 3 we investigate the special situation of U/T, where T is the standard maximal torus of U consisting of diagonal matrices. The cohomology of the flag manifold U/T is most conveniently presented as a quotient of $H^*(BT;Z)$. That is, $H^*(U/T;Z) = Z[x_1, \ldots, x_n] / I(n)$, where the polynomial variables x_i all have degree 2 and I(n) is the ideal in the polynomial ring generated by the symmetric polynomials in the x_i. We

notice that T fixes the coordinate axes $L_1, \ldots, L_n$ in C^n. If we use L_i we obtain the obvious map $\pi_i : U/T \longrightarrow CP^{n-1}$ by setting $\pi_i(uT) = uL_i$. The classes x_i mod I(n) are precisely $\pi_i^*(y)$ for a suitable generator y in $H^2(CP^{n-1};Z)$. In particular it follows that for $i=1,\ldots,n$ we have x_i^n in I(n). We find a surprising converse:

<u>Theorem</u> 6. If $u \in Z[x_1,\ldots,x_n]$ is an element of grade 2 and $u^n \in I(n)$, then $u \equiv ax_i$ mod I(n) for some $a \in Z$ and some i with $1 \leq i \leq n$.

The proof involves first replacing the integers Z by the field of rationals Q and using the derivation $D = \frac{\partial}{\partial x_1} + \ldots + \frac{\partial}{\partial x_n}$ to give an inductive proof. The interested reader is referred to [7] for details. Ian Macdonald pointed out to us that this result appeared in D.Monk's thesis [13]. This reassures us that the result is correct, especially in view of the fact that the techniques of proof are quite different.

We now wish to relate the case U/T to U/H. We may as well take $T \subset H$ and assume that $L_1,\ldots,L_k$ are precisely all the lines in C^n fixed by H. We obtain two-dimensional cohomology classes $y_1,\ldots,y_k$ (from our canonical maps $U/H \longrightarrow CP^{n-1}$ constructed from $L_1,\ldots,L_k$) such that under the standard map $k : U/T \longrightarrow U/H$ we have $k^*y_i = x_i$ mod I(n).

<u>Corollary</u> 7. If $u \in H^2(U/H;Z)$ is a class with $u^n = 0$, then $u = ay_i$ for some a in Z and some i with $1 \leq i \leq k$.

The idea of the proof is simple. The map $k^*: H^*(U/H;Z) \longrightarrow H^*(U/T;Z)$ is a monomorphism, [4] and elements in the image of k^* come from elements invariant under the action of the Weyl group of H on $H^*(BT;Z)$. See [7] for details.

We will now launch ourselves into the proof of Theorem 3 proper. We notice that the group $W = N/H$ acts transitively on the classes

$y_1, \ldots, y_k$ (indeed, more is true - W contains all permutations of these classes).

Let $X = U/H \times B$ be a friendly space and $f: X \longrightarrow X$ a homotopy equivalence. To fix notation, let us pick $L=L_1$ and let $y = c_1(h)$, the first Chern class of the Hopf bundle over CP^{n-1}. The map $r: X \longrightarrow CP^{n-1}$ has the property that $r^*y = y_1$. Consider $v = f^*y_1$. Since B is simply connected we have $v = u + b$, where $u \in H^2(U/H;Z)$, $b \in H^2(B;Z)$. Since $v^n = 0$, it follows that $u^n = 0$, so by Corollary 7 $u = ay_i$ for some i with $1 \leq i \leq k$ and some a in Z. Expanding $v^n = 0$ we obtain $0 = na^{n-1}y_i^{n-1}b$ + terms in other components. Since $y_i^{n-1} \neq 0$ and $H^2(B;Z)$ is torsion-free, we must have either $b = 0$ or $a = 0$. We show that $a = 0$ is impossible, for then $v = f^*y_1 = b$, $v^n = b^n = 0$, so $b^{n-1} = 0$ (remember, B is far from CP^{n-1}) or $v^{n-1} = 0$, which is ridiculuous, because f^* is an isomorphism and $y_1^{n-1} \neq 0$. This means that we must have the other alternative: $f^*y_1 = ay_i$, and $a = 1$ or -1 since f^* is an isomorphism. Let w in W be any element which takes y_i to y_1, then $f(w \times 1)$ takes $y_1 = c_1(r^*h)$ to either y_1 or $-y_1$. Noticing that $-y_1 = c_1(r^*h^{-1})$ and that nonequivariant line bundles are classified by their first Chern class we have the proof of Theorem 3.

References

1. M.F.Atiyah and G.B.Segal, Lectures on equivariant K-theory, Mimeographed notes, Oxford, 1965.

2. M.F.Atiyah and D.O.Tall, Group representations, λ-rings and the J-homomorphism, Topology 8 (1969), 253-297.

3. A.Back, Homotopy rigidity for Grassmannians (to appear). Preprint, University of Chicago, 1978.

4. A.Borel, Sur la cohomologie des espacès fibrés principaux et des espaces homogenes de groupes de Lie compacts, Ann. of Math., 57 (1953), 115-207.

5. G.Bredon, Introduction to compact transformation groups, Academic Press, London and New York, 1972.

6. F.P.Cass and V.P.Snaith, On C*-algebra extensions of Lie groups, Lecture at the Waterloo Algebraic Topology Conference, June 1978.

7. J.Ewing and A.Liulevicius, Homotopy rigidity of linear actions on friendly homogeneous spaces (to appear).

8. C.N.Lee and A.Wasserman, On the groups JO(G), Mem. Amer. Math. Soc. no. 159 (1975).

9. A.Liulevicius, Homotopy rigidity of linear actions: characters tell all, Bull. Amer. Math. Soc. 84 (1978), 213 - 221.

10. ____________, Homotopy rigidity of sturdy spaces, Proceedings of the Aarhus Algebraic Topology Symposium, August 1978 (to appear).

11. J.McLeod, The Künneth formula in equivariant K-theory, Proceedings of the Waterloo Algebraic Topology Conference, June 1978 (to appear).

12. A.Meyerhoff and T.Petrie, Quasi-equivalences of G-modules, Topology 15 (1976), 69-75.

13. J.D.Monk, The geometry of flag manifolds, Proceedings London Math. Soc.(3) 9 (1959), 253-286.

14. H.V.Pittie, Homogeneous vector bundles on homogeneous spaces, Topology 11 (1972), 199-203.

15. G.B.Segal, Equivariant K-theory, Inst. Hautes Etudes Sci. Publ. Math. No. 34 (1968), 129-151.

16. ____ _____, Cohomology of topological groups, Symposia Mathematica, vol. IV (INDAM, Rome, 1968/69), 377-387.

17. R.Steinberg, On a theorem of Pittie, Topology 14 (1975), 173-177.

The University of Chicago
Chicago, Illinois

July 1978

The Kunneth Formula in Equivariant K-theory

John McLeod*

§1: Most topological spaces which arise naturally in mathematics are to some extent symmetrical. To take full advantage of this extra structure the algebraic topologist should work in the category of spaces X which are acted upon by a symmetry group G, and symmetry preserving maps. An appropriate cohomology theory for this is the "equivariant K-theory" $K_G^*(X)$ of Atiyah and Segal [6; 13]. However, the computation of $K_G^*(X)$ tends to be rather difficult, partly because of the complexity of the coefficient ring, which is R(G), the character ring of G.

This paper extends substantially the range of techniques available for calculating $K_G^*(X)$ when G is a compact connected Lie group. For example it will be shown (in §4) that if T is a maximal torus of G and W is the Weyl group then $K_G^*(X) \cong K_T^*(X)^W$, which is a generalisation of the classical result $R(G) \cong R(T)^W$.

The motivation for this work is the Künneth formula spectral sequence first studied by Hodgkin [9; 10] whose object is to calculate $K_G^*(X \times Y)$ in terms of $K_G^*(X)$ and $K_G^*(Y)$. In particular (setting Y = G/H) it can be used to compute $K_H^*(X)$ from $K_G^*(X)$ when H is a closed subgroup of G. Hodgkin constructed a spectral sequence

$$E_2^{**} = \mathrm{Tor}_{R(G)}^{**}(K_G^*(X),\ K_G^*(Y)) \Longrightarrow F_G^*(X,Y) \qquad (1.1)$$

but only in certain special cases, for example when $\pi_1(G)$ is

---

* This article was prepared by Victor Snaith from the previously unpublished writings of the author.

torsion free and X or Y is a free G-space, did he prove $F_G^*(X,Y)$ is equal to $K_G^*(X \times Y)$. However in [16] Snaith showed that in order to establish $F_G^*(X,Y) \cong K_G^*(X \times Y)$ for all X and Y when $\pi_1(G)$ is torsion free it is sufficient to do so in the generic case $X = Y = G/T$, the space of left cosets of a maximal torus T in G. This amounts to proving the following result, which is our

<u>Main Theorem</u>

When $\pi_1(G)$ is torsion free

$$R(T) \underset{R(G)}{\otimes} R(T) \cong K_T^*(G/T).$$

In addition to the Künneth theorem, some consequences of the main theorem are (see §4)

1) A new proof of Hodgkin's theorem on $K^*(G)$.

2) A new direct proof that the map $\alpha(G,T) : R(T) \to K^*(G/T)$ defined by Atiyah and Hirzebruch in their foundation paper [5] on K-theory is onto.

3) $K_G^*(X) \cong K_T^*(X)^W$ ($\pi_1(G)$ may be allowed torsion here) if $K_T^*(X)$ is R(T)-free.

It should be remarked that Hodgkin set up a general theory of Künneth type spectral sequences and so other convergence results (which say nothing about the K_G^* case) have been obtained [14].

We must now begin work. The reader is assumed to be familiar K_G^* and with the Stiefel diagram approach to Lie groups. For further details see [6; 13] and [1] respectively. This paper is a substantial part of the author's thesis [12]. I would like to thank Vic Snaith for suggesting the problem and John Conway for lending me his spherical blackboard. I would like also to thank the National Research Council for their financial support.

1.2: Henceforth let G be a compact connected Lie group with Weyl group W and T a maximal torus. Let R(G), R(T) be their character rings. Preliminary to the proof of the main theorem I will establish the following commutative diagram of W-equivariant maps of Z/2-graded R(T)-algebras.

$$\begin{array}{ccc} R(T) \underset{R(G)}{\otimes} R(T) & \xrightarrow{\lambda} & K_G^*(G/T) \\ \Big\downarrow \underset{w\in W}{\oplus} \varepsilon w & & \Big\downarrow i^* \\ \underset{w\in W}{\oplus} R(T) & \xrightarrow{\cong} & K_T^*(W) \end{array} \qquad (1.3)$$

The algebras on the left of (1.3) are concentrated in the zero grading.

1.4: Explanation of (1.3)

(a) T acts on the left of G/T by t(gT) = tgT. gT is a fixed point if and only if $g \in N$, the normaliser of T (i.e. the fixed point set is W). W acts on the right of G/T by (gT)(nT) = gnT ($n \in N$; $nT \in W$). The inclusion $i : W \longrightarrow G/T$ is T-W equivariant so it induces the W-equivariant map, i^*, in (1.3). W acts on the left of the R(T)-algebras $K_T^*(G/T)$ and $K_T^*(W)$. If $a \in \underset{W}{\oplus} R(T) = K_T^*(W)$ let a_w denote its w-th coordinate. Then if $v \in W$ the action on $\underset{W}{\oplus} R(T)$ is clearly given by $(v^*a)_w = a_{wv}$.

(b) Consider $R(T) \underset{R(G)}{\otimes} R(T)$ as an R(T)-algebra by action through the left hand factor. Define $\lambda : R(T) \underset{R(G)}{\otimes} R(T) \longrightarrow K_T^0(G/T)$ by

$\lambda(1 \otimes [V]) = G \underset{T}{\times} V$ (T acts on left of G)

$\lambda([V] \otimes 1) = G/T \times V$ (T acts diagonally)

where V is a T-representation. If V is a G-representation then $G \times_T V \cong G/T \times V$ so λ is well defined.

W acts on the left of R(T) as follows. If V is a representation of T then w(V) is V with new action $t_{\#}v = n^{-1}tnv$ $(w = nT \in W)$. Let W act on the left of $R(T) \otimes_{R(G)} R(T)$ by $w(\sum a_i \otimes b_i) = \sum a_i \otimes wb_i$.

We now check that λ is W-equivariant in (1.3). Firstly $\lambda w([V] \otimes 1) = \lambda([V] \otimes 1) = G/T \times V = w^*(G/T \times V) = w^*\lambda([V] \otimes 1)$ and $\lambda w(1 \otimes [V]) = \lambda(1 \otimes w[V]) = G/T \times w(V)$ while $w^*\lambda(1 \otimes [V]) = w^*(G \times_T V)$. Define an isomorphism $\psi : G \times_T w(V) \longrightarrow w^*(G \times_T V)$ by $\psi[g,v] = (gT, [gn,v])$ where $w = nT \in W$, $v \in V$, $g \in G$.

Hence λ is W-equivariant as claimed.

(c) Define $\oplus\varepsilon w : R(T) \otimes_{R(G)} R(T) \longrightarrow \oplus_W R(T)$ of (1.3) by the formula

$$(\oplus\, \varepsilon w(\sum a_i \otimes b_i))_v = \varepsilon v(\sum a_i \otimes b_i) = \sum a_i(vb_i).$$

It is W-equivariant because

$$\begin{aligned}(u^*(\oplus\, \varepsilon w(\sum a_i \otimes b_i)))_v &= (\oplus\, \varepsilon w(\sum a_i \otimes b_i))_{vu} \\ &= \sum a_i(vub_i) \\ &= (\oplus\, \varepsilon w(\sum a_i \otimes ub_i))_v \\ &= (\oplus\, \varepsilon w(u(\sum a_i \otimes b_i)))_v.\end{aligned}$$

(d) The square (1.3) commutes. For $i^*\lambda([V] \otimes 1)$ and $\oplus\, \varepsilon w([V] \otimes 1)$ both have [V] in all coordinates while

$$i^{*}\lambda(1\otimes[V]) = [nT \underset{T}{\times} V] = w(V)$$

which is $(\oplus\varepsilon w(1\otimes[V]))_{w}$.

1.5: Definition of the subalgebra A

Let A be the subalgebra of $\bigoplus_{W} R(T)$ consisting of those elements a such that for every root θ of G and every $w \in W$ $(1-e^{\theta})$ divides $a_w - a_{r_\theta w}$. Here e^{θ} is the 1-dimensional T-representation associated with the root θ and r_θ is reflection in the Weyl group corresponding to θ.

Theorem 1.6

In (1.3) i^* is an isomorphism of $K_T^*(G/T)$ onto A.

Theorem 1.7

If $\pi_1(G)$ is torsion free in (1.3) $\oplus\varepsilon w$ is an isomorphism of $R(T) \underset{R(G)}{\otimes} R(T)$ onto A.

1.8: The main theorem is an immediate consequence of §§1.4(d), 1.6, 1.7.

§2: In this section, in a series of lemmas, we will establish Theorem 1.6.

2.1: G/T is equivariantly homeomorphic to G^c/B where G^c is a complex affine algebraic group and B is a Borel subgroup [7]. G^c/B has a T-equivariant decomposition into affine cells. The cells are in one-one correspondence with the elements of the Weyl group W. Choose a positive root system for G. If $w \in W$ then the corresponding cell V_w is $\oplus V_\theta$ summed over $\theta > 0$ such that $w^{-1}\theta < 0$ (where V_θ is the 1-dimensional representation with character e^{θ}). The origin of V_w is the fixed point $w \in W \subset G/T$ [8, p.347]. Choose a total ordering $(w_1,\ldots,w_N)$ of W such that $r \leq s$ implies $\dim V_{w_r} \leq \dim V_{w_s}$. Let $A_n = V_{w_1} \cup \ldots \cup V_{w_n} \subseteq G/T$. Hence $w_1 = 1$, $A_{w_1} = V_{w_1}$ is a point and $A_N = G/T$. Each A_n is a left T-space.

<u>Lemma 2.2</u>

In (1.3) $i^* : K_T^*(G/T) \longrightarrow K_T^*(W)$ is injective.

<u>Proof</u>

We have exact sequences

$$\begin{array}{ccccccccccc} 0 & \longrightarrow & K_T^1(A_{n+1}) & \longrightarrow & K_T^1(A_n) & \longrightarrow & K_T^0(V_{w_{n+1}}) & \longrightarrow & K_T^0(A_{n+1}) & \longrightarrow & K_T^0(A_n) \longrightarrow 0 \\ & & & & & & \Big\downarrow \cong & & & & \\ & & & & & & R(T) & & & & \end{array}$$

$A^1 = *$ so, by induction on n, $K_T^1(A_n) = 0$ for all n and in particular $K_T^1(G/T) = 0$. Now we have a commutative diagram

$$\begin{array}{ccccccccc} 0 & \longrightarrow & K_T^0(V_{w_{n+1}}) & \xrightarrow{j_{n+1}} & K_T^0(A_{n+1}) & \xrightarrow{k_{n+1}} & K_T^0(A_n) & \longrightarrow & 0 \\ & & \Big\downarrow i^* & & \Big\downarrow i^* & & \Big\downarrow i^* & & \\ 0 & \longrightarrow & K_T^0(V_{w_{n+1}} \cap W) & \longrightarrow & K_T^0(A_{n+1} \cap W) & \longrightarrow & K_T^0(A_n \cap W) & \longrightarrow & 0. \end{array}$$

By the Thom isomorphism the left-hand i^* is injective, and $A_1 = A_1 \cap W$, so by induction $i^* : K_T^0(G/T) \longrightarrow K_T^0(W)$ is injective.

<u>Lemma 2.3</u>

$i^*(K_T^*(G/T)) \subseteq A$.

Proof

Let θ be a root of G. Let C by the identity component of the centraliser of $\ker(\theta : T \longrightarrow \mathbb{C})$. The Weyl group of C is $\{1, r_\theta\}$ and C/T is made up of two cells —— a point, e, and V_θ. Thus we have the following diagram.

$$\begin{array}{ccccccccc} 0 & \longrightarrow & K_T^0(V_\theta) & \longrightarrow & K_T^0(C/T) & \longrightarrow & K_T^0(e) & \longrightarrow & 0 \\ & & \downarrow & & \downarrow & & \downarrow & & \\ 0 & \longrightarrow & K_T^0(r_\theta) & \longrightarrow & K_T^0(\{1,r_\theta\}) & \longrightarrow & K_T^0(\{1\}) & \longrightarrow & 0 \\ & & \downarrow \cong & & \downarrow \cong & & \downarrow \cong & & \\ & & R(T) & & R(T) \oplus R(T) & & R(T) & & \end{array}$$

As a basis for $K_T^0(C/T)$ over R(T) take the class of the trivial line (which maps to 1 in $K_T^0(e)$) and the image of the Thom class in $K_T^0(V_\theta)$. The image of the former in $K_T^0(\{1,r_\theta\})$ is (1,1) and that of the latter is $(1 - e^\theta, 0)$. Any element of $K_T^0(C/T)$ is a linear combination of these so the difference of coordinates of the image in $K_T^0(\{1,r_\theta\})$ is always divisible by $1 - e^\theta$.

However the composition

$K_T^0(G/T) \xrightarrow{i^*} K_T^0(W) \longrightarrow K_T^0(\{1,r_\theta\})$ equals

$K_T^0(G/T) \longrightarrow K_T^0(C/T) \longrightarrow K_T^0(\{1,r_\theta\})$ so if

$x \in K_T^0(G/T)$ $(i^*x)_1 - (i^*x)_{r_\theta}$ is divisible by $1 - e^\theta$. Also if $w \in W$

then

$$\begin{aligned} (i^*wx)_1 - i^*(wx)_{r_\theta} &= (w^*i^*x)_1 - (w^*i^*x)_{r_\theta} \\ &= (i^*x)_w - (i^*x)_{r_\theta w} \end{aligned}$$

is divisible by $1 - e^\theta$. Since θ is an arbitrary root $i^*(K_T^*(G/T)) \subseteq A$.

2.4: Interlude

If $w \in W$ let $N(w) = \dim_{\mathbb{C}} V_w$, the number of positive roots θ such that $w^{-1}\theta < 0$. If $\underline{\ell}$ is a vector in the interior of the fundamental Weyl chamber then $\langle\theta,\underline{\ell}\rangle > 0$ and $\langle\theta, w\underline{\ell}\rangle < 0$ or equivalently $\theta > 0$ and $\ker\theta$ separates $\underline{\ell}$ from $w\underline{\ell}$. Hence $N(w)$ is the number of hyperplanes in the Stiefel diagram crossed by the line from $\underline{\ell}$ to $w\underline{\ell}$. Now let $r_\theta \in W$ be the reflection corresponding to a positive root θ with $w^{-1}\theta < 0$. Since $r_\theta\theta = -\theta < 0$ both $r_\theta\underline{\ell}$ and $w\underline{\ell}$ are on the opposite side of $\ker\theta$ from $\underline{\ell}$ (note $r_\theta = r_\theta^{-1}$). Hence $\underline{\ell}$ and $r_\theta w\underline{\ell}$ are on the same side of $\ker\theta$. Consider the straight line from $\underline{\ell}$ to $w\underline{\ell}$. This crosses $\ker\theta$ at a single point, P. By reflecting the part of the line from P to $w\underline{\ell}$ in $\ker\theta$ we get a (bent) line from $\underline{\ell}$ to $r_\theta w\underline{\ell}$. Since the Stiefel diagram is symmetrical about $\ker\theta$ the new straight line (from $\underline{\ell}$ to $r_\theta w\underline{\ell}$) crosses less hyperplanes than the old one because in particular it does not cross $\ker\theta$ — in other words $N(w) > N(r_\theta w)$. (Notice that if $\theta' > 0$ and $w^{-1}\theta' > 0$ the same reasoning yields $N(w) < N(r_{\theta'} w)$ in this case.)

2.5: Completion of proof of Theorem 1.6

We must show $i^*(K_T^*(G/T)) \supseteq A$. We have the following maps induced by inclusions.

$$\begin{array}{ccccccccc}
K_T(w_N) & & K_T(w_{N-1}) & & & & K_T(w_2) & & K_T(w_1) \\
\uparrow & & \uparrow & & & & \uparrow & & \uparrow \\
K_T(G/T) & \xrightarrow{k_N} & K_T^*(A_{N-1}) & \xrightarrow{k_{N-1}} & \cdots & \longrightarrow & K_T^*(A_2) & \xrightarrow{k_2} & K_T^*(A_1) \\
\uparrow{\scriptstyle j_N} & & \uparrow{\scriptstyle j_{N-1}} & & & & \uparrow{\scriptstyle j_2} & & \uparrow \\
K_T^*(V_{w_N}) & & K_T^*(V_{w_{N-1}}) & & & & K_T^*(V_{w_2}) & & K_T^*(V_{w_1})
\end{array}$$

By the Thom isomorphism $K_T^*(V_{w_n})$ is a free $R(T)$-module on one generator g_n which restricts in $K_T^0(w_n) = R(T)$ to $\prod_{\theta>0, w^{-1}\theta<0} (1-e^\theta)$. All the k_i are onto so choose $f_n \in K_T^0(G/T)$ such that $k_{n+1}k_{n+2}\cdot\cdot\cdot k_N(f_n) = j_n(g_n)$. Then we have $(i^*f_n)_{w_n} = \prod_{\theta>0, w^{-1}\theta<0} (1-e^\theta)$ and $(i^*f_n)_{w_r} = 0$ if $r < n$ since $k_n j_n = 0$.

Let $a \in A$. We prove by downward induction on n that if $a_{w_r} = 0$ for $r \leq n$ then $a \in i^*(K_T^*(G/T))$. This will complete the proof of Theorem 1.6. For $n = N$ the claim is obvious because $a = 0$.

Here is the induction step. Assume $a_{w_r} = 0$ for $r \leq n-1$. Suppose θ is a positive root such that $w_n^{-1}\theta < 0$. Then $N(r_\theta w_n) < N(w_n)$, by §2.4, and so $a_{r_\theta w_n} = 0$ (by definition of the ordering on W). But since $a \in A$ $(1-e^\theta)$ divides $a_{w_n} - a_{r_\theta w_n} = a_{w_n}$ so that $\prod_{\theta>0, w_n^{-1}\theta<0} (1-e^\theta)$ divides a_{w_n} since $R(T)$ is a unique factorisation domain and the factors are coprime. Let $a_n = c_n \prod_{\theta>0, w_n^{-1}\theta<0} (1-e^\theta)$ $(c_n \in R(T))$ then $a - i^*(c_n f_n) \in A$ by Lemma 2.3. Also $(a - i^*(c_n f_n))_{w_r} = 0$ if $r \leq n$. By induction there exists $q \in K_T^*(G/T)$ such that $i^*(q + c_n f_n) = a$.

§3: In this section, in a series of lemmas, we will establish Theorem 1.7.

Lemma 3.1

Let $a \in R(T)$ and let θ be a root of G. Then $(1-e^{\theta})$ divides $a - r_{\theta}a$.

Proof

A general element of R(T) is $\sum m_i e^{\theta_i}$ where $m_i \in Z$, θ_i are weights and e^{θ_i} the corresponding T-representations. Since $r_{\theta}(\theta_i) = \theta_i - n_i\theta$ where $n_i = \frac{2<\theta,\theta_i>}{<\theta,\theta>}$ we have

$$\sum m_i e^{\theta_i} - r_{\theta}(\sum m_i e^{\theta_i}) = \sum m_i (e^{\theta_i} - e^{r_{\theta}\theta_i})$$

$$= \sum m_i e^{\theta_i}(1 - e^{n_i\theta})$$

which is clearly divisible by $(1-e^{\theta})$.

Lemma 3.2

$$\underset{w}{\oplus} \varepsilon w (R(T) \underset{R(G)}{\otimes} R(T)) \subseteq A$$

Proof

Let $\sum a_i \otimes b_i \in R(T) \underset{R(G)}{\otimes} R(T)$. Then

$$(\oplus \varepsilon w(\sum a_i \otimes b_i))_u - (\oplus \varepsilon w(\sum a_i \otimes b_i))_{r_{\theta}u}$$

$$= \sum a_i u b_i - \sum a_i r_{\theta} u b_i$$

$$= \sum a_i (u b_i - r_{\theta} u b_i)$$

which is divisible by $(1-e^{\theta})$ by Lemma 3.1.

3.3: For the rest of this section (except for §3.8) we assume $\pi_1(G)$ is torsion free. In this case $\sum_{\theta>0} e^{\theta/2}$ (summed over positive roots, θ) is a weight. Hence $\prod_{\theta>0} e^{\theta/2}$ is a unit of R(T).

Define $\delta \in R(T)$ by the equation

$$\delta = \prod_{\theta>0} (e^{\theta/2} - e^{-\theta/2}) = (\prod_{\theta>0} e^{\theta/2})^{-1} \prod_{\theta>0} (e^{\theta} - 1).$$

Lemma 3.4

If $\pi_1(G)$ is torsion free there exists $x \in R(T) \otimes_{R(G)} R(T)$ such that

$$\varepsilon\, w(x) = \begin{cases} \delta & \text{if } w = 1 \\ 0 & \text{otherwise.} \end{cases}$$

Proof

Let $\{e_w \mid w \in W\}$ be the R(G)-basis for R(T) given in [18]. It will suffice to find elements $c_w \in R(T)$ such that

$$\sum_{v \in W} c_v u e_v = \begin{cases} \delta, & \text{if } u = 1 \\ 0 & \text{if } 1 \neq u \in W. \end{cases} \tag{3.5}$$

For then we set $x = \sum c_w \otimes e_w$. Let M_w be the matrix with entries

$$(M_w)_{uv} = ue_v \quad (v \neq w),$$

$$(M_w)_{uw} = 0 \quad (u \neq 1),$$

and $(M_w)_{1w} = 1.$

Then the simultaneous solution of (3.5) is possible if and only if $\det(ue_v)$ divides $\delta \det M_w$ for all $w \in W$, the solution being

$c_w = \dfrac{\delta \det M_w}{\det(ue_v)}$. But by the proof of [18, Lemma 2.4] $\det(ue_v) = \delta^{\frac{|W|}{2}}$.

Moreover for each root $\theta > 0$ $(e^\theta - 1)$ divides the difference between the u-th row and the $r_\theta u$-th row of M_w (by Lemma 3.1) provided neither u nor $r_\theta u$ is 1. There are $\frac{|W|}{2} - 1$ such disjoint pairs $\{u, e_\theta u\}$ in W so $(e^\theta - 1)^{\frac{|W|}{2} - 1}$ divides $\det M_w$. Hence $\delta^{\frac{|W|}{2} - 1}$ divides $\det M_w$ since if $\theta \neq \pm \phi$ then $(e^\theta - 1)$ and $(e^\phi - 1)$ are co-primes in the U.F.D., R(T).

<u>Lemma 3.6</u>

If $\pi_1(G)$ is torsion free then $\underset{w}{\oplus}\varepsilon w(R(T) \underset{R(G)}{\otimes} R(T)) \supseteq A$.

<u>Proof</u>

Choose x as in Lemma 3.4 and suppose $x = \sum a_i \otimes b_i$. Let $x_w = \sum w a_i \otimes b_i$ then

$$\oplus\varepsilon v x_w = \begin{cases} 0 & \text{if } w \neq v, \\ w(\delta) & \text{if } w = v. \end{cases}$$

Also $w(\delta) = (\det w)\delta$ where $\det w = (-1)^{N(w^{-1})}$. Let $a \in A$ and let $y = \sum_{w \in W} (\det w) a_w x_w$. Then $\underset{w}{\oplus}\varepsilon w(y) = \delta a$. Let $\theta > 0$ be a root. The expression $\sum_w (\det w) a_w x_w$ splits into pairs of terms.

$$(\det u) a_u x_u + (\det r_\theta u) a_{r_\theta u} x_{r_\theta u}$$

$$= \det u((a_u - a_{r_\theta u}) x_u + a_{r_\theta u}(x_u - x_{r_\theta u})).$$

Now $x_u - x_{r_\theta u} = \sum (u a_i - r_\theta u a_i) \otimes b_i$ which is divisible by $(e^\theta - 1)$ by

Lemma 3.1. Also $a_u - a_{r_\theta u}$ is divisible by $(e^\theta - 1)$ because $a \in A$.

Hence y is divisible by $(e^\theta - 1)$ for all $\theta > 0$ and therefore by δ.

Hence $\delta^{-1}y \in R(T) \underset{R(G)}{\otimes} R(T)$ and $\underset{W}{\oplus}\varepsilon w(\delta^{-1}y) = a$.

3.7: Completion of the proof of Theorem 1.7

It suffices to show $\underset{W}{\oplus}\varepsilon w$ is injective when $\pi_1(G)$ is torsion free. Here are two proofs: -

Proof I: From [16] λ is injective in (1.3) and so is i^*, by Lemma 2.2.

Proof II: Let $R(T)_{(0)}$ be the field of fractions of $R(T)$. We have a diagram

$$\begin{array}{ccc}
R(T) \underset{R(G)}{\otimes} R(T) & \xrightarrow{\ \oplus\varepsilon w\ } & \underset{W}{\oplus} R(T) \\
\Big\downarrow \alpha & & \Big\downarrow \beta \\
R(T)_{(0)} \underset{R(G)}{\otimes} R(T) & \xrightarrow{\ (\oplus\varepsilon w)_{(0)}\ } & \underset{W}{\oplus} R(T)_{(0)}
\end{array}$$

$R(T) \underset{R(G)}{\otimes} R(T)$ is free over $R(T)$ so α is injective (and so is β).

Since δ is a unit $(\oplus\varepsilon w)_{(0)}$ is onto and therefore is an isomorphism, by dimensions.

3.8: Proposition

Without any assumption on $\pi_1(G)$

$$\oplus\varepsilon w\colon R(T) \underset{R(G)}{\otimes} R(T) \longrightarrow \underset{W}{\oplus} R(T)$$

is injective.

Proof

In this case $G = H/\Gamma$ where Γ is a finite central subgroup of H and $\pi_1(H)$ is torsion free. Also $T = T'/\Gamma$ for T' a maximal torus of H. Also $R(T'/\Gamma) \underset{R(H/\Gamma)}{\otimes} R(T'/\Gamma) \subseteq R(T') \underset{R(H)}{\otimes} R(T')$ and $\underset{W}{\oplus} R(T'/\Gamma) \subseteq \underset{W}{\oplus} R(T')$. Thus $\underset{W}{\oplus} \varepsilon w$ is injective by comparing the (T,G)-case with the (T',H)-case to which the proof of §3.7 applies.

§4: Consequences of the Main Theorem

4.1: Theorem

The Hodgkin spectral sequence [10] takes the form of a strongly convergent spectral sequence

$$E_2^{*,*} = \mathrm{Tor}^{**}_{R(G)}(K_G^*(X),\ K_G^*(Y)) \Longrightarrow K_G^*(X \times Y)$$

where G is compact, connected and $\pi_1(G)$ is torsion free. (Here $X, Y \in A_G$, the category of G-spaces used in [10].)

Proof

The Main Theorem proves the hypothesis to which §4.1 was reduced in [16].

4.2: Hodgkin's Theorem [2; 3; 11]

If G is a compact, connected Lie group of rank m with $\pi_1(G)$ torsion free then $K^*(G)$ is an exterior algebra over Z on m (primitive) generators.

Proof

The spectral sequence of §4.1 with $X = Y = G$ takes the form $E_2 = \mathrm{Tor}_{R(G)}(Z,Z) \Longrightarrow K^*(G)$. The structure of $R(G)$ is the tensor product of a polynomial ring on n generators with a Laurent polynomial ring on m-n generators. The Koszul resolution shows E_2 is the required exterior algebra generated by Tor^{-1}. Hence the spectral sequence collapses. The images of the generators of E_2 can be described geometrically (see [9; 10]).

4.3: A direct proof that the Atiyah-Hirzebruch map is onto.

The associated vector bundle construction gives a homomorphism $\alpha(G,T) : R(T) \longrightarrow K^0(G/T)$ [5] which fits into a diagram.

$$\begin{array}{ccc}
R(T) \underset{R(G)}{\otimes} R(T) & \xrightarrow{\ \lambda\ } & K^0_T(G/T) \\
\downarrow{\scriptstyle \eta\otimes 1} & & \downarrow{\scriptstyle \psi} \\
Z \underset{R(G)}{\otimes} R(T) & \xrightarrow{\ \lambda'\ } & K^0(G/T) \\
{\scriptstyle \xi}\downarrow & \nearrow{\scriptstyle \alpha(G,T)} & \\
R(T) & &
\end{array}$$

In [5] it was conjectured that $\alpha(G,T)$ was onto. This was first proved independently by Seymour [15] and Snaith [17]. They proved that λ' was an isomorphism by showing that $R(T)^\wedge$ is free over $R(G)^\wedge$ and using an $I(G)$-adically completed version of Hodgkin's spectral sequence ($(\_)^\wedge$ denotes $I(G)$-adic completion).

It is fairly easy to see that λ' is injective. We will show it is onto by the geometry used in §2. Consider (notation as in §2) the diagram

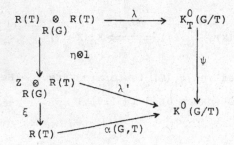

$$\begin{array}{ccccccccc} 0 & \longrightarrow & K_T^0(V_{w_{r+1}}) & \longrightarrow & K_T^0(A_{r+1}) & \longrightarrow & K_T^0(A_r) & \longrightarrow & 0 \\ & & \downarrow \chi_{r+1} & & \downarrow \psi_{r+1} & & \downarrow \psi_r & & \\ 0 & \longrightarrow & K^0(V_{w_{r+1}}) & \longrightarrow & K^0(A_{r+1}) & \longrightarrow & K^0(A_r) & \longrightarrow & 0 \end{array}$$

which has exact rows because $K^1(A_s) = 0$ for all s. Since χ_{r+1} is onto, by the Thom isomorphism theorem, induction on r shows that ψ_r is onto for all r. In particular $\psi = \psi_N$ is onto and so $\alpha(G,T)$ is also onto.

Theorem 4.4

Let $X \in A_G$ and let G have Weyl group W and maximal torus T. Then the restriction homomorphism induces an isomorphism

$$j^* : K_G(X) \xrightarrow{\cong} K_T^*(X)^W,$$

the subring of W-invariants, if $K_T^*(X)$ is R(T)-free.

Proof

If $\pi_1(G)$ is torsion free the spectral sequence yields $R(T) \underset{R(G)}{\otimes} K_G^*(X) \cong K_T^*(X)$ from which the result follows. If $\pi_1(G)$ has torsion let H,T' be as in §3,8 and consider

$$\begin{array}{ccc} K_G^*(X) & \xrightarrow{p^*} & K_H^*(X) \\ j^* \downarrow \uparrow j_* & & k_* \uparrow \downarrow k^* \\ K_T^*(X) & \xrightarrow{q^*} & K_{T'}^*(X) \end{array}$$

in which j_*, k_* are left inverses to j^*, k^* respectively and are given by an index construction [4, §4.9 et seq.]. From [16] we know that

$R(\Gamma) \otimes K_T^*(X) \cong K_{T'}^*(X)$ and that q^* is injective. We know that q^* commutes with the W-action and $\mathrm{im}(k^*)$ equals the W-invariants. The result follows by chasing using $q^* j^* = k^* p^*$ and $p^* j_* = k_* q^*$. Here we have used that if $K_T^*(X)$ is $R(T)$-free then $K_{T'}^*(X)$ is $R(T')$-free.

Remark 4.5:

Theorem 4.4 can probably be improved to require only that $K_T^*(X)$ has no p-torsion when p divides $|W|$. However the following example shows that some condition is required on $K_T^*(X)$.

Take RP^2 with $SU(2)$ acting trivially. Then $K_{SU(2)}^*(RP^2) = R(SU(2)) \otimes K^*(RP^2)$ and $K^*(RP^2) \cong Z \oplus Z/2$ concentrated in even degree. Hence

$$K_{S^1}^{\alpha}(SU(2) \times RP^2) \cong (R(S^1) \otimes_{R(SU(2))} Z) \otimes K^{\alpha}(RP^2)$$

and

$$R(S^1) \otimes_{R(SU(2))} Z \cong Z[\alpha,\alpha^{-1}]/(\alpha+\alpha^{-1}-2).$$

The Weyl group is $Z/2$ whose generator interchanges α and α^{-1}. Thus

$$K_{S^1}^0(SU(2) \times RP^2)^W \cong Z/2 \oplus Z/2 \oplus Z.$$

However

$$K_{SU(2)}^0(SU(2) \times RP^2) \cong K^0(RP^2).$$

REFERENCES

1. J. F. Adams, Lectures on Lie Groups, Benjamin, 1967.

2. S. Araki, Hopf structures attached to K-theory: Hodgkin's Theorem, Ann. Math. 85(1967), 508-525.

3. M. F. Atiyah, On the K-theory of compact Lie Groups, Topology 4(1965), 95-99.

4. M. F. Atiyah, Bott periodicity and the index of elliptic operators, Q. J. Math. (Oxford), (2)19(1968), 113-140.

5. M. F. Atiyah & F. Hirzebruch, Vector bundles and homogeneous spaces, Proc. Symp. Pure Math., Vol. 3 Differential geometry, A.M. Soc., (1961), 7-38.

6. M. F. Atiyah & G. B. Segal, Equivariant K-theory, Univ. of Warwick lecture notes, 1965.

7. A. Borel, Kählerian coset spaces of semi-simple Lie Groups, Proc. Nat. Acad. Sci., 40(1954), 1147-1151.

8. A. Borel, Linear algebraic groups, Benjamin, 1969.

9. L. Hodgkin, An equivariant Künneth formula in K-theory, University of Warwick preprint, 1968.

10. L. Hodgkin, The equivariant Künneth theorem in K-theory, Lecture Notes in Maths. 496, pp.1-100, Springer-Verlag, 1975.

11. L. Hodgkin, On the K-theory of Lie Groups, Topology 6(1967), 1-36.

12. J. McLeod, Thesis, Cambridge University, (1975).

13. G. B. Segal, Equivariant K-theory, I.H.E.S. Pub. Math., 34(1968) 129-151.

14. R. M. Seymour, On the convergence of the Eilenberg-Moore spectral sequence, Math. Proc. Cambs. Phil. Soc., (83)1(1978).

15. R. M. Seymour, Thesis, Warwick University, (1970).

16. V. P. Snaith, On the Künneth formula spectral sequence in equivariant K-theory, Proc. Camb. Phil. Soc., 72(1972), 167-177.

17. V. P. Snaith, On the K-theory of homogeneous spaces and conjugate bundles of Lie Groups, Proc. L. M. Soc., (3)22(1971), 562-584.

18. R. Steinberg, On a theorem of Pittie, Topology, 14(1975), 173-177.

Trinity College
Cambridge
England

ISOTOPY CLASSES OF PERIODIC DIFFEOMORPHISMS ON SPHERES

Reinhard Schultz

One of the most important themes in the topology of manifolds is the recoverability of geometrical information from algebraic data. This is perhaps most strikingly illustrated in the theory of surfaces, where algebra reflects geometry in an extremely faithful fashion. The following rather detailed example of this close connection can serve as a mathematical starting point for the present paper:

THEOREM (Nielsen [25], corrected by Fenchel [13] and Macbeath [22]). *Let* M^2 *be an oriented surface, and let* $f: M^2 \to M^2$ *be a homotopy equivalence with* f^n *homotopic to the identity. Then* f *is homotopic to a diffeomorphism* g *with* $g^n = 1$.

Examples due to F. Raymond and L. Scott [27] show that the analogous statement in every higher dimension is false. On the other hand, work of W. Heil and J. Tollefson [15] shows that an analogous partial statement *is* true for certain "nice" 3-manifolds - namely, sufficiently large oriented Seifert fiber spaces. Results of this sort are one source of motivation for the following question:

Suppose $f: M^n \to M^n$ *is a* (DIFF, PL, TOP) *automorphism with* f^p *isotopic to the identity, where* $p > 1$. *Is* f *isotopic to a similar automorphism* g *such that either*

(a) g *has period* p, *or*

(b) g *is periodic (hence its period is divisible by* p)?

As F. Raymond has observed, there are relatively simple examples where the answer is negative (I am grateful to him for bringing this to my attention in a letter). Namely, if $n \geq 5$ diffeomorphisms of the torus T^n exist with f homotopic but not isotopic to the identity and f^2 isotopic to the identity. Some examples follow from the work of Kirby and Siebenmann ([18]; by the surgery theoretic construction given there the PL homeomorphisms are approximations to diffeomorphisms - compare [39, §15]); many more examples follow from pseudo-isotopy

---

Partially supported by NSF Grants MCS 76-08794 and MCS 78-02913.

theory (compare [14,16]; D. Burghelea has independently calculated $\pi_o(\mathrm{Diff}^+T^n)$ by similar methods). Since T^n is aspherical an action with fixed points induces a monomorphism $G \to \mathrm{Aut}(\pi_1)$ [10]. Further considerations eliminate the fixed point free case also.

In this paper we shall consider manifolds that are spherical concentrating mainly on the standard sphere S^n itself. Furthermore, for the sake of simplicity we shall also assume throughout that p is an odd prime; similar considerations hold for prime powers (and even $p = 2$), but the technical problems are considerably greater. Our results may be summarized as follows:

THEOREM. (i) Let $p \geq 5$ be a fixed odd prime. Then for infinitely many values of n there is a class in $\pi_o(\mathrm{Diff}^+S^n)$ having order p and represented by a diffeomorphism of period p. If $p = 3$, at least some nontrivial classes are realizable in this manner (compare Thm. 1.8).

(ii) Let p be an odd prime such that $2^{2k+1} \equiv 1 \bmod p$ for some k. Then for at least one value of n there is a class in $\pi_o(\mathrm{Diff}^+S^n)$ of order p not represented by a period p diffeomorphism (Theorem 4.7).

(iii) Let p be an irregular prime that does not satisfy the condition of (ii). Then for infinitely many value of n there are classes in $\pi_o(\mathrm{Diff}^+S^n)$ of order p not represented by period p diffeomorphism (Theorem 4.5).

REMARKS 1. Diff^+S^n denotes the groups of orientation preserving diffeomorphisms with a suitable C^r topology, $r \geq 1$.

2. There are infinitely many odd primes satisfying the conditions in (ii) and (iii) respectively (if $p \equiv 7 \bmod 8$, then (ii) is true, while the case of (iii) is treated in Section 4).

3. We shall explain later how one can replace "period p diffeomorphisms" by "periodic diffeomorphisms" in the conclusions of (ii) and (iii).

4. According to work completed by J. Cerf [9], the group $\pi_o(\mathrm{Diff}^+S^n)$ is diffeomorphic to the Kervaire-Milnor abelian group Θ_{n+1} of homotopy spheres if $n \geq 5$, the map being given by taking a class f to the exotic sphere $\Sigma_f^{n+1} = D_+^{n+1} \cup_f D_-^{n+1}$.

Aside from the previously described motivation, this work was also motivated by an effort to understand an extra obstruction - theoretic term that arises in [29,§3] and is dismissed as "undetermined" there. In fact, the central part of this paper is the modification of ideas from [29] and its yet unpublished sequels to give a precise interpretation of the "undetermined" extra term. The present version of our central technical theorem (Theorem 2.1) was partly the result of questions by Lowell Jones, a fact that I acknowledge cheerfully.

Finally, I would like to thank John Ewing for his remarks about the number-theoretic problems arising in parts (ii) and (iii) of the above theorem; these saved me a great deal of work.

1. Realizing nontrivial isotopy classes

We shall begin by giving a procedure for finding free $\mathbb{Z}_p$ actions representing certain isotopy classes of diffeomorphisms on S^{2n-1}. Let $\mathbb{Z}_p$ act freely on S^{2n-1} via restriction of the free linear S^1 action, let L^{2n-1} be the corresponding lens space, and let $\pi: S^{2n-1} \to L^{2n-1}$ be the orbit space projection. Consider the set $\text{Top-S}(L^{2n-1}, S^{2n-1})$ of all pairs $(\mathcal{S}, \mathcal{T})$ where $\mathcal{S}$ is a topological smoothing of L^{2n-1} in the sense of [19] and $\mathcal{T}$ is an identification of the pulled back smoothing $\pi^*\mathcal{S}$ (on S^{2n-1}) with the standard smoothing of the sphere. The following is then immediate from (say) the methods and results of [19]:

PROPOSITION 1.1. <u>Assuming</u> $n \geq 2$, <u>the above set</u> $\text{Top-S}(L^{2n-1}, S^{2n-1})$ <u>corresponds bijectively to the group</u>

$$[L^{2n-1} \cup_\pi e^{2n}, \text{Top}/0] \quad \blacksquare$$

Given a representative for an element in $\text{Top-S}(L^{2n-1}, S^{2n-1})$, we get a free smooth action of $\mathbb{Z}_p$ on S^{2n-1} by covering transformations such that the quotient is L^{2n-1} <u>with smooth structure given by</u> $\mathcal{S}$. If we make the convention that $\exp(2\pi i/p) \in \mathbb{Z}_p$ is the Standard Generator, then the isotopy class of the free action's Standard Generator determines an element of $\pi_0(\text{Diff}^+ S^{2n-1})$. Let I be this isotopy class map.

Since the codomain of I is an abelian group by construction and domain is also one by Proposition 1.1, it is difficult to avoid asking whether I is a homomorphism.

The next result answers this question affirmatively and carries still further valuable information:

THEOREM 1.2. <u>Let</u> L^{2n+1} <u>be the</u> $(2n+1)$<u>-dimensional lens space constructed in analogy with</u> L^{2n-1}, <u>so that we may write</u>

$$L^{2n+1} = (L^{2n-1} \cup_\pi e^{2n}) \cup_h e^{2n+1}.$$

<u>Then the following diagram is commutative</u>:

$$\begin{array}{ccc}
\text{Top-S}(L^{2n-1}, S^{2n-1}) & \rightarrow & [L^{2n-1} \cup_\pi e^{2n}, \text{Top}/O] \\
{\scriptstyle -I}\downarrow & & \\
\pi_o(\text{Diff}^+ S^{2n-1}) & & \downarrow{\scriptstyle h^*} \\
\downarrow{\scriptstyle \cong} & & \\
\Theta_{2n} & \xrightarrow{\cong} & \pi_{2n}(\text{Top}/O).
\end{array}$$

PROOF. As in [29] we may thicken up $L^{2n-1} \cup_\pi e^{2n}$ into a regular neighborhood

$$\mathcal{R}_o = S^{2n-1} \times_{\mathbb{Z}_p} D^2 \cup D^{2n} \times [-\varepsilon, \varepsilon],$$

where $S^{2n-1} \times [-\varepsilon, \varepsilon]$ is identified with $S^{2n-1} \times$ [small closed arc] in $S^{2n-1} \times_{\mathbb{Z}_p} S^1 \cong S^{2n-1} \times (S^1/\mathbb{Z}_p) = S^{2n-1} \times S^1$. Let $\mathcal{R}$ be $\mathcal{R}_o$ with the corners suitably rounded. Then the smoothing $\mathcal{S}$ gives us a smoothing of $S^{2n-1} \times_{\mathbb{Z}_p} D^2$, and the trivialization $\mathcal{T}$ allows one to extend this smoothing to $\mathcal{R}$. In effect, one has smoothings of two pieces of $\mathcal{R}$ that coincide on the overlap, and these two smoothings are simply pasted together. By this construction, the following triangle commutes:

$$\begin{array}{ccc}
\text{Top-S}(L^{2n-1}, S^{2n-1}) & \xrightarrow{\cong} & \\
\downarrow{\scriptstyle \text{thickening and pasting}} & & [L^{2n-1} \cup_\pi e^{2n}, \text{Top}/O]. \\
\text{Top-S}(\mathcal{R}) & \xrightarrow[\lambda]{\cong} &
\end{array}$$

(λ is the isomorphism given by [19]).

On the other hand, the following diagram is also commutative by direct inspection of the defining maps:

$$\begin{array}{ccc}
\text{Top-S}(\mathcal{R}) & \xrightarrow[\cong]{\lambda} & [L^{2n-1} \cup_\pi e^{2n}, \text{Top}/O] \\
{\scriptstyle \text{take boundary}}\downarrow & & \downarrow{\scriptstyle h^*} \\
\text{Top-S}(\partial\mathcal{R} = S^{2n}) & \xrightarrow[\cong]{\lambda} & \pi_{2n}(\text{Top}/O)
\end{array}$$

Because of these two commutative diagrams, the proof of Theorem 1.2 reduces to showing that the following square is also commutative:

$$\begin{array}{ccc} \text{Top-S}(L^{2n-1}, S^{2n-1}) & \xrightarrow{\;\cong\;} & \text{Top-S}(\mathcal{R}) \\ {\scriptstyle -I}\downarrow & & \downarrow{\scriptstyle \partial} \\ \Theta_{2n} & \xrightarrow{\;\cong\;} & \text{Top-S}(\partial\mathcal{R}=S^{2n}). \end{array}$$

To see this, observe that $\partial\mathcal{R}$ may be constructed by surgery as follows: In the new smoothing of $\mathcal{R}$, the submanifold $S^{2n-1} \times_{\mathbb{Z}_p} S^1$ is transformed into the mapping torus for the Standard Generator γ of the induced free $\mathbb{Z}_p$ action on S^{2n-1}. Moreover, $\partial\mathcal{R}$ is obtained from this mapping torus by performing surgery along a fiber. Given the canonical identification between the mapping torus of γ and $S^{2n-1} \times S^1 \# \Sigma_\gamma^{2n}$ (compare [5]), it is fairly routine to verify that $\partial\mathcal{R}$ is diffeomorphic to Σ_γ^{2n}. The minus sign appears because a reversal of orientation occurs when we identify the mapping torus and the connected sum (compare the minus sign in [29, Thm. 3.4]) ■

By Theorem 2, the search for nontrivial isotopy classes representing free $\mathbb{Z}_p$ actions reduces to looking for nontrivial classes in the image of h^*. Examples of this sort are given by the following homotopy-theoretic result:

PROPOSITION 1.3. *Let* p *be an odd prime, and let* $\alpha_i, \beta_k \in \pi_{*(p)}$ *be the usual elements* (compare [24], [38]).

Let $j: \pi_{*(p)} \to \pi_*(F/O)_{(p)}$ *be the usual projection, and let* i *satisfy* $1 \le i \le p-1$. *Then there are classes*

$$B_i \in [L^{2n(i)-1} \cup_\pi e^{2n(i)}, Top/O]$$

with $n(i) = \frac{1}{2}\text{Stem}(\beta_i\beta_1^{p-i})$, $1 \le i \le p-1$ *and the following further properties:*

(a) B_i *is nullhomotopic on the* $[\text{Stem}(\beta_i\beta_1^{p-i-1}\alpha_1)-1]$*-skeleton of* L, *and the obstruction to nullhomotoping it on the next skeleton is* $j(\beta_i\beta_1^{p-i-1}\alpha_1)$.

(b) $h^*B_i = j(\beta_i\beta_1^{p-i})$ ■

NOTE. The element on the right hand side of (b) are non zero by Toda's calculations [38, Theorem 4.15].

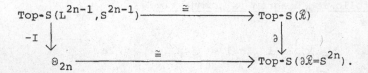

The proof essentially amounts to verifying that an element such as B_i actually exists for the complex in question, observing that the action of P^1 in $H^*(L^{2n+1};\mathbb{Z}_p)$ tells us that h^*B_i is obtained by inserting the obstruction into the bracket $\langle -,\alpha_1,\ldots,\alpha_1\rangle$ (p terms), and using the formula $\beta_1 = \langle \alpha_1,\ldots,\alpha_1\rangle$ (p terms). Similar calculations have been described in papers by G. Brumfiel [7,8].

COROLLARY 1.4. <u>The classes</u> $j(\beta_i\beta_1^{p-1}) \in \pi_{2*}(\mathrm{Top}/0)$ <u>are representable by free</u> $\mathbb{Z}_p$ <u>actions on spheres</u>.

There are two features of this that merit improvement. In the first place, for each odd prime we only get finitely many isotopy classes represented. Furthermore, no isotopy classes on even dimensional spheres are realized. We can take care of these problems by giving a modified version of the original construction for <u>semifree</u> $\mathbb{Z}_p$ actions. The key point is that all resmoothing must be done <u>away from the fixed point set</u>.

For this purpose we consider a set

$$\mathrm{Top}\text{-}S(L^{2n-1}xD^{k+1};\ S^{2n-1}xD^{k+1},\ L^{2n-1}xS^k)$$

consisting of (a) topological smoothings on $L^{2n-1}xD^{k+1}$ and (b) identifications of the induced smoothings of $S^{2n-1}xD^{k+1}$ and $L^{2n-1}xS^k$ that are consistent on $S^{2n-1}xS^k$ (the induced smoothing of $S^{2n-1}xD^{k+1}$ is given by taking universal coverings, and that of $L^{2n-1}xS^k$ is given by restricting to the boundary). As in the previous case, smoothing theory [19] gives a homotopy theoretic interpretation of this set.

PROPOSITION 1.5. <u>If</u> $k + 2n \geq 5$, <u>then there is an isomorphism</u>

$$\mathrm{Top}\text{-}S(L^{2n-1}xD^{k+1};\ S^{2n-1}xD^{k+1},\ L^{2n-1}xS^k)\cong[L^{2n-1}\cup e^{2r},\Omega^{k+1}\ \mathrm{Top}/0].$$

Given a class in this set, we may form a smooth $\mathbb{Z}_p$ action on S^{k+2n} as follows: Let W be the underlying smooth manifold homeomorphic to $L^{2n-1}xD^{k+1}$, and let $\widetilde{W}$ be its universal covering with the induced $\mathbb{Z}_p$ action. By the extra conditions imposed, we know that the action on $\partial\widetilde{W}$ is smoothly equivalent to the standard linear action on $S^{2n-1}xS^k$ in a prescribed way. Therefore we may form a smooth $\mathbb{Z}_p$ action on the manifold

$$\Sigma^{k+2n} = W \cup_{\tilde{f}} D^{2n}xS^k \ .$$

Since $\widetilde{W}$ is identified (nonequivariantly) with $S^{2n-1} \times D^{k+1}$ by an extension of $\widetilde{f}$ (a condition on Top S), a relatively standard cutting and pasting result (compare [28, 1.1]) implies that Σ^{k+2n} is nonequivariantly diffeomorphic to S^{k+2n}. As before we have an isotopy class mapping

$$I: \text{Top-}S(\cdots) \to \pi_o(\text{Diff}^+ S^{k+2n}) \cong \Theta_{k+2n+1} \quad ,$$

and Theorem 1.2 generalizes in the expected way:

THEOREM 1.6. <u>Under the isomorphism of 1.5, the map</u> $-I$ <u>corresponds to the homomorphism</u>

$$h^*: [L^{2n-1} \cup e^{2n}, \Omega^{k+1} \text{Top}/0] \to \pi_{2n}(\Omega^{k+1}(\text{Top}/0)) \quad ■$$

Actually, this theorem is a parameterized version of 1.2.

The point of 1.3 and 1.4 was that certain classes in $\pi_*(\text{Top}/0)$ could be found in the image of h^* - specifically, classes that could be written as $\langle x, \alpha_1, \ldots, \alpha_1 \rangle$ (p terms) had this property. Here is another simpler criterion:

PROPOSITION 1.7. <u>Let</u> $n \geq p$, $n \not\equiv -1 \mod p$. <u>Then the image of</u> h^* <u>contains</u> $j(\alpha_1 \cdot \pi_{2n+k+4-2p})$.

IDEA OF PROOF. The key point is that the Steenrod operation P^1 hits the top class in $H^*(L^{2n+1}; \mathbb{Z}_p)$. Given this, one proceeds again as in [8]. ■

<u>Remark</u>. Since α_1 annihilates π_{odd}, this result is totally useless for finding free $\mathbb{Z}_p$ actions with exotic isotopy classes.

We can now find an infinite number of nontrivial classes in $\Theta_* = \pi_o(\text{Diff}^+ S^{*-1})$ that have periodic representatives if $p \geq 5$. I am grateful to H. Miller for explaining how these examples and still further information could be recovered from [24].

THEOREM 1.8. (i) <u>Suppose</u> $p \geq 5$. <u>Let</u> t <u>be a positive integer such that</u> $t \not\equiv 0, -1 \mod p$ <u>or</u> $t \in \{p-1, p\}$. <u>Then</u> $j(\alpha_1 \beta_i)$ <u>is a nonzero class with a periodic representative</u>.

(ii) <u>Suppose</u> $p = 3$ <u>and</u> $t = 1, 2, 3,$ <u>or</u> 6. <u>Then</u> $j(\alpha_1 \beta_t)$ <u>admits a periodic representative</u>.

PROOF. In the first place, $\alpha_1\beta_t \notin$ Image J for $t \leq p$ by the results of [38] for $t < p$ and [26, Thm. 7.9] plus [34] for $t = p$. Secondly, if $p = 3$, then $\alpha_1\beta_6 \notin$ Image J by computations of M. Tangora and D. Ravenel (see [37] for example). Thus it remains to check that $\alpha_1\beta_t \neq 0$ in $\mathrm{Ext}^{3,*}$ of the Adams-Novikov spectral sequence for $\pi_{*(p)}$ if $t \not\equiv 0, -1 \bmod p$ and $p \geq 5$; if this is true, it is immediate that $\alpha_1\beta_t$ is a nontrivial permanent cycle since no non zero differential can hit it. But the nonvanishing of these classes is true by [24, 2.13].

2. Isotopy classes and knot invariants

In the previous section we gave sufficient conditions for the existence of periodic representatives in $\pi_0(\mathrm{Diff}^+ S^n)$, and in this section we shall develop machinery to give necessary conditions. Technically speaking, the results of this section complement those of [29,§3] and constitute a further development of the techniques presented there.

Let Φ be a smooth action of $\mathbb{Z}_p$ on S^{n+2k}; denote the fixed point set by F^n, the normal representation at a fixed point by V, and the equivariant normal bundle of F^n by ξ. The fundamental data in [29] was an equivariant map ρ' from the unit sphere bundle $S(\xi)$ into $S(V)$ whose restriction to a typical fiber $S(\xi)_* \cong S(V)$ had some possibly large degree $d \equiv 1 \bmod p$. This in turn allowed us to form a $\mathbb{Z}_{(p)}$ homology equivalence h from $S(\xi)/G = L(\xi)$ into $S^n \times L(V)$ using ρ'/G on the second factor and

$$L(\xi) \xrightarrow{\text{proj}} F^n \xrightarrow{\text{deg } 1} S^n$$

on the first (recall that F^n is a $\mathbb{Z}_{(p)}$ homology sphere by P.A. Smith theory). In fact, by taking joins we also constructed homology equivalences $h_M: L(\xi \oplus M) \to S^n \times L(V \oplus M)$ with M an arbitrary free G module.

The homology equivalence h itself is $\mathbb{Z}_{(p)}$ homologically h-cobordant to $1 \times d: S^n \times L(V)$, where $d: L(V)\circlearrowleft$ is a map of degree $d = \deg(\rho'|S(V))$. This fact is implicit in the proof of [29, Prop.3.1]. Let M be the free 2-dimensional G-module given by restriction from the free 2-dimensional S^1-module (in contrast with [29], an explicit choice is important here because we are especially concerned with the action of the Standard Generator for G).

The standard isotopy class $\gamma = \gamma(\Phi)$ of the action Φ is the image of $\exp(2\pi i/p)$ under the map $\Phi_*: Z_p = \pi_0(Z_p) \to \pi_0(\mathrm{Diff}+S^{n+2K}) = \Theta_{n+2K+1}$. Since the codomain is abelian, this is an isomorphism invariant of Φ.

THEOREM 2.1. Assume the above notation. Then h_M is $\mathbb{Z}_{(p)}$-homologically h-cobordant to $(1\times d)\ \#\Delta(\gamma)$, where "$\#\Delta(\gamma)$" denotes taking connected sums with the homotopy sphere γ.

PROOF. We shall adopt the setting used to prove [29, Thm. 3.4]. Let W be the homology h-cobordism from $S^n \times S^{2k-1}$ to $S(\xi)$ constructed in [29, Prop. 3.1] from the G-actions, and let $\mathfrak{W}$ be the manifold

$$L(V\oplus M)\times S^n \times [0,\varepsilon] \cup W\times_G D(M) \cup S(\xi\oplus M/G\times[1-\varepsilon,1]$$

constructed in [29, p.117] with rounded corners (see Figure 1 on that page). Recall that a submanifold $P_o = S^n \times D^{2k} \cup W \cup D(\xi)$ was embedded in the "third component" of

$$\partial\mathfrak{W} = L(V\oplus M)\times S^n \cup S(\xi\oplus M)/G \cup X_o,$$

where

$$X_o = \{D(V)\times S^1 \times S^n\} \cup [W\text{-open collar}]\times_G S(M) \cup \{S(\xi)\times_G S^1\}.$$

Both W and $\mathfrak{W}$ have reference maps to $S^n \times L(V\oplus M)$ that are nearly homology equivalences ([29, §3])
Recall from [29,3.5] that the angle-straightened version P of P_o is diffeomorphic to $S^n \times S^{2k}$ because we are assuming the G-action is on S^{n+2k}. In fact, one can also give a reasonably brief description of the straightened manifold X obtained from X_o.

(2.2) ASSERTION. There is a smooth G action on $P = S^n \times S^{2k}$ such that

(a) The action on $P - \mathrm{Int}\, D^n_- \times D^{2k}_-$ is equivalent to the linear action of G on $S^n \times S(V\oplus \mathbb{R})$

(b) X is equivalent to $P\times_G S(M)$ ∎

Part (a) is true because the surgery done in proving [29,3.5] involves an equivariant embedding of $S(V)\times D^{n+1}$, so that one can fill in with $D(V)\times S^n$; the linearity portion is true because the embedding factored through a linear disk $D(V)\times D^n \subseteq S^{n+2k}$. Given this information and the above decomposition of X_o, part (b) of the assertion is a routine exercise.

As noted in [29, p. 118], the identification $P \cong S^n \times S^{2k}$ allows us to paste an extra piece onto $\mathfrak{W}$ so that it looks more like a $\mathbb{Z}_{(p)}$-homology h-cobordism. Unfortunately, there is a misprint on line 14, however; the manifolds listed on that line should be $I \times S^n \times D^{2k+1}$ and $I \times P$ respectively. At any rate, one pastes on a copy of $I \times S^n \times D^{2k+1}$ along $I \times S^n \times S^{2k} \subseteq S^1 \times_G P = X$. Since the restriction of $\mathfrak{W} \to S^n \times L(V \oplus M)$ to X factors through $S^n \times (D(V) \times S^1) \subseteq S^n \times L(V \oplus M)$, there is no difficulty in extending the map to the enlarged manifold $\mathfrak{W}_1$. We now have

$$\partial \mathfrak{W}_1 = L(V \oplus M) \times S^n \cup S(\xi \oplus M)/G \cup Y,$$

where $Y = S^o \times S^n \times D^{2k+1} \cup_\delta I \times S^n \times S^{2k}$ and δ is the identity on one copy of $S^n \times S^{2k}$ and the Standard Generator of G (acting on $P = S^n \times S^{2k}$) on the other. Actually, the linearity of the G-action off $D^n_- \times D^{2k}_+$ tells us that Y is diffeomorphic to $S^n \times S^{2k+1} \# -\Sigma_\gamma^{n+2k+1}$, and one can next perform a smooth surgery along $D^n \times S^{2k+1}$ to get a new manifold $\mathfrak{W}_3$ with

$$\mathfrak{W}_3 = L(V \oplus M) \times S^n \cup S(\xi \oplus M)/G \cup -\Sigma_\gamma^{n+2k+1}$$

(and a reference maps $\mathfrak{W}_3 \to S^n \times L(V \oplus M)$ extending the original map). The negative sign comes from the choice of orientation conventions (compare [29, §3, esp. p. 120]. Topologically one could go a step further and form $\mathfrak{W}_4'$ (with reference map) by coning off the last boundary component, and a straightforward calculation would imply that $\mathfrak{W}_4'$ is a homological h-cobordism between $L(V \oplus M) \times S^n$ and $S(\xi \oplus M)/G$. Smoothly, however, one must join two points in $L(V \oplus M) \times S^n$ and Σ by an embedded arc that meet the boundary only at end points, take a closed tubular neighborhood, and let $\mathfrak{W}_4$ be $\mathfrak{W}_3$ minus the interior of this tubular neighborhood. Then $\mathfrak{W}_4$ is again a homological h-cobordism; however, the bottom is no longer $S^n \times L(V \oplus W)$ but rather $S^n \times L(V \oplus W) \# \Sigma_\gamma^{n+2k+1}$.

The construction of Theorem 2.1 allows us to interpret the undetermined term γ' occuring in [29, Thm. 3.4]:

COROLLARY 2.3. Let Φ be the original G-action on S^{n+2k}, let $\omega(\Phi) \in \pi_n(F_G(V)/C_G(V))_{(p)}$ be the knot invariant, as defined in [29], let

$$q_\circledast : \pi_n(F_G(V \oplus M)/C_G(V \oplus M)) \to [S^n \times L(V \oplus M), F/O]_{(p)}$$

be defined as in [29, §2], let $\gamma \in \pi_n(F/0)_{(p)}$ be the Pontrjagin-Thom invariant of F^n, and let δ be the isotopy class of a Standard Generator. Then

$$q\otimes[\omega(\Phi)\oplus M] \oplus\gamma = e^*(j(\delta))$$

where $e: S^n\times L(V\oplus M) \to S^{n+2k+1}$ has degree 1 and $j: \pi_o(\mathrm{Diff}^+S^{n+2k}) \to \pi_{n+2k+1}(F/0)_{(p)}$ is the usual homomorphism.

PROOF. This follows upon taking the p-local normal invariants of the two homology equivalences discussed in 2.1 (the normal invariants are those defined in [29, §1]) and applying [29, Cor. 2.5]. ■

In fact, an analog of Corollary 2.3 is true for G actions on arbitrary $\mathbb{Z}_{(p)}$ homology spheres. To state this, we construct a homomorphism $\varphi:\pi_o(\mathrm{Diff}^+\Sigma^m) \to \pi_{m+1}(F/0)_{(p)}$ as follows: Let g represent a typical class in the domain, and let $d: \Sigma^m \to S^m$ have degree 1. Then dg also has degree 1 and consequently is homotopic to d; choose a homotopy. Using d and this homotopy one can construct a homology equivalence $h: \Sigma \times_G S^1 \to S^m\times S^1$. The normal invariant of h lies in

$$[S^m\times S^1,F/0]_{(p)} = \pi_m(F/0)_{(p)}\oplus\pi_{m+1}(F/0)_{(p)},$$

and we take $\varphi(g)$ to be the second component of this normal invariant. There is a theoretical indeterminacy caused by the choice of homotopy, and it turns out that the indeterminacy is the image of a homomorphism from $\pi_1(F(\Sigma^m,S^m),d)$. On the other hand, a check of the spectral sequence for $\pi_*(F(\Sigma^m,S^m),d)$ shows that π_1 is finite of order prime to p, and therefore the indeterminacy is in fact zero. Given this, we may state the generalization we want:

PROPOSITION 2.4. Let Φ be a smooth G-action on the $\mathbb{Z}_{(p)}$- homology sphere Σ^{n+2k} with notation as before. Let

$$c: S^n\times L(V\oplus M) \to S^n\times L(V\oplus M)/S^n\times L(V) \simeq S^{n+2k}\vee S^{n+2k+1}$$

be the collapsing map as in [29, §3]. Then, in the notation of [29, Thm. 3.4], we have

$$q\otimes[\omega(\Phi)\oplus M] +\gamma = c^*(-q(\Sigma)\oplus\varphi(g_\Phi)),$$

where g is the isotopy class determined by a Standard Generator of G ■

The proof follows along the lines of [29,§3] and the present section but requires a more delicate treatment. A typical complication is that Σ might not be almost diffeomorphic to the boundary of a $\mathbb{Z}_{(p)}$ acyclic manifold and it might be necessary to take a connected sum of Σ with it self many times (the number prime to p, however) in order to fulfill this condition. Since we shall not need this result later in this paper, the proof is not given.

3. Some non-h-cobordant manifolds

For our purposes the most important consequence of Theorm 2.1 is that a G-action yields a $\mathbb{Z}_{(p)}$ homology h cobordism between a lens space bundle and the connected sum of a trivial bundle with the homotopy sphere that carries the isotopy invariant. Therefore, if we can find homotopy spheres for which the connected sum and the appropriate lens space bundles are not homo logically h-cobordant, we shall have examples of isotopy classes without periodic representatives. The following result gives a convenient place to start looking for examples:

THEOREM 3.1. Let F^n be a $\mathbb{Z}_{(p)}$-homology sphere with a weakly almost complex structure, let $W^{2\ell}$ be a free G-module, and let ξ be a G-vector bundle over F (trivial G-action on F) with fiber W. Write $\xi = \Sigma\xi_j \otimes T_j$ where T_j runs over all irreducible G-modules and the ξ_j are complex vector bundles. Assume that the rational Chern classes of F and each ξ_j are zero. Finally, let $n + 2\ell = 4s \geq 8$, and suppose M^{4s-1} is a homotopy sphere bounding a parallelizable manifold with $L(\xi) = S(\xi)/G$ and $S^n \times L(W) \# M$ both $\mathbb{Z}_{(p)}$-homologically h-cobordant. Then M is a 2-primary element of bP_{4s}.

Remark 3.2. Given a smooth G action on S^{4s-2}, we get such an F^n and ξ unless $n = 2$ and $2^{odd} = 1 \bmod p$. Take F^n to be the fixed point set of G and $\xi = \nu \oplus M$ where ν is the equivariant normal-bundle. The the canonical stable framing on $D(\nu) \times M \subseteq S^{4s-2} \times M$ plus the complex structure on ν induce a weakly almost complex structure on F^n. By Ewing's results [11] the rational Chern classes of ξ all vanish, and thus by construction of the weak almost complex structure on F^n <u>its</u> rational Chernclass is also trivial.

We shall use this to find explicit negative examples in the next section.

PROOF OF 3.1. Since the integral homology of $BU_{n_1}\times\ldots\times BU_{n_q}$ is torsion free and concentrated in even dimensions, it follows that the Atiyah-Hirzebruch spectral sequence for the complex bordism collapses. Suppose that ξ is the given G vector bundle and $U_{n_1}\times\ldots\times U_{n_q}$ corresponds to its structural group (i.e., ξ_j has complex dimension n_j). Since elements in the free MU_* module $MU_*(BU_{n_1}\times\ldots\times BU_{n_q})$ are detected by rational characteristic numbers (this of course uses the structure theorems for MU_* as in [35]), the rational Chern class assumptions tell us that (F^n,ξ) is a weakly almost complex boundary. Let $(\mathfrak{W}^{n+1},\Xi)$ be the cobounding manifold with vector bundle, and let $L(\Xi)$ denote the associated lens space bundle.

Suppose that V is a $\mathbb{Z}_{(p)}$ homology h-cobordism between $L(\xi)$ and $S^n\times L(W)\#M$, and let P^{4s} be a (2s-1)-connected parallelizable manifold with $\partial P^{4s} = M$. Form the closed oriented 4s-manifold

$$\mathfrak{U} = L(\Xi)\cup_{L(\xi)} V\cup_{S\times L\#M} D^{n+1}\times L(W)\#P.$$

The rational cohomology and Pontrjagin classes are now easy to determine. In the first place, the rational Serre spectral sequence for $L(\Xi)$ collapses because the associated vector bundle $\Xi = \Sigma\xi_j$ has zero rational Euler class; this is true because the rational Chern class of each ξ_j is trivial and the Euler class of is a monimial in these Chern classes. Therefore by the Mayer-Vietoris sequence for $\mathfrak{U} = L(\Xi)\cup\overline{(\mathfrak{U}-L(\Xi))}$ we see that the kernel of the restriction map

$$i^*\colon H^*(\mathfrak{U};Q)\to H^*(L(\Xi);Q)$$

splits into two pieces. One is 1-dimensional in degrees $n+1$ and $4s$ and zero elsewhere, and it is the image of the map

$$\psi^*\colon H^*(D^{n+1}\times L,S^n\times L)\underset{\text{excision}}{\cong} H^*(\mathfrak{U},L(\Xi)\cup V\cup P)\to H^*(\mathfrak{U}).$$

The other is a piece that maps monomorphically to $H^{2s}(P)$ by restriction. This information gives us a very firm hold on the rational Pontrjagin numbers as follows:

LEMMA 3.3. <u>All decomposable rational Pontrjagin numbers of</u> $\mathfrak{U}$ <u>are zero</u>.

PROOF. Consider the rational Pontrjagin classes $p_k(L(\boxminus))=i^*p_k(\mathfrak{U})$. We claim these lie in the image of

$$\pi^*: H^*(\mathfrak{W}) \to H^*(L(\boxminus)).$$

But suppose we look instead at the sphere bundle $S(\boxminus)$ that finitely covers $L(\boxminus)$; then some well known vector bundle identities imply that the tangent bundle of $S(\boxminus)$ is induced stably via projection onto $\mathfrak{W}$. Since $S(\boxminus) \to L(\boxminus)$ is a rational isomorphism in cohomology, it follows that the rational Pontrjagin classes of $L(\boxminus)$ must come from $H^*(\mathfrak{W})$ as claimed. In fact, these classes can be lifted to $H^*(\mathfrak{W},\partial\mathfrak{W})$ because the rational Pontrjagin classes of $\partial\mathfrak{W} = F^n$ are zero (by the Hirzebruch signature theorem).

It follows that $p_k(L(\boxminus))$ may be expressed as $j^*(EXC)^*(\pi,\partial\pi)^*\alpha_k$, where $\alpha_k \in H^{4k}(\mathfrak{W},\partial\mathfrak{W})$ and $j^*,(EXC)^*,(\pi,\partial\pi)^*$ are given in the following diagram:

$$\begin{array}{ccc}
H^*(\mathfrak{W},\partial\mathfrak{W}) & \xrightarrow{\ \ell^*\ } & H^*(\mathfrak{U}) \\
\Big\downarrow (\pi,\partial\pi)^* & & \Big\uparrow j^* \\
H^*(L(\boxminus),L(\xi)) & \xrightarrow[\text{excision}]{(EXC)^*} & H^*(\mathfrak{U},\overline{\mathfrak{U}-L(\boxminus)}).
\end{array}$$

Putting the above observations together with the parallelizability of P, it follows that $p_k(\mathfrak{U}) = \ell^*\alpha_k + \psi^*\beta_k$ for suitable α_k and β_k. Suppose now that we look at a 4s-dimensional monomial

$$p_{k_1}\cdots p_{k_r} = (\ell^*\alpha_{k_1} + \psi^*\beta_{k_1})\cdots(\ell^*\alpha_{k_r} + \psi^*\beta_{k_r}).$$

Of course we want to show this vanishes unless there is only one factor. The expression breaks up into a sum of monomials, most of which contain both α and β type factors and one each containing solely α or β type factors, adn we shall verify that each such monomial is zero.

The mixed type monomials vanish because each $\ell^*\alpha_j\ \psi^*\beta_k$ is zero. To see this, notice that $\ell^*\alpha_j$ comes from $H^*(\mathfrak{U},\overline{\mathfrak{U}-L(\boxminus)})$ and $\psi^*\beta_k$ comes from $H^*(\mathfrak{U},L(\boxminus)\cup V\cup P)$, so that the product comes from $H^*(\mathfrak{U},\mathfrak{U}) = 0$. Next, each $\psi^*\beta_j\psi^*\beta_k$ is zero because all cup products in $H^*(D^{n+1}\times L, S^n\times L)$ are zero. Finally, the pure α-monomial $\ell^*\alpha_{k_1}\cdots\ell^*\alpha_{k_r}$ is the image of a class in $H^{4s}(\mathfrak{W},\partial\mathfrak{W})$. But $4s>\dim\mathfrak{W}= n+1$ by construction, and therefore the α-monomial is also zero ■

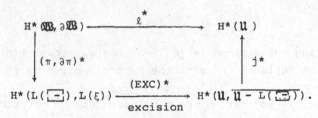

PROOF of 3.1 CONCLUDED. Since all decomposable Pontrjagin numbers of $\mathfrak{U}$ vanish, the Hirzebruch signature theorem implies that the signature and $\hat{A}$ genus can be related to each other in a very explicit way (compare [4]). If we knew that $\mathfrak{U}$ was a spin manifold, then one could use the integrality theorem for the $\hat{A}$ genus to conclude that the signature of P was divisible by 8 order bP_{4s} as in a paper of W. Browder [4], and from this it would follow that the homotopy sphere M^{4s-1} must be the standard one. However, we do not know if $\mathfrak{U}$ is a spin manifold, and therefore we must be content with a weaker result of Borel-Hirzebruch [2,(25.5)] stating that the $\hat{A}$ genus of $\mathfrak{U}$ lies in the ring $\mathbb{Z}\left[\frac{1}{2}\right]$. Having <u>this</u> at our disposal we may argue as in [4] that the order of M in bP_{4s} is a power of 2 ∎

<u>Complements to 3.1</u> A. Using the methods and results of a paper by G. Brumfiel [6], one can say a little more. Namely if

$$f: \Theta_{4s-1} \to bP_{4s}$$

is the splitting map defined in [6], one can say that f(M) is 2-primary. One replaces P with the manifold given by [6, Thm. 1.5].

B. It is possible to sharpen the conclusion on M of f(M) and show that it actually is zero. By the Atiyah-Hirzebruch integrality theorem [1], it is enough to show that $\mathfrak{U}$ or something similar can be chosen to be a Spin^c manifold. This can be done using the Anderson-Brown-Peterson decomposition of $M\text{Spin}^c$; since we do not need this sharpening, we shall pursue this point any further.

4. <u>Finding non-realizable isotopy classes</u>.

We shall say that a smooth $G = \mathbb{Z}_p$ action on S^{n+2k} with fixed point set F^n is <u>unexceptional</u> if, given the splitting of the equivariant normal bundle as $\nu = \Sigma\xi_j \otimes T_j$, the rational Chern classes of the bundles ξ_j are all trivial. With this notation, we may summarize the results of Sections 2 and 3 as follows:

THEOREM 4.1. <u>Let</u> p <u>be an odd prime, and suppose</u> $x \in bP_{4s}$ <u>has order</u> p. <u>Then the isotopy class of</u> x <u>in</u> $\pi_0(\text{Diff}^+ S^{4s-2})$ <u>is not representable by an unexceptional period</u> p <u>diffeomorphism</u>.

This follows from Theorems 2.1 and 3.1.

The usefulness of 4.1 depends on two things - finding classes of order p for a given prime p and showing that all G-actions on the appropriate sphere are unexceptional. Here are some basic conditions under which these things happen:

(4.2) Classes of order p exist in some bP_{4s} if and only if p is an irregular prime (as defined in [3]) or 2 has odd multiplicative order mod p (e.q., see [36, Ch.6]).

(4.3) All G-actions on spheres are unexceptional if and only if 2 has even multiplicative order mod p [11, 32].

By these results, if we can find irregular primes such that 2 has even multiplicative order mod p, then we shall have examples of isotopy classes with no period p representatives. But suppose $p \equiv \pm 3 \bmod 8$; then 2 is not a perfect square mod p [20,p.68] and hence 2 must have even multiplicative order mod p (the unit group of $\mathbb{Z}_p$ is isomorphic to $\mathbb{Z}_{p-1}$). Hence the following result of T. Metsänkylä [23] gives us an infinite number of primes for which (4.2) and (4.3) are valid:

(4.4) METSÄNKYLÄ'S THEOREM. If $n \geq 6$ there are infinitely many irregular primes $\not\equiv \pm 1 \bmod n$ ∎

If $n = 8$ the first such prime is 37, which is also the first irregular prime.

We then have the following conclusion:

THEOREM 4.5. Let p be an irregular prime congruent to ±3 mod 8 (by (4.4) there are infinitely many such primes). Then for infinitely many values of s there are classes in $bP_{4s} \subseteq \pi_0(\mathrm{Diff}^+ S^{4s-2})$ of order p that admit no representatives with period p.

PROOF. By 2.1, 3.1, and 4.4, it is merely necessary to show that for such a prime there are infinitely many groups bP_{4s} with elements of order p. But the order of bP_{4s} is $2^{x(s)}(2^{2s-1})\ \mathrm{num}\,(B_s/4s)$, where B_s denotes the s-th Bernoulli number [17], and classical results of analytic number theory (compare [3, Ch. 5, §§5,8]) imply that p divides infinitely many of the numerators $\mathrm{num}\,B_s/4s$ if p is irregular ∎

COMPLEMENT 4.5A. The classes described in 4.5 admit no representatives with arbitrary (finite) period.

IDEA OF PROOF. The main step is to show there are no representatives of period p^r with $r \geq 2$. If $G = \mathbb{Z}_{p^r}$ acts semifreely on S^{n+2k} with fixed point set F^n, this may be done exactly as for $\mathbb{Z}_p$. In general, however, one needs the concepts used to study ultrasemifree actions as in [30]; this is formally parallel to the semifree case, being based upon the fact that G acts freely of the fixed point set F^n of $\mathbb{Z}_p$. One complication is that the fixed point set must be viewed as a manifold with (possibly ineffective) action of $G/\mathbb{Z}_p$. If we let U denote the tangent space to F^n at a fixed point of the whole group, then in analogy with Theorem 2.1 we get that the manifold $S(\nu \oplus M)/G$ is $\mathbb{Z}_{(p)}$-homologically h-cobordant to $S(V\oplus M) \times_G S(U \oplus \mathbb{R}) \# \Sigma_\gamma^{n+2k+1}$, where $\gamma \in \Theta_{n+2k+1}$ is the isotopy class of the Standard Generator for G (compare [30, §3]. In further analogy with 3.1, we have that these two manifolds are not homologically h-cobordant if the equivariant tangent bundle of S^{n+2k} has finite order in $\widetilde{KO}_G(S^{n+2k})$ and $\Sigma \in bP_{4*}$ has order p. Since an extension of Ewing's calculations in [11] show that ν always has finite order if 2 has even multiplicative order mod p, (compare [12], [33]), in analogy with 4.5 we see that no representatives with period p^r exist.

It is now easy to exclude all other possible periods. If there were a representaive of period q, say f, we could construct a representative g of order a power of p as follows: Write $q = q'p^r$ where $r \geq 1$ and q' is prime to p, and choose s' so that $s'q \equiv 1 \bmod p$. Then $g = f^{s'q}$ would be isotopic to f and have order p^r. Therefore the classes in 4.5 admit no periodic representatives ■

If 2 has odd multiplicative order mod p, it is still possible to give at least one example of a non representable isotopy class mod p. To do this, we must extend our previous definition of unexceptionality to actions of $G = \mathbb{Z}_{p^r}$. Namely, we shall require that equivariant tangent bundle of S^{n+2k} have finite order in $\widetilde{KO}_G(S^{n+2k})$. In analogy with [21], this condition is equivalent to the previous one if $G = \mathbb{Z}_p$.

THEOREM 4.6. <u>Suppose that</u> 2 <u>has multiplicative order</u> $2s+1 \bmod p$. <u>Then all smooth</u> G-<u>actions on spheres of dimension</u> $\leq 4s+3$ <u>are unexceptional</u>.

This will be proved in a forthcoming paper [33]. At any rate, we can now proceed exactly as in the case of irregular $p \equiv \pm 3 \bmod 8$ to obtain the following result:

THEOREM 4.7. <u>Suppose that 2 has multiplicative order</u> $2s+1 \bmod p$. <u>Then there is a class of order</u> p <u>in</u> $bP_{4s+4} \subseteq \pi_0(\mathrm{Diff}^+ S^{4s+2})$ <u>containing no periodic representatives</u> ■

<u>Remarks</u> 1. The class exists because, as noted earlier, the order of bP_{4s+4} is a power of 2 times $(2^{2s+1}-1)\cdot \mathrm{num}\,(B_{s+1}/4(s+1))$.

2. The first examples for which 4.7 applies are quite simple; namely, they are the classes in $bP_8 = \Theta_7 \cong \pi_0(\mathrm{Diff}^+ S^6) \cong \mathbb{Z}_{28}$ of order 7 (since $2^3 \equiv 1 \bmod 7$).

DEPARTMENT OF MATHEMATICS
PURDUE UNIVERSITY
WEST LAFAYETTE, INDIANA 47907, U.S.A.

REFERENCES

1. M. Atiyah and F. Hirzebruch, Riemann-Roch theorems for differentiable manifolds, Bull. Amer. Math. Soc 65 (1959), 276-281.

2. A. Borel and F. Hirzebruch, Characteristic classes and homogeneous spaces. II, Amer. J. Math. 81 (1959), 315-382.

3. Z. I. Borevich and I. R. Shafarevich, Number Theory (Transl. from Russian by N. Greenleaf), Pure and Applied Mathematics Vol. 20. Academic Press, New York, 1966.

4. W. Browder, On the action of Θ^n $(\partial\pi)$, Differential and Combinatorial Topology (A Symposium in Honor of M. Morse), Princeton Mathematical Senes No. 27, 23-56. Princeton University Press, Princeton, 1965.

5. ______________ , Diffeomorphisms of 1-connected manifolds, Trans. Amer. Math. Soc. 128 (1967), 155-163.

6. G. Brumfiel, On the homotopy groups of BPL and PL/O, Ann. of Math. 88 (1968) 291-311.

7. ______________ , Homotopy equivalences of almost closed manifolds, Comment. Math. Helv. 46 (1971), 381-407; ibid., Proc. Sympos. Pure Math. 22, 73-79. American Mathematical Society, Providence, 1971.

8. ______________ , [Free] differentiable S^1 actions on homotopy spheres, preprint, University of California - Berkeley, 1968.

9. J. Cerf, La stratification naturelle des espaces de fonctions différentiables réeles et la théorème de la pseudo-isotopie, Inst. Hautes Études Sci. Pub. Math. 39 (1970), 5-173.

10. P. Conner and F. Raymond, Manifolds with few periodic homeomorphisms, Proc. Second Conference on Compact Transofrmation Groups (Univ. of Mass., Amherst, 1971) Lecture Notes in Mathematics Vol. 299, 1-75 Springer, New York, 1972.

11. J. Ewing, Spheres as fixed point sets, Quart. J. Math. Oxford (2) 27, (1976) 445-455.

12. ______________ , Semifree actions of finite groups on homotopy spheres, preprint, Indiana University and University of Virginia, 1977.

13. W. Fenchel, Bemaerkninger om endlige grupper af afbildningsklasser, Mat. Tidsskr. B, Årg. 1950, 90-95.

14. A. Hatcher, The second obstruction for pseudo-isotopies, Astérisque 6, 239-275. Société Mathématique de France, Paris, 1973.

15. W. Heil and J. L. Tollefson, Deforming homotopy involutions of 3-manifolds to involutions, Topology, to appear.

16. W.-C. Hsiang and R. W. Sharpe, Parameterized surgery and pseudo-isotopy, Pac. J. Math. 67 (1976), 401-459.

17. M. Kervaire and J. Milnor, Groups of homotopy spheres, Ann. of Math. 78 (1973), 514-537.

18. R. C. Kirby and L. Siebenmann, On the triangulation of manifolds and the Hauptvermutung, Bull. Amer. Math. Soc. 75 (1969), 742-749.

19. R. Lashof, The immersion approach to triangulation and smoothing, Proc. Sympos. Pur Math. 22, 131-164. American Mathematical Society, Providence, 1971.

20. W. J. LeVeque, Topics in Number Theory, Vol. I. Addison-Wesley, Reading, Mass. 1956.

21. P. Löffler, Über die G-Rahmbarigkeit von G-Homotopiesphären, Arch. Math. (Basel) 29, (1977), 629-634.

22. A. M. Macbeath, On a theorem of J. Nielsen, Quart. J. Math. Oxford (2) 13 (1962), 235-236.

23. T. Metsänkylä, Note on the distribution of irregular primes, Ann. Acad. Sci. Fenn. Ser. A I No. 492 (1971),7pp.

24. H. Miller, D. Ravenel, and W. S. Wilson, Periodic phenomena in the Adams-Novikov spectral sequence, Ann. of Math. 106 (1977), 469-516.

25. J. Nielsen, Abbildungsklassen endlicher Ordnung, Acta. Math. 75 (1943), 23-115.

26. S. Oka, The stable homotopy groups of spheres I, Hiroshima Math. J. 1 (1971), 305-337.

27. F. Raymond and L. Scott, The failure of Nielsen's theorem in higher dimension, Arch. Math. (Basel), to appear.

28. R. Schultz, Composition constructions on diffeomorphisms of $S^p \times S^q$, Pac. J. Math. 42 (1972), 739-754.

29. ____________ , Differentiable group actions on homotopy spheres: I, Invent. Math. 31 (1975), 105-128.

30. ____________, *ibid*. II, to appear.

31. ____________, Smooth actions of small groups on exotic spheres, Proc. Sympos. Pure Math. 32, 491-496. American Mathematical Society, Providence, 1978.

32. ____________, Spherelike G-manifolds with exotic equivariant tangent bundles, to appear in Studies in Alg.Top.(Academic Press).

33. ____________, Finding equivariantly framed group actions on homotopy spheres, to appear.

34. L. Smith, Realizing complex bordism modules IV, Amer. J. Math. 99 (1977), 418-436.

35. R. Stong, Notes on Cobordism Theory, Mathematical Notes No. 7. Princeton University Press, Princeton, 1968.

36. D. Sullivan, Geometric TopologyI-Localization, Periodicity, and Galois Symmetry, multi-copied notes, MIT, 1970.

37. M. Tangora, Some homotopy groups mod 3, Reunion Sobre Teoria de Homotopia (Northwestern 1974), Notas de Matemáticas y Simposia No. 1, 227-246. Sociedad Matemática Mexicana, México, D. F., 1975.

38. H. Toda, p-primary components of homotopy groups IV., Mem. Coll. Sci. Kyoto Univ. 32 (1959), 297-332.

39. C. T. C. Wall, Surgery on Compact Manifolds, London Math. Soc. Monographs No. 1. Academic Press, New York, 1970.

ORIGINAL BROWN-PETERSON SPECTRA

J. M. BOARDMAN

(Research supported by NSF Grant MCS76-23466)

Introduction

Let p be a fixed prime. The ring spectrum known as BP was constructed by Quillen in [9] as a summand of the localization MU_p of the Thom spectrum MU at p. It has often escaped notice (including mine until S. Rosen pointed it out to me) that BP is not the same as the spectrum, call it X, constructed by Brown and Peterson in [6], which we shall call an original Brown-Peterson spectrum; the homotopy groups are visibly different. The purpose of this note is to study original Brown-Peterson spectra in terms of BP. Most of this material appeared in the preprint [3].

We find easily in §1 that the localization X_p of X is BP. This suggests that the stable localization theory developed in [2] should be a useful tool. We use it freely and find that it does indeed give us far more precise information about X.

The construction of X is inductive and involves countably many choices; the authors were careful to allow for this indeterminacy by using the indefinite article throughout. As a consequence, the properties of X are rather obscure. Our first result will be clear from §1 and §5.

Theorem A. There is a canonical original Brown-Peterson ring spectrum.

One would obviously like to know the effect of varying the choices in the construction of X.

Theorem B. If p is an odd prime, there are uncountably many distinct original Brown-Peterson spectra. But if p=2, there is only one.

We prove the first statement in §2 and §3, and the second in §4.

In §5 we study ring structures. It is by no means clear from [6] that any of the X admit a ring structure.

Theorem C. Every original Brown-Peterson spectrum admits uncountably many different ring structures.

When X is a ring spectrum, X_p is also, and Theorem B of [4] yields a ring spectrum isomorphism $X \cong BP$, so that $\pi_*(X)$ becomes a subring of $\pi_*(BP)$.

Theorem D. Suppose given an original Brown-Peterson ring spectrum X, so that $\pi_*(X)$ is a subring of $\pi_*(BP)$. If p is an odd prime, there are uncountably many different ring structures on X that all yield the same subring $\pi_*(X)$ of $\pi_*(BP)$. But if $p=2$, the subring $\pi_*(X)$ determines the ring spectrum structure on X uniquely up to isomorphism.

The moral appears to be that the prime 2 really is different.

1. The localization square

Take any original Brown-Peterson spectrum X, and write $G = \pi_*(X)$, which is a free abelian group. After some reorganization and translation into our stable category $\underset{\sim}{S}_h$, the construction yields a filtered spectrum

$$\ldots X(3) \subset X(2) \subset X(1) \subset X(0) = X,$$

whose quotients $X(n)/X(n+1)$ are GEM-spectra (graded or generalized Eilenberg-MacLane spectra) having free abelian homotopy groups. (In fact, the procedure is to construct $X/X(n)$ by induction on n.) One could regard it as a geometric realization of an explicit resolution of $H^*(X;Z/p) = A/(\beta)$ by non-free A-modules, where A denotes the mod p Steenrod algebra and β the Bockstein element. If X is not known in advance, such a realization is not generally available; but in this case the use of a large amount of additional data makes it possible. Fortunately, we do not need to be specific about the details.

If we localize at p, $X_p = BP$, because $\pi_*(X_p) = G_p$ is $Z_{(p)}$-free of the correct rank and $H^*(X_p;Z/p)$ vanishes in odd degrees. (For $H_*(X_p)$ must be of finite type over $Z_{(p)}$, and can have no p-torsion by the universal coefficient theorem, so that we have enough information to apply Theorem A of [4].)

On the other hand, if we localize away from p to get $X[p^{-1}]$, the construction breaks up completely because the invariants used in assembling the tower all have order some power of p, so that $X[p^{-1}]$ must be just the GEM-spectrum $K(G[p^{-1}])$. We therefore obtain the localization square

$$(1.1)\qquad \begin{array}{ccc} X & \longrightarrow & X_p = BP \\ \downarrow & & \downarrow \\ X[p^{-1}]=K(G[p^{-1}]) & \longrightarrow & X_\emptyset = K(G_\emptyset), \end{array}$$

which neatly summarizes X from our point of view. If we regard G as embedded as a subgroup of $\pi_*(BP)$, this square shows by Theorem 2.11 of [2] that X is completely determined up to isomorphism by the subgroup G of $\pi_*(BP)$. Conversely, given any free abelian subgroup G of $\pi_*(BP)$ such that $G_p = \pi_*(BP)$, we can synthesize uniquely a corresponding spectrum X and square (1.1). We need to know that we have not inadvertently generalized the situation.

<u>1.2 LEMMA</u>. <u>Every free abelian subgroup</u> G <u>of</u> $\pi_*(BP)$ <u>such that</u> $G_p = \pi_*(BP)$ <u>arises from some original Brown-Peterson spectrum.</u>

<u>Proof</u>. Fix a particular filtered original Brown-Peterson spectrum X, and let $G = \pi_*(X)$. We obtain a localization square (1.1) of <u>filtered</u> spectra, and also (see [6]) a filtration of G by free abelian subgroups $G(n) = \pi_*(X(n))$. Our approach is to synthesize different filtered spectra Y by varying the filtered spectrum $X[p^{-1}]$, without disturbing the filtered spectra X_p or $X_\emptyset$.

Take any free abelian subgroup F of $G_p = \pi_*(BP)$ such that $F_p = G_p$. We filter F by the subgroups $F(n) = F \cap G(n)_p \subset G_p$, then filter the spectrum $K(F[p^{-1}])$ in the obvious way by subspectra $K(F(n)[p^{-1}])$. For each n we synthesize a spectrum $Y(n)$ by Theorem 2.11 of [2] by the square

$$\begin{array}{ccc} Y(n) & \longrightarrow & X(n)_p \\ \downarrow & & \downarrow \\ K(F(n)[p^{-1}]) & \longrightarrow & K(F(n)_\emptyset) = X(n)_\emptyset. \end{array}$$

The same theorem also provides a sequence of maps

$$\ldots\ Y(n+1) \to Y(n) \to \ldots \to Y(2) \to Y(1) \to Y(0) = Y,$$

which we may without loss of generality regard as a filtered spectrum. We assert that this filtered spectrum is obtainable by the Brown-Peterson construction. First, $Y(n)/Y(n+1)$ is indeed a GEM-spectrum having the correct homotopy groups

F(n)/F(n+1), by considering its localization square. Second, if we examine all the other data used in the construction (see [6]) we find it is all p-local information consisting of conditions on $Y(n)_p = X(n)_p$ rather than $Y(n)$ itself, and therefore valid by the choice of X.]]]

Let us consider the homotopy and homology pullback localization squares induced by (1.1):

$$\begin{array}{ccc} \pi_*(X) = G & \longrightarrow & \pi_*(X_p) = G_p = \pi_*(BP) \\ \downarrow & & \downarrow \\ \pi_*(X[p^{-1}]) = G[p^{-1}] & \longrightarrow & \pi_*(X_\emptyset) = G_\emptyset \end{array} \tag{1.3}$$

and

$$\begin{array}{ccc} H_*(X) & \longrightarrow & H_*(X_p) = H_*(BP) \\ \downarrow & & \downarrow \\ H_*(X[p^{-1}]) = H_*(K(G[p^{-1}])) & \longrightarrow & H_*(X_\emptyset) \end{array} \tag{1.4}$$

The homotopy square (1.3) requires no further comment. The homology square (1.4), however, is more complicated. Let us write H for the image of $H_*(X)$ in $H_*(BP)$, so that $H_p = H_*(BP)$. Then $H_*(X)$ consists of two parts - plenty of torsion prime to p corresponding to the torsion in $H_*(K(G[p^{-1}]))$, and a torsion-free part H which must be given by $H = H_*(BP) \cap G[p^{-1}]$. (We are of course using the Hurewicz homomorphism to embed $\pi_*(BP)$ in $H_*(BP)$ etc., and taking intersections in the rational vector space $G_\emptyset$.) We easily recover G from H by

$$H \cap \pi_*(BP) = H_*(BP) \cap G[p^{-1}] \cap \pi_*(BP) = G[p^{-1}] \cap G_p = G.$$

Let us summarize this section so far:

1.5 LEMMA. There are canonical 1-1 correspondences between

(a) original Brown-Peterson spectra X equipped with an isomorphism $X_p \cong BP$;

(b) free abelian subgroups G of $\pi_*(BP)$ such that $G_p = \pi_*(BP)$;

(c) free abelian subgroups H of $H_*(BP)$ such that $H_p = H_*(BP)$.]]]

This is the precise description we seek of original Brown-Peterson spectra. To classify the spectra X themselves, all we have to do is allow for the different choices of isomorphism $X_p \cong BP$.

<u>The canonical example</u>. The ring $\pi_*(BP)$ is well known to be the polynomial ring $Z_{(p)}[v_1,v_2,v_3,\ldots]$ on the Hazewinkel generators [7] v_n in degrees $2(p^n-1)$. As in [4] we write V for the subring $Z[v_1,v_2,v_3,\ldots]$ over the <u>integers</u>. Similarly in homology, $H_*(BP) = Z_{(p)}[w_1,w_2,w_3,\ldots]$ where w_n (usually written m_{p^n-1}) also has degree $2(p^n-1)$, and we again write $W = Z[w_1,w_2,w_3,\ldots]$. For the canonical original Brown-Peterson spectrum we take $G = V$, which corresponds to $H = W$. Its ring structure will be obvious from §5. The only fact we need about the Hurewicz homomorphism of BP is that

$$v_n = pw_n \mod p^2W.$$

2. Automorphisms of BP

According to Lemma 1.5, all we have to do to establish Theorem B is count the number of orbits in the action of the automorphism group of BP on the set of all free abelian subgroups G of $\pi_*(BP)$ such that $G_p = \pi_*(BP)$. Alternatively, we may work in homology with subgroups H of $H_*(BP)$, where the action is much more accessible. From Quillen [9], after some rewriting (see §2 of [4]), there is a 1-1 correspondence between maps $f{:}BP \to BP$ and additive homomorphisms $u{:}T \to V_p$ of degree zero, where $T = Z[t_1,t_2,t_3,\ldots]$ is another polynomial ring on generators t_n in degrees $2(p^n-1)$, given in homology by

$$f_*{:}W_p \xrightarrow{r_*} T\otimes W_p \xrightarrow{u\otimes 1} V_p\otimes W_p \xrightarrow{\text{mult}} W_p, \tag{2.1}$$

in which r_* is the ring homomorphism defined on generators by

$$r_*w_n = t_n \otimes 1 + \Sigma_{i=1}^{i=n-1}\, t_{n-i}^{p^i} \otimes w_i + 1\otimes w_n . \tag{2.2}$$

Moreover, f will be an automorphism provided $u(1)$ is not divisible by p in

$Z_{(p)}$, the component of V_p in degree 0 (see Lemma 2.10 of [4]).

In particular, we have scalar multiplication by any element of $Z_{(p)}$. In degree zero, H is free on one generator, which must be some unit c of $Z_{(p)}$; after scalar multiplication by c^{-1} we have H = W in degree zero. The only automorphism of $Z_{(p)}$ that leaves Z fixed is -1, but since multiplication by -1 preserves every subgroup of $H_*(BP)$, it is of no interest. Therefore from now on, we consider only subgroups H that are exactly $Z \subset Z_{(p)}$ in degree zero, and automorphisms f that induce the identity map of $H_0(BP)$, or equivalently u(1)=1.

2.3 LEMMA. For any such automorphism f of BP, $f_* = 1: H_*(BP) \to H_*(BP)$ mod p.

Proof. Apart from degree zero, every element of V_p lies in pW_p. Hence, from (2.1) and (2.2), only the value u(1) = 1 is relevant mod p.]]]

This suggests some arithmetic invariants. Since $H \subset W_p$ is free abelian and $H_p = W_p$, H and W must have the same rank in each degree, and we can compare them by introducing an abstract additive isomorphism $\phi: W \cong H$. Then the localization $\alpha = \phi_p$ is an additive automorphism of $W_p = H_p = H_*(BP)$.

2.4 DEFINITION. We write $\det_N(\alpha) \in Z_{(p)}$ for the determinant of α in degree N.

We can, of course, compose ϕ with any automorphism of W in degree N, which will have determinant ±1, so that H determines $\det_N(\alpha)$ only up to sign. In the case $H \subset W$, which we shall see in Lemma 2.6 is not really special at all, there is a simple interpretation. In each degree N, the subgroup H has finite index in W, namely $|\det_N(\alpha)|$. For arbitrary H, this generalizes to the quotient of the index of $H \cap W$ in W by the index of $H \cap W$ in H.

Proof of Theorem B, Case $p \geq 5$. We have just defined a collection of numbers $\det_N(\alpha)$ for each positive multiple N of 2(p-1), and by Lemma 2.3 these are invariants of the original Brown-Peterson spectrum X up to sign and modulo p. That is, they take values in the quotient $(Z/p)^*/\pm 1$ of the multiplicative group $(Z/p)^*$ of units of Z/p. This quotient group is non-trivial as soon as $p \geq 5$. It is evident that the values $\det_N(\alpha)$ may be chosen arbitrarily, thus giving rise to uncountably many different X.]]]

If $p=3$ or $p=2$, this proof fails because the quotient group of values is trivial. We take up the case $p=3$ in §3.

We shall need a good supply of automorphisms of BP.

2.5 LEMMA. Suppose θ is any additive automorphism of $H_N(BP)$ such that $\theta = 1 \bmod p^k$. Then if k is large enough (in fact $k \geq N/2(p-1)$) there exists an automorphism f of BP such that in homology

(a) $f_* = 1$ in degrees less than N;

(b) $f_* = \theta$ in degree N;

(c) $f_* = 1 \bmod p^k$ in all degrees.

Proof. We take f as the automorphism induced as in (2.1) by a homomorphism $u:T \to V_p$ that is zero in all degrees except 0 and N. Of course we require $u(1) = 1$. As w^σ runs through the monomials of W in degree N we find $f_* w^\sigma = w^\sigma + u(t^\sigma)$. If k is large enough we have $p^k W_p \subset V_p$ in degree N, so that the choices $u(t^\sigma) = \theta(w^\sigma) - w^\sigma$ are legitimate. Then (c) holds in all degrees.]]]

A generalization of this allows us to reduce the number of groups H to be considered.

2.6 LEMMA. Let H be any free abelian subgroup of W_p such that $H_p = W_p$.

(a) There is an automorphism f of BP such that $f_* H \subset W$.

(b) There is an automorphism f of BP such that $f_* H$ is a subring of W_p, and there are uncountably many nonisomorphic possible rings $f_* H$.

Proof. (a) As in Lemma 2.5, f is induced by a homomorphism $u:T \to V_p$, which we construct by induction on degree. Let $\{x_i\}$ be a Z-base of H in degree N, which is therefore also a $Z_{(p)}$-base of W_p in degree N. Then from (2.1) and (2.2) we have

$$f_* x_i = u(y_i) + z_i,$$

where the elements z_i are determined by the choice of u in lower degrees and the y_i form a $Z_{(p)}$-base of T_p in degree N. So to choose the $u(y_i) \in V_p \subset W_p$ is to choose u in degree N. We can always make $f_* x_i$ lie in W because

$W_p = V_p + W$.

(b) We construct similarly an automorphism g such that $g_*L \subset H$ and take $f = g^{-1}$, so that $L \subset f_*H$. Here L is also constructed by induction on degree, large enough to contain the group D generated by products of elements of f_*H of lower degree, to ensure that f_*H is a subring of W_p. It is possible to choose L free abelian because D is finitely generated in each degree. Except in degrees of the form $2(p^n-1)$, the "group of indecomposables" of the ring f_*H, which is f_*H/D or H/g_*D, is finite but arbitrarily large (by choosing L/D arbitrarily large); this assures uncountably many distinct rings f_*H.]]]

(We are not claiming to satisfy (a) and (b) with the same automorphism.)

3. The case p=3

In this section we prove Theorem B for the case $p=3$.

We again consider the numbers $\det_N(f_*)$ for the automorphism f of BP induced as in (2.1) by a homomorphism $u:T \to V_3$, where of course we continue to assume $u(1) = 1$. However, we now have to work mod 9 in order to have a nontrivial quotient group $(Z/9)^*/\pm 1$. Lemma 2.3 is definitely false with p replaced by p^2.

In degree N let us write $f_* = 1+\theta : H_N(BP) \to H_N(BP)$. Then Lemma 2.3 does show that θ is divisible by 3, so that in the expansion of $\det_N(f_*)$ any term involving two or more factors from θ makes no contribution mod 9. This leaves only the terms

$$\det_N(f_*) = 1 + \text{trace}\ \theta \qquad \text{mod } 9.$$

The only monomials in V that are nontrivial mod $9W$ are 1 and the generators v_n, so that the only values of u that can make any contribution to $\det_N(f_*)$ mod 9 are the values $u(t^\sigma)$ where t^σ is any monomial in the t_i of degree $2(3^n-1)$ for some n. A simple counting argument is not enough to establish the theorem because there are far too many such monomials t^σ.

We have to be more explicit. To compute the trace of θ we use the obvious base of $H_*(BP)$ consisting of all the monomials $w^\sigma = w_1^{\sigma(1)} w_2^{\sigma(2)} \dots$. Let us define elements $C(w^\rho, w^\sigma)$ of T by the identity

$$r_* w^\sigma = \Sigma_\rho \, C(w^\rho, w^\sigma) \otimes w^{\sigma-\rho},$$

or as zero for inappropriate values of ρ. Then we can rewrite (2.1) as

$$f_* w^\sigma = \Sigma_\rho \, u(C(w^\rho, w^\sigma)) w^{\sigma-\rho}. \tag{3.1}$$

We recall that $v_n = 3w_n \bmod 9$. Thus the only important terms for trace θ mod 9 are those for which v_n appears in $u(C(w^\rho, w^\sigma))$ for some n and $w^\sigma = w^{\sigma-\rho} w_n$; that is, w^ρ must be w_n. In other words,

$$\det_N(f_*) = 1 + \text{trace } \theta = 1 + \Sigma_\sigma \Sigma_n 3c_{n,\sigma} \bmod 9, \tag{3.2}$$

where $c_{n,\sigma}$ denotes the coefficient of v_n in $u(C(w_n, w^\sigma))$ and we sum over all positive n and all monomials w^σ in degree N.

We need a good method for computing the coefficients $C(w_n, w^\sigma)$ in T. That is, we seek the coefficient of

$$w_1^{\sigma(1)} w_2^{\sigma(2)} \dots w_{n-1}^{\sigma(n-1)} w_n^{\sigma(n)-1} w_{n+1}^{\sigma(n+1)} \dots$$

in

$$r_* w^\sigma = \Pi_k (r_* w_k)^{\sigma(k)} = \Pi_k (1 \otimes w_k + \Sigma_{i<k} t_{k-i}^{3^i} \otimes w_i)^{\sigma(k)},$$

where we interpret w_0 as 1. The only way to obtain enough powers of all the w_k for $k > n$ is to take the leading term $1 \otimes w_k$ from every factor $r_* w_k$ of the product in this range; in other words, the numbers $\sigma(k)$ for $k > n$ are irrelevant. For $k=n$ we have

$$(r_* w_n)^{\sigma(n)} = 1 \otimes w_n^{\sigma(n)} + \Sigma_i \, \sigma(n) \, t_{n-i}^{3^i} \otimes w_n^{\sigma(n)-1} w_i + \text{lower terms},$$

so that to find exactly $\sigma(n)-1$ powers of w_n we must choose some i from the sum. With i chosen, we need to pick out $w_i^{-1} \cdot \Pi_{k=1}^{k=n-1} w_k^{\sigma(k)}$ from the remaining factors $\Pi_{k=1}^{k=n-1} r_* w_k^{\sigma(k)}$. This leads to the inductive formula

$$C(w_n, w^\sigma) = \Sigma_{i<n} \, \sigma(n) \, t_{n-i}^{3^i} \, C(w_i, w^\sigma) \qquad (\text{if } n > 0), \tag{3.3}$$

which together with the obvious $C(w_0,w^\sigma) = 1$ serves to compute $C(w_n,w^\sigma)$ in general. The answer, by induction on n, is

$$C(w_n,w^\sigma) = \Sigma\, \sigma(i_1)\sigma(i_2)\ldots\sigma(i_s)\, t_{i_1}\, t_{i_2-i_1}^{3^{i_1}} \ldots t_{i_s-i_{s-1}}^{3^{i_{s-1}}}, \tag{3.4}$$

where we sum over all sequences $0 < i_1 < i_2 < \ldots < i_s = n$.

Thus by (3.2) $u(t^\beta)$ contributes to $\det_N(f_*)$ mod 9 only if t^β is a monomial of the familiar form

$$t^\beta = t_{i_1}\, t_{i_2-i_1}^{3^{i_1}}\, t_{i_3-i_2}^{3^{i_2}} \ldots t_{i_r-i_{r-1}}^{3^{i_{r-1}}},$$

where of course N must be not less than the degree of t^β, which is $2(3^{i_r}-1)$. So in the range of degrees $2(3^n-1) \leq N < 2(3^{n+1}-1)$, the values $\det_N(f_*)$ in $(Z/9)^*/\pm 1 = Z/3$ depend linearly on the $u(t^\beta)$, where t^β runs through all monomials of the above form with $0 < i_1 < i_2 < \ldots\ i_s \leq n$, of which there are just 2^n. But the number of degrees N in this range with $H_N(BP) \neq 0$ is $\frac{1}{2}(3^{n+1}-3^n) = 3^n$. It follows that for additive automorphisms α of $H_*(BP)$ as in Definition 2.4, there must be at least 3^n-2^n independent linear combinations of the $\det_N(\alpha)$ in this range of degrees that are invariant under the action of the automorphism group of BP (although for us one will suffice). As n varies we obtain infinitely many invariants, all of them arbitrary, which again yield uncountably many distinct original Brown-Peterson spectra.

4. The case $p=2$

In this section we establish the remaining case of Theorem B. Although much of the theory is quite similar to that for $p=3$ and the first three lemmas hold (with changes) for odd primes, the conclusions are very different.

As before, let H be a free abelian subgroup of $H_*(BP) = W_2$ such that $H_2 = W_2$, and choose an additive automorphism α of $H_*(BP)$ such that $\alpha W = H$. We show first that the determinants $\det_N(\alpha)$ are in effect the only invariants we need to consider.

4.1 LEMMA. *Suppose* $\det_N(\alpha) = \pm 1 \bmod 2^k$. *Then if* k *is large enough* $(k \geq N/2)$, *there is an automorphism* f *of* BP *such that*

(a) $f_* = 1$ *in degrees less than* N;

(b) $f_* W = \alpha W = H$ *in degree* N;

(c) $\det_d(f_*) = 1 \bmod 2^k$ *for all* d.

Proof. By changing the sign of α on one base element of W if necessary, we may assume $\det_N(\alpha) = 1 \bmod 2^k$. We use the obvious base of W (over Z) or of W_2 (over $Z_{(2)}$) consisting of the appropriate monomials w^σ; suppose its rank is q. It is standard (e.g. Bass [1, Corollary 5.2]) that the epimorphism of rings $Z \to Z/2^k$ induces an epimorphism of groups $SL(q,Z) \to SL(q,Z/2^k)$, so that the image of α in $SL(q,Z/2^k)$, which we have by hypothesis, lifts to some θ in $SL(q,Z)$. Then $\alpha\theta^{-1}W = \alpha W = H$; but by Lemma 2.5 $\alpha\circ\theta^{-1}$ is realizable in degree N by some geometric automorphism f_* as required.]]]

We need to iterate this over a range of degrees.

4.2 LEMMA. *Suppose* $\det_d(\alpha) = \pm 1 \bmod 2^k$ *in all degrees* d *such that* $M \leq d \leq N$. *Then if* k *is large enough* $(k \geq N/2)$ *there is an automorphism* f *of* BP *such that*

(a) $f_* = 1$ *in degrees less than* M;

(b) $f_* W = H$ *in all degrees* d *such that* $M \leq d \leq N$;

(c) $\det_d(f_*) = 1 \bmod 2^k$ *for all* d.

Proof. By induction on N, starting from the case N=M, which is Lemma 4.1. If the result holds for N-1, yielding a geometric automorphism g of BP, we apply Lemma 4.1 to the additive automorphism $g_*^{-1} \circ \alpha$ to obtain an automorphism h of BP, and deduce the result for N by taking $f = g \circ h$.]]]

It is well known (e.g. Jacobson [8,p.114, Theorem 4]) that for any $k \geq 3$ the group $(Z/2^k)^*/\pm 1$ is cyclic of order 2^{k-2} on the generator 5 (as can be seen by expanding $5^{2^{k-3}} = (1+4)^{2^{k-3}}$ by the binomial theorem and observing that the group has order 2^{k-2}). The final ingredient we need for the main proof is an automorphism that realizes a generator of this group.

We first need the existence of some useful elements of $H_*(BP)$.

4.3 LEMMA. In every even positive degree $2N$, $W = Z[w_1,w_2,w_3,\ldots]$ contains exactly one monomial of one of the following forms:

(a) $w_{i_1}w_{i_2}\ldots w_{i_s}$ or (b) $w_{i_1}^2 w_{i_2}\ldots w_{i_s}$

where $0 < i_1 < i_2 < \ldots < i_s$ and $s > 0$. Further, $i_s = n$ where $2^n-1 \le N < 2^{n+1}-1$.

Proof. Given N, we define $n=i_s$ by the inequality, and proceed by induction on n. If $N = 2^n-1$ the desired monomial is w_n. If $N = 2^{n+1}-2$, it is w_n^2. Otherwise we write $N = 2^n-1+M$, so that $0 < M < 2^n-1$, and use the monomial xw_n where x is the monomial chosen in degree $2M$.

To verify uniqueness all we need do is check that no other choice for the last factor is possible, and use induction on n. If $i_1=a$ and $i_s=b$, then for fixed a and b the largest possible degree is that of the monomial $w_a^2 w_{a+1}w_{a+2}\ldots w_{b-1}w_b$, which is twice

$$2(2^a-1)+(2^{a+1}-1)+\ldots+(2^b-1) = 2^{b+1}-(b-a+2) \le 2^{b+1}-2. \;]]]$$

4.4 LEMMA. For every even positive integer $2N$, there is an automorphism f of BP such that

(a) $f_* = 1$ in all degrees less than $2(2^n-1)$, where $2^n-1 \le N < 2^{n+1}-1$;

(b) $\det_d(f_*) = 1 \bmod 8$ in all degrees $d < 2N$;

(c) $\det_{2N}(f_*) = 5 \bmod 8$.

Proof. The proof divides into two cases according to Lemma 4.3.

Case (a) Suppose that in degree $2N$, W contains the monomial $w_{i_1}w_{i_2}\ldots w_{i_s}$, where $i_s=n$. We define the monomial t^β in T in degree $2(2^n-1)$ as

$$t^\beta = t_{i_1}\, t_{i_2-i_1}^{2^{i_1}} \cdots t_{i_s-i_{s-1}}^{2^{i_{s-1}}},$$

and take f as the automorphism of BP induced as in (2.1) by the homomorphism $u:T \to V$ defined by $u(1) = 1$, $u(t^\beta) = 2v_n$, and zero on all other monomials of T.

We note that $u(t^{\beta}) = 4w_n \bmod 8$. The analog of (3.2) is

$$\det_d(f_*) = 1 + \Sigma_{\sigma}\, 4c_{\sigma} \bmod 8, \tag{4.5}$$

where c_{σ} denotes the coefficient of t^{β} in $C(w_n, w^{\sigma})$ and we sum over all monomials w^{σ} in degree d. Corresponding to (3.4) we have

$$C(w_n, w^{\sigma}) = \Sigma\, \sigma(j_1)\sigma(j_2)\ldots\sigma(j_r)t_{j_1} t_{j_2-j_1}^{2^{j_1}} \ldots t_{j_r-j_{r-1}}^{2^{j_{r-1}}} \tag{4.6}$$

where we sum over all choices of the j's satisfying $0 < j_1 < j_2 < \ldots < j_r = n$. The only difficulty in recognizing the occurrence of t^{β} in $C(w_n, w^{\sigma})$ is that there may be repetitions among the suffixes $j_t - j_{t-1}$; but this is easily resolved by writing every monomial in T uniquely as a product, without repetition, of factors of the form $t_j^{2^m}$, from which it is clear that t^{β} occurs only if $r=s$ and $j_t = i_t$ for all t, and then with coefficient $\sigma(i_1)\sigma(i_2)\ldots\sigma(i_s)$. Thus $c_{\sigma}=1$ for exactly one monomial w^{σ} in degree 2N, namely the monomial provided by Lemma 4.3, and is zero for all other monomials in degree $\leq$ 2N.

<u>Case (b)</u> Suppose instead we have the monomial $w_{i_1}^2 w_{i_2} w_{i_3} \ldots w_{i_s}$ in degree 2N. We write $m = i_1$ and $n = i_s$. This time we define

$$t^{\beta} = t_{i_1}^2\, t_{i_2-i_1}^{2^{i_1}}\, t_{i_3-i_2}^{2^{i_2}} \ldots t_{i_s-i_{s-1}}^{2^{i_{s-1}}}$$

which has degree $2(2^m-1+2^n-1)$, and define the automorphism f of BP as induced by the homomorphism $u: T \to V$ given by $u(1) = 1$, $u(t^{\beta}) = v_m v_n$, and zero on all other monomials of T. Then $u(t^{\beta}) = 4w_m w_n \bmod 8$.

Just as before we find

$$\det_d(f_*) = 1 + \Sigma_{\sigma}\, 4c_{\sigma} \bmod 8,$$

except that c_{σ} now denotes the coefficient of t^{β} in $C(w_m w_n)$. (We write $C(w_m w_n, w^{\sigma})$ simply as $C(w_m w_n)$ for convenience, suppressing the dependence on σ.) The technique for computing $C(w_m w_n)$ is essentially the same as for $C(w_n)$, except that we need to consider more terms:

$$(r_*w_k)^q = 1 \otimes w_k^q + \Sigma_{i<k}\, q\, t_{i-k}^{2^i} \otimes w_i w_k^{q-1} + \Sigma_{i<k} \binom{q}{2}\, t_{k-i}^{2^{i+1}} \otimes w_i^2 w_k^{q-2}$$

$$+ \Sigma_{i<j<k}\, q(q-1)\, t_{k-i}^{2^i} t_{k-j}^{2^j} \otimes w_i w_j w_k^{q-2}$$

$$+ \text{terms with lower powers of } w_k.$$

Since $q(q-1)$ is always even, the last sum stated may be ignored, and the required inductive formulae are, where we interpret w_0 as 1:

$$\text{(4.7)} \qquad C(w_j w_k) = \Sigma_{i<k}\, \sigma(k)\, t_{k-i}^{2^i}\, C(w_j w_i) \qquad \text{if } 0 \leq j < k,$$

and

$$\text{(4.8)} \qquad C(w_k^2) = \Sigma_{i<k} \binom{\sigma(k)}{2} t_{k-i}^{2^{i+1}}\, C(w_i^2) \qquad \text{if } k > 0$$

together with the obvious $C(1) = 1$. The first is a generalization of the analog of (3.3).

In this case detecting the occurrence of t^β is more complicated. The monomial t^β itself can take on four different appearances, depending on the values of i_1 and i_2 and s:

$$\text{(4.9)} \quad \begin{cases} \text{(i)} & t^\beta = t_m^2\, x^{2^m} = t_m^2\, y^4 & \text{if } i_1 > 1; \\ \text{(ii)} & t^\beta = t_m^2\, t_{i_2-m}^2\, x^4 & \text{if } i_1 = 1 \text{ and } i_2 > 2; \\ \text{(iii)} & t^\beta = x^4 & \text{if } i_1 = 1 \text{ and } i_2 = 2; \\ \text{(iv)} & t^\beta = t_m^2 & \text{if } s = 1. \end{cases}$$

Here x and y stand for some monomials in T. Consider the sequence of induction steps required to produce each term of $C(w_m w_n)$ with t^β. We assert that to obtain t^β, we must use (4.8) exactly once, as the last step, to pass from $C(w_m^2)$ to $C(w_0^2)$, for which we require $\sigma(m) \geq 2$. Once this is established, it is clear that the previous steps must have been to use (4.7) to pass from $C(w_m w_n)$ to $C(w_m w_{j_2})$ to $C(w_m w_{j_3})$ etc., until we get to $C(w_m^2)$. The only values that will produce t^β are $j_2=i_{s-1}$, $j_3=i_{s-2}$, ..., $j_s=i_1=m$, with coefficient $\binom{\sigma(m)}{2}\sigma(i_2)\sigma(i_3)\dots\sigma(i_s)$, and the result follows as in Case (a).

It is clear that once (4.8) is used, we have to keep on using it. Suppose first that (4.8) is not used at all, so that the last step was to pass from $C(w_0w_a)$ to $C(w_0^2)$ by (4.7). One of the previous steps must have been to pass from $C(w_bw_c)$ to $C(w_bw_0)$ by (4.7), where $0 < a \leq b < c$. Then the term of $C(w_mw_n)$ will have the form $\lambda t_a t_c x^2$, (for some number λ), which by (4.9) is never a term with t^β.

Suppose next that (4.8) is used twice, to pass from $C(w_b^2)$ to $C(w_a^2)$ to $C(w_0^2)$. We observe that in (4.7) and (4.8), $C(w_iw_j)$ is always multiplied by a monomial of the form $x^{2^{\min(i,j)}}$, and it follows by induction that the same is true of $C(w_iw_j)$ in the expansion of $C(w_mw_n)$. In this case we therefore have a term of $C(w_mw_n)$ of the form $\lambda t_a^2 t_{b-a}^{2^{a+1}} x^{2^b}$, where $0 < a < b$, which again can never be a term in t^β. Therefore (4.8) must be used exactly once, to pass from $C(w_a^2)$ to $C(w_0^2)$. If this is the only step in the induction the term is λt_a^2, which can only correspond to t^β as in case (iv) of (4.9), and then only if $a=m$.

Otherwise, the previous step must have been to pass from $C(w_aw_b)$ to $C(w_a^2)$ by (4.7), where $0 < a < b$. Then the term of $C(w_mw_n)$ has the form $\lambda t_a^2 t_{b-a}^{2^a} x^{2^b}$. If $a > 1$ we can write this as $\lambda t_a^2 y^4$, which can only correspond to t^β as in case (i) of (4.9). If $a=1$ and $b > 2$ we have $\lambda t_1^2 t_{b-1}^2 x^4$, with $b-1 \neq 1$; this corresponds to case (ii) of (4.9). Finally, if $a=1$ and $b=2$ the term reduces to $\lambda t_1^4 x^4$, which corresponds to case (iii). Thus in all cases we must have $a=m$ to obtain a term in t^β, and our assertion is justified.]]]

<u>Proof of Theorem B.</u> We first work in the range of degrees $2(2^n-1) \leq d < 2(2^{n+1}-1)$, for some fixed n. From Lemma 4.4, by induction on d, the values $\det_d(f_*)$ may be chosen arbitrarily in $(Z/8)^*/\pm 1 = Z/2$ in this range, where f is an automorphism of BP such that $f_* = 1$ in degrees less than $2(2^n-1)$. If we consider the determinants as taking values in the cartesian product L of copies

of $(Z/2^k)^*/\pm 1$, one for each (even) d in our range of degrees, we have in effect just stated that every element of $L/2L$ can be realized. It follows from the structure of L that every element of L itself can be realized.

Suppose $H(n)$ is a free abelian subgroup of W_2 such that $H(n) = W$ in degrees less than $2(2^n-1)$ and $H(n)_2 = W_2$, and choose an additive automorphism α of $H_*(BP)$ such that $\alpha W = H(n)$. We have just seen that there is an automorphism f of BP such that $\det_d(f_* \circ \alpha) = \det_d(f_*).\det_d(\alpha) = \pm 1 \bmod 2^k$ for each d in our range. By Lemma 4.2, there is another automorphism g of BP such that $g_*W = f_*\alpha W = f_*H(n)$ in our range of degrees. Let $e(n) = g^{-1} \circ f$, another automorphism of BP, and let $H(n+1) = e(n)_*H(n)$; then $H(n+1) = W$ in degrees less than $2(2^{n+1}-1)$ and we have our induction step.

Now take any free abelian subgroup $H = H(1)$ of W_2 that agrees with W in degree zero. By induction on n we have subgroups $H(n)$ of W_2 and automorphisms $e(n)$ of BP such that $e(n)_*H(n) = H(n+1)$. There is no difficulty (as is easily seen algebraically from (2.1)) in forming the infinite composite $e = \ldots \circ e(3) \circ e(2) \circ e(1)$, since by construction $e(n)_* = 1$ in degrees less than $2(2^n-1)$. Then $e_*H = W$, as required.]]]

5. Ring structures

By ring spectrum we understand a spectrum E equipped with a commutative and associative multiplication map $\mu: E \wedge E \longrightarrow E$ with unit $i: S^0 \longrightarrow E$ (working entirely in the homotopy category $\underline{S}_h$), so that $\pi_*(E)$ is a ring (with 1). It is clear that any localization of E is automatically also a ring spectrum.

The converse is not automatic. Given ring spectra E_p and $E[p^{-1}]$ that localize to the same ring structure on $E_\emptyset$, we seek a ring structure on E that localizes to E_p and $E[p^{-1}]$. The rational ring spectrum $E_\emptyset$ is completely described by the ring $\pi_*(E_\emptyset)$. It is trivial to construct i; we have the (strict) pullback square of rings

$$(5.1)\qquad \begin{array}{ccc} \pi_*(E) & \longrightarrow & \pi_*(E_p) \\ \downarrow & & \downarrow \\ \pi_*(E[p^{-1}]) & \longrightarrow & \pi_*(E_\emptyset). \end{array}$$

For μ we need instead the exact sequence (2.10 of [2])

$$\ldots\{E_\emptyset\wedge E_\emptyset,E_\emptyset\}_1 \longrightarrow \{E\wedge E,E\}\rightarrow \{E_p\wedge E_p,E_p\}\oplus\{E[p^{-1}]\wedge E(p^{-1}],E[p^{-1}]\}$$

$$\longrightarrow \{E_\emptyset\wedge E_\emptyset,\ E_\emptyset\}\rightarrow\ldots$$

which shows that there is a map $\mu:E\wedge E\longrightarrow E$ that localizes correctly, possibly many, but fails to guarantee commutativity etc.

However, if X is an original Brown-Peterson spectrum, we have $\{X_\emptyset\wedge X_\emptyset,X_\emptyset\}_1=0$ for trivial dimensional reasons, so that μ is unique. The exact sequence shows also that μ will be commutative, and similar proofs show that μ is associative and has i as unit. Thus rationally equivalent ring structures on X_p and $X[p^{-1}]$ do yield a well-defined ring structure on X. Further, Theorem B of [4] shows that the ring structure on $X_p = BP$ is unique up to isomorphism (essentially because there is a good universal property).

Proof of Theorem C. In the localization square (1.1), if G is a subring of $\pi_*(BP) = V_p$, $X[p] = K(G[p^{-1}])$ is canonically a ring spectrum and therefore yields a ring structure on X. By Lemma 2.6 the condition on G can be arranged (recall $G = H\cap V_p$ in W_p), in uncountably many ways and the resulting spectra X are distinguished by the rings $H_*(X)/\text{torsion}$.]]]

Proof of Theorem D. Given the subring G of $\pi_*(BP)$, Theorem 1.4 of [5] shows $K(G[p^{-1}])$ admits an essentially unique ring structure if $p=2$, or uncountably many distinct ones if p is odd.]]]

REFERENCES

[1] H. Bass, K-theory and stable algebra, Publ. Math. I.H.E.S. 22 (1964)5-60.

[2] J.M. Boardman, Stable homotopy theory, Appendix C, Localization theory (mimeograph)The Johns Hopkins University, August 1975.

[3] ___________, Appendix D, Localization and splittings of MU (mimeograph), The Johns Hopkins University, February 1976.

[4] ___________, Splittings of MU and other spectra, Lecture Notes in Mathematics (Springer-Verlag) 658 (1978)27-79.

[5] ___________, Graded Eilenberg-MacLane ring spectra (preprint).

[6] E.H. Brown, Jr., and F.P. Peterson, A spectrum whose Z_p cohomology is the algebra of reduced p-th powers, Topology 5 (1966)149-154.

[7] M.Hazewinkel, A universal formal group and complex cobordism, Bull. Amer. Math. Soc. 81 (1975)930-933.

[8] N.Jacobson, Lectures in abstract algebra, III, Van Nostrand (1964).

[9] D.Quillen, On the formal group laws of unoriented and complex cobordism theory, Bull. Amer. Math. Soc. 75 (1969)1293-1298.

September 1978

Department of Mathematics
The Johns Hopkins University
Baltimore, Maryland 21218

BP-OPERATIONS AND MAPPINGS OF STUNTED COMPLEX PROJECTIVE SPACES

Donald M. Davis
Lehigh University, Bethlehem, Pa.
Northwestern University, Evanston, Ill.

1. INTRODUCTION

Let p be a prime number. Spaces X and Y are stably p-equivalent if there exist integers r and s and a p-equivalence $f: \Sigma^r X \to \Sigma^s Y$, i.e. f induces an isomorphism in Z_p-cohomology. Let $CP_n^{n+k} = CP^{n+k}/CP^{n-1}$ denote the stunted complex projective space.

In [4] we determined which stunted complex projective spaces with at most $3p - 2 + (-1)^p$ cells are stably p-equivalent. We saw that there are more stable p-equivalences than one might expect from naively localizing the result of [6] on stable homotopy type of stunted CP's. This enabled us to calculate lower bounds for the stable genus sets $SG(CP_n^{n+k})$ with $k \leq 4$.

As originally defined in [9], SG(X) is the set of stable homotopy types stably p-equivalent to X for all primes p. For example, we showed $SG(CP_2^4) \supset \{CP_2^4, CP_{10}^{12}\}$, $SG(CP_7^9) \supset \{CP_7^9, CP_{19}^{21}\}$, and $SG(CP_8^{12}) \supset \{CP_{8+48i}^{12+48i} : 0 \leq i < 60,\ i \not\equiv 4(5)\}$. This raises the question of whether the stable genus set of a stunted CP can contain stable homotopy types which do not contain any stunted CP's. In order to facilitate the study of this question, we mimic the work of Zabrodsky ([12]) to prove

<u>Theorem 1.1</u> $SG(CP_{n+1}^{n+k}) \approx (Z_t^*/\pm 1)^k/M$, where t is a product of primes $\leq k$, Z_t^* is the group of units in the ring of integers mod t, $M = M(CP_{n+1}^{n+k}) = \{(\bar{a}_1,\ldots,\bar{a}_k)\in(Z_t^*/\pm 1)^k$ such that there exists $f\colon CP_{n+1}^{n+k} \to CP_{n+1}^{n+k}$ with $f^*x^{n+i} = a_i x^{n+i}\}$.

In Section 3 we will prove the generalization of this theorem for any $SG(X)$.

The question of determining those cohomology homomorphisms induced by self-maps of stunted CP's was first posed by Gitler in [7]. The techniques which we use to determine necessary conditions for stable p-equivalences are also useful in determining necessary conditions for cohomology homomorphisms to be induced by self-maps. In the case of stunted CP's with three cells it is easy to check that these necessary conditions are also sufficient. This together with Theorem 1.1 implies that the lower bounds calculated above for $SG(CP_n^{n+2})$ are sharp.

<u>Theorem 1.2</u> There is a stable map $f\colon CP_n^{n+2} \to CP_n^{n+2}$ with $f^*x^{n+i} = a_i x^{n+i}$ if and only if the following four conditions are satisfied:

$$n(a_0 - a_2) \equiv 0(3)$$
$$(n+1)(a_1 - a_2) \equiv 0(2)$$
$$n(a_0 - a_1) \equiv 0(2)$$
$$(2n + \tbinom{n}{2})(a_2 - a_0) + n(n+1)(a_0 - a_1) \equiv 0(4).$$

<u>Theorem 1.3</u> $SG(CP_2^4) = SG(CP_{10}^{12}) = \{CP_2^4, CP_{10}^{12}\}$

$$SG(CP_7^9) = SG(CP_{19}^{21}) = \{CP_7^9, CP_{19}^{21}\}$$

$$SG(CP_n^{n+2}) = \{CP_n^{n+2}\} \text{ for } n \in \{0,1,3,4,5,6,8,9,12,21\}.$$

According to [5] each CP_n^{n+2} belongs to exactly one of the 14 stable homotopy types listed in Theorem 1.3.

1.2 and 1.3 and their proofs suggest the conjectures that BP primary operations detect all necessary conditions for cohomology homomorphisms induced by self-maps of stunted CP's and that the stable genus of stunted CP's consists only of stunted CP's. The fact that Feder and Gitler obtained complete results for stable homotopy type using ch suggests that only primary K-theoretic or BP operations are relevant for stunted CP's -- that the indeterminancy for attaching maps contains all maps not in the image of J. That primary BP operations should give complete results for stable p-equivalences of stunted CP's is compatible with this observation.

In [4] we used the coaction $BP_*(CP) \to BP_*(BP) \otimes_{BP_*} BP_*(CP)$ to determine the necessary conditions for stable p-equivalences (p odd). The extremely complicated formulas which arise there make it very unlikely that the coaction might be used to give general formulas for CP's with arbitrarily many cells. The operations in $BP^*(\ )$ are somewhat more promising for this end because they allow the possibility of calculating with one operation at a time. Because the coaction is dual to the action of all the r_I operations of Adams and Quillen ([2],[11]), it is not surprising that the same results (for questions about coefficients in maps of stunted CP's) are obtained when using all the r_I's as when using the coaction in $BP_*(\ )$.

Recently, Bendersky ([3]) introduced operations ρ_I in $BPQ^*(\ ) = BP^*(\ ) \otimes Q$, where Q denotes the rational numbers, which are somewhat easier to calculate on CP than are the r_I. He calculates a linear relationship between the two sets of operations, so again it is not surprising that the ρ_I's yield the same results as the r_I's on the questions considered in this paper.

Novikov ([10]) introduced a type of Adams operation ψ^q in $BPQ^*(\ )$, for any q not divisible by the prime p which underlies BP. Although the properties of the ψ^q are quite different from those of the ρ_I (for example ψ^q is multiplicative while ρ_I satisfies a Cartan formula), in our range of calculation the same results are obtained using all ψ^q's as using all ρ_I's. This can be explained by relating the ψ^q's and the ρ_I's.

<u>Theorem 1.4</u> $\psi^q = \sum_I q^I m^I \rho_I$, where I ranges over all finite sequences of nonnegative integers, $m^{(i_1,i_2,\ldots)} = m_1^{i_1} m_2^{i_2}\cdots \in BPQ^{-2i_1(p-1)-2i_2(p^2-1)\cdots}$ and $q^{(i_1,i_2,\ldots)} = q^{i_1(p-1)+i_2(p^2-1)+\cdots}$.

In the range of calculation of this paper ($\ll 2(p^2-1)$) only the ρ_i's occur, and Theorem 1.4 is invertible, enabling expression of ρ_i in terms of the ψ^q. For example, $\psi^q = q^{p-1}m_1\rho_1 + q^{2p-2}m_1^2\rho_2$ on a k-dimensional class in a complex of dimension $< k + 6(p - 1)$. If

$p = 2$, let $q = 3$ and 5 and use Cramer's rule to obtain (on such a class) $m_1\rho_1 = \frac{5}{6}\psi^3 - \frac{3}{10}\psi^5$ and $m_1^2\rho_2 = \frac{1}{10}\psi^5 - \frac{1}{6}\psi^3$. However, when ρ_{01} enters into play, Theorem 1.4 can no longer be inverted because the vectors of coefficients of ρ_{01} and ρ_{p+1} are not linearly independent. Thus it is plausible that the ρ_I's might give more information than the ψ^q's.

BP operations, since they distinguish the role of a certain prime, are particularly well-suited to studying stable p-equivalences; however, for necessary conditions for cohomology homomorphisms of self-maps of stunted CP's, we do not care to distinguish the roles of the primes. Thus perhaps the Landweber-Novikov operations in $MU^*(\ )$ or ch would be a more effective tool. The necessary conditions obtained using ch are easy to state in a general form; they involve numbers $c_{n,i}$ = coefficient of x^i in $(\frac{e^x-1}{x})^n$. These will be discussed in Section 5.

This paper has benefited from conversations with Bill Dwyer, Sam Gitler, and Clarence Wilkerson. The work was supported by an NSF grant.

2. BP OPERATIONS

In this section we show how to compute ρ_i and ψ^q in $BPQ^*(CP)$ and how to use these to determine necessary conditions for stable p-equivalences and for cohomology homomorphisms of self-maps. We also prove Theorem 1.4 expressing ψ^q in terms of the ρ_I's.

Recall that BPQ^* is a polynomial algebra generated by

$m_i \in BPQ^{-2(p^i-1)}$ with $pm_1 = v_1$ an integral class, and that $BPQ^*(CP^n) \approx BPQ^*[u]/u^{n+1}$ with $\deg(u) = 2$. It follows readily from [3; Theorem 3.2] that the generating function $Y(z) = \sum_{i\geq 0} \rho_i(u)z^i$ satisfies $zY(z)^p + Y(z) - \log u = 0$, where $\log u = \sum_{i\geq 0} m_i u^{p^i}$. From this one calculates $\rho_i(u) = (-1)^i c_i(\log u)^{i(p-1)+1}$, where $\sum_{i\geq 1} c_i x^{i-1} = (1 + \sum_{i\geq 1} c_i x^i)^p$. Thus $c_i = \sum (i_1, i_2, \ldots, p-\Sigma i_j) c_2^{i_2} c_3^{i_3} \cdots$ where the sum is taken over all tuples such that $i-1 = i_1 + 2i_2 + 3i_3 + \cdots$, and $(i_1, i_2, \ldots)$ denotes the multinomial coefficient. For example $c_0 = c_1 = 1$, $c_2 = p$, $c_3 = p^2 + \binom{p}{2}$. By the Cartan formula we now obtain

$$\rho_0(u^k) = (\log u)^k$$

$$\rho_1(u^k) = -k(\log u)^{k+p-1}$$

$$\rho_2(u^k) = (kp + \binom{k}{2})(\log u)^{k+2(p-1)}$$

The action of ρ_i on BPQ^* is given in [3] by

$$\rho_i(m^I) = \begin{cases} 1 & \text{if } m^I = m_1^i \\ 0 & \text{if not.} \end{cases}$$

Now suppose $f\colon CP_k^{k+p-1} \to \Sigma^{2k-2\ell} CP_\ell^{\ell+p-1}$ is a stable p-equivalence. Then $f^* u^{\ell+p-1} = a_1 u^{k+p-1}$ and $f^* u^\ell = a_0 u^k + bm_1 u^{k+p-1}$ with $\nu_p(a_i) = 0$, where $\nu_p(\ )$ denotes the exponent of p. $\nu_p(b) \geq 1$ since $f^* u^\ell \in BP^*(CP)$ so that $bm_1 = cv_1 = cpm_1$.

$0 = \rho_1 f^* u^\ell - f^* \rho_1 u^\ell = (b - a_0 k)u^{k+p-1} - (-a_1 \ell u^{k+p-1})$. Thus

$b = a_0k - a_1\ell$, so $\nu_p(a_0k-a_1\ell) \geq 1$, which implies that we cannot have exactly one of k and ℓ divisible by p. This is exactly the condition obtained in this case in [4]. The above merely serves to illustrate a simple case of how to calculate with ρ_i.

The Adams operations ψ^q satisfy

i) $\psi^q u = \frac{1}{q}\exp(q \log u)$, where exp is the series inverse to log

ii) $\psi^q(x\cdot y) = \psi^q x\cdot\psi^q y$

iii) $\psi^q(m_i) = q^{p^i-1}m_i$.

From (i) we calculate $\mathrm{mod}(m_2,\ldots)$

$$\psi^q(u) = u - c_q v_1 u^p + q^{p-1}c_q v_1^2 u^{2p-1}$$

$$-q^{p-1}c_q(c_q \frac{3p-1}{2} + 1)v_1^3 u^{3p-2} + \cdots$$

where $c_q = \frac{q^{p-1}-1}{p}$. It is interesting to note that $\psi^q u \in BP^*(CP) \subset BPQ^*(CP)$; this may have some bearing on current investigation by Landweber and others as to the extent to which ψ^q are integral operations.

Suppose $f\colon CP_n^{n+2} \to CP_n^{n+2}$ satisfies $f^*x^{n+i} = a_i x^{n+i}$ in $H^*(\ )$. In $BP^*(\ )$ with $p = 3$ we have $f^*u^n = a_0u^n + bv_1u^{n+2}$ and $f^*u^{n+2} = a_2u^{n+2}$, with the a_i the same as in $H^*(\ )$.

$$0 = \psi^q f^* u^n - f^*\psi^q u^n = c_q((a_2 - a_0)n - 3b).$$

Consequently $(a_2-a_0)n \equiv 0(3)$. In $BP^*(\ )$ with $p = 2$ we have

$$f^*u^n = a_0u^n + d_1v_1u^{n+1} + d_2v_1^2u^{n+2}$$

$$f^*u^{n+1} = a_1u^{n+1} + d_3v_1u^{n+2}$$

$$f^*u^{n+2} = a_2u^{n+2}.$$

Then

$$\begin{aligned} 0 &= \text{coefficient of } v_1^2u^{n+2} \text{ in } (\psi^q f^* u^n - f^* \psi^q u^n) \\ &= c_q^2[(a_0-a_2)(2n + \tbinom{n}{2})-2d_1(n+1) + 4d_2] \\ &\quad + c_q[(a_0-a_2)n-d_1(n+1) + 4d_2 + nd_3]. \end{aligned}$$

Choosing values of q which give distinct c_q implies

$$0 = (a_0-a_2)(2n + \tbinom{n}{2}) - 2d_1(n+1) + 4d_2$$

$$0 = (a_0-a_2)n - d_1(n+1) + 4d_2 + nd_3.$$

The d's can be eliminated by utilizing the coefficient of v_1u^{n+1} in $\psi^q f^*u^n - f^*\psi^q u^n$ and the coefficient of v_1u^{n+2} in $\psi^q f^* u^{n+1} - f^*\psi^q u^{n+1}$, eventually yielding the conditions of Theorem 1.2.

This is clearly not a very efficient way of obtaining these conditions, although the final results are strong. In Section 5 we shall see that ch gives necessary conditions more efficiently, and quite likely gives results just as strong as those obtained from BP.

<u>Proof of Theorem 1.4</u> Bendersky defined $\{\rho_I\}$ to be dual to $\{\eta_R(m^I)\}$ under the pairing $\langle\ ,\ \rangle: BP^*(BP) \otimes_{BP_*} BP_*(BP) \to BP^*$.

Thus $\langle \Sigma q^I m^I \rho_I, \eta_R m^F \rangle = q^F m^F$. Therefore it suffices to show $\langle \psi^q, \eta_R m^F \rangle = q^F m^F$. But this follows from [1; Prop. 2, p. 72], which implies $\langle \psi^q, \eta_R m_i \rangle = \psi^q m_i = q^{p^i - 1} m_i$.

3. THE RELATIONSHIP BETWEEN STABLE GENUS AND SELF-MAPS

In this section we dualize the main theorem of [12], which relates the genus of an H_0-space to its self-maps. Theorem 1.1 is an immediate corollary of the following more general result.

<u>Theorem 3.1</u> Suppose X is a finite CW complex of dimension d, and t is divisible by every torsion prime in $\pi^s_\eta(X)$ for $\eta \leq d$ and by the product of the orders of the cokernels of the Hurewicz homomorphisms of X modulo torsion. Let ℓ denote the number of nonzero groups $\pi_\eta(X)$/torsion, $\eta \leq d$. Let $[X,X]_*$ denote the group of stable self-maps of X which induce isomorphisms in $(\pi^s_\eta(\ )/\text{torsion}) \otimes \mathbb{Z}_t$, $\eta \leq d$. Then there is an exact sequence

$$[X,X]_t \xrightarrow{\alpha} (\mathbb{Z}_t^*/\pm 1)^\ell \xrightarrow{\xi} SG(X) \to 0,$$

where α is given by $\langle \det((\pi^s_\eta(\ )/\text{torsion}) \otimes \mathbb{Z}_t)\rangle_\eta$.

<u>Proof</u>. Let s_i denote the rank of the ith nonzero $\pi^s_{\eta_i}(X)$/torsion, $\eta_i \leq d$. Let $\mathcal{M}_t(s_i)$ denote the group of $(s_i \times s_i)$-matrices in $\mathbb{Z}$ whose determinants are prime to t. Let $\mathcal{M}_t(\bar{s}) = \mathcal{M}_t(s_1) \oplus \cdots \oplus \mathcal{M}_t(s_\ell)$, and similarly define $GL(\mathbb{Z},\bar{s})$ and $GL(\mathbb{Z}_t,\bar{s})$. We define a function $\xi'\colon \mathcal{M}_t(\bar{s}) \to SG(X)$ as follows. Let $S^{\bar{\eta}} = (S^{\eta_1})^{s_1} \times \cdots \times (S^{\eta_\ell})^{s_\ell}$.

Let $h_0 \colon S^{\overline{\eta}} \to X$ induce an isomorphism in $\pi^s_\eta(\ )/\text{torsion}$, $\eta \leq d$. This is possible since we are working in the stable category where $\times$ and $\vee$ are equivalent. If $A \in \mathcal{M}_t(\overline{s})$, let $q_A \colon S^{\overline{\eta}} \to S^{\overline{\eta}}$ induce A in $\pi^s_*(\ )/\text{torsion}$. Define $\xi'(A)$ to be the space Y in the following pushout diagram:

$$\begin{array}{ccc} S^{\overline{\eta}} & \xrightarrow{\ q_A\ } & S^{\overline{\eta}} \\ {\scriptstyle h_0}\downarrow & & \downarrow{\scriptstyle h'_A} \\ X & \xrightarrow{\ f_A\ } & Y \end{array}$$

Let $\mathbb{P}_t = \{p \in \mathbb{P} \colon p \mid t\}$. Then q_A is a $\mathbb{P}_t$-equivalence, for its homology matrices are invertible mod p for $p \in \mathbb{P}_t$, and h_0 is a $(\mathbb{P}-\mathbb{P}_t)$-equivalence (since primes not dividing t will not divide the homology of cofibre (h_0)). As in [12; proof of 2.3], using [8; Cor. 7.10, p. 97], this implies $Y \in SG(X)$.

Reduction mod t defines $\mathcal{M}_t(\overline{s}) \xrightarrow{\ \rho_t\ } GL(\mathbb{Z}_t, \overline{s})$. By Corollary 3.3 below, $\xi'(A) = \xi'(A + tB)$ so that ξ' induces $\overline{\xi} \colon GL(\mathbb{Z}_t, \overline{s}) \to SG(X)$.

To see that ξ', and hence $\overline{\xi}$, is surjective, let $Y \in G(X)$. For each $p_i \mid t$ there is a p_i-equivalence $f_i \colon X \to Y$. Let $g_i \colon Y \to Y$ be a p_i-equivalence such that $H^*(g_i; \mathbb{Z}_{p_j}) = 0$, $j \neq i$. The composite f:

$$X \to X \vee \cdots \vee X \xrightarrow{\ \vee g_i f_i\ } Y \vee \cdots \vee Y \to Y$$

is a $\mathbb{P}_t$-equivalence. Let the matrix A represent $\pi^s_*(f)/\text{torsion}$ with respect to prebases $\overline{x}$, $\overline{y'}$. After change of basis, we may assume A is diagonal $= (d_1, \ldots, d_r)$ with $(t, d_i) = 1$. Then

$f_*x_i = d_i y_i' + \theta_i$ with θ_i a torsion element. Order $(\theta_i)|t$ by assumption so there exists θ_i' such that $d_i\theta_i' = \theta_i$. Let $y_i = y_i' + \theta_i'$. Then

$$\begin{array}{ccc} S^{\bar\eta} & \xrightarrow{\;d_1,\ldots,d_r\;} & S^{\bar\eta} \\ {\scriptstyle h_0}\downarrow & & \downarrow{\scriptstyle \bar y_i} \\ X & \xrightarrow{\;\;f\;\;} & Y \end{array}$$

commutes. h_0 and $\bar y_i$ are $(\mathbb{P}-\mathbb{P}_t)$-equivalences, and f and $(d_1,\ldots,d_r)$ are $\mathbb{P}_t$-equivalences. By the dual of [13; 1.6], the diagram is a pushout and $Y = \xi'(A)$.

If $B \in GL(\mathbb{Z},\bar s)$, then $q_B\colon S^{\bar\eta} \to S^{\bar\eta}$ is a homeomorphism. If $h\colon S^{\bar\eta} \to Y$ is any map, and

$$\begin{array}{ccc} S^{\bar\eta} & \xrightarrow{\;q_B\;} & S^{\bar\eta} \\ {\scriptstyle h}\downarrow & & \downarrow \\ Y & \xrightarrow{\;g\;} & Z \end{array}$$

is a pushout diagram, then g is a homeomorphism. Thus $\xi'(BA) = \xi'(A)$ if $B \in GL(\mathbb{Z},\bar s)$. Note that $GL(\mathbb{Z},\bar s) \subset \mathcal{M}_t(\bar s)$ and let $\mathcal{A} = \rho_t(GL(\mathbb{Z},\bar s)) \subset GL(\mathbb{Z}_t,\bar s)$. The above argument shows that $\bar\xi$ factors to give $\xi\colon GL(\mathbb{Z}_t,\bar s)/\mathcal{A} \to SG(X)$. As in [13; 1.8], det defines an isomorphism

$$GL(\mathbb{Z}_t,\bar s)/\mathcal{A} \approx (\mathbb{Z}_t^*/\pm 1)^{\ell},$$

which produces the surjection ξ of the theorem.

Taking the matrix in $\pi_*^s(\ )$/torsion gives a homomorphism $\varphi\colon [X,X]_t \to \mathscr{M}_t(\mathbb{Z},\bar{s})$. We shall complete the proof of 3.1 by showing $\xi'(A) = \xi'(B)$ if and only if $\det(A^{-1}B) = \det(\varphi(f))$ for some f.

If $\det(A^{-1}B) = \det(\varphi(f))$, then the pushout diagram

$$\begin{array}{ccccc}
S^{\bar\eta} & \xrightarrow{q_{\varphi(f)}} & S^{\bar\eta} & \xrightarrow{q_A} & S^{\bar\eta} \\
\downarrow{\scriptstyle h_0} & & \downarrow{\scriptstyle h_0} & & \downarrow{\scriptstyle h_A} \\
X & \xrightarrow{f} & X & \longrightarrow & Y = \xi'(A)
\end{array}$$

implies $\xi'(A\varphi(f)) = Y = \xi'(A)$. Since $\det(A\varphi(f)) = \det B$, we obtain $\xi'(B) = \xi'(A\varphi(f)) = \xi'(A)$.

Conversely, suppose $\xi'(A) = \xi'(B) = Y$. Let $\tilde{A}\in\mathscr{M}_t(\bar{s})$ satisfy $\rho_t(A\tilde{A}) = \rho_t(\tilde{A}A) = I$. The pushout diagram

$$\begin{array}{ccccc}
S^{\bar\eta} & \xrightarrow{q_A} & S^{\bar\eta} & \xrightarrow{q_{\tilde{A}}} & S^{\bar\eta} \\
\downarrow{\scriptstyle h_0} & & \downarrow{\scriptstyle h_1} & & \downarrow{\scriptstyle h'} \\
X & \xrightarrow{f_A} & Y & \xrightarrow{f'} & Z
\end{array}$$

implies $Z = \xi'(\tilde{A}A) = \xi'(I) = X$. Let $f = f'\circ f_B$. Then $\det(\varphi(f)) = \det(A^{-1}B)$. This final step uses

i) h_1 determines a basis for $\pi_*(Y)$/torsion (as in [12; 2.3]);

ii) $\pi_*(Y)/\text{torsion} \approx \pi_*(X)/\text{torsion}$ (since $Y\in G(X)$);

iii) if φ is a homomorphism of isomorphic finitely generated free abelian groups, then $|\det \varphi|$ is independent of choice of bases.

The following lemma is dual to [12; 1.7 and 1.8].

Lemma 3.2 Suppose X and t are as above, $f: X \to Y$ and $g: W \to Y$ are stable maps, $\{x_i\}$ a prebasis for $\pi_*^s(X)$/torsion, and $\{z_i\} \subset \pi_*^s(W)$ arbitrary elements of the same gradation as the x_i's. Then there exists $f': X \to Y$ such that $(f'_* - f_*)(x_i) = t \cdot g_*(z_i)$.

Proof. Let $S = S^{\bar{\eta}}$ be as above. It suffices to construct a stable map $d: X \to X \vee S$ such that

$$\begin{array}{ccc} S & \xrightarrow{\text{pinch}} & S \vee S \\ \downarrow{\scriptstyle h_0} & & \downarrow{\scriptstyle h_0 \vee t} \\ X & \xrightarrow{d} & X \vee S \end{array}$$

commutes, and then let f' be the composite

$$X \xrightarrow{d} X \vee S \xrightarrow{f \vee \bar{z}} Y \vee W \xrightarrow{1 \vee g} Y \vee Y \xrightarrow{\text{fold}} Y.$$

d is constructed inductively on the skeleta of the relative CW complex (X,S), using a CW decomposition obtained by adjoining Moore spaces. The construction is dual to that of [12; 1.7]. For example, the first two steps are given by the diagram

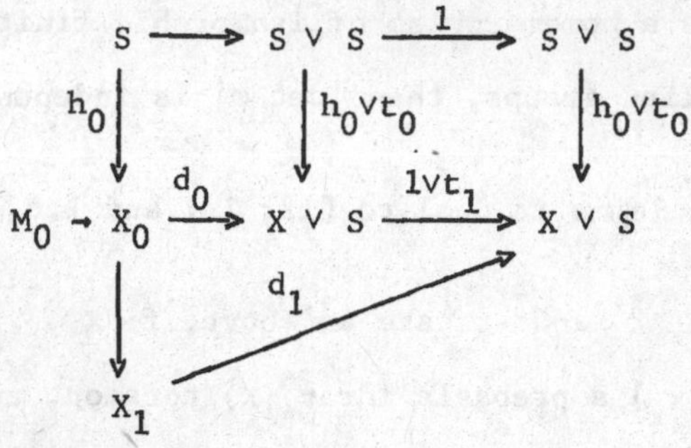

The map d_1 exists because t_1 annihilates $H_*(\text{cofibre } t_0)$. ∎

__Corollary 3.3__ Suppose X and t are as above, $A \in \mathcal{M}_t(\bar{s})$, and $B \in \mathcal{M}(\bar{s})$ is any integer matrix. Then $\xi'(A) = \xi'(A + tB)$.

__Proof__. Apply 3.2 with $f = f_A$, $W = S^{\bar\eta}$, $g = h_A$, and $\bar{z}$ corresponding to the rows of B. The resulting maps form a commutative diagram

$$
\begin{array}{ccc}
S^{\bar\eta} & \xrightarrow{\;q_{A+tB}\;} & S^{\bar\eta} \\
{\scriptstyle h_0}\downarrow & & \downarrow{\scriptstyle h_A} \\
X & \xrightarrow{\;f'\;} & Y
\end{array}
$$

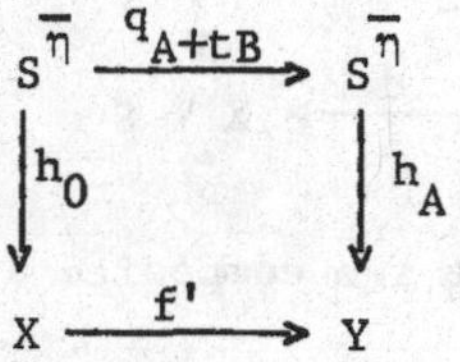

showing $Y = \xi'(A + tB)$.

4. STABLE GENUS AND SELF MAPS OF CP_n^{n+2}

In this section we prove Theorems 1.2 and 1.3 by showing that the estimates for $SG(CP_n^{n+2})$ and $M(CP_n^{n+2})$, the cohomology homomorphisms induced by stable self maps, are sharp.

We begin by tabulating the necessary conditions for cohomology homomorphisms implied by Theorem 1.2 for the 14 stable homotopy types of CP_n^{n+2}.

n(mod 24)	necessary conditions for CP_n^{n+2}	
0	$a_1 \equiv a_2(2)$	
$1 \sim 17$	$a_0 \equiv a_1(2)$	$a_0 \equiv a_2(6)$
$2 \sim 22$	$a_1 \equiv a_2(2)$	$a_0 \equiv a_2(12)$
$3 \sim 15$	$a_0 \equiv a_1(2)$	$a_0 \equiv a_2(4)$
$4 \sim 20$	$a_1 \equiv a_2(2)$	$a_0 \equiv a_2(6)$
$5 \sim 13$	$a_0 \equiv a_1(2)$	$a_0 \equiv a_2(3)$
$6 \sim 18$	$a_1 \equiv a_2(2)$	$a_0 \equiv a_2(4)$
$7 \sim 11$	$a_0 \equiv a_1(2)$	$a_0 \equiv a_2(12)$
$8 \sim 16$	$a_1 \equiv a_2(2)$	$a_0 \equiv a_2(3)$
9	$a_0 \equiv a_1(2)$	$a_0 \equiv a_2(2)$
$10 \sim 14$	$a_1 \equiv a_2(2)$	$a_0 \equiv a_2(12)$
12	$a_0 \equiv a_1(2)$	$a_0 \equiv a_2(2)$
$19 \sim 23$	$a_1 \equiv a_2(2)$	$a_0 \equiv a_2(12)$
21	$a_0 \equiv a_1(2)$	

To check that any tuple satisfying these conditions may be realized by a self-map requires a brute force study of the cell structure of these cell complexes. The burden is eased somewhat by the use of S-duality.

Consider for example the case $n \equiv 4(24)$, in which the stable type of CP_n^{n+2} is $(S^0 \vee S^2) \cup_h e^4$, with attaching map $h = (\alpha_1 + 2\nu, \eta)$. For any $a_0, a_1 \in \mathbf{Z}$ and $\epsilon \in \mathbf{Z}_2$ we may form a map

$(a_0 \vee \epsilon\eta^2, a_1): S^0 \vee S^2 \xrightarrow{g} S^0 \vee S^2$. g can be extended to a self-map of CP_n^{n+2} of degree a_2 on the top cell if and only if $a_2 h = g \circ h \epsilon \pi_3^s(S^0 \vee S^2) \approx \mathbb{Z}_{24} \oplus \mathbb{Z}_2$. $g \circ h = ((2a_0 + 4\epsilon)\nu + a_0\alpha_1, a_1\eta)$ because $\eta^3 = 4\nu$. Thus there is a self-map of CP_n^{n+2} of degree (a_0, a_1, a_2) if and only if $a_2(\alpha_1 + 2\nu) = (2a_0 + 4\epsilon)\nu + a_0\alpha_1 \epsilon \pi_3^s$ (for some ϵ) and $a_2\eta = a_1\eta\epsilon\pi_1^s$. This is equivalent to $a_2 \equiv a_0(3)$, $a_2 \equiv a_0(2)$, and $a_2 \equiv a_1(2)$, showing that the necessary conditions of 1.2 are also sufficient in this case. Note we have also derived the necessary conditions without any apparent use of BP operations, except to the extent they are used to determine attaching maps. However, in cases of many cells, operations are clearly a more effective method of determining necessary conditions.

An S-dual of CP_4^6 is CP_{17}^{19}. Thus the necessary conditions for cohomology homomorphisms of self-maps of CP_n^{n+2} with $n \equiv 17$ are also sufficient by duality from the case $n \equiv 4$. Other cases follow similarly, completing the proof of 1.2.

We now use Theorem 1.1 and the above determination of $M(CP_n^{n+2})$ to prove

$$|SG(CP_n^{n+2})| = \begin{cases} 2 & \text{if } n \equiv 2,7,10,19(24) \\ 1 & \text{for the other 10 stable homotopy types,} \end{cases} \tag{4.2}$$

which implies Theorem 1.3. Let $G = (\mathbb{Z}/2^A 3^B)^* = \{a \equiv \pm 1(6)\}$ and $\overline{G} = G/\pm 1$. (4.2) follows from Theorem 1.1, (4.1), and the following

<u>Proposition 4.3</u> $|\overline{G}^3/\{(\overline{a}_0,\overline{a}_1,\overline{a}_2)\colon a_0 \equiv a_1(2),\ a_0 \equiv a_2(n)$ with $n = 2,3,4$, or $6\}| = 1$;

$$|\overline{G}^3/\{(\overline{a}_0,\overline{a}_1,\overline{a}_2)\colon a_0 \equiv a_1(2),\ a_0 \equiv a_2(12)\}| = 2.$$

<u>Proof</u>. The first part follows from the fact that $a_0, a_1 \in G$ implies either $a_0 + a_1 \equiv 0(4)$ or $a_0 - a_1 \equiv 0(4)$ and either $a_0 + a_1 \equiv 0(6)$ or $a_0 - a_1 \equiv 0(6)$. The generators of the second quotient are $(\overline{1},\overline{1},\overline{1})$ and $(\overline{1},\overline{1},\overline{7})$.

5. THE USE OF ch IN STUDYING SELF-MAPS OF CP

<u>Definition 5.1</u>. Let $c_{n,i}$ = coefficient of x^i in $(\frac{e^x-1}{x})^n$. $c_{n,i} = \frac{n!}{(n+i)!} S(n+i,n)$, where $S(\ ,\ )$ is the Stirling number of the second kind, numbers of much interest to combinatorialists.

Naturality of the Chern character ch yields necessary conditions for the coefficient homomorphisms in self-maps of stunted CP's, which can be expressed in terms of the $c_{n,i}$. However, because closed form expressions for $c_{n,i}$ become unwieldy, these necessary conditions seem to be nearly as complicated as those obtained from BP. It is conceivable that the necessary conditions implied by ch might also be sufficient. The work of [6] on stable homotopy type provides some evidence for this.

<u>Theorem 5.2</u> If $f\colon CP_n^{n+k} \to CP_n^{n+k}$ satisfies $f^*x^{n+i} = a_i x^{n+i}$, then $R^{(\ell)}_{n+\ell,j} \in \mathbf{Z}$ for $1 \leq j \leq k - \ell$, $0 \leq \ell \leq k - 1$, where

$$R_{m,i}^{(\ell)} = (a_{i+\ell} - a_\ell)c_{m,i} - \sum_{j=1}^{i-1} (a_{j+\ell} - a_\ell)c_{m,j}E_{m,i,i-j},$$

and $E_{m,i,r}$ is defined recursively by

$$E_{m,i,r} = c_{m+i-r,r} - \sum_{\ell=1}^{r-1} E_{m,i-\ell,r-\ell}c_{m+i-\ell,\ell}.$$

Example 5.3 $E_{m,i,1} = c_{m+i-1,1}$

$$E_{m,i,2} = c_{m+i-2,2} - c_{m+i-2,1}c_{m+i-1,1}$$

$$R_{m,1}^{(\ell)} = (a_{\ell+1} - a_\ell)c_{m,1}$$

$$R_{m,2}^{(\ell)} = (a_{\ell+2} - a_\ell)c_{m,2} - (a_{\ell+1} - a_\ell)c_{m,1}c_{m+1,1}$$

$$R_{m,3}^{(\ell)} = (a_{\ell+3} - a_\ell)c_{m,3} - (a_{\ell+2} - a_\ell)c_{m,2}c_{m+2,1}$$

$$-(a_{\ell+1} - a_\ell)c_{m,1}(c_{m+1,2} - c_{m+1,1}c_{m+2,1}).$$

Proposition 5.4 If $f: CP_n^{n+3} \to CP_n^{n+3}$ satisfies $f^*x^{n+i} = a_ix^{n+i}$, then the following six expressions are integers:

$(a_1-a_0)c_{n,1}$

$(a_2-a_0)c_{n,2}-(a_1-a_0)c_{n,1}c_{n+1,1}$

$(a_3-a_0)c_{n,3}-(a_2-a_0)c_{n,2}c_{n+2,1}-(a_1-a_0)c_{n,1}(c_{n+1,2}-c_{n+1,1}c_{n+2,1})$

$(a_2-a_1)c_{n+1,1}$

$(a_3-a_1)c_{n+1,2}-(a_2-a_1)c_{n+1,1}c_{n+2,1}$

$(a_3-a_2)c_{n+2,1}$

Substituting $c_{n,1} = \frac{n}{2}$, $c_{n,2} = \frac{n}{6} + \frac{1}{4}\binom{n}{2} = \frac{n(3n+1)}{24}$, and $c_{n,3} = \frac{n}{24} + \frac{n(n-1)}{12} + \frac{1}{8}\binom{n}{3}$ into Proposition 5.4 yields explicit necessary conditions for the coefficients.

Proof of Theorem 5.2 Suppose $\bar{x}^{n+i}$ are the generators of $\widetilde{KU}(CP_n^{n+k})$ and $f^! \bar{x}^n = \Sigma\, b_i \bar{x}^{n+i}$. Then $f^* ch\bar{x}^n = \sum_{i=0}^{k} a_i c_{n,i} x^{n+i}$ and $ch\, f^! \bar{x}^n = \sum_{i=0}^{k} x^{n+i} \sum_{j=0}^{i} b_j c_{n+j,i-j}$. Equating coefficients yields $a_i c_{n,i} = \sum_{j=0}^{i} b_j c_{n+j,i-j}$, or

$$b_0 = a_0$$

$$b_i = (a_i - b_0)c_{n,i} - \sum_{j=1}^{i-1} b_j c_{n+j,i-j}.$$

Since the b_i must be integers, the relations $R_{n,i}^{(0)}$ follow. $R_{n+\ell,i}^{(\ell)}$ is obtained by applying a similar technique to $\bar{x}^{n+\ell}$.

REFERENCES

1. J. F. Adams, Lectures on generalized cohomology, Lecture Notes in Mathematics, Springer-Verlag, 99 (1969), 1-138.

2. J. F. Adams, "Quillen's work on formal groups and complex cobordism," in Stable homotopy and generalized homology, Univ. of Chicago Press, 1974.

3. M. Bendersky, "Rational cobordism operations," Proc. Amer. Math. Soc., 69 (1978), 193-198.

4. D. M. Davis, "Stable p-equivalences of stunted complex projective spaces," to appear in Indiana Univ. Math. Jour.

5. S. Feder and S. Gitler, "Stable homotopy types of stunted complex projective spaces," Proc. Camb. Phil. Soc. 73 (1973), 431-438.

6. S. Feder and S. Gitler, "The classification of stunted projective spaces by stable homotopy type," Trans. Amer. Math. Soc. 225 (1977), 59-82.

7. S. Gitler, "Algunas problemas sobre espacios proyectivos," Bol. Soc. Mat. Braz. (1973), 173-181.

8. P. Hilton, G. Mislin, and J. Roitberg, Localization of nilpotent groups and spaces, North-Holland, American Elsevier, New York, 1975.

9. G. Mislin, "The genus of an H-space," Symposium on Algebraic Topology, Lecture Notes in Mathematics, Springer-Verlag, 249 (1971), 75-83.

10. S. P. Novikov, "The methods of algebraic topology from the viewpoint of cobordism theories," Math. U.S.S.R.-Izvestija 1 (1967), 827-913.

11. D. G. Quillen, "On the formal group laws of unoriented and complex cobordism theories," Bull. Amer. Math. Soc. 75 (1969), 1293-1298.

12. A. Zabrodsky, "p-equivalences and homotopy type," Lecture Notes in Mathematics, Springer-Verlag, 418 (1974), 161-171.

13. A. Zabrodsky, "On the genus of finite CW-H-spaces," Comm. Math. Helv. 49 (1974), 48-64.

On the Stable Homotopy of Symplectic Classifying and Thom Spaces

Stanley O. Kochman and Victor P. Snaith[1]

1. Introduction. In this paper we compute the first 26 stable homotopy groups of MSp(n) and BSp(n) for all n. By [11], [12] the stable homotopy type of BSp(n) is equivalent to that of $\bigvee_{k\leq n} MSp(k)$, so it suffices to compute the groups $\pi_j^S(MSp(k))$. To do this we study the Atiyah-Hirzebruch spectral sequences:

$$(1.1)\quad {}_kE^2_{u,v} = \tilde{H}_u(MSp(k);\pi^S_v) \Longrightarrow \pi^S_{u+v}(MSp(k)) \qquad (k\geq 1)$$

When k>>i, $\pi^S_{4k+i}(MSp(k)) \cong \pi_i(\underline{MSp})$ which is known for i≤100 [3, Part III]. We use only the first 23 of these groups, which are tabulated in §7, Table B_1. This knowledge enables us to determine the differentials in ${}_kE^r_{*,*}$ for k>>0. The differentials in the other spectral sequences then follow easily. The details of this strategy are explained in detail below. The resulting extension problems are resolved by classical techniques in §5. The results of our calculations are listed in §7, Table G.

Our methods also apply to MU(n), BU(n) and BO(2n). M. Mahowald [unpublished], using Adams spectral sequences, has also determined these groups. R. Mosher [7], using the Atiyah-Hirzebruch spectral sequence, computes the first 19 stable homotopy groups of CP^∞. These results are used in [12, Part I] to

---

[1]This research was partially supported by the National Research Council of Canada.

determine the low dimensional homotopy groups of MU(n) and BU(n). It is considerably easier to study all the $\pi_*^S(MU(k))$ simultaneously, however, we will not do that here since the symplectic case illustrates the method adequately. K. Li [5] has studied $\pi_*^S(BO(2n))$ using our method.

We will use the following notation. Let $b_k \in H_{4k}(HP^\infty)$ be a generator. Then $H_*(BSp) = Z[b_1, b_2, \ldots]$ and $H_*(BSp(n))$ is the subgroup with basis $(b_{i_1} b_{i_2} \cdots b_{i_t};\ 0 \le t \le n)$. In view of the stable splitting of BSp(n) cited above, $H_*(MSp(n))$ is the subgroup with basis $(b_{i_1} \cdots b_{i_n})$. Let B_i be the image of b_{i+1} under the canonical homomorphism $H_{4i+4}(MSp(1)) \longrightarrow H_{4i}(\underline{MSp})$. Then $H_*(\underline{MSp}) = Z[B_1, B_2, \ldots]$. The structure map $\Sigma^4 MSp(n) \longrightarrow MSp(n+1)$ induces multiplication by b_1 in homology. To abbreviate the notation, when no confusion results, factors of b_1 have been omitted from polynomials in the b_i. For example the element $b_3 - b_2^2$ in ${}_N E^8_{4N+8,0}$ is an abbreviation for $b_3 b_1^{N-1} - b_2^2 b_1^{N-2}$. Analogously there are generators $a_k \in H_{2k}(CP^\infty)$ which determine elements $A_{k-1} \in H_{2k-2}(\underline{MU})$ such that $H_*(\underline{MU}) = Z[A_1, A_2, \ldots]$. Throughout this paper N will denote a large integer.

Now we will discuss the method in detail. The spectral sequences (1.1) admit the following three structures.

(a) The product in the $\underline{MSp}$-spectrum $\{MSp(h) \wedge MSp(k) \longrightarrow MSp(h+k)\}$ induces pairings $\{{}_h E^r_{u,v} \otimes {}_k E^r_{u',v'} \longrightarrow {}_{h+k} E^r_{u+u',v+v'};\ r \ge 2\}_{h \ge 1, k \ge 1}$ which are the canonical ones when r=2. In particular the differentials commute with multiplication by $b_1 \in {}_1 E^\infty_{4,0}$. This allows us to make deductions about differentials in ${}_kE^r_{*,*}$ from those in ${}_{k+1}E^r_{*,*}$ (c.f. §3).

(b) There is a π^S_*-module structure induced by the composition pairing of homotopy groups. This simplifies the determination of differentials. The ring π^S_*, in dimensions less than or equal to 22, is tabulated in §7, Table A.

(c) A Landweber-Novikov operation of degree t induces a map of spectral sequences ${}_NE^r_{u,v} \longrightarrow {}_{N+t}E^r_{u,v}$ when $N >> t+u+v$. The definition and properties of these operations is given in §2. These operations are used repeatedly in §§3,6 to determine infinite families of differentials from the lowest dimensional differentials.

To assist in determining differentials originating in ${}_kE^r_{4u,0}$ we use the following information about Hurewicz homomorphisms. D. Segal [10] proves that

$$\text{Image}\left[h : \pi^S_{4j}(HP^\infty) \longrightarrow H_{4j}(HP^\infty)\right] = \begin{cases} Z(2j)!b_j & j \text{ even} \\ Z\frac{1}{2}(2j)!b_j & j \text{ odd} \end{cases}$$

From [2; Part III, §6], Image $\left[h : \pi_{2i}(\underline{MU}) \longrightarrow H_{2i}(\underline{MU})\right]$ may be computed. In §§4,6 we use this information to deduce five transgressive differentials. From [3, Part III], we know Image$\left[h : \pi_{4i}(\underline{MSp}) \longrightarrow H_{4i}(\underline{MSp})\right]$ for $i \le 5$. This information is tabulated in §7, Table B_2.

This paper is organized as follows. All differentials localized at the prime two, except four of the transgressive differentials mentioned above, are computed in §3. In §3 we use all the information described above, in particular we use the Landweber-Novikov operations of §2. The results of §§3,4 allow us to analyze the spectral sequences (1.1) localized at the prime two. This information is tabulated in §7, Tables C, D, E. To pass from these tables to stable homotopy groups localized at the prime two, there are several extension problems concerning $\pi_*^S(HP^\infty)_{(2)}$ and $\pi_*^S(MSp(2))_{(2)}$ which are solved in §5. Our methods apply when localizing at odd primes as well. However, in our range of dimensions the computations are few and easy. The results can be found in §6 and are tabulated in §7, Tables F, G.

Perhaps we should point out that our method of calculation is not algorithmic. On the other hand, it does not break down at the point where we stopped.

2. Landweber-Novikov Operations in the Atiyah-Hirzebruch Spectral Sequence.

The Landweber-Novikov operations s_I, [4] and [8], are induced by self-maps of the spectrum MSp. Hence s_I induces a map of spectral sequences from ${}_NE^r_{k,p}$ to ${}_{N+i}E^r_{k,p}$ for deg $s_I = 4i$ and $N>>i+k+p$. These maps are the canonical ones when $r=2$. In Section 3 we will use these operations to deduce an infinite sequence of d^r differentials from the d^r differentials of least degree. This procedure is analogous to the one used in [3, §5,6,8] to determine differentials in the mod two Adams spectral sequence for $\pi_*(\text{MSp})$. The theory of this section clearly generalizes by replacing MSp by any ring spectrum and $\pi_*^S(\ )$ by any generalized homology theory.

The following theorem lists the basic properties of the s_I on ${}_NE^r_{k,p}$, $\pi_t^S(\text{MSp}(N))$ and $H_t(\text{MSp}(N);R)$ with R a commutative ring. These operations will be defined in the proof of the theorem. Let $k\Delta_r$ denote the sequence $(0,\ldots,0,k,0,\ldots)$ with k the rth entry.

THEOREM 2.1 Let $I=(i_1,\ldots,i_q)$ be a sequence of natural numbers with $i=i_1+\ldots+i_q$. Let N be large compared to i, t, k and p. Then there are Landweber-Novikov operations:

$$s_I : H_t(\text{MSp}(N);R) \longrightarrow H_t(\text{MSp}(N+i);R)$$

$$s_I : \pi_t^S(\text{MSp}(N)) \longrightarrow \pi_t^S(\text{MSp}(N+i))$$

$$s_I : {}_NE^r_{k,p} \longrightarrow {}_{N+i}E^r_{k,p}$$

These operations have the following properties.

(a) The s_I are group homomorphisms.

(b) The s_I commute with the Hurewicz homomorphism

$$h : \pi_t^S(\text{MSp}(n)) \longrightarrow H_t(\text{MSp}(n)).$$

(c) The action of s_I on $\pi_*^S(MSp(N))$ modulo the Atiyah-Hirzebruch filtration gives the action of s_I on ${}_N E_{*,*}^{\infty}$.

(d) $s_I \circ d^r = d^r \circ s_I$

(e) The s_I defined above correspond to the usual s_I under the isomorphisms $H_{4N+j}(MSp(N)) \cong H_j(\underline{MSp})$ and $\pi_{4N+j}^S(MSp(N)) \cong \pi_j(\underline{MSp})$ for $0 \le j \le 4N-2$.

(f) Assume that $x_j \in H_*(MSp(N_j);R)$, $x_j \in \pi_*^S(MSp(N_j))$ or $x_j \in {}_{N_j}E_{*,*}^r$ such that $s_I(x_j)$ is defined for $j=1,2$. Then

$$s_I(x_1 x_2) = \sum_{I=J+K} s_J(x_1) s_K(x_2)$$

(g) The action of s_I on ${}_N E_{*,*}^r$ is a homomorphism of π_*^S-modules.

(h) The action of s_I on $H_*(MSp(N);R)$ is given by the Cartan formula (f) and the following two facts.

(i) $s_I(b_k b_1^{N-1}) = 0$ if $i \ge 2$.

(ii) $s_{\Delta_t}(b_k b_1^{N-1})$ is the summand of $b_1^{N-1}(b_1+\ldots+b_k)^{t+1}$ of degree $4k+4N-4$.

(i) The action of s_I on ${}_N E_{*,*}^2 = H_*(MSp(N)) \otimes \pi_*^S$ is $s_I \otimes 1$ where the latter s_I is given by (h).

Proof. $s_I \in MSp^i(\underline{MSp})$ is given by a sequence of compatible maps $s_I^{q,N}: \Sigma^{4q} MSp(N)^{(4q+8N-2)} \longrightarrow MSp(q+n+i)^{(4q+8N-2)}$ for q,N large compared to i. For q small compared to N, the structure map $\Sigma^{4q} MSp(N+i) \longrightarrow MSp(q+N+i)$ is a $(4q+8N-2)$-equivalence with "inverse" $G_i^{q,N}: MSp(q+N+i)^{(4q+8N-2)} \longrightarrow \Sigma^{4q} MSp(N+i)$. A straightforward argument shows that the maps $G_i^{q,N} \circ s_I^{q,N} : \Sigma^{4q}MSp(N)^{(4q+8N-2)} \longrightarrow \Sigma^{4q}MSp(N+i)$ induce three well-defined maps s_I as described in the first part of this theorem. Properties (a)-(e), (g) and (i) are easy to verify. (f) follows from (e), the Cartan formula for the action of s_I on $H_*(\underline{MSp})$ and the following observation. Let x be an element of $H_*(MSp(N);R)$, $\pi_*^S(MSp(N))$ or ${}_NE_{*,*}^r$ such that $s_I(x)$ is defined. If $\deg s_H \leq \deg s_I$ then $s_H(x)$ is defined too. Now (h) follows from (e) and [4, Lemma 5.6].

3. <u>Differentials</u>. Throughout this section all spectral sequences will be localized at the prime two. In this section we compute the differentials which occur in the first 26 degrees of the spectral sequences (1.1). Thus tables C, D and E of §7 as well as the row of Table B_1 containing $\pi^S_{k+4N}(MSp(N))_{(2)}$ can be filled in from the information in this section. There are four transgressive differentials whose values we record here although we postpone their verification to §4. We do this to avoid interrupting our discussion with a lengthy computation of images under the Hurewicz homomorphism. Many of the results here give information in all degrees. All of the differentials of this section are valid in any of the spectral sequences (1.1) in which they make sense by adding factors of b_1 to each summand of the equation.

The following theorem is the simplest illustration of our basic method. We determine $d^4(b_2)$ from the knowledge of $\pi_3(\underline{MSp})$. Then by Landweber-Novikov operations we determine $d^4(b_k)$ for all $k \geq 1$.

THEOREM 3.1 For $k \geq 1$, $d^4(b_k) = (k-1)\nu b_{k-1}$.

<u>Proof</u>. $\pi^S_7(HP^\infty) \cong \pi_3(\underline{MSp}) = 0$. Thus the infinite cycle $\nu \in {}_1E^4_{4,3}$ must bound. The only possibility is $d^4(b_2)$, and we can define ν so that $d^4(b_2) = \nu$. For $k \geq 2$, ${}_1E^4_{4k-4,3} = Z_8 \nu b_{k-1}$, $s_{\Delta_{k-2}}(\nu b_{k-1}) = \nu$ and $s_{\Delta_{k-2}} d^4(b_k) = d^4 s_{\Delta_{k-2}}(b_k) = (k-1)d^4(b_2) = (k-1)\nu$. Thus $d^4(b_k) = (k-1)\nu b_{k-1}$.

In the next theorem we determine all the d^8-differentials on ${}_kE^8_{4u,0}$ by Landweber-Novikov operations from the knowledge of

$d^8(b_3-b_2^2)$ and $d^8(4b_3)$.

THEOREM 3.2 (a) For $k\geq 1$, $d^8(b_{8k+1}) = (2k+1)(k+1)\sigma b_{8k-1}$.

(b) For $k\geq 0$, $d^8(2b_{8k+5}) = (8k+2)\sigma b_{8k+3}$.

(c) For $k\geq 0$, $d^8(4b_{4k+3}) = 8(k-1)\sigma b_{4k+1}$.

(d) For $k\geq 1$, $d^8(8b_{2k}) = 8(k-1)\sigma b_{2k-2}$.

(e) $d^8(b_3-b_2^2) = \sigma$

(f) $d^8(8b_2^3) = 8\sigma b_2$

(g) $d^8(4b_4+4b_3b_2) = 4\sigma b_2$

(h) $d^8(3b_3b_2-2b_2^3-b_4) = \sigma b_2$

(i) $d^8(2b_3^2) = 8\sigma b_3+8\sigma b_2^2$

(j) $d^8(4b_4b_2-b_3^2-b_5) = 15\sigma b_3$

(k) $d^8(2b_2^4) = 12\sigma b_2^2$

(l) $d^8(4b_3b_2^2) = 4\sigma b_3+8\sigma b_2^2$

(m) $d^8(2b_3b_2^2-b_3^2-b_2^4) = 14\sigma b_3+2\sigma b_2^2$

(n) $d^8(8b_5b_2) = d^8(8b_4b_3) = 8\sigma b_3b_2$

(o) $d^8(2b_6 - 2b_5b_2) = 4\sigma b_4+14\sigma b_3b_2$

(p) $d^8(b_6-5b_5b_2+2b_4b_3-6b_3^2b_2) = 14\sigma b_4+9\sigma b_3b_2+8\sigma b_2^3$

(q) $d^8(8b_4b_2^2) = 8\sigma b_4+8\sigma b_2^3$

(r) $d^8(b_4b_3-b_4b_2^2-3b_3^2b_2+5b_3b_2^3-2b_2^5) = \sigma b_4+12\sigma b_3b_2+3\sigma b_2^3$

(s) $d^8(8b_3b_2^3) = 8\sigma b_3b_2$

(t) $d^8(4b_3^2b_2+4b_4b_3) = 8\sigma b_4+4\sigma b_3b_2$

(u) $d^8(8b_2^5) = 0$

Proof. $\pi^S_{4N+7}(MSp(N)) \cong \pi_7(\underline{MSp}) = 0$. Thus the infinite cycle $\sigma\epsilon_N E^8_{4N,7}$ must bound. The only possiblities are $d^8(b_3-b_2^2)$ and $d^8(4b_3)$. However, $\eta\sigma\neq 0$ in $_N E^8_{4N,8}$ and $\eta d^8(4b_3) = d^8(4\eta b_3) = 0$. Thus we can define σ so that $d^8(b_3-b_2^2) = \sigma$. By D. Segal [10], $4b_3$ is not an infinite cycle while $8b_3$ is an infinite cycle. Hence $d^8(4b_3) = 8\sigma$. In §4 we will give an alternate proof of this fact. We now apply the method of the proof of Theorem 3.1. For (a) we use $s_{\Delta_{8k-2}}$; for (b) $s_{\Delta_{8k+2}}$; for (c) $s_{\Delta_{4k}}$; for (d) $s_{\Delta_{2k-3}}$; for (f)-(h) s_{Δ_1}; for (i)-(m) s_{Δ_2}, $s_{2\Delta_1}$ and for (n)-(u) s_{Δ_3}, $s_{\Delta_2+\Delta_1}$, $s_{3\Delta_1}$.

We determine all the d^8-differentials on $_k E^8_{4u,1}$ by Landweber-Novikov operations from the knowledge of $d^8(\eta b_3)$ and $d^8(\eta b_2^2)$.

THEOREM 3.3 (a) For $k\geq 1$, $d^8(\eta b_{2k+1}) = (k\varepsilon+(k-1)\eta\sigma)b_{2k-1}$.

(b) For $k\geq 1$, $d^8(\eta b_{2k}) = (k-1)(\eta\sigma+\varepsilon)b_{2k-2}$.

(c) For $k\geq 5$, $d^8(\eta b_k b_2) = (k-1)(\eta\sigma+\varepsilon)b_{k-1} + k\varepsilon b_{k-2}b_2$.

(d) $d^8(\eta b_2^2) = \eta\sigma+\varepsilon$

(e) $d^8(\eta b_3 b_2) = \varepsilon b_2$

(f) $d^8(\eta b_2^3) = (\eta\sigma+\varepsilon)b_2$

(g) $d^8(\eta b_4 b_2) = (\eta\sigma+\varepsilon)b_3 + (\eta\sigma+\varepsilon)b_2^2$

(h) $d^8(\eta b_3^2) = d^8(\eta b_2^4) = 0$

(i) $d^8(\eta b_3 b_2^2) = (\eta\sigma+\varepsilon)b_3 + \varepsilon b_2^2$

(j) $d^8(\eta b_4 b_3) = \varepsilon b_4 + (\eta\sigma+\varepsilon)b_3 b_2$

(k) $d^8(\eta b_4 b_2^2) = (\eta\sigma+\varepsilon)b_4 + (\eta\sigma+\varepsilon)b_2^3$

(l) $d^8(\eta b_3 b_2^3) = (\eta\sigma+\varepsilon)b_3 b_2 + \varepsilon b_2^3$

(m) $d^8(\eta b_3^2 b_2) = d^8(\eta b_2^5) = 0$

<u>Proof</u>. $\pi_{4N+8}(MSp(N)) \cong \pi_8(\underline{MSp})$ is torsion-free. Thus $d^8 : {}_N E^8_{4N+8,1} \longrightarrow {}_N E^8_{4N,8}$ is an isomorphism. $\eta d^8(\eta b_2^2) = d^8[(\eta b_2)^2] = 2(\eta b_2)d^8(\eta b_2) = 0$. Thus $d^8(\eta b_2^2) = \eta\sigma+\varepsilon$. By Theorem 3.2(e), $d^8(\eta b_3) = \eta d^8(b_3-b_2^2) + d^8(\eta b_2^2) = \eta\sigma + (\eta\sigma+\varepsilon) = \varepsilon$. We now apply the method of the proof of Theorem 3.1. For (a) we use $s_{\Delta_{2k-2}}$; for (b) $s_{\Delta_{2k-3}}$; for (c) $s_{\Delta_{k-2}}$, $s_{\Delta_{k-3}+\Delta_{k-1}}$; for (e),(f) s_{Δ_1}; for (g)-(i) s_{Δ_2}, $s_{2\Delta_1}$ and for (j)-(m) s_{Δ_3}, $s_{\Delta_2+\Delta_1}$, $s_{3\Delta_1}$.

We determine all the d^{12}-differentials on ${}_k E^{12}_{4u,0}$ by Landweber-Novikov operations from the knowledge of $d^{12}(16b_4)$, $d^{12}(8b_4-8b_2^3)$ and $d^{12}(8b_3b_2)$.

THEOREM 3.4 (a) $d^{12}(16b_4) = \zeta$

(b) $d^{12}(8b_4-8b_2^3) = (4c+2)\zeta$

(c) $d^{12}(8b_3b_2) = (2e+1)\zeta$

(d) $d^{12}(16b_5) = d^{12}(8b_2^4) = 2\zeta b_2$

(e) $d^{12}(4b_3^2) = (4e+2)\zeta b_2$

(f) $d^{12}(4b_3b_2^2+2b_3^2+4b_5) = (4c+6e+2)\zeta b_2$

(g) $d^{12}(6b_3^2-8b_3b_2^2+4b_2^4) = (2e+1)\zeta b_2$

(h) $d^{12}(8b_4b_2-2b_3^2) = d^{12}(6b_3^2 - 12b_3b_2^2 - 12b_5) = (4c+2e)\zeta b_2$

(i) $d^{12}(8b_2^5) = \zeta b_2^2$

(j) $d^{12}(8b_6) = (4c+4e+6)\zeta b_3$

Proof. By D. Segal [10], $128b_4$ is an infinite cycle in ${}_1E^{12}_{16,0}$, but $64b_4$ is not an infinite cycle. Thus we can define ζ so that $d^{12}(16b_4) = \zeta$. In Theorem 4.3 we will prove that $d^{12}(8b_4-8b_2^3) = (4c+2)\zeta$ and $d^{12}(8b_3b_2) = (2e+1)\zeta$. We now apply the method of the proof of Theorem 3.1. For (d)-(h) we use s_{Δ_1}, and for (i),(j) we use s_{Δ_2}, $s_{2\Delta_1}$.

The following theorem completes our computations through stable degree 14.

THEOREM 3.5 (a) $d^8(\sigma b_3) = 0$

(b) $d^8(\sigma b_2^2) = \sigma^2$

(c) $d^8(\sigma b_3b_2) = 0$

(d) $d^8(\sigma b_4) = d^8(\sigma b_2^3) = \sigma^2 b_2$

(e) $d^8(\sigma b_3^2) = d^8(\sigma b_2^4) = 0$

(f) $d^8(\sigma b_5) = d^8(\sigma b_3b_2^2) = \sigma^2 b_3$

(g) $d^8(\sigma b_4b_2) = \sigma^2 b_3+\sigma^2 b_2^2$

(h) $d^8(2\nu b_4) = 0$

(i) $d^8(\nu b_4 b_3+\nu b_3^2 b_2) = 0$

(j) $d^{12}(2\nu b_4) = \kappa$

(k) $d^{12}(\nu b_4 b_3+\nu b_3^2 b_2) = \kappa b_3+\kappa b_2^2$

(l) $d^8(\varepsilon b_3) = d^8(\varepsilon b_2^2) = \eta\kappa$

(m) $d^8(\varepsilon b_4) = d^8(\varepsilon b_3 b_2) = d^8(\varepsilon b_2^3) = \eta\kappa b_2$.

Proof. $\pi_{14+4N}(MSp(N)) \cong \pi_{14}(\underline{MSp}) = Z_2\eta\phi_2 + Z_2\eta\phi_1 q_0$. The infinite cycle $\eta(b_4+b_2^3), \mu, \eta b_2$ represents $\phi_2, \eta q_0, \phi_1$, respectively. Thus $\eta^2(b_4+b_2^3)$, $\eta\mu b_2$ represents $\eta\phi_2$, $\eta\phi_1 q_0$, respectively, Hence the infinite cycles κ and σ^2 are boundaries. Now $d^8(2\nu b_4) = 0$ and $d^8(\sigma b_3) = d^8 d^8(b_5+b_3^2-4b_4 b_2) = 0$ by Theorem 3.2(j). Thus the only possibilities for the two boundaries above are $d^8(\sigma b_2^2)$ and $d^{12}(2\nu b_4)$. Observe that $\eta d^8(\sigma b_2^2) = \sigma d^8(\eta b_2^2) = \sigma(\eta\sigma+\varepsilon) = 0$. Thus $d^8(\sigma b_2^2) = \sigma^2$ and $d^{12}(2\nu b_4) = \kappa$. We now use s_{Δ_1}, s_{Δ_2} and $s_{2\Delta_1}$ to prove (c)-(g), (i) and (k). Observe that $0 = \eta d^{12}(2\nu b_4) = \eta\kappa$ in ${}_1E^{12}_{*,*}$ while $\eta\kappa\neq 0$ in ${}_1E^8_{*,*}$. Thus $\eta\kappa$ must be a d^8-boundary. The only possibilities are $d^8(\varepsilon b_3)$ and $d^8(\varepsilon b_2^2)$. $d^8(\varepsilon b_3)+d^8(\varepsilon b_2^2) = d^8 d^8(\eta b_3 b_2^2)+\sigma d^8(\eta b_3) = 0$. Hence $d^8(\varepsilon b_3) = d^8(\varepsilon b_2^2) = \eta\kappa$. We now use s_{Δ_1} to prove (m).

THEOREM 3.6 (a) $d^{16}(8b_2^4-16b_5) = \rho$

(b) $d^{16}(64b_5) = 16\rho$

(c) $d^{16}(64b_4b_2+32b_3^2) = 24\rho$

(d) $d^{16}(32b_6) = 8\rho b_2$

(e) In ${}_NE^{16}_{*,*}$, ρb_2 is a d^{16}-boundary.

(f) $d^8(\mu b_3) = \eta\rho$

(g) $d^8(\mu b_2^2) = 0$

(h) $d^8(\mu b_4) = d^8(\mu b_2^3) = 0$

(i) $d^8(\mu b_3 b_2) = \eta\rho b_2$

Proof. $q_0^2 \in \pi_{16}(\underline{MSp}) \cong \pi_{4N+16}(MSp(N))$ is represented by $2^8 b_2^4$. In addition, ρ must be a d^{16}-boundary because $\pi_{4N+15}(MSp(N)) \cong \pi_{15}(\underline{MSp}) = 0$. Thus we can define ρ so that $d^{16}(8b_2^4-16b_5) = \rho$. By D. Segal [10], $64b_5$ is not an infinite cycle while $128b_5$ is an infinite cycle. Thus $d^{16}(64b_5) = 16\rho$. In Theorem 4.3 we will prove that $d^{16}(64b_4b_2+32b_3^2) = 24\rho$. We now use s_{Δ_1} to prove (d). Note that ρb_2 is nonzero in ${}_NE^{16}_{4N+4,15}$ and $\pi_{4N+19}(MSp(N)) \cong \pi_{19}(\underline{MSp}) = 0$. Thus ρb_2 must be a d^{16}-boundary. Observe that $\eta d^8(\mu b_2^2) = \mu d^8(\eta b_2^2) = \mu(\eta\sigma+\varepsilon) = 0$ and $\eta d^8(\mu b_3) = \mu d^8(\eta b_3) = \mu\varepsilon = \eta^2\rho$. Multiplication by η is a monomorphism from ${}_NE^8_{4N,16}$ to ${}_NE^8_{4N,17}$. Thus $d^8(\mu b_2^2) = 0$ and $d^8(\mu b_3) = \eta\rho$. We now use s_{Δ_1} to prove (h) and (i).

The following theorem completes our calculations through stable degree 19.

THEOREM 3.7 (a) $d^{16}(\eta b_2^4) = \eta^*$

(b) $d^{16}(\eta b_3^2) = 0$

(c) $d^{16}(\eta b_6) = d^{16}(\eta b_3^2 b_2) = 0$

(d) $d^{16}(\eta b_2^5) = \eta^* b_2$

(e) $d^{16}(\eta^2 b_4 b_2) = \eta\eta^*$

(f) $d^8(\zeta b_3) = 4\nu^*$

(g) $d^8(\zeta b_2^2) = 0$

(h) $d^8(\zeta b_4) = 0$

(i) $d^8(\zeta b_3 b_2) = d^8(\zeta b_2^3) = 4\nu^* b_2$

(j) $d^{12}(2\sigma b_4) = \nu^* + f\eta\bar{\mu}$

(k) $d^{12}(2\sigma b_5) = 2\nu^* b_2$

<u>Proof</u>. $\pi_{4N+16}(MSp(N)) \cong \pi_{16}(\underline{MSp})$ is torsion-free. Hence the infinite cycle η^* must bound. The only possibilities are $d^{16}(\eta b_3^2)$ and $d^{16}(\eta b_2^4)$. From [3, Part III] we see that $\eta^2 b_3^4 \in {}_N E^{\infty}_{4N+32,2}$ represents $\phi_0^2[V_{0,1}^4] = [\phi_0 V_{0,1}^2]^2 + x$. Here x is an element of $\pi_{34}(\underline{MSp})$ of Adams filtration five and an element of $\pi_{34+4N}(MSp(N))$ of Atiyah-Hirzebruch filtration 4N+28. Hence ηb_3^2 is an infinite cycle, and $d^{16}(\eta b_2^4) = \eta^*$. We now use s_{Δ_1} to prove (c) and (d).

$\pi^S_{4N+17}(MSp(N)) \cong \pi_{17}(\underline{MSp}) = Z_2[\phi_0 V_{0,1}^2] + Z_2 \eta R(0,1) + Z_2\, \eta q_0^2$. We showed above that $\eta b_3^2 \in {}_N E^{\infty}_{4N+16,1}$ represents $[\phi_0 V_{0,1}^2]$. Since $\mu \in {}_N E^{\infty}_{4N,9}$

represents ηq_0, we see that $\overline{\mu} \in {}_N E^{\infty}_{4N,17}$ represents ηq_0^2. $\pi^S_{4N+18}(MSp(N)) \cong \pi_{18}(\underline{MSp}) = Z_2\eta[\Phi_0 V^2_{0,1}] + Z_2\eta^2 q_0^2 + Z_2\phi_1\phi_2 + Z_2\phi_1^2 q_0$. Now $\eta[\Phi_0 V^2_{0,1}]$ is represented by $\eta^2 b_3^2 \in {}_N E^{\infty}_{4N+16,2}$, $\eta^2 q_0^2$ is represented by $\eta\overline{\mu} \in {}_N E^{\infty}_{4N,10}$, $\phi_1\phi_2$ is represented by $(\eta b_2)(\eta b_4 + \eta b_2^3)$ $= \eta^2 b_4 b_2 + \eta^2 b_2^4 \in {}_N E^{\infty}_{4N+16,2}$ and $\phi_1^2 q_0$ is represented by $(\eta b_2)(\mu b_2)$ $= \eta\mu b_2^2 \in {}_N E^{\infty}_{4N+8,10}$. Thus $d^{16}(\eta^2 b_4 b_2 + \eta^2 b_2^4) = 0$, and $d^{16}(\eta^2 b_4 b_2)$ $= \eta d^{16}(\eta b_2^4) = \eta\eta^*$. In addition ν^* must be a multiple of $\eta\overline{\mu}$ in ${}_N E^{\infty}_{*,*}$. $\pi^S_{4N+19}(MSp(N)) \cong \pi_{19}(\underline{MSp}) = 0$. Thus $d^{12}(2\sigma b_4)$ is a nonzero element of order four. Hence $4\nu^* = 0$ in ${}_N E^{12}_{*,*}$, and we can define ν^* so that $d^{12}(2\sigma b_4) = \nu^* + f\eta\overline{\mu}$. Thus $4\nu^*$ is a d^8-boundary. The only possibilities are $d^8(\zeta b_3)$ and $d^8(\zeta b_2^2)$. By Theorem 3.4(i), $d^8(\zeta b_2^2) = d^8 d^8(8b_2^5) = 0$. Hence $d^8(\zeta b_3) = 4\nu^*$. We use s_{Δ_1} to prove (h) and (i).

Next we complete our computations through stable degree 22.

THEOREM 3.8 (a) $d^8(\nu^3 b_4) = d^8(\nu^3 b_2^3) = 0$

(b) $d^8(\kappa b_3) = d^8(\kappa b_2^2) = \eta\overline{\kappa}$

(c) $d^{12}(\nu^3 b_4) = 2\overline{\kappa}$

(d) $d^{20}(\eta b_3^2 b_2) = d^{20}(\eta b_6) = \overline{\kappa}$

(e) $d^{20}(128 b_6) = \overline{\zeta}$

(f) $d^{12}(\eta\sigma b_4) = \overline{\sigma} + 4g\overline{\zeta}$

<u>Proof</u>. $\eta d^8(\nu^3 b_4) = d^8(\eta\nu^3 b_4) = 0$ and $\eta d^8(\nu^3 b_2^3) = 0$.

Multiplication by η is a monomorphism from ${}_N E^8_{4N+4,16}$ to ${}_N E^8_{4N+4,17}$.

Thus $d^8(\nu^3 b_4) = d^8(\nu^3 b_2^3) = 0$. $\pi^S_{4N+20}(MSp(N)) \cong \pi_{20}(\underline{MSp})$ is torsion-free, $\pi^S_{4N+21}(MSp(N)) \cong \pi_{21}(\underline{MSp}) = Z_2\phi_3 + Z_2\phi_1 R(0,1) + Z_2\phi_1 q_0^2 + Z_2\phi_2 q_0$ and $\pi^S_{4N+22}(MSp(N)) \cong \pi_{22}(\underline{MSp}) = \eta\pi_{21}(\underline{MSp}) + Z_2\phi_1[\phi_0 V^2_{0,1}]$. (Note that multiplication by η is a monomorphism from $\pi_{21}(\underline{MSp})$ to $\pi_{22}(\underline{MSp})$.)

From N. Ray [9] we see that $s_{\Delta_2}(\phi_3) = \phi_2$ which is represented by $\eta(b_4+b_2^3) \in {}_N E^\infty_{4N+12,1}$, $s_{2\Delta_2}(\phi_3) = \phi_1$ which is represented by $\eta b_2 \in {}_N E^\infty_{4N+4,1}$ and $s_{\Delta_1}(\phi_3) = 0$. Hence ϕ_3 is represented by $\eta(b_6+b_3^2 b_2) \in {}_N E^\infty_{4N+20,1}$. In proving Theorem 3.7 we showed that $\phi_0 R(0,1)$ is represented by $\bar{\mu} \in {}_N E^\infty_{4N,17}$. Since $s_{\Delta_1}(\phi_1 R(0,1)) = \phi_0 R(0,1)$, $\phi_1 R(0,1)$ is represented by $\bar{\mu} b_2 \in {}_N E^\infty_{4N+4,17}$. Note that $\phi_0 q_0^2$, $\phi_1 q_0$, $\phi_0 q_0$ is represented by $\mu b_2^2 \in {}_N E^\infty_{4N+8,9}$, $\mu b_2 \in {}_N E^\infty_{4N+4,9}$, $\mu \epsilon {}_N E^\infty_{4N,9}$ respectively. Hence $\phi_1 q_0^2$ is represented by $\mu b_2^3 + s\nu^3 b_4 \in {}_N E^\infty_{4N+12,9}$ and $\phi_2 q_0$ is represented by $\mu b_4 + \mu b_2^3 + (s+t)\nu^3 b_4 \in {}_N E^\infty_{4N+12,9}$. Thus $\bar{\kappa} \in {}_N E^8_{4N,20}$, $\eta\bar{\kappa} \in {}_N E^8_{4N,21}$, $\kappa b_3 \in {}_N E^8_{4N+8,14}$, $\kappa b_2^2 \in {}_N E^8_{4N+8,14}$, $\eta\sigma b_4 \in {}_N E^{12}_{4N+12,8}$, $\nu^3 b_4 \in {}_N E^{12}_{4N+12,9}$ and $\eta b_3^2 b_2 \in {}_N E^{16}_{4N+20,1}$ are boundaries or have nonzero differentials. Hence $d^{12}(\nu^3 b_4) = 2\bar{\kappa}$. By Theorem 3.5(m) $d^8(\kappa b_3) = d^8(\kappa b_2^2)$ must equal $\eta\bar{\kappa}$. Note that $s_{2\Delta_2}(\phi_1[\phi_0 V^2_{0,1}]) = \phi_1\phi_0$ is represented by $\eta^2 b_2 \in {}_N E^\infty_{4N+4,2}$. Hence

$\phi_1[\phi_0 V_{0,1}^2]$ is represented by $\eta^2 b_3^2 b_2 \in {}_N E_{4N+20,2}^{\infty}$. Thus $d^{20}(\eta b_3^2 b_2) = \overline{\kappa}$. Since $\eta b_6 + \eta b_3^2 b_2$ is an infinite cycle, $d^{20}(\eta b_6) = \overline{\kappa}$. By D. Segal [10], $2^{10} b_6$ is an infinite cycle but $2^9 b_6$ is not an infinite cycle. Thus $d^{20}(2^7 b_6)$ is a nonzero element of order eight. Thus we can define $\overline{\zeta}$ so that $d^{20}(2^7 b_6) = \overline{\zeta}$. Since $d^{12}(\eta\sigma b_4)$ is nonzero and $\overline{\zeta}$ has order eight in ${}_N E_{19,0}^{20}$, $d^{12}(\eta\sigma b_4) = \overline{\sigma} + 4g\overline{\zeta}$.

4. Images in $H_*(\underline{MU})$ under the Hurewicz Homomorphism. Throughout this section all groups and spectral sequences will be localized at the prime two. In this section we derive four transgressive differentials $d^{4k}(x)$ in ${}_NE^{4k}_{4N,4k-1}$ with N large. These differentials were assumed in Section 3. Such a differential is the last possible nonzero differential on x. Consider the following commutative diagram in which all the maps are the canonical ones and H is the composite of the maps in the bottom row.

$$\begin{array}{ccccccccc}
\pi_{4k}(\underline{MU}) & \xleftarrow{\alpha_*} & \pi_{4k}(\underline{MSp}) & \rightarrowtail & \pi^S_{4k+4N}(MSp(N)) & \longrightarrow\!\!\!\!\rightarrow & & & {}_NE^{\infty}_{4k+4N,0} \\
\downarrow h_U & & \downarrow h_{Sp} & & & & & & \downarrow \\
H_{4k}(\underline{MU};Z_{(2)}) & \xleftarrow{\alpha_*} & H_{4k}(\underline{MSp};Z_{(2)}) & \twoheadleftarrow & H_{4k+4N}(MSp(N);Z_{(2)}) & \twoheadleftarrow & {}_NE^{2}_{4k+4N,0} & \leftarrowtail & {}_NE^{4k}_{4k+4N,0}
\end{array}$$

$$H: {}_NE^{4k}_{4k+4N,0} \longrightarrow H_{4k}(\underline{MU};Z_{(2)})$$

Thus if $d^{4k}(x) = 0$ then $H(x) \in$ Image h_U. Conversely, we will see that if $d^{4k}(x)$ is nonzero then the determination of the coset $H(x)$ + Image h_U can be used to deduce $d^{4k}(x)$.

The following theorem is a consequence of [2, Part II §6]. In particular it is helpful to use the formula of [12, 6.12].

THEOREM 4.1 Image h_U is a polynomial subalgebra of $H_*(\underline{MU};Z_{(2)}) = Z_{(2)}[A_n|n\geq 1]$ with a generator g_n in each even degree 2n. The first six of these generators can be taken as follows.

$g_1 = 2A_1$

$g_2 = 3A_2-2A_1^2$

$g_3 = 6A_3-6A_2A_1+2A_1^3$

$g_4 = 5A_4-14A_3A_1-6A_2^2+9A_2A_1^2-10A_1^4$

$g_5 = 15A_5-35A_4A_1-12A_3A_2+42A_3A_1^2+15A_2^2A_1-23A_2A_1^3+12A_1^5$

$g_6 = 7A_6-32A_5A_1-26A_4A_2-12A_3^2+84A_4A_1^2+124A_3A_2A_1+21A_2^3+115A_2A_1^4-168A_3A_1^3$
$\quad -160A_2^2A_1^2-134A_1^6$

We want to compute the transgressions $d^8(4b_3)$, $d^{12}(8b_4-8b_2^3)$, $d^{12}(8b_3b_2)$ and $d^{16}(64b_4b_2+32b_3^2)$. We thus determine the Image h_U cosets of these elements in the following theorem.

THEOREM 4.2 (a) $H(4b_3)$ + Image $h_U = 8A_1^4$ + Image h_U.

(b) $H(8b_4-8b_2^3)$ + Image $h_U = 32A_1^6$ + Image h_U.

(c) $H(8b_3b_2)$ + Image $h_U = 48A_1^6$ + Image h_U.

(d) $H(64b_4b_2+32b_3^2)$ + Image $h_U = 192A_1^8$ + Image h_U

Proof (a) Under the four maps whose composite is H we have that $4b_3 \longmapsto 4b_3 \longmapsto 4b_3 \longmapsto 4B_2 \longmapsto 4(2A_4-2A_3A_1+A_2^2)$. Now $8A_4-8A_3A_1+4A_2^2 \equiv 8g_4+2g_3g_1+8A_1^4$ mod (16), and (16) ⊂ Image h_U in degree eight. Thus $H(4b_3) \equiv 8A_1^4$ modulo Image h_U.

(b) $H(8b_4-8b_2^3) \equiv -16g_6+14g_3^2+8g_5g_1+16g_4g_2+4g_4g_1^2-8g_2^3-3g_3g_1^3-20g_3g_2g_1$
$+10g_2^2g_1^2-g_2g_1^4+32A_1^6$ mod (64). In degree twelve, (64) ⊂ Image h_U. Hence $H(8b_4-8b_2^3) \equiv 32A_1^6$ modulo Image h_U.

(c) $H(8b_3b_2) \equiv 32g_4g_2+8g_3g_2g_1-48g_2^3-4g_4g_1^2+g_3g_1^3-2g_2^2g_1^2+48A_1^6$ (mod 64)
Thus $H(8b_3b_2) \equiv 48A_1^6$ modulo Image h_U.

(d) $H(64b_4b_2+32b_3^2) \equiv 128g_4^2+32g_6g_1^2+16g_5g_1^3+32g_4g_2g_1^2+32g_3^2g_2+32g_2^4$ $+12g_3^2g_1^2+128g_4g_2^2+32g_3g_2^2g_1-8g_3g_2g_1^3-16g_2^3g_1^2+8g_4g_1^4+2g_3g_1^5+4g_2^2g_1^4+2g_2g_1^6$ $-64A_1^8 \bmod (256)$. In degree sixteen, $(256) \subset \text{Image } h_U$. Thus $H(64b_4b_2+32b_3^2) \equiv 192A_1^8$ modulo Image h_U.

We can now compute the required transgressions.

THEOREM 4.3 (a) $d^8(4b_3) = 8\sigma$.

(b) $d^{12}(8b_4-8b_2^3) = (4c+2)\zeta$.

(c) $d^{12}(8b_3b_2) = (2e+1)\zeta$.

(d) $d^{16}(64b_4b_2+32b_3^2) = 24\rho$.

Proof. (a) Let $d^8(4b_3) = \alpha\sigma$. From the degree eight entries in Table B_2, we see that α is divisible by 8. Since $d^8(b_3-b_2^2) = \sigma$, $4b_3-\alpha(b_3-b_2^2)$ is an infinite cycle. Hence $H[4b_3-\alpha(b_3-b_2^2)] \in \text{Image } h_U$. By Theorem 4.2(a), $\alpha \equiv 8 \bmod 16$. Thus $d^8(4b_3) = 8\sigma$.

(b) Let $d^{12}(8b_4-8b_2^3) = \alpha\zeta$. Since $d^{12}(16b_2^3) = (2a+1)\zeta$, $8b_4-8b_2^3 + 16(2s+1)\alpha b_2^3$ is an infinite cycle. Hence $H[8b_4-8b_2^3+16(2s+1)\alpha b_2^3] \in \text{Image } h_U$. By Theorem 4.2(b), $\alpha \equiv 2 \bmod 4$. Thus $d^{12}(8b_4-8b_2^3)$ is 2ζ or 6ζ.

(c) Let $d^{12}(8b_3b_2) = \alpha\zeta$. Then $H[8b_3b_2+16(2t+1)\alpha b_2^3] \in \text{Image } h_U$. By Theorem 4.2(c), α must be odd.

(d) Let $d^{16}(64b_4b_2+32b_3^2) = \alpha\rho$. Since $d^{16}(8b_2^4-16b_5) = \rho$, $H[64b_4b_2+32b_3^2-\alpha(8b_2^4-16b_5)] \in \text{Image } h_U$. By Theorem 4.2(d), $\alpha \equiv 24 \bmod 32$.

5. Extension problems. Throughout this section all groups and spectral sequences will be localised at the prime two. In this section we shall discuss the extension problems which must be solved in order to pass from the E^{∞}-terms of the spectral sequences to the stable homotopy groups tabulated in §7. These occur only for HP^{∞} and $MSp(2)$. Let $[x]\in\pi_*^S(X)$ denote an element represented by an E^2-term x in one of the spectral sequences.

We begin with some simple observations which we will use repeatedly. The proof is omitted.

PROPOSITION 5.1 (a) Let $0\neq\eta\in\pi_1^S$. If $x\in {}_jE^2_{p,q}$ is an infinite cycle then $[\eta x]\in\pi^S_{p+q+1}(MSp(j))$ may be chosen to be of order two.

(b) If $x\in {}_jE^2_{p,q}$ is an infinite cycle and $0\neq[\eta x]\in {}_jE^{\infty}_{p,q+1}$ then $[x]\in\pi^S_{p+q}(MSp(j))$ is not divisible by two.

(c) The natural map $\alpha:\Sigma^{4t}MSp(n)\longrightarrow MSp(n+t)$ induces multiplication by b_1^t, ${}_nE^2_{*,*}\longrightarrow {}_{n+t}E^2_{*,*}$. Consequently if $a\in {}_nE^{\infty}_{p,q}$ and $b\in {}_nE^{\infty}_{p+s,q-s}$ $(s>0)$ satisfy $2[b]=[a]\in\pi^S_{p+q}(MSp(n))$ then $2[b_1^tb]=[b_1^ta]\in\pi^S_{p+q+4t}(MSp(n+t))$.

By means of Proposition 5.1 we easily determine the following extensions.

THEOREM 5.2 (a) $\pi^S_{18}(HP^{\infty})_{(2)}\cong Z_2[\sigma^2 b_1]\oplus Z_2[\mu\eta b_2]$

(b) $\pi^S_{22}(HP^{\infty})_{(2)}\cong Z_2[\eta\bar{\mu}b_1]\oplus Z_2[\kappa b_2]$

(c) $\pi^S_{18}(MSp(2))_{(2)}\cong Z_2[\eta\mu b_1^2]\oplus Z_2[\eta^2 b_2^2]$

(d) $\pi_{22}^{S}(MSp(2))_{(2)} \cong Z_2[\eta\mu b_2 b_1] \oplus Z_2[\nu^2 b_2^2]$

(e) $\pi_{24}^{S}(MSp(2))_{(2)} \cong Z_2[\eta^* b_1^2] \oplus Z_2[\varepsilon(b_3 b_1 - b_2^2)] \oplus Z_2[\sigma\eta b_2^2] \oplus 3Z_{(2)}$.

(f) $\pi_{26}^{S}(HP^\infty)_{(2)} \cong Z_2[\eta\bar{\mu} b_2] \oplus Z_2[\eta\mu b_4] \oplus Z_2[\eta^2 b_6]$.

THEOREM 5.3 $\pi_{19}^{S}(HP^\infty)_{(2)} \cong Z_{32}[\zeta b_2] \oplus Z_2[\sigma b_3]$

Proof. The composition series for $\pi_{19}^{S}(HP^\infty)_{(2)}$ looks as follows.

$$Z_{16}\rho b_1 \subset F_1 \subset \pi_{19}^{S}(HP^\infty)_{(2)}, \quad \pi_{19}^{S}(HP^\infty)_{(2)}/F_1 \cong Z_2[\sigma b_3].$$

From [7] we know that in $\pi_{19}^{S}(CP^\infty)$ there is a class $[\zeta a_4]$ represented by $\zeta a_4 \in E_{8,11}^{\infty} \cong Z_4$ such that $[\zeta a_4]$ generates a Z_{64}. Also $[2\rho a_2]$ generates a Z_{16} in this Z_{64} so that $4[\zeta a_4] = (2s+1)[2\rho a_2]$. The canonical map $i:CP^\infty \longrightarrow HP^\infty$ in homology satisfies $i_*(a_{2k}) = b_k$. Immediately we see that $F_1 \cong Z_{32}$. From §7, Table E and Proposition 5.1(c) with n=1=t, p=4, q=15 and s=4 it follows that $2[\sigma b_3] = 0$.

THEOREM 5.4 $\pi_{23}^{S}(HP^\infty)_{(2)} \cong Z_8[\rho b_2] \oplus Z_2[8\sigma b_4]$.

Proof. Here the extension looks like

$$0 \longrightarrow Z_8[\rho b_2] \longrightarrow \pi_{23}^{S}(HP^\infty)_{(2)} \longrightarrow Z_2[8\sigma b_4] \longrightarrow 0.$$

Let $\pi: HP^\infty \longrightarrow HP^\infty/S^4$ be the collapsing map. Set $W = HP^\infty/S^4$. It will suffice to show that $\pi_*[\rho b_2] \in \pi_{23}^{S}(W)_{(2)}$ is not divisible by two. Such divisibility is expressed by the existence of the following diagram of cofibration sequences ($n >> 0$).

$$\begin{array}{ccccccc}
S^{8n+15} & \xrightarrow{\ \rho\ } & S^{8n} & \xrightarrow{\ i\ } & C_\rho & \xrightarrow{\ j\ } & S^{8n+16} \\
{\scriptstyle 2}\big\downarrow & & \big\downarrow{\scriptstyle b_2} & & \big\downarrow{\scriptstyle b_2'} & & \big\downarrow{\scriptstyle 2} \\
S^{8n+15} & \xrightarrow{\ \rho'\ } & \Sigma^{8n-8}W & \xrightarrow{\ i'\ } & C_{\rho'} & \xrightarrow{\ j'\ } & S^{8n+16}
\end{array} \qquad (5.5)$$

We will show that (5.5) cannot exist by means of an e-invariant computation in K-theory. We assume that the reader is familiar with [1, esp. §7].

Let ψ^k denote the k-th Adams operation. $KU^0(HP^\infty) = Z[[w]]$ where $\psi^3(w) = w^3+6w^2+9w$ and $\psi^2(w) = w^2+4w$. Since the ψ^k are ring homomorphisms we obtain

$$\psi^3(w^2) = w^6+12w^5+54w^4+108w^3+81w^2,$$

$$\psi^3(w^3) = w^9+18w^8+\ldots+1458w^4+729w^3, \qquad (5.6)$$

$$\psi^2(w^2) = w^4+8w^3+6w^2 \quad \text{and}$$

$$\psi^2(w^3) = w^6+12w^5+48w^4+64w^3.$$

Let $\beta\in\widetilde{KU}^0(S^2) \cong \widetilde{KU}^{-2}(S^0)$ be a generator. Recall that β is invertible and $\psi^k(\beta) = k\beta$. Also β^m generates $\widetilde{KU}^0(S^{2m})$. Now $\pi^*:\widetilde{KU}^4(\Sigma^{8n-8}W) \longrightarrow \widetilde{KU}^4(\Sigma^{8n-8}HP^\infty)$ is injective with image generated by $\{\beta^{4n-6}w^t;\ t\geq 2\}$. Consider the exact sequence

$$0 \longrightarrow \widetilde{KU}^4(S^{8n+16}) \cong Z \xrightarrow{\ j'^*\ } \widetilde{KU}^4(C_{\rho'}) \xrightarrow{\ i'^*\ } \widetilde{KU}^4(\Sigma^{8n-8}W) \longrightarrow 0.$$

We may choose $\xi_2,\xi_3\in KU^4(C_{\rho'})$ and then define $\xi_k = w^{k-2}\xi_2$ for $k\geq 4$ so as to satisfy $i'^*(\xi_k) = \beta^{4n-6}w^k$ for $k\geq 2$. From (5.6) we obtain the following equations in which $A = j'^*(\beta^{4n+6})$.

$$\psi^3(\xi_2) = 3^{4n-6}(\xi_6+\ldots+108\xi_3+81\xi_2)+c_3A,$$
$$\psi^2(\xi_2) = 2^{4n-6}(\xi_4+8\xi_3+16\xi_2)+c_2A, \tag{5.7}$$
$$\psi^3(\xi_3) = 3^{4n-6}(\xi_9+18\xi_8+\ldots+1458\xi_4+729\xi_3)+d_3A \quad \text{and}$$
$$\psi^2(\xi_3) = 2^{4n-6}(\xi_6+12\xi_5+48\xi_4+64\xi_3)+d_2A.$$

Now we compute $\psi^2\psi^3(\xi_k)$ and $\psi^3\psi^2(\xi_k)$ $(k=2,3)$ and equate the coefficients of A. Observe that if $k\geq 4$ then $\psi^t(\xi_k) = \psi^t(w)^{k-2}\psi^t(\xi_2)$ is a linear combination of the ξ_k since $w^2A=0$. Thus from (5.7) we obtain the following integral equations.

$$c_3(2^{4n+6}-2^{4n-2})+4d_2 3^{4n-3} = c_2(3^{4n+6}-3^{4n-2})+2^{4n-3}d_3 \quad \text{and}$$
$$d_2(3^{4n+6}-3^{4n}) = d_3(2^{4n+6}-2^{4n}). \tag{5.8}$$

Next we observe as follows that d_3 is even. There is an exact sequence

$$0 \longrightarrow \widetilde{KO}^4(S^{8n+16}) \cong Z \xrightarrow{j'^*} \widetilde{KO}^4(C_{\rho'}) \xrightarrow{i'^*} \widetilde{KO}^4(\Sigma^{8n-8}W) \longrightarrow 0.$$

Also $w^3 \in \mathrm{Image}[c:\widetilde{KO}^4(HP^\infty) \longrightarrow \widetilde{KU}^4(HP^\infty)]$. Thus we may choose $\xi_3 = c(\xi_3')$ for $\xi_3' \in \widetilde{KO}^4(C_{\rho'})$. Then $\psi^3(\xi_3') = y+d_3'j'^*(B)$ where $c(y) = 3^{4n-6}(\xi_9+18\xi_8+\ldots+729\xi_3)$ and $B\in\widetilde{KO}^4(S^{8n+16})$ is a generator. Since $c(B) = 2\beta^{4n+6}$ we obtain $d_3 = 2d_3'$ as required. Therefore (5.8) implies that 2^{4n-7} divides c_2.

Now consider (5.5) and the induced maps on $\widetilde{KU}^4$. Observe that $v = b_2^*(\beta^{4n-4}w^2) \in \widetilde{KU}^4(S^{8n})$ is a generator and that $b_2^*(\beta^{4n-4}w^k) = 0$ for $k\geq 3$. Thus we obtain the following equations.

$$b_2'^*(\xi_2) = v, \quad b_2'^*(\xi_3) = aA$$
$$b_2'^*(\xi_k) = b_2'^*(w^{k-2}\xi_2) = b_2'^*(w^{k-2})v = 0 \quad \text{for } k\geq 4 \quad \text{and} \tag{5.9}$$
$$\psi^2(v) = 2^{4n-2}v + (2c_2+2^{4n-3}a)A.$$

However the last equation of (5.9) measures the K-theory e-invariant of $\rho \in \pi_{15}^S$ which is computed in [1, §7]. In fact the calculation given there assures us that the 2-exponent of $2c_2 + 2^{4n-3}a$ is exactly $4n-7$ and not $4n-6$ as the existence of (5.5) implies.

The following general result will be used to determine $\pi_{19}^S(MSp(2))_{(2)}$ and $\pi_{23}^S(MSp(2))_{(2)}$.

THEOREM 5.10 Let $Z_{2^u}x \subset {}_2E^{\infty}_{4k+8,4r-1}$ and let $Z_{2^r}y = {}_2E^{\infty}_{4k+4,4r+3}$. Assume that $2^u x \neq 0$ in ${}_2E^{4r}_{4k+8,4r-1}$. Also assume that there is $n>2$ such that $x\, b_1^{n-2} = 0$ in ${}_nE_{4n+4k,4r-1}$ and $y\, b_1^{n-2} \neq 0$ ${}_nE^{4r}_{4n+4k-4,4r+3}$. Let $z \in {}_2E^{4r}_{4k+4r+8,0}$ and $w \in {}_nE^{4r}_{4n+4k+4r,0}$ be any two elements such that $d^{4r}(z) = 2^u x$ and $d^{4r}(w) = x$. Then $d^{4r}(2^u w - z b_1^{n-2}) = 0$ and $d^{4r+4}(2^u w - z b_1^{n-2}) = \alpha y\, b_1^{n-2}$. Moreover, there are representatives $[x]$, $[y]$ in $\pi^S_{4k+4r+7}(MSp(2))$ of x, y, respectively, such that:

$$2^u[x] \equiv \alpha[y] \text{ modulo } F_{4k}\pi^S_{4k+4r+7}(MSp(2)).$$

<u>Proof</u>. Let X_k, Y_k denote the k-skeleton of $MSp(2)$, $MSp(n)$, respectively.

$$\begin{array}{ccccc}
 & & & & \pi^S_{4k+4r+7}(MSp(2)) \\
 & & & & \Uparrow \\
 & & & & \pi^S_{4k+4r+7}(X_{4k+4r+8}) \\
 & & & & \uparrow g_1 \\
{}_2E^4_{4k+4r+8,0} & = & \pi^S_{4k+4r+8}(X_{4k+4r+8}/X_{4k+4r+4}) & \xrightarrow{\ \partial\ } & \pi^S_{4k+4r+7}(X_{4k+4r+4}) \\
 & & & & \uparrow g_2 \\
{}_2E^4_{4k+8,4r-1} & = & \pi^S_{4k+4r+7}(X_{4k+8}/X_{4k+4}) & \xleftarrow{\ f_1\ } & \pi^S_{4k+4r+7}(X_{4k+8}) \\
 & & & & \uparrow g_3 \\
{}_2E^4_{4k+4,4r+3} & = & \pi^S_{4k+4r+7}(X_{4k+4}/X_{4k}) & \xleftarrow{\ f_2\ } & \pi^S_{4k+4r+7}(X_{4k+4})
\end{array}$$

Since x, y are infinite cycles there are $\{x\}\in\pi^S_{4k+4r+7}(X_{4k+8})$ and $\{y\}\in\pi^S_{4k+4r+7}(X_{4k+4})$ such that $f_1\{x\}=x$, $f_2\{y\}=y$ and we can define $[x]\equiv g_1g_2\{x\}$, $[y]=g_1g_2g_3\{y\}$. Let $2^u[x]=\gamma[y]$ modulo $F_{4k}\pi^S_{4k+4r+7}(MSp(2))$. Then $g_1g_2(2^u\{x\}-\gamma g_3\{y\})=0$. Hence $g_2(2^u\{x\}-\gamma g_3\{y\})=\partial(A)$ where $d^{4r}(A)=f_1(2^u\{x\}-\gamma g_3\{y\})=2^u x$. Thus $A=z+z_0$ where $z_0\in {}_2E^{4r}_{4k+4r+8,0}$ is a d^{4r}-cycle. Hence

$$\partial(z+z_0) = 2^u g_2\{x\} - \gamma g_2 g_3\{y\}. \qquad (5.11)$$

$$\begin{array}{ccccc}
{}_nE^2_{4n+4k+4r,0} = \pi^S_{4n+4k+4r}(Y_{4n+4k+4r}/Y_{4n+4k+4r-4}) & \xrightarrow{\partial'} & \pi^S_{4n+4k+4r-1}(Y_{4n+4k+4r-4}) \\
 & & \uparrow g_2' \\
{}_nE^2_{4n+4k,4r-1} = \pi^S_{4n+4k+4r-1}(Y_{4n+4k}/Y_{4n+4k-4}) & \xleftarrow{f_1'} & \pi^S_{4n+4k+4r-1}(Y_{4n+4k}) \\
 & & \uparrow g_3' \\
{}_nE^2_{4n+4k-4,4r+3} = \pi^S_{4n+4k+4r-1}(Y_{4n+4k-4}/Y_{4n+4k-8}) & \xleftarrow{f_2'} & \pi^S_{4n+4k+4r-1}(Y_{4n+4k-4})
\end{array}$$

The canonical map $\Sigma^{4n-8}MSp(2) \longrightarrow MSp(n)$ induces a map from the first diagram to the second one. The induced map on E^4-terms is multiplication by b_1^{n-2}. Since xb_1^{n-2} is a d^{4r}-boundary, there is $B\in {}_nE^{4r}_{4n+4k+4r,0}$ such that $\partial'(B)=g_2'\{x\}'$ where $f_1'\{x\}'=d^{4r+4}(B)$. Thus $B=w+w_0$ where $w_0\in {}_nE^{4r}_{4n+4k+4r,0}$ with $d^{4r+4}(w_0)=0$. Hence $\partial'(w+w_0)=g_2'\{x\}'$ and

$$\partial'(2^u w + 2^u w_0) = 2^u g_2'\{x\}'. \qquad (5.12)$$

Subtracting (5.11)' from (5.12) gives

$$\partial'(2^u w+2^u w_0-zb_1^{n-2}-z_0b_1^{n-2}) = \gamma g_2' g_3'\{y\}'.$$

Hence $d^{4r+4}(2^u w - zb_1^{n-2} - z_0 b_1^{n-2}) = \gamma f_2'\{y\}' = \gamma y b_1^{n-2}$. $d^{4r+4}(z_0)$ is divisible by 2^v. Hence $d^{4r+4}(2^u w - zb_1^{n-2}) = (\gamma - 2^v c) y b_1^{n-2}$. Thus $\alpha \equiv \gamma - 2^v c \bmod 2$. Hence $\alpha[x] \equiv \gamma[x] - 2^v c[x] \equiv \gamma[x] \equiv [y]$ modulo $Z(2[y]) + F_{4k}\pi^S_{4k+4r+7}(MSp(2))$. Thus we can modify $[x]$ if necessary so that $\alpha[x] \equiv [y]$ modulo $F_{4k}\pi^S_{4k+4r+7}(MSp(2))$.

THEOREM 5.13 (a) $\pi^S_{19}(MSp(2))_{(2)} \cong Z_8[\nu b_2^2]$

(b) $\pi^S_{23}(MSp(2))_{(2)} \cong Z_{128}[2\sigma b_2^2]$

Proof. Both parts of this theorem are applications of Theorem 5.10 in which the indeterminacy is zero. For part (a) in Theorem 5.10 we take $n=3$, $u=1$, $r=1$, $v=2$, $x=\nu b_2^2$, $y=\sigma b_2 b_1$, $z=9b_3b_2-3b_4b_1$ and $w=3b_2^3$. The required differentials are given by Theorems 3.1 and 3.2(d),(g),(h). For part (b) the composition series has the following form.

$$Z_{16}[\rho b_1^2] \subset F_1 \subset \pi^S_{23}(MSp(2))_{(2)},$$
$$F_1/[\rho b_1^2] \cong Z_2[\zeta b_2 b_1],$$
$$\pi^S_{23}(MSp(2))_{(2)}/F_1 \cong Z_4[2\sigma b_2^2].$$

From Theorem 5.3 we see that $F_1 \cong Z_{32}$. Now apply Theorem 5.10 with $n>>0$, $r=2$, $u=2$, $v=5$, $y=\zeta b_2 b_1$, $x=2\sigma b_2^2$ and z, w chosen according to Theorem 3.2. Then $4w-zb_1^{n-2}$ has an odd multiple of $4b_2^4$ as a summand. Hence by Theorem 3.4 α will be odd.

THEOREM 5.14 (a) $\pi^S_{25}(HP^\infty)_{(2)} \cong Z_2[\bar{\mu} b_2] \oplus Z_2[\eta\eta^* b_2] \oplus Z_2[(\mu+s\nu^3)b_4]$

(b) There is an exact sequence

$$\begin{array}{c} 0 \\ \downarrow \\ Z_2[\bar{\mu}_1 b_1^2] \oplus Z_2[\eta\eta^* b_1^2] \oplus Z_2[\nu^3 b_2^2] \oplus Z_2[\mu b_2^2] \oplus Z_2[\eta\varepsilon b_2^2] \\ \downarrow \\ \pi_{25}^S(MSp(2))_{(2)} \\ \downarrow \\ Z_2[\eta b_3^2] \\ \downarrow \\ 0 . \end{array}$$

<u>Proof</u>. (a) We must show $2[(\mu+s\nu^3)b_4] = 0$. If $\eta[1024b_6] = [(\mu+s\nu^3)b_4]$ we are finished. Let us suppose otherwise so that $\eta\pi_{24}^S(HP^\infty)$ consists of multiples of $[\eta\eta^* b_2]$. By Theorem 5.4 we may form the Toda bracket [13] $<[8\sigma b_4],2,\eta> \subset \pi_{25}^S(HP^\infty)$ which consists of elements of the form $x + \lambda[\eta\eta^* b_2]$ with $\lambda \in Z_2$. We claim x may be taken as $[(\mu+s\nu^3)b_4]$. For consider the collapsing map $\pi: HP^\infty \longrightarrow HP^\infty/HP^3$. In the spectral sequence for $\pi_*^S(HP^\infty/HP^3)$ the differentials of §3 easily show that $\mu b_4 \in E_{16,9}^\infty$ represents a non-trivial element of order two. However

$$\begin{aligned} &\pi_*<[8b_4\sigma],2,\eta> \\ &\equiv <8\sigma,2,\eta>b_4 \quad \text{mod } \eta\pi_{24}^S(HP^\infty/HP^3) \\ &\equiv \mu b_4 \quad \text{mod } \eta\pi_{24}^S(HP^\infty/HP^3) \end{aligned}$$

since $\mu \in <8\sigma,2,\eta>$ by [13, p.189]. Finally $2[(\mu+s\nu^3)b_4]$ will be represented by $2<[8\sigma b_4],2,\eta> \subset <[8\sigma b_4],2,2\eta> \subset [8\sigma b_4]\pi_2^S(HP^\infty) = 0$.

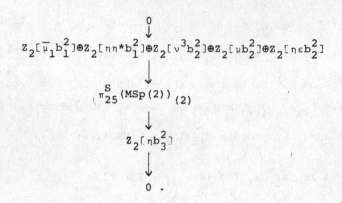

(b) The composition series looks like

$$Z_2[\bar{\mu}b_1^2]\oplus Z_2[\eta\eta^* b_1^2] \subset F_1 \subset \pi_{25}^S(MSp(2))_{(2)},$$

$$F_1/(Z_2\oplus Z_2) \cong Z_2[\nu^3 b_2^2]\oplus Z_2[\mu b_2^2]\oplus Z_2[\eta\varepsilon b_2^2],$$

$$\pi_{25}^S(MSp(2))_{(2)}/F_1 \cong Z_2[\eta b_3^2].$$

We will show $F_1 \cong 5Z_2$. Proposition 5.1(a) quickly reduces the task to showing that $2[\mu b_2^2] = 0$. Consider the following map $HP^\infty \xrightarrow{\Delta} HP^\infty \wedge HP^\infty \xrightarrow{m} MSp(2)$. Here Δ is the diagonal map and m is the MSp-multiplication. By part (a) $[(\mu+s\nu^3)b_4]\in\pi_{25}^S(HP^\infty)_{(2)}$ has order two. However $m_*\Delta_*[(\mu+s\nu^3)b_4] = [(\mu+s\nu^3)m_*\Delta_* b_4] = [(\mu+s\nu^3)b_2^2] = [\mu b_2^2] + s[\nu^3 b_2^2]$.

6. Homotopy at Odd Primes. In this section we analyze the Atiyah-Hirzebruch spectral sequences (1.1) localized at an odd prime p. In total degree less than or equal to 26 these spectral sequences are nontrivial only for p=3,5,7,11. We will analyze the case p=3 in detail. Analogous but easier arguments show that ${}_kE^{\infty}_{u,v}=0$ when $k\geq 1$, $26\geq u+v$, $v>0$ and $p\in\{5,7,11\}$. Thus for the remainder of this section p=3 and all the ${}_kE^r_{u,v}$ are localized at the prime 3. The differentials which we compute in this section determine the ${}_kE^r_{u,v}$ for $u+v\leq 26$. The ${}_1E^r_{u,v}$ are tabulated in §7, Table F. We leave the tabulation of the other ${}_kE^r_{u,v}$ to the reader. The conclusion is that ${}_2E^{\infty}_{12,7} = Z_3\alpha_2 b_2$ and ${}_kE^{\infty}_{u,v}=0$ for $k\geq 2$, $v>0$, $26\geq u+v$ and $(k,u,v)\neq(2,12,7)$.

THEOREM 6.1 For $k\geq 2$, $d^4(b_k)=(k-1)\alpha_1 b_{k-1}$.

Proof. By D. Segal [10], $3b_2$ is an infinite cycle, but b_2 is not an infinite cycle. Thus we can define α_1 so that $d^4(b_2)=\alpha_1$. We now use $s_{\Delta_{k-2}}$ to show that $d^4(b_k)=(k-1)\alpha_1 b_{k-1}$ for $k\geq 3$.

THEOREM 6.2 (a) $d^8(3b_3) = \alpha_2$

(b) $d^8(2b_3+b_2^2) = 0$

(c) For $k\geq 2$, $d^8(3b_{3k}) = \alpha_2 b_{3k-2}$

(d) For $k\geq 1$, $d^8(b_{3k+1}) = (k-1)\alpha_2 b_{3k-1}$

(e) For $k\geq 0$, $d^8(3b_{3k+2}) = 0$

(f) $d^8(3b_3 b_2) = 0$

(g) $d^8(b_2^3) = \alpha_2 b_2$

(h) $d^8(b_4 b_2 - b_5) = 2\alpha_2 b_3$

(i) $d^8(3b_3^2) = 2\alpha_2 b_3 + \alpha_2 b_2^2$

(j) For $k \geq 1$, $d^8(\alpha_1 b_{3k}) = (3k-2)\beta_1 b_{3k-2}$

(k) $d^8(\alpha_1 b_3^2) = 2\beta_1 b_3 + \beta_1 b_2^2$

Proof. By D. Segal [10] $9b_3$ is an infinite cycle, but $3b_3$ is not an infinite cycle. Thus we can define α_2 so that $d^8(3b_3) = \alpha_2$. If $d^8(2b_3+b_2^2) = \alpha_2$ then $8b_3+b_2^2$ is an infinite cycle in ${}_NE^8_{4N+8,0}$, and if $d^8(2b_3+b_2^2) = 2\alpha_2$ then $5b_3+b_2^2$ is an infinite cycle in ${}_NE^8_{4N+8,0}$. Let H denote the map of §4 with $Z_{(2)}$ replaced by $Z_{(3)}$. Then

$H(8b_3+b_2^2)$ + Image $h_U = 15A_2^2$ + Image h_U and

$H(5b_3+b_2^2)$ + Image $h_U = 21A_2^2$ + Image h_U. However, neither $15A_2^2$ nor $21A_2^2$ is in Image h_U. Thus $d^8(2b_3+b_2^2) = 0$. We prove (c) with $s_{\Delta_{3k-3}}$; (d) with $s_{\Delta_{3k-2}}$; (e) with $s_{\Delta_{3k-1}}$; (f),(g) with s_{Δ_1} and (h),(i) with $s_{\Delta_2}, s_{2\Delta_1}$. Since $\pi_*(\underline{MSp})$ has no odd torsion [6], $\beta_1 \in {}_NE^8_{4N,10}$ must be a d^8-boundary. Thus we can define β_1 so that $d^8(\alpha_1 b_3) = \beta_1$. We now use $s_{\Delta_{3k-3}}$ to prove (j) and s_{Δ_2}, $s_{2\Delta_1}$ to prove (k).

THEOREM 6.3 (a) $d^{12}(b_4) = \alpha'_3$

(b) $d^{12}(3b_3b_2) = \frac{1}{2}(h-6)\alpha'_3$

(c) $d^{12}(3b_5) = h\alpha'_3 b_2$ with $h \not\equiv 0 \bmod 3$.

(d) $d^{12}(3b_2^3) = i\alpha'_3$ with $i \not\equiv 0 \bmod 3$.

(e) $d^{12}(9b_6) = 3i\alpha'_3 b_3$

(f) $d^{16}(27b_5) = \alpha_4$

(g) $d^{16}(27b_6) = j\alpha_4 b_2$ with $j \neq 0 \mod 3$.

(h) $d^{20}(81b_6) = \alpha_5$

Proof. By D. Segal [10] $9b_4$, $81b_5$ and $243b_6$ are infinite cycles while $3b_4$, $27b_5$ and $81b_6$ do not survive to ${}_N E^{\infty}_{*,*}$. Thus we can define α_3' so that $d^{12}(b_4) = \alpha_3'$. In addition $d^{12}(3b_5)$ must be a nonzero element of order nine. $s_{\Delta_1}(3b_5) = 6b_4+2(3b_3b_2)$. Thus $d^{12}(3b_3b_2) = \frac{1}{2}(h-6)\alpha_3'$. $d^{12}(9b_6)$ must have order at least three, and $s_{\Delta_2}(9b_6) = 27b_4+18(3b_3b_2)+3(3b_2^3)$. Thus $d^{12}(3b_2^3)$ has order nine and (d),(e) are valid. $d^{16}(27b_5)$ must be nonzero, so we define α_4 as $d^{16}(27b_5)$. $d^{16}(27b_6)$ must be non-zero of order 3 which proves (g). $d^{20}(81b_6)$ must also be a nonzero element of order 3 so we define α_5 as $d^{16}(81b_6)$.

7. Tables. This section contains seven sets of tables labled A through G. Tables A, B contain known information which we use in our computations while Tables C through G contain the results of our calculations. Before proceeding to the tables we say a few words on how to interpret each of them. In all of these tables, omitted groups are zero.

Table A lists the first 22 stable homotopy groups of spheres as well as all the multiplicative relations in this range. The first 19 dimensions are taken from H. Toda [13]. We thank M. Mahowald for the information on dimensions 20, 21 and 22.

Table B_1 lists the first 23 homotopy groups of MSp. The $\phi_n \in \pi_{8n-3}(\underline{MSp})$ are defined by N. Ray [9]. The remaining notation as well as the data in Table B is taken from [3, Part III]. We also list the elements of $\pi_{k+4N}(MSp(N))_{(2)}$, in the notation of ${}_NE_{*,*}$, which correspond to the elements of $\pi_k(\underline{MSp})$. For each k the group generators in Table B_1 have been ordered so that the ith generator of $\pi_k(\underline{MSp})$ corresponds to the ith generator of $\pi_{k+4N}(MSp(N))$ for all i. In Table B_2 we give the Hurewicz images of 12 elements of $\pi_*(\underline{MSp})$. All other algebra generators of $\pi_*(MSp)$ in the first 23 dimensions map to zero under the Hurewicz homomorphism. The integers f,...,m in Table B_2 are not used elsewhere in this paper.

In Table C we list the groups ${}_NE^r_{4N+u,v}$ localized at the prime two for large N, all r and $u+v \le 26$. In each of the tables C_2 through C_5, a portion of the table of ${}_NE^r_{*,*}$ is enclosed by double lines. If the group ${}_NE^r_{u,v}$ is enclosed by double lines then ${}_NE^r_{u,v} = {}_NE^\infty_{u,v}$ and ${}_NE^s_{u,v}$ is omitted from the tabulation of ${}_NE^s_{*,*}$

for $s>r$. The letters c, e, s, t in Tables B, C and D are defined in Theorems 3.4 and 3.8.

In Table D, Table F we list the groups ${}_1E^r_{u,v}$ for all r and $u+v\le 26$ localized at the prime two, three, respectively. In all three of these tables we use the same convention as in Table C on groups ${}_1E^r_{u,v}$ enclosed by double lines. In Table E we display ${}_2E^\infty_{u,v}$ for $u+v\le 26$ localized at the prime two. For $k\ge 3$ we do not display ${}_kE^\infty_{u,v}$ localized at the prime two because ${}_kE^\infty_{4k+u,v} \cong {}_NE^\infty_{4N+u,v}$ for $N>>k\ge 3$ and $26\ge u+v$.

In Table G we list the groups $\pi^S_k(MSp(t))$ for $k\le 26$. There is one ambiguous extension in $\pi^S_{25}(MSp(2))$ which is either a direct sum of 6 copies of Z_2 or a direct sum of $Z_4[\eta b_3^2]$ with 4 copies of Z_2. There are no other ambiguities in the table. Recall that by the stable splitting of $BSp(n)$, $\pi^S_k(BSp(n)) = \bigoplus_{i=1}^{k} \pi^S_k(MSp(i))$ and $\pi^S_k(BSp) = \bigoplus_{i=1}^{\infty} \pi^S_k(MSp(i))$. Thus $\pi^S_k(BSp(n))$ and $\pi^S_k(BSp)$ for $k\le 26$ are also given by Table G. Note that all the upper blank boxes in Table G represent zero groups.

n	π_n^S	Relations
0	Z	
1	$Z_2\eta$	
2	$Z_2\eta^2$	
3	$Z_8\nu+Z_3\alpha_1$	$\eta^3=4\nu$
6	$Z_2\nu^2$	
7	$Z_{16}\sigma+Z_3\alpha_2+Z_5\alpha_{1,5}$	
8	$Z_2\eta\sigma+Z_2\varepsilon$	
9	$Z_2\nu^3+Z_2\eta\varepsilon+Z_2\mu$	$\eta^2\sigma+\eta\varepsilon=\nu^3$
10	$Z_2\eta\mu+Z_3\beta_1$	$\eta^2\varepsilon=0,\ \nu\sigma=0,\ \alpha_1\alpha_2=0$
11	$Z_8\zeta+Z_9\alpha_3'+Z_7\alpha_{1,7}$	$\eta^2\mu=4\zeta,\ \nu\varepsilon=0$
13	$Z_3\alpha_1\beta_1$	
14	$Z_2\sigma^2+Z_2\kappa$	$\nu\zeta=0$
15	$Z_{32}\rho+Z_2\eta\kappa+Z_3\alpha_4+Z_5\alpha_{2,5}$	$\eta\sigma^2=0,\quad \sigma\varepsilon=0$
16	$Z_2\eta\rho+Z_2\eta^*$	$\eta^2\kappa=0,\ \sigma\mu=\eta\rho,\ \varepsilon^2=0,\ \eta\nu\varepsilon=0$
17	$Z_2\eta\eta^*+Z_2\nu\kappa+Z_2\eta^2\rho+Z_2\bar{\mu}$	$\varepsilon\mu=\eta^2\rho=\eta\sigma\mu$
18	$Z_8\nu^*+Z_2\eta\bar{\mu}$	$\eta^2\eta^*=4\nu^*,\ \nu\rho=0,\ \sigma\zeta=0,\ \mu^2=\eta\bar{\mu}$
19	$Z_2\bar{\sigma}+Z_8\bar{\zeta}+Z_3\alpha_5+Z_{11}\alpha_{1,11}$	$\eta^2\bar{\mu}=4\bar{\zeta},\ \eta^*\nu=0,\ \nu\eta^*=0,\ \varepsilon\zeta=0,\ \eta\nu^*=0$
20	$Z_8\bar{\kappa}+Z_3\beta_1^2$	$\eta\bar{\sigma}=0,\ \eta\bar{\zeta}=0,\ \nu^2\kappa=4\bar{\kappa},\ \nu\bar{\mu}=0,\ \mu\zeta=0$
21	$Z_2\sigma^3+Z_2\eta\bar{\kappa}$	$\nu\nu^*=\sigma^3,\ \sigma\kappa=0$
22	$Z_2\nu\bar{\sigma}+Z_2\varepsilon\kappa$	$\eta^2\bar{\kappa}=\varepsilon\kappa,\ \nu\bar{\zeta}=0,\ \sigma\rho=0,\ \zeta^2=0$

Table A: Stable Homotopy of Spheres

k	0	1	2	3	4	5	6	7	8
$\pi_k(\underline{MSp})$	Z	$Z_2\phi_0$	$Z_2\phi_0^2$	0	$ZQ(0)$	$Z_2\phi_1$	$Z_2\phi_0\phi_1$	0	$Z[h_0^2V_{0,1}]+Zq_0$
$\pi^S_{k+4N}(MSp(N))_{(2)}$	$Z_{(2)}$	$Z_2\eta$	$Z_2\eta^2$	0	$Z_{(2)}$	$Z_2[\eta b_2]$	$Z_2\eta[\eta b_2]$	0	$Z_{(2)}+Z_{(2)}$

k	9	10	11	12
$\pi_k(\underline{MSp})$	$Z_2\phi_0q_0$	$Z_2\phi_1^2+Z_2\phi_0^2q_0$	0	$ZQ(1)+Z[Q(0)V_{0,1}]+ZQ(0)q_0$
$\pi^S_{k+4N}(MSp(N))_{(2)}$	$Z_2[\mu]$	$Z_2[\eta^2b_2^2]+Z_2\eta[\mu]$	0	$Z_{(2)}+Z_{(2)}+Z_{(2)}$

k	13	14	15	16
$\pi_k(\underline{MSp})$	$Z_2\phi_2+Z_2\phi_1q_0$	$Z_2\phi_0\phi_2+Z_2\phi_0\phi_1q_0$	0	$Z[h_0V_{0,1}^2]+Z[h_0^2V_{0,2}]+Zq_0^2$ $+Z[h_0^2V_{0,1}]q_0+ZR(0,1)$
$\pi^S_{k+4N}(MSP(N))_{(2)}$	$Z_2[\eta b_4+\eta b_2^3]+Z_2[\mu b_2]$	$Z_2\eta[\eta b_4+\eta b_2^3]+Z_2\mu[\mu b_2]$	0	$Z_{(2)}+Z_{(2)}+Z_{(2)}+Z_{(2)}+Z_{(2)}$

k	17	18	19
$\pi_k(\underline{MSp})$	$Z_2[\phi_0 V_{0,1}^2]+Z_2\phi_0 q_0^2$ $+Z_2\phi_0 R(0,1)$	$Z_2\phi_1\phi_2+Z_2\phi_0[\phi_0 V_{0,1}^2]$ $+Z_2\phi_1^2 q_0+Z_2\phi_0^2 q_0^2$	0
$\pi_{k+4N}^S(MSp(N))_{(2)}$	$Z_2[\eta b_3^2]+Z_2\bar{\mu}+Z_2[\mu b_2^2]$	$Z_2\eta[\eta b_4 b_2+\eta b_2^4]+Z_2\eta[\eta b_3^2]$ $+Z_2\eta[\mu b_2^2]+Z_2\eta\bar{\mu}$	0

k	20	21	22	23
$\pi_k(\underline{MSp})$	$Z[h_0^2 V_{1,2}]+Z[Q(0)V_{0,1}^2]$ $+Z[Q(1)V_{0,1}]+Z[Q(0)V_{0,2}]$ $+Z[Q(0)V_{0,1}]q_0+ZQ(1)q_0$ $+ZQ(0)q_0^2$	$Z_2\phi_3+Z_2\phi_1 R(0,1)$ $+Z_2\phi_1 q_0^2+Z_2\phi_2 q_0$	$Z_2\phi_0\phi_3+Z_2\phi_0\phi_1 R(0,1)$ $+Z_2\phi_0\phi_1 q_0^2+Z_2\phi_0\phi_2 q_0$ $+Z_2\phi_1[\phi_0 V_{0,1}^2]$	0
$\pi_{k+4N}^S(MSp(N))_{(2)}$	$Z_{(2)}+Z_{(2)}+Z_{(2)}+Z_{(2)}$ $+Z_{(2)}+Z_{(2)}+Z_{(2)}$	$Z_2[\eta b_6+\eta b_3^2 b_2]$ $+Z_2[\mu b_2^2+t\nu^3 b_4]+Z_2[\bar{\mu}b_2]$ $+Z_2[\mu b_4+s\nu^3 b_4]$	$Z_2\eta[\eta b_6+\eta b_3^2 b_2]$ $+Z_2[\eta\mu b_2^2]+Z_2\eta[\bar{\mu}b_2]$ $+Z_2[\eta\mu b_4]+Z_2[\eta^2 b_3^2 b_2]$	0

Table B_1: Homotopy of $\underline{MSp}$

X	h(X)
$Q(0)$	$8B_1$
q_0	$16B_1^2$
$[h_0^2 v_{0,1}]$	$4B_2-8B_1^2$
$Q(1)$	$8B_3-8(5+8c)B_1^3$
$[Q(0)v_{0,1}]$	$8B_2B_1-16(5+2e)B_1^3$
$[h_0 v_{0,1}^2]$	$2B_2^2+4f$
$[h_0^2 v_{0,2}]$	$4B_4+4B_2B_1^2+8g$
$R(0,1)$	$16B_3B_1-16B_1^4+32h$
$[h_0^2 v_{1,2}]$	$4B_5+4B_4B_1+4B_3B_2+4B_2B_1^3+8i$
$[Q(0)v_{0,1}^2]$	$8B_2^2B_1+16j$
$[Q(1)v_{0,1}]$	$8B_3B_2+8B_2B_1^3+16k$
$[Q(0)v_{0,2}]$	$8B_4B_1+8B_2B_1^3+16m$

Table B_2: Image of the Hurewicz

Homomorphism $h: \pi_*(\underline{MSp}) \longrightarrow H_*(\underline{MSp}; Z_{(2)})$

22	$Z_2\varepsilon\kappa$				
21	$Z_2\eta\bar{\kappa}$				
20	$Z_4\bar{\kappa}$				
19	$Z_2\bar{\sigma}+Z_8\bar{\zeta}$				
18	$Z_8\nu^*+Z_2\eta\bar{\mu}$	$Z_4(2\nu^*b_2)+Z_2\eta\bar{\mu}b_2$			
17	$Z_2\eta\eta^*+Z_2\eta^2\rho+Z_2\bar{\mu}$	$Z_2\eta\eta^*b_2+Z_2\eta^2\rho b_2+Z_2\bar{\mu}b_2$			
16	$Z_2\eta\rho+Z_2\eta^*$	$Z_2\eta\rho b_2+Z_2\eta^*b_2$			
15	$Z_{32}\rho+Z_2\eta\kappa$	$Z_{32}\rho b_2+Z_2\eta\kappa b_2$			
14	$Z_2\sigma^2+Z_2\kappa$	$Z_2\sigma^2b_2$	$Z_2\sigma^2b_3+Z_2\kappa b_3-Z_2\sigma^2b_2^2+Z_2\kappa b_2^2$		
11	$Z_8\zeta$	$Z_8\zeta b_2$	$Z_8\zeta b_3+Z_8\zeta b_2^2$		
10	$Z_2\eta\mu$	$Z_2\eta\mu b_2$	$Z_2\eta\mu b_3+Z_2\eta\mu b_2^2$	$Z_2\eta\mu b_4+Z_2\eta\mu b_3b_2+Z_2\eta\mu b_2^3$	
9	$Z_2\eta\varepsilon+Z_2\mu$	$Z_2\nu^3b_2+Z_2\eta\varepsilon b_2+Z_2\mu b_2$	$Z_2\eta\varepsilon b_3+Z_2\mu b_3+Z_2\eta\varepsilon b_2^2+Z_2\mu b_2^2$	$Z_2\nu^3b_4+Z_2\eta\varepsilon b_4+Z_2\mu b_4+Z_2\eta\varepsilon b_3b_2$ $+Z_2\mu b_3b_2+Z_2\nu^3b_2^3+Z_2\eta\varepsilon b_2^3+Z_2\mu b_2^3$	
8	$Z_2\eta\sigma+Z_2\varepsilon$	$Z_2\eta\sigma b_2+Z_2\varepsilon b_2$	$Z_2\eta\sigma b_3+Z_2\varepsilon b_3+Z_2\eta\sigma b_2^2+Z_2\varepsilon b_2^2$	$Z_2\eta\sigma b_4+Z_2\varepsilon b_4+Z_2\eta\sigma b_3b_2$ $+Z_2\varepsilon b_3b_2+Z_2\eta\sigma b_2^3+Z_2\varepsilon b_2^3$	
7	$Z_{16}\sigma$	$Z_{16}\sigma b_2$	$Z_{16}\sigma b_3+Z_{16}\sigma b_2^2$	$Z_{16}\sigma b_4+Z_{16}\sigma b_3b_2+Z_{16}\sigma b_2^3$	$Z_{16}\sigma b_5+Z_{16}\sigma b_3^2+Z_{16}\sigma b_4b_2+Z_{16}\sigma b_3b_2^2$ $+Z_{16}\sigma b_2^4$

3	$Z_2\nu(b_4b_3+b_3^2b_2)$
2	$Z_2\eta^2b_6+Z_2\eta^2b_5b_2+Z_2\eta^2b_4b_3+Z_2\eta^2b_3^2b_2+Z_2\eta^2b_4b_2^2+Z_2\eta^2b_3b_2^3+Z_2\eta^2b_2^5$
1	$Z_2\eta b_6+Z_2\eta b_5b_2+Z_2\eta b_4b_3+Z_2\eta b_3^2b_2+Z_2\eta b_4b_2^2+Z_2\eta b_3b_2^3+Z_2\eta b_2^5$
0	$Z_{(2)}(b_6-5b_5b_2+2b_4b_3-6b_3^2b_2)+Z_{(2)}(8b_5b_2)+Z_{(2)}(8b_4b_2^2)$ $+Z_{(2)}(b_4b_3-b_4b_2^2-3b_3^2b_2+5b_3b_2^3-2b_2^5)+Z_{(2)}(4b_3^2b_2+4b_4b_3)$ $+Z_{(2)}(8b_3b_2^3)+Z_{(2)}(8b_2^5)$
	4N+20

3	0	0	0	$Z_2(2\nu b_4)$	0
2	$Z_2\eta^2$	$Z_2\eta^2 b_2$	$Z_2\eta^2 b_3+Z_2\eta^2 b_2^2$	$Z_2\eta^2 b_4+Z_2\eta^2 b_3 b_2+Z_2\eta^2 b_2^3$	$Z_2\eta^2 b_5+Z_2\eta^2 b_3^2+Z_2\eta^2 b_4 b_2+Z_2\eta^2 b_3 b_2^2$ $+Z_2\eta^2 b_2^4$
1	$Z_2\eta$	$Z_2\eta b_2$	$Z_2\eta b_3+Z_2\eta b_2^2$	$Z_2\eta b_4+Z_2\eta b_3 b_2+Z_2\eta b_2^3$	$Z_2\eta b_5+Z_2\eta b_3^2+Z_2\eta b_4 b_2+Z_2\eta b_3 b_2^2$ $+Z_2\eta b_2^4$
0	$Z_{(2)}$	$Z_{(2)}(8b_2)$	$Z_{(2)}(b_3-b_2^2)+Z_{(2)}(4b_3)$	$Z_{(2)}(8b_4)+Z_{(2)}(8b_2^3)$ $+Z_{(2)}(3b_3b_2-2b_2^3-b_4)$	$Z_{(2)}(2b_5)+Z_{(2)}(2b_3^2)+Z_{(2)}(2b_2^4)$ $+Z_{(2)}(4b_4b_2-b_3^2-b_5)+Z_{(2)}(2b_3b_2^2-b_3^2-b_2^4)$
	4N	4N+4	4N+8	4N+12	4N+16

Table C_1: ${}_N E^8_{u,v}$, N large

22	0		
21	0		
20	$Z_4\bar{\kappa}$		
19	$Z_2\bar{\sigma}+Z_8\bar{\zeta}$		
18	$Z_4\nu^*+Z_2\eta\bar{\mu}$	$Z_2(2\nu^*b_2)+Z_2\eta\bar{\mu}b_2$	
17	$Z_2\eta\eta^*+Z_2\bar{\mu}$	$Z_2\bar{\mu}b_2+Z_2\eta\eta^*b_2$	
16	$Z_2\eta^*$	$Z_2\eta^*b_2$	
15	$Z_{32}\rho$	$Z_{32}\rho b_2$	
14	$Z_2\kappa$	0	$Z_2(\kappa b_3+\kappa b_2^2)$
11	$Z_8\zeta$	$Z_8\zeta b_2$	$Z_4(2\zeta b_3)+Z_8\zeta b_2^2$

10	$Z_2\eta\mu$	$Z_2\eta\mu b_2$	$Z_2\eta\mu b_2^2$
9	$Z_2\mu$	$Z_2\mu b_2$	$Z_2\mu b_2^2$
2	$Z_2\eta^2$	$Z_2\eta^2 b_2$	$Z_2\eta^2 b_2^2$
1	$Z_2\eta$	$Z_2\eta b_2$	0
0	$Z_{(2)}$	$Z_{(2)}(8b_2)$	$Z_{(2)}(8b_3)+Z_{(2)}(4b_3-8b_2^2)$
	4N	4N+4	4N+8

10	$Z_2\eta\mu b_4+Z_2\eta\mu b_2^3$	
9	$Z_2\mu b_4+Z_2\mu b_2^3+Z_2\nu^3 b_4$	
8	$Z_2\eta\sigma b_4$	
7	$Z_4(2\sigma b_4)$	
6	0	0
3	$Z_2(2\nu b_4)$	0
2	$Z_2\eta^2(b_4+b_2^3)$	$Z_2\eta^2 b_3^2+Z_2\eta^2 b_2^4+Z_2\eta b_4 b_2$
1	$Z_2\eta(b_4+b_2^3)$	$Z_2\eta b_3^2+Z_2\eta b_2^4$
0	$Z_{(2)}(16b_2^3)+Z_{(2)}(8b_3b_2)$ $+Z_{(2)}(8b_4-8b_2^3)$	$Z_{(2)}(16b_5)+Z_{(2)}(8b_2^4)+Z_{(2)}(8b_4b_2-2b_3^2)$ $+Z_{(2)}(8b_2^4-12b_3b_2^2+6b_3^2-12b_5)$ $+Z_{(2)}(6b_3^2-8b_3b_2^2+4b_2^4)$
	4N+16	

3	$Z_2\nu(b_4b_3+b_3^2b_2)$
2	$Z_2\eta^2b_6+Z_2\eta^2b_3^2b_2+Z_2\eta^2b_2^5$
1	$Z_2\eta b_6+Z_2\eta b_3^2b_2+Z_2\eta b_2^5$
0	$Z_{(2)}+Z_{(2)}+Z_{(2)}+Z_{(2)}+Z_{(2)}+Z_{(2)}+Z_{(2)}$
	4N+20

Table C_2: ${}_N E^{12}_{u,v}$, N large

22	0	
20	$Z_2\overline{\kappa}$	
19	$Z_8\overline{\zeta}$	
18	$Z_2\eta\overline{\mu}$	$Z_2\eta\overline{\mu}b_2$
17	$Z_2\eta\eta^*+Z_2\overline{\mu}$	$Z_2\eta\eta^*b_2+Z_2\overline{\mu}b_2$
16	$Z_2\eta^*$	$Z_2\eta^*b_2$
15	$Z_{32}\rho$	$Z_{32}\rho b_2$

6	0	
3	0	0
2	$Z_2\eta^2b_3^2+Z_2\eta^2b_2^4+Z_2\eta^2b_4b_2$	$Z_2\eta^2b_6+Z_2\eta^2b_3^2b_2+Z_2\eta^2b_2^5$
1	$Z_2\eta b_3^2+Z_2\eta b_2^4$	$Z_2\eta b_6+Z_2\eta b_3^2b_2+Z_2\eta b_2^5$
0	$Z_{(2)}+Z_{(2)}+Z_{(2)}+Z_{(2)}+Z_{(2)}$	$Z_{(2)}+Z_{(2)}+Z_{(2)}+Z_{(2)}+Z_{(2)}+Z_{(2)}$ $+Z_{(2)}$
	4N+16	4N+20

14	0	0	0
11	0	0	0
	4N	4N+4	4N+8

10	$Z_2\eta\mu b_4+Z_2\eta\mu b_2^3$
9	$Z_2(\mu+s\nu^3)b_4+Z_2(\mu b_2^3+t\nu^3b_4)$
2	$Z_2\eta^2(b_4+b_2^3)$
1	$Z_2\eta(b_4+b_2^3)$
0	$Z_{(2)}(128b_2^3)+Z_{(2)}[8b_3b_2-16(5+2e)b_2^3]+Z_{(2)}[8b_4-8(5+8c)b_2^3]$
	4N+12

Table C_3: ${}_NE^{16}_{u,v}$, N large

22	0
20	$Z_2\bar{\kappa}$
19	$Z_8\bar{\zeta}$

18	$Z_2\eta\bar{\mu}$	$Z_2\eta\bar{\mu}b_2$
17	$Z_2\bar{\mu}$	$Z_2\bar{\mu}b_2$
15	0	0
	4N	4N+4

6	0
2	$Z_2\eta^2b_3^2+Z_2\eta^2(b_4b_2+b_2^4)$
1	$Z_2\eta b_3^2$
0	$Z_{(2)}+Z_{(2)}+Z_{(2)}+Z_{(2)}+Z_{(2)}$
	4N+16

3	0
2	$Z_2\eta^2b_6+Z_2\eta^2b_3^2b_2$
1	$Z_2\eta b_6+Z_2\eta b_3^2b_2$
0	$Z_{(2)}+Z_{(2)}+Z_{(2)}+Z_{(2)}+Z_{(2)}+Z_{(2)}+Z_{(2)}$
	4N+20

Table C_4: ${}_NE^{20}_{u,v}$, N large

22	0		
19	0		
	4N	3	0
		2	$Z_2\eta^2 b_6+Z_2\eta^2 b_3^2 b_2$
		1	$Z_2\eta(b_6+b_3^2 b_2)$
		0	$Z_{(2)}+Z_{(2)}+Z_{(2)}+Z_{(2)}+Z_{(2)}+Z_{(2)}+Z_{(2)}$
			4N+20

Table C_5: ${}_N E^{\infty}_{u,v}$, N large

22	$Z_2\varepsilon\kappa$					
21	$Z_2\eta\bar{\kappa}$					
20	$Z_4\bar{\kappa}$					
19	$Z_2\bar{\sigma}+Z_8\bar{\zeta}$					
18	$Z_8\nu^*+Z_2\eta\bar{\mu}$	$Z_4(2\nu^*b_2)+Z_2\eta\bar{\mu}b_2$				
17	$Z_2\eta\eta^*+Z_2\eta^2\rho+Z_2\bar{\mu}$	$Z_2\eta\eta^*b_2+Z_2\bar{\mu}b_2+Z_2\eta^2\rho b_2$				
16	$Z_2\eta\rho+Z_2\eta^*$	$Z_2\eta\rho b_2+Z_2\eta^*b_2$				
15	$Z_{32}\rho+Z_2\eta\kappa$	$Z_{32}\rho b_2+Z_2\eta\kappa b_2$				
14	$Z_2\sigma^2+Z_2\kappa$	$Z_2\sigma^2b_2$	$Z_2\kappa b_3+Z_2\sigma^2b_3$			
11	$Z_8\zeta$	$Z_8\zeta b_2$	$Z_8\zeta b_3$			
10	$Z_2\eta\mu$	$Z_2\eta\mu b_2$	$Z_2\eta\mu b_3$	$Z_2\eta\mu b_4$		
9	$Z_2\eta\varepsilon+Z_2\mu$	$Z_2\nu^3b_2+Z_2\mu b_2+Z_2\eta\varepsilon b_2$	$Z_2\mu b_3+Z_2\eta\varepsilon b_3$	$Z_2\nu^3b_4+Z_2\mu b_4+Z_2\eta\varepsilon b_4$		
8	$Z_2\eta\sigma+Z_2\varepsilon$	$Z_2\eta\sigma b_2+Z_2\varepsilon b_2$	$Z_2\eta\sigma b_3+Z_2\varepsilon b_3$	$Z_2\eta\sigma b_4+Z_2\varepsilon b_4$		
7	$Z_{16}\sigma$	$Z_{16}\sigma b_2$	$Z_{16}\sigma b_3$	$Z_{16}\sigma b_4$	$Z_{16}\sigma b_5$	
3	0	0	0	$Z_2(2\nu b_4)$	0	0
2	$Z_2\eta^2$	$Z_2\eta^2b_2$	$Z_2\eta^2b_3$	$Z_2\eta^2b_4$	$Z_2\eta^2b_5$	$Z_2\eta^2b_6$
1	$Z_2\eta$	$Z_2\eta b_2$	$Z_2\eta b_3$	$Z_2\eta b_4$	$Z_2\eta b_5$	$Z_2\eta b_6$
0	$Z_{(2)}$	$Z_{(2)}(8b_2)$	$Z_{(2)}(4b_3)$	$Z_{(2)}(8b_4)$	$Z_{(2)}(2b_5)$	$Z_{(2)}(8b_6)$
	4	8	12	16	20	24

Table D_1: ${}_1E^8_{u,v}$

v \ u	4	8	12	16	20	24
22	0					
20	$Z_4\bar{\kappa}$					
19	$Z_2\bar{\sigma}+Z_8\bar{\zeta}$					
18	$Z_4\nu^*+Z_2\eta\bar{\mu}$	$Z_2(2\nu^*b_2)+Z_2\eta\bar{\mu}b_2$				
17	$Z_2\bar{\mu}+Z_2\eta\eta^*$	$Z_2\eta\eta^*b_2+Z_2\bar{\mu}b_2$				
16	$Z_2\eta^*$	$Z_2\eta^*b_2$				
15	$Z_{32}\rho$	$Z_{32}\rho b_2$				
14	$Z_2\sigma^2+Z_2\kappa$	$Z_2\kappa b_2$	0			
11	$Z_8\zeta$	$Z_8\zeta b_2$	$Z_4(2\zeta b_3)$			
10	$Z_2\eta\mu$	$Z_2\eta\mu b_2$	0	$Z_2\eta\mu b_4$		
9	$Z_2\mu$	$Z_2\eta\varepsilon b_2+Z_2\mu b_2$	0	$Z_2\mu b_4+Z_2\nu^3b_4$		
8	$Z_2\eta\sigma$	$Z_2\eta\sigma b_2$	0	$Z_2\eta\sigma b_4$		
7	$Z_8\sigma$	$Z_8\sigma b_2$	$Z_2\sigma b_3$	$Z_8(2\sigma b_4)$		
6				0	0	0
3				$Z_2(2\nu b_4)$	0	0
2	$Z_2\eta^2$	$Z_2\eta^2b_2$	0	0	0	$Z_2\eta^2b_6$
1	$Z_2\eta$	$Z_2\eta b_2$	0	0	0	$Z_2\eta b_6$
0	$Z_{(2)}$	$Z_{(2)}(8b_2)$	$Z_{(2)}(8b_3)$	$Z_{(2)}(16b_4)$	$Z_{(2)}(16b_5)$	$Z_{(2)}(8b_6)$

Table D_2: ${}_1E^{12}_{u,v}$

22	0	
20	$Z_2\bar{\kappa}$	
19	$Z_8\bar{\zeta}$	
18	$Z_2\eta\bar{\mu}$	$Z_2\eta\bar{\mu}b_2$
17	$Z_2\bar{\mu}+Z_2\eta\eta^*$	$Z_2\eta\eta^*b_2+Z_2\bar{\mu}b_2$
16	$Z_2\eta^*$	$Z_2\eta^*b_2$
15	$Z_{32}\rho$	$Z_{32}\rho b_2$

14	$Z_2\sigma^2$	$Z_2\kappa b_2$	0
11	0	$Z_2\zeta b_2$	0
	4	8	12

10	$Z_2\eta\mu b_4$
9	$Z_2(\mu+s\nu^3)b_4$
7	$Z_2(8\sigma b_4)$
0	$Z_{(2)}(128b_4)$
	16

6	0	0
2	0	$Z_2\eta^2b_6$
1	0	$Z_2\eta b_6$
0	$Z_{(2)}(64b_5)$	$Z_{(2)}(32b_6)$
	20	24

Table D_3: ${}_1E^{16}_{u,v}$

22	0
20	$Z_2\bar{\kappa}$
19	$Z_8\bar{\zeta}$

18	$Z_2\eta\bar{\mu}$	$Z_2\eta\bar{\mu}b_2$
17	$Z_2\bar{\mu}+Z_2\eta\eta^*$	$Z_2\bar{\mu}b_2+Z_2\eta\eta^*b_2$
16	$Z_2\eta^*$	$Z_2\eta^*b_2$
15	$Z_{16}\rho$	$Z_8\rho b_2$
	4	8

6	0
0	$Z_{(2)}(128b_5)$
	20

6	0
2	$Z_2\eta^2b_6$
1	$Z_2\eta b_6$
0	$Z_{(2)}(128b_6)$
	24

Table D_4: ${}_1E^{20}_{u,v}$

22	0
19	0
	4

6	0
2	$Z_2\eta^2b_6$
1	0
0	$Z_{(2)}(1024b_6)$
	24

[illegible] $^{\infty}$

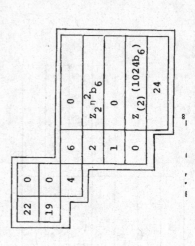

18	$Z_2\eta\bar{\mu}$				
17	$Z_2\bar{\mu}+Z_2\eta\eta^*$				
16	$Z_2\eta^*$				
15	$Z_{16}\rho$				
14	0	0			
11	0	$Z_2\zeta b_2$			
10	$Z_2\eta\mu$	$Z_2\eta\mu b_2$	$Z_2\eta\mu b_2^2$		
9	$Z_2\mu$	$Z_2\mu b_2$	$Z_2\mu b_2^2+Z_2\nu^3b_2^2+Z_2\eta\varepsilon b_2^2$		
8	0	0	$Z_2\varepsilon(b_3-b_2^2)+Z_2\eta\sigma b_2^2$		
7	0	$Z_4\sigma b_2$	$Z_4(2\sigma b_2^2)$		
6	0	0	$Z_2\nu^2b_2^2$		
3	0	0	$Z_2\nu b_2^2$	0	
2	$Z_2\eta^2$	$Z\eta^2b_2$	$Z_2\eta^2b_2^2$	0	$Z_2\eta^2b_3^2$
1	$Z_2\eta$	$Z_2\eta b_2$	0	0	$Z_2\eta b_3$
0	$Z_{(2)}$	$Z_{(2)}(8b_2)$	$Z_{(2)}(8b_3)+Z_{(2)}(4b_3-8b_2^2)$	$Z(128b_4)+$ $+Z_{(2)}[8b_3b_2-16(2e+1)b_4]$	$Z_{(2)}+Z_{(2)}+Z_{(2)}$
	8	12	16	20	24

Table E: ${}_2E^{\infty}_{u,v}$

20	$Z_3\beta_1^2$					
19	$Z_3\alpha_5$					
15	$Z_3\alpha_4$	$Z_3\alpha_4 b_2$				
13	0	0	$Z_3\alpha_1\beta_1 b_3$			
11	$Z_9\alpha_3'$	$Z_9\alpha_3' b_2$	$Z_9\alpha_3' b_3$			
10	$Z_3\beta_1$	0	0	$Z_3\beta_1 b_4$		
7	$Z_3\alpha_2$	$Z_3\alpha_2 b_2$	$Z_3\alpha_2 b_3$	$Z_3\alpha_2 b_4$		
3	0	0	$Z_3\alpha_1 b_3$	0	0	$Z_3\alpha_1 b_6$
0	$Z_{(3)} b_1$	$Z_{(3)}(3b_2)$	$Z_{(3)}(3b_3)$	$Z_{(3)} b_4$	$Z_{(3)}(3b_5)$	$Z_{(3)}(3b_6)$
	4	8	12	16	20	24

Table F_1: ${}_1E^8_{u,v}$ localized at 3

20	0		
19	$Z_3\alpha_5$		
15	$Z_3\alpha_4$	$Z_3\alpha_4 b_2$	
11	$Z_9\alpha_3'$	$Z_9\alpha_3' b_2$	$Z_9\alpha_3' b_3$

10	0	0	0
7	0	$Z_3\alpha_2 b_2$	$Z_3\alpha_2 b_3$
0	$Z_{(3)}$	$Z_{(3)}(3b_2)$	$Z_{(3)}(9b_3)$
	4	8	12

10	0		
3	0	0	0
0	$Z_{(3)}(3b_4)$	$Z_{(3)}(3b_5)$	$Z_{(3)}(9b_6)$
	16	20	24

Table F_2: ${}_1E^{12}_{u,v}$ localized at 3

20	0	
19	$Z_3\alpha_5$	
15	$Z_3\alpha_4$	$Z_3\alpha_4b_2$

11	0	0	$Z_3\alpha_3'b_3$
	4	8	12

10	0
0	$Z_{(3)}(9b_4)$
	16

3	0	0
0	$Z_{(3)}(27b_5)$	$Z_{(3)}(27b_6)$
	20	24

Table F_3: ${}_1E^{16}_{u,v}$ localized at 3

20	0	
15	0	0
	4	8

3	0	0
0	$Z_{(3)}(81b_5)$	$Z_{(3)}(243b_6)$
	20	24

Table F_4: ${}_1E^{\infty}_{u,v}$ localized at 3

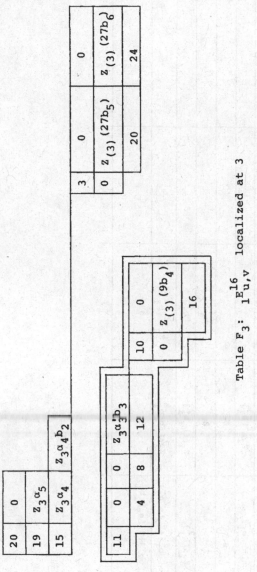

k	$\pi_k^S(HP^\infty)$	$\pi_k^S(MSp(2))$	$\pi_k^S(MSp(3))$	$\pi_k^S(MSp(4))$	$\pi_k^S(MSp(5))$	$\pi_k^S(MSp(6))$
0	0					
4	$Z[b_1]$					
5	$Z_2\eta[b_1]$					
6	$Z_2\eta^2[b_1]$					
8	Z	$Z[b_1^2]$				
9	$Z_2[\eta b_2]$	$Z_2\eta[b_1^2]$				
10	$Z_2\eta[\eta b_2]$	$Z_2\eta^2[b_1^2]$				
11	$Z_8\sigma[b_1]$	0				
12	$Z_2\eta\sigma[b_1]+Z$	Z	$Z[b_1^3]$			
13	$Z_2\mu[b_1]$	$Z_2[\eta b_2b_1]$	$Z_2\eta[b_1^3]$			
14	$Z_2\eta\mu[b_1]$	$Z_2\eta[\eta b_2b_1]$	$Z_2\eta^2[b_1^3]$			
15	$Z_8[\sigma b_2]+Z_3[\alpha_2b_2]$	0	0			
16	$Z_2\eta[\sigma b_2]+Z$	$Z+Z$	Z	$Z[b_1^4]$		
17	$Z_2[\mu b_2]+Z_2\varepsilon[\eta b_2]$	$Z_2\mu[b_1^2]$	$Z_2[\eta b_2b_1^2]$	$Z_2\eta[b_1^4]$		
18	$Z_2\sigma^2[b_1]+Z_2\eta[\mu b_2]$	$Z_2\eta[\mu b_1^2]+Z_2[\eta^2b_2^2]$	$Z_2\eta[\eta b_2b_1^2]$	$Z_2\eta^2[b_1^4]$		
19	$Z_{32}[\zeta b_2]+Z_2[\sigma b_3]$ $+Z_2[\alpha_2b_3]$	$Z_8[\nu b_2^2]+Z_3[\alpha_2b_2b_1]$	0	0		

20	$Z_2\eta^*[b_1]+Z$	Z+Z	Z+Z	Z	$Z[b_1^5]$	
21	$Z_2\bar{\mu}[b_1]+Z_2\eta\eta^*[b_1]$	$Z_2[\mu b_2 b_1]$	$Z_2\mu[b_1^3]$	$Z_2[\eta b_2 b_1^3]$	$Z_2\eta[b_1^5]$	
22	$Z_2\eta\bar{\mu}[b_1]+Z_2[\kappa b_2]$	$Z_2\eta[\mu b_2 b_1]+Z_2\nu[\nu b_2^2]$	$Z_2\eta\mu[b_1^3]+Z_2[\eta^2 b_2^2 b_1]$	$Z_2\eta[\eta b_2 b_1^3]$	$Z_2\eta^2[b_1^5]$	
23	$Z_8[\rho b_2]+Z_2[8\sigma b_4]$ $+Z_3[\alpha_3' b_3]$	$Z_{128}[2\sigma b_2^2]$	0	0	0	
24	$Z_2[\eta^* b_2]+Z$	$Z_2\eta^*[b_1^2]+Z_2[\eta\sigma b_2^2]$ $+Z_2[\varepsilon b_3 b_1-\varepsilon b_2^2]$ $+Z+Z+Z$	Z+Z+Z	Z+Z	Z	$Z[b_1^6]$
25	$Z_2[\bar{\mu} b_2]+Z_2\eta[\eta^* b_2]$ $+Z_2[(\mu+s\nu^3)b_4]$	$\left[Z_2[\eta b_3^2]?(Z_2\eta\eta^*[b_1^2]\right.$ $+Z_2\eta[\varepsilon b_3 b_1-\varepsilon b_2^2]$ $\left.+Z_2[\nu^3 b_2^2])\right]+$ $Z_2[\mu b_2^2]+Z_2\bar{\mu}[b_1^2]$	$Z_2[\mu b_2 b_1^2]$ $+Z_2[\eta b_4 b_1^2+\eta b_2^3]$	$Z_2\mu[b_1^4]$	$Z_2[\eta b_2 b_1^4]$	$Z_2[\eta b_1^6]$
26	$Z_2\eta[\bar{\mu} b_2]+Z_2[\eta^2 b_6]$ $+Z_2\eta[(\mu+s\nu^3)b_4]$	$Z_2\eta[\mu b_2^2]+Z_2\eta[\bar{\mu} b_1^2]$ $+Z_2\eta[\eta b_3^2]$	$Z_2\eta[\mu b_2 b_1^2]$ $+Z_2\eta[\eta b_4 b_1^2-\eta b_2^3]$	$Z_2\eta\mu[b_1^4]$ $+Z_2[\eta^2 b_2^2 b_1^2]$	$Z_2\eta[\eta b_2 b_1^4]$	$Z_2\eta^2[b_1^6]$

Table G: Summary of Results

Bibliography

1. J. F. Adams, "On the Groups J(X)-IV," Topology 5(1966), 21-27.

2. J. F. Adams, "Stable Homotopy and Generalised Homology", The U. of Chicago Press, Chicago, Ill. 1974.

3. S. O. Kochman, "The Symplectic Cobordism Ring", Memoirs A.M.S. (to appear).

4. P. S. Landweber, "Cobordism Operations and Hopf Algebras", Trans. A.M.S. 129(1967), 94-110.

5. K. Li, Thesis, The U. of Western Ontario (to appear).

6. J. Milnor, "On the Cobordism Ring Ω_* and a Complex Analogue", Amer. J. Math. 82(1960), 505-521.

7. R. E. Mosher, "Some Stable Homotopy of Complex Projective Space," Topology 7(1968), 179-193.

8. S. P. Novikov, "The Methods of Algebraic Topology from the Viewpoint of Cobordism Theories", Izv. Akad. Nauk. S.S.S.R., Seriia Mat. 31(1967), 855-951 (translation in Math. U.S.S.R.-Izvestiia 1 (1967), 827-913).

9. N. Ray, "Indecomposables in Tors MSp_*", Topology 10(1971), 261-270.

10. D. Segal, "On the Stable Homotopy of Quaternionic and Complex Projective Spaces", Proc. A.M.S. 25(1970), 838-841.

11. V. P. Snaith, "Towards Algebraic Cobordism", Bull. A.M.S. 83(1977), 384-385.

12. V. P. Snaith, "Algebraic Cobordism and K-Theory", Memoirs A.M.S. (to appear).

13. H. Toda, "Composition Methods in Homotopy Groups of Spheres", Annals of Math. Studies No. 49, Princeton U. Press, Princeton, N.J., 1962.

The University of Western Ontario

NEW APPLICATIONS OF COMMUTATIVE ALGEBRA TO BROWN-PETERSON HOMOLOGY

Peter S. Landweber[1]
Rutgers University
New Brunswick, N.J. 08903

1. <u>Introduction and results</u>. My purpose here is to prove the main results of Johnson and Yosimura's paper "Torsion in Brown-Peterson Homology and Hurewicz Homomorphisms" [3] in a more algebraic and conceptual manner, and at the same time to prove several new results about BP_*BP-comodules.

Earlier applications of commutative algebra to complex bordism MU and to BP [6,7,9] were based on the notion of the set Ass(M) of the <u>associated</u> prime ideals of a module M, the prime annihilator ideals of elements of M, and the convenience that for coherent (= finitely presented) modules over BP_* one can carry over techniques from the Noetherian case, especially primary decomposition. In this paper we make use of the set $Ass_w(M)$ of <u>weakly associated</u> prime ideals of a module M [1], the prime ideals minimal among prime ideals containing Ann(x) for some $x \in M$. For BP_*BP-comodules, <u>these are precisely the radicals of annihilator ideals</u>, and are shown to be <u>invariant ideals</u> in BP_*.

Let $\mathcal{BC}$ denote the category of (associative) BP_*BP-comodules [3,9] and $\mathcal{BC}_0$ the subcategory of comodules which are coherent as BP_*-modules. For a finite complex X, $BP_*(X) \in \mathcal{BC}_0$, and for any CW-spectrum X, $BP_*(X)$ lies in $\mathcal{BC}$. A comodule structure is determined by a structure map

$$\psi : M \to BP_*BP \otimes_{BP_*} M$$

given by $\psi(x) = \sum_E c(t^E) \otimes r_E(x)$, where c denotes the canonical antiautomorphism of BP_*BP. We prefer to deal with the Quillen

[1]Supported in part by a grant from the National Science Foundation.

operations $\{r_E\}$ rather than ψ; notice that for $x \in M$, $r_E x = 0$ for all but a finite number of exponent sequences.

We also recall [4,9] that $BP_* = Z_{(p)}[v_1, v_2, \ldots]$ and that (with $v_0 = p$) the invariant prime ideals in BP_* are $I_n = (v_0, \ldots, v_{n-1})$ for $0 \leq n \leq \infty$. In particular, $I_0 = (0)$ and $I_\infty = (p, v_1, \ldots, v_n, \ldots)$.

We now collect the principal results of the paper, and some of their immediate consequences.

Theorem 1: Let $M \in \mathcal{BP}$ and $x \neq 0$ in M. Then

$$\sqrt{\mathrm{Ann}(x)} = \{\lambda \in BP_* : \lambda^k x = 0 \text{ with } k > 0\}$$

is an invariant prime ideal in BP_*.

Here are an alternate version and an immediate corollary.

Theorem 1′: If $M \in \mathcal{BP}$, then each weakly associated prime ideal of M is invariant.

Corollary 1: For $M \in \mathcal{BP}$, $\mathrm{Ass}_w(M)$ consists of the radicals $\sqrt{\mathrm{Ann}(x)}$ for nonzero elements $x \in M$.

Theoren 1 ⟹ Theorem 1′: If $\sqrt{\mathrm{Ann}(x)} = I_n$, then I_n is the only minimal prime ideal containing $\mathrm{Ann}(x)$, and it is invariant.

Theorem 1′ ⟹ Theorem 1: It's generally true that $\sqrt{\mathrm{Ann}(x)}$ is the intersection of the minimal prime ideals P_α containing $\mathrm{Ann}(x)$. If each P_α is invariant, hence some I_n, there is clearly just one minimal prime ideal containing $\mathrm{Ann}(x)$ and it is invariant, so that $\sqrt{\mathrm{Ann}(x)}$ is both invariant and prime. QED

In the MU case, with invariant prime ideals $I_n(p)$ [6], we have the following analogue.

Theorem 1_{MU}: Let M be a MU_*MU-comodule and $x \neq 0$ in M. Then

(a) $\sqrt{\mathrm{Ann}(x)}$ is invariant;

(b) each minimal prime ideal containing $\mathrm{Ann}(x)$ is invariant;

(c) there are only a finite number of minimal primes containing $\sqrt{\mathrm{Ann}(x)}$, and $\sqrt{\mathrm{Ann}(x)}$ is their intersection.

We shall stay with the BP-case for the rest of this report; many of the techniques work equally well for the MU case and in the purely algebraic setting of [7].

Corollary 2 (Johnson-Yosimura [3]): If $M \in \mathcal{BP}$ and $x \in M$ is v_n-torsion ($v_n^t x = 0$ with $t > 0$) then x is v_{n-1}-torsion.

Proof: $v_n^t = 0 \Longrightarrow v_n \in \sqrt{\mathrm{Ann}(x)} = I_m \Longrightarrow m \geq n+1$ $\Longrightarrow v_{n-1} \in \sqrt{\mathrm{Ann}(x)} \Longrightarrow v_{n-1}^s x = 0$ for some $s > 0$.

Recall that for $M \in \mathcal{BP}$, an element $x \in M$ is primitive [3,6] if $r_E x = 0$ for all $E \neq 0$.

Theorem 2: If $M \in \mathcal{BP}$ and $I_n = \sqrt{\mathrm{Ann}(x)}$ with $x \in M$ and $n < \infty$, then $I_n = \mathrm{Ann}(y)$ for some primitive element y in M.

Corollary 3 (Johnson-Yosimura [3]): Let $M \in \mathcal{BP}$. If all primitives of M are v_n-torsion, then M is a v_n-torsion module.

Proof: If M were not v_n-torsion, then for some $x \in M$ we have $v_n \notin \sqrt{\mathrm{Ann}(x)}$, hence $\sqrt{\mathrm{Ann}(x)} = I_m$ with $m \leq n$. By Theorem 2, there is a primitive $y \in M$ with $I_m = \mathrm{Ann}(y)$, so y is not v_n-torsion, a contradiction. QED

Corollary 4 (Johnson-Yosimura [3]): Let $M \in \mathcal{BP}$. If no nonzero primitive is v_n-torsion, then M is v_n-torsion free.

Proof: If M is not v_n-torsion free, then by Theorem 1 and Lemma 2.9, $N = \{x \in M : x \text{ is } v_n\text{-torsion}\}$ is a nonzero invariant submodule of M, hence it contains a nonzero primitive element (any nonzero element of lowest degree in a finitely-generated invariant submodule of N will do) which is v_n-torsion. QED

Whereas Theorems 1 and 2 allow one to deduce results of Johnson and Yosimura [3], our remaining results are consequences of the following algebraic analogue of a result of [3]:

Theorem 3: Let $M \in \mathcal{BP}$ and put $E(n)_* = Z_{(p)}[v_1, \ldots, v_n, v_n^{-1}]$. Then $v_n^{-1}M = 0$ iff $E(n)_* \otimes_{BP_*} M = 0$.

In view of the exact functor theorem [9], we obtain an immediate corollary.

Corollary 5 (Johnson-Yosimura [3]): *For a CW-spectrum* X, $v_n^{-1}BP_*(X) = 0$ *iff* $E(n)_*(X) = 0$. *Hence* $v_n^{-1}BP_*(\ )$ *and* $E(n)_*(\ )$ *have the same acyclic spaces.*

The next two results follow from Theorem 3.

Corollary 6: *Let* J *be an invariant ideal in* BP_*. *Then* $v_n^{-1}J$ *is a finitely generated ideal in* $v_n^{-1}BP_*$, *i.e., there is a finitely generated (invariant) ideal* $J_0 \subset J$ *such that* $v_n^{-1}J_0 = v_n^{-1}J$.

Theorem 4: *Let* $M \in \mathcal{BP}$ *be a finitely generated* BP_**-module. Then* M *is coherent if and only if there exists* $n \geq 0$ *such that* $v_n : M \to M$ *is injective.*

For a BP_*-module M, let $p \dim M$ denote the *projective dimension* of M, and $w \dim_{\mathcal{BP}} M$ denote the *$\mathcal{BP}$-weak dimension* of M [12], the largest t such that $\mathrm{Tor}_t^{BP_*}(A,M) \neq 0$ with $A \in \mathcal{BP}$.

Corollary 7: *Let* $M \in \mathcal{BP}$ *be a finitely generated* BP_**-module. The following are equivalent:*

(a) M *is coherent;*

(b) $p \dim M < \infty$;

(c) $w \dim_{\mathcal{BP}} M < \infty$;

(d) $v_n : M \to M$ *is injective for some* $n \geq 0$.

For a BP_*-module M, we call M *pseudocoherent* iff each finitely generated submodule N of M is finitely presented.

Corollary 8: *If* $M \in \mathcal{BP}$, *then* M *is pseudocoherent iff, for each finitely generated submodule* N *of* M, *there exists* $n \geq 0$ *with* $v_n : M \to M$ *injective.*

Corollary 9: *If* $M \in \mathcal{BP}$ *and* $w \dim_{\mathcal{BP}} M < \infty$, *then* M *is pseudocoherent.*

Corollary 8 is immediate from Theorem 4, while Corollary 9 will be proved at the same time as Corollary 7.

For a BP_*-module M, let $T_\infty M$ denote the submodule of elements which are v_n-torsion for all n, and put $\bar{M} = M/T_\infty M$.

Theorem 5: *If* $M \in \mathcal{BP}$, *then* $\bar{M} = M/T_\infty M$ *is pseudocoherent iff each finitely generated submodule* N *of* M *has* $|\mathrm{Ass}_w(N)| < \infty$.

In §2 we prove Theorems 1, 2 and 3. The remaining results are proved in §3.

These results and techniques are useful for the application of the homology theories $E(n)_*(\ )$ to stable homotopy theory, a program initiated by D. C. Ravenel [11].

2. Weakly associated prime ideals and primitive elements

2.1. *Preliminaries*. We begin with a bit of commutative algebra. Let A be a commutative ring, and M an A-module. A prime ideal P is *associated* to M if $P = \mathrm{Ann}(x)$ for some $x \in M$. This is a good notion when A is Noetherian. A prime ideal P is *weakly associated* to M if it is minimal among prime ideals containing $\mathrm{Ann}(x)$ for some $x \in M$. Let $\mathrm{Ass}(M)$ = all associated prime ideals, and $\mathrm{Ass}_w(M)$ = all weakly associated prime ideals. Thus $\mathrm{Ass}(M) \subset \mathrm{Ass}_w(M)$.

2.2. *Lemma*: *If* $P \in \mathrm{Ass}_w(M)$ *and* P *is finitely generated, then* $P \in \mathrm{Ass}(M)$.

We refer to [10] for the proof, which is based on the proof of Theorem 86 in Kaplansky [5].

2.3. *Lemma*: *If* N *is a submodule of* M, *then*

$$\mathrm{Ass}_w(N) \subset \mathrm{Ass}_w(M) \subset \mathrm{Ass}_w(N) \cup \mathrm{Ass}_w(M/N) .$$

This is part of an exercise in Bourbaki's Algèbre Commutative [1, Ch. 4, §1, EX. 17].

2.4. *Proof of Theorem 1′ for* $M \in \mathcal{BP}_0$. By the *prime filtration theorem* [7,9], the coherent comodule M has a filtration in $\mathcal{BP}_0$, $0 = M_0 \subset M_1 \subset \ldots \subset M_n = M$, such that M_i/M_{i-1} is stably isomorphic to BP_*/P_i with P_i a finitely generated invariant prime ideal in

BP_*. Hence 2.3 implies that each element of $Ass_w(M)$ is one of the P_i, since

$$Ass_w(A/P) = \{P\}$$

for any prime ideal P in a ring A. Thus each $P \in Ass_w(M)$ is invariant. QED

2.5. Remarks. a) Using 2.2, it follows that $Ass_w(M) = Ass(M)$ for $M \in \mathcal{BP}_0$.

b) The prime filtration theorem depends on knowing that each $P \in Ass(M)$ is invariant for $M \in \mathcal{BP}_0$, which depends on primary decomposition [7]. One can also prove 2.4 directly by primary decomposition, using the fact [1, Ch. 4, §2, EX. 20] that, when a submodule N has a reduced primary decomposition in M, $Ass_w(M/N)$ coincides with the associated set of prime ideals.

2.6. Proof of Theorem 1′: Let $P \in Ass_w(M)$, $M \in \mathcal{BP}$. Then P is a minimal prime ideal containing $Ann(x)$ for some $x \in M$. Replace M by the smallest invariant submodule N containing x; N is finitely generated over BP_*, and has a filtration in $\mathcal{BP}$, $0 = N_0 \subset N_1 \subset \ldots \subset N_k = N$ such that N_i/N_{i-1} is stably isomorphic to BP_*/J_i with J_i an invariant ideal. By 2.3, it suffices to show that if J is an invariant ideal in BP_*, then each $P \in Ass_w(BP_*/J)$ is invariant. Thus let $x \in BP_*$ represent $\bar{x} \in BP_*/J$; it will suffice to show that $\sqrt{Ann(\bar{x})} = I_n$ for some n, $0 \leq n \leq \infty$.

Write $J = \cup J_k$, where the J_k $(k = 1,2,\ldots)$ are finitely generated invariant ideals and $J_k \subset J_{k+1}$ for all k. Let x represent $x_k \in BP_*/J_k$. Then one verifies rapidly that $Ann(\bar{x}) = \cup Ann(x_k)$, and so $\sqrt{Ann(\bar{x})} = \bigcup \sqrt{Ann(x_k)}$. By 2.4, $\sqrt{Ann(x_k)} = I_n$ for some n depending on k; since $\sqrt{Ann(x_k)} \subset \sqrt{Ann(x_{k+1})}$, it is evident that $\sqrt{Ann(\bar{x})}$ is also some I_n. QED

2.7. Proof of Theorem 2. We will reduce Theorem 2 to the following special case:

2.8. Proposition: If $M \in \mathcal{BP}$ and I_n is the smallest element of $\mathrm{Ass}_w(M)$, $n < \infty$, then $I_n = \mathrm{Ann}(y)$ for some primitive $y \in M$.

2.9. Lemma. If $M \in \mathcal{BP}$, $0 \leq n < \infty$, then both submodules $\{x \in M : I_n x = 0\}$ and $\{x \in M : I_n^k x = 0$ for some $k > 0\}$ are invariant.

This follows immediately from [7, Lemma 2.3].

2.10. 2.8 $\Longrightarrow$ Theorem 2. Let $I_n \in \mathrm{Ass}_w(M)$, $n < \infty$. Put $N = \{x \in M; I_n^k x = 0$ for some $k > 0\}$. Then $I_n = \sqrt{\mathrm{Ann}(x)}$ with $x \in M$; clearly x lies in N, so $I_n \in \mathrm{Ass}_w(N)$, and clearly I_n is the smallest element of $\mathrm{Ass}_w(N)$.

2.11. Proof of 2.8. By 2.2, $I_n = \mathrm{Ann}(x)$ for some $x \in M$. Put $N = \{x \in M : I_n x = 0\}$; then $x \in N$, and so we can replace M by N. So assume from the start that $I_n M = 0$.

Now suppose we don't have $I_n = \mathrm{Ann}(y)$ for any primitive $y \in M$. Then each primitive element in M is v_n-torsion. By assumption and 2.2, $I_n = \mathrm{Ann}(x)$ for some $x \in M$. Hence

$$K = \{m \in M : m \text{ is } v_n\text{-torsion}\}$$

is a proper invariant submodule of M, i.e., $M/K \neq 0$. Choose $x \in M$ representing a nonzero primitive element $\bar{x}$ of M/K. Then $r_E x \in K$ for all $E \neq 0$. Since $r_E x \neq 0$ for only finitely many E's, we have $v_n^t r_E x = 0$ for some $t > 0$ and all $E \neq 0$. Since $I_n M = 0$, $v_n : M \to M$ is a morphism of comodules, hence $r_E(v_n^t x) = 0$ for all $E \neq 0$. Thus $v_n^t x$ is primitive, hence x is v_n-torsion, so $x \in K$ and $\bar{x} = 0$. This is a contradiction, completing the proof of 2.8 and Theorem 2. QED

2.12. Proof of Theorem 3. It is evident that if $v_n^{-1} M = 0$ for $M \in \mathcal{BP}$, then also $E(n)_* \otimes_{BP_*} M = 0$ since the ring homomorphism $BP_* \to E(n)_* = Z_{(p)}[v_1, \ldots, v_n, v_n^{-1}]$ factors throught $v_n^{-1} BP_*$.

Now assume $E(n)_* \otimes_{BP_*} M = 0$. We want to show that each element of M is v_n-torsion. If this is <u>not</u> true, the smallest weakly associated prime ideal of M is I_m, $m \leq n$. By Theorem 2, M contains a primitive element x such that $\mathrm{Ann}(x) = I_m$. Thus M contains an invariant submodule N stably isomorphic to BP_*/I_m. Since $m \leq n$,

$$BP_*/I_m \otimes_{BP_*} Z_{(p)}[v_1,\ldots,v_n,v_n^{-1}] \cong BP_*[v_m,\ldots,v_n,v_n^{-1}] \neq 0 .$$

By the exact functor theorem [9], extended from $\mathcal{BP}_0$ to $\mathcal{BP}$ since Tor_i commutes with direct limits, the injection $0 \to N \to M$ remains exact when tensored with $E(n)_*$. Hence our assumption that $E(n)_* \otimes_{BP_*} M = 0$ implies that $E(n)_* \otimes_{BP_*} N = 0$, and this is a contradiction. QED

3. <u>Consequences of Theorem 3</u>

3.1. <u>Proof of Corollary 6</u>: Let J be an invariant prime ideal in BP_*, and consider the ring homomorphisms

$$BP_* \to v_n^{-1}BP_* \to E(n)_* .$$

Since $E(n)_*$ is Noetherian, J extends to a finitely generated ideal in $E(n)_*$. Applying the exact functor theorem to $E(n)_*$, we see that $E(n)_* \otimes_{BP_*} J$ is finitely generated over $E(n)_*$, hence we can choose a finitely generated ideal $J_0 \subset J$ for which

$$E(n)_* \otimes_{BP_*} J_0 = E(n)_* \otimes_{BP_*} J .$$

Thus $E(n)_* \otimes_{BP_*} (J/J_0) = 0$, and so Theorem 3 implies that $v_n^{-1}(J/J_0) = 0$, i.e., $v_n^{-1}J = v_n^{-1}J_0$. QED

3.2. <u>Proof of Theorem 4</u>: Let $M \in \mathcal{BP}$ be a finitely generated BP_*-module, and assume $v_n : M \to M$ is injective. Choose an exact sequence

$$0 \to K \to F \to M \to 0$$

in $\mathcal{BP}$ with F finitely generated and free [9, Prop. 2.4]. Our aim is to show that K is finitely generated.

Tensor with $E(n)_*$ and apply the exact functor theorem to obtain the exact sequence

$$0 \to E(n)_* \otimes_{BP_*} K \to E(n)_* \otimes_{BP_*} F \to E(n)_* \otimes_{BP_*} M \to 0 .$$

The advantage is that $E(n)_*$ is Noetherian, so $E(n)_* \otimes_{BP_*} K$ is finitely generated over $E(n)_*$.

Let $\bar{k}_1, \ldots, \bar{k}_r \in E(n)_* \otimes_{BP_*} K$ be generators over $E(n)_*$. Multiplying them by a power of v_n, we may assume that $\bar{k}_i = 1 \otimes k_i$ with $k_i \in K$. Let K_0 be the BP_*-submodule of K generated by all $r_E(k_i)$; then K_0 is a finitely generated invariant submodule of K for which $E(n)_* \otimes_{BP_*} (K/K_0) = 0$. As in the previous argument, the key step is supplied by Theorem 3: we can conclude that $v_n^{-1}(K/K_0) = 0$.

Next consider the diagram

$$\begin{array}{ccccccccc}
 & & & & & & 0 & & \\
 & & & & & & \downarrow & & \\
0 & \to & K/K_0 & \to & F/K_0 & \to & M & \to & 0 \\
 & & \downarrow & & \downarrow & & \downarrow & & \\
0 & \to & v_n^{-1}K/K_0 & \to & v_n^{-1}F/K_0 & \to & v_n^{-1}M & \to & 0 \\
 & & \| & & & & & & \\
 & & 0 & & & & & &
\end{array}$$

and observe that

$$K/K_0 \cong \operatorname{Ker}(F/K_0 \to v_n^{-1}(F/K_0)) .$$

Since $F/K_0 \in \mathcal{BP}_0$, there is a finitely generated module N over $Z_{(p)}[v_1, \ldots, v_\ell]$ with $\ell \geq n$ such that

$$F/K_0 \cong BP_* \otimes_{Z_{(p)}[v_1, \ldots, v_\ell]} N .$$

It is then elementary that

$$\operatorname{Ker}(F/K_0 \to v_n^{-1}(F/K_0)) \cong BP_* \otimes_{Z_{(p)}[v_1, \ldots, v_\ell]} (\operatorname{Ker}(N \to v_n^{-1}N))$$

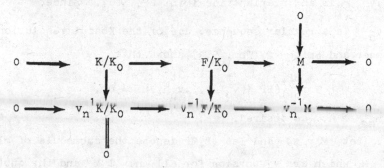

and this implies that K/K_0 is finitely generated. Since K_0 is also finitely generated, we conclude that K is finitely generated, as desired. QED

3.3. Proof of Corollary 7. We have just proved that (d) $\Longrightarrow$ (a). It is well-known that (a) $\Longrightarrow$ (b) and (b) $\Longrightarrow$ (c). That (c) $\Longrightarrow$ (d) results from the next lemma, which also suffices for the proof of Corollary 9.

3.4. Lemma: If $M \in \mathcal{BP}$ and $w \dim_{\mathcal{BP}} M \leq n < \infty$, then $v_n : M \to M$ is injective. I.e., if M has nontrivial v_n-torsion, then $w \dim_{\mathcal{BP}} M > n$.

Proof. Let $x \neq 0$ in M with $v_n^t x = 0$, $t > 0$. Hence $I_n \subset \sqrt{\mathrm{Ann}(x)}$, and so for sufficiently large t we have $v_k^t x = 0$ for $0 \leq k \leq n$. Choose exponents $t_0, \ldots, t_n \geq t$ so that $(v_0^{t_0}, \ldots, v_n^{t_n})$ is an invariant ideal in BP_* [8]. Since $v_0^{t_0}, \ldots, v_n^{t_n}$ is a regular sequence, use of the Koszul resolution as in Conner and Smith [2, Thm. 5.3] shows that

$$\mathrm{Tor}_{n+1}^{BP_*}(BP_*/(v_0^{t_0}, \ldots, v_n^{t_n}), M) \neq 0 ,$$

so $w \dim_{\mathcal{BP}} M > n$, as desired. QED

3.5. Let $M \in \mathcal{BP}$ and let $T_\infty M$ denote the submodule of elements of M which are v_n-torsion for all n. $T_\infty M$ and the quotient $\bar{M} = M/T_\infty M$ are in $\mathcal{BP}$.

3.6. Lemma: If $M \in \mathcal{BP}$, then

$$\mathrm{Ass}_w(\bar{M}) = \mathrm{Ass}_w(M) - \{I_\infty\} .$$

Proof. By 2.3 we have the inclusion

$$\mathrm{Ass}_w(M) \subset \mathrm{Ass}_w(T_\infty M) \cup \mathrm{Ass}_w(\bar{M}) .$$

Since $\mathrm{Ass}_w(T_\infty M) \subset \{I_\infty\}$, it will suffice to show that

$$\mathrm{Ass}_w(\bar{M}) \subset \mathrm{Ass}_w(M) - \{I_\infty\} .$$

If $I_n \in Ass_w(\bar{M})$, then $n < \infty$ and we can find $x \in M$ representing $\bar{x} \in \bar{M}$ so that $I_n = \sqrt{Ann(\bar{x})}$. We claim that $I_n = \sqrt{Ann(x)}$. Since $Ann(x) \subset Ann(\bar{x})$, it suffices to show that x is v_k-torsion for $k < n$; since $\bar{x}$ is v_k-torsion for $k < n$, we have $v_k^t x \in T_\infty M$ which certainly implies that x is also v_k-torsion. QED

3.7. Proof of Theorem 5. Assume that $\bar{M}$ is pseudocoherent, and let N be a f.g. invariant submodule of M. Since $N \subset T_\infty M + N$, it suffices to show that

$$|Ass_w(T_\infty M + N)| < \infty .$$

Since $\bar{M}$ is pseudocoherent, its f.g. submodule $(T_\infty M + N)/T_\infty M$ is coherent, and so the desired conclusion follows from Lemma 2.3.

Conversely, assume that $|Ass_w(N)| < \infty$ for each f.g. submodule N of M, and let $\bar{N}$ be a f.g. submodule of $\bar{M}$. Since $\bar{N}$ lies in a f.g. invariant submodule of $\bar{M}$, we shall assume that $\bar{N}$ is invariant. In order to show that $\bar{N}$ is coherent, it suffices to show that $|Ass_w(\bar{N})| < \infty$, in view of Theorem 4 and the fact that $I_\infty \notin Ass_w(\bar{M})$. Choose a f.g. invariant submodule N of M which projects onto $\bar{N}$, so that $T_\infty M + N$ is the inverse image of $\bar{N}$ in M. By 2.3 we see that $|Ass_w(T_\infty M + N)| < \infty$, and then 3.6 implies that also $|Ass_w(\bar{N})| < \infty$. QED

3.8. Remark. Further corollaries on invariant ideals and primary decomposition follow easily from Theorem 4 and the techniques of [6,7,10]. Notably, each invariant I_n-primary ideal Q in BP_* with $n < \infty$ (i.e., $Ass_w(BP_*/Q) = \{I_n\}$) is finitely generated, and every invariant finitely generated ideal is a finite intersection of invariant finitely generated primary ideals. More generally, an invariant submodule N of a comodule $M \in \mathcal{BP}$ is a finite intersection of invariant primary submodules if and only if $|Ass_w(M/N)| < \infty$.

REFERENCES

[1] N. Bourbaki, Algèbre Commutative, Ch. 4, Hermann, Paris, 1967.

[2] P. E. Conner and L. Smith, On the complex bordism of finite complexes, Publ. Math. IHES 37 (1969), 117-221.

[3] D. C. Johnson and Z. Yosimura, Torsion in Brown-Peterson homology and Hurewicz homomorphisms, Duke Math. J. (to appear).

[4] D. C. Johnson and W. S. Wilson, BP operations and Morava's extraordinary K-theories, Math. Z. 144 (1975), 55-75.

[5] I. Kaplansky, Commutative Rings, Allyn and Bacon, Boston, 1970.

[6] P. S. Landweber, Annihilator ideals and primitive elements in complex bordism, Illinois J. Math. 17 (1973), 273-284.

[7] ———, Associated prime ideals and Hopf algebras, J. Pure and Applied Algebra 3 (1973), 43-58.

[8] ———, Invariant regular ideals in Brown-Peterson homology, Duke Math. J. 42 (1975), 499-505.

[9] ———, Homological properties of comodules over MU_*MU and BP_*BP, Amer. J. Math. 98 (1976), 591-610.

[10] P. Landweber and W. Parry, When does a primary decomposition exist?, Rutgers University preprint.

[11] D. C. Ravenel, Localization with respect to certain periodic homology theories, U. of Washington preprint.

[12] Z. Yosimura, to appear.

THE SIGNATURE OF SYMPLECTIC AND SELF-CONJUGATE MANIFOLDS

Peter S. Landweber[1]
Rutgers University
New Brunswick, N.J. 08903

1. Introduction

This will be a report on several problems concerning the Hirzebruch signature of manifolds, from the point of view of cobordism theory. First I will survey results on the signature of SU and Spin manifolds, due to Oshanin [13], and then summarize the results of Jones [8, 9] on the signature of symplectic manifolds. The final topic is the solution of the corresponding problem for self-conjugate cobordism. The results are displayed in the table below. The main challenge was to give complete answers to 2-primary problems, even though the 2-primary structure of Ω_*^{Sp} and Ω_*^{SC} are not known.

We may view the signature as a ring homomorphism $\sigma : \Omega_*^{SO} \to Z$, since it is multiplicative and vanishes on boundaries of oriented manifolds. Thus we can ask about its values on Ω_{4n}^{G} for each of the bordism groups in the diagram

$$\begin{array}{ccccccc}
\Omega_{4n}^{Sp} & \longrightarrow & \Omega_{4n}^{SU} & \longrightarrow & \Omega_{4n}^{Spin} & & \\
\downarrow & & \downarrow & & \downarrow & & \\
\Omega_{4n}^{SC} & \longrightarrow & \Omega_{4n}^{U} & \longrightarrow & \Omega_{4n}^{SO} & \xrightarrow{\ \sigma\ } & Z
\end{array}$$

The image of $\Omega_*^G \xrightarrow{\sigma} Z$ is now known in all these cases with the generators given in the following table for the cases of interest:

---
[1] Supported in part by a grant from the National Science Foundation.

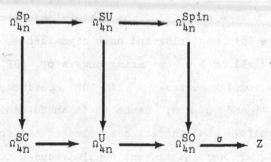

	4	8	12	16	20	24	28	32	
Spin	16	1	...						
SU	16	2	16	1 ...					
Sp	16	4	16	2	16	4	16	1	...
SC	8	2	16	1 ...					

In particular, in these cases $\mathrm{Sign}(\Omega^G_{4n})$ repeats with periodicity 8, 16 or 32 in the dimension 4n. One faces no problem for oriented and unitary bordism since CP^{2n} has signature 1.

The determination of the values of the signature on symplectic cobordism was done by Les Jones [8, 9] in his thesis which I directed. Partial results on the relation between the Kervaire invariant and the signature of Spin and SU-manifolds were obtained by Janey Daccach [4] in his thesis.

Bob Stong pointed out the futility of searching in Ω^{SC}_{12} for an example with signature 8, which put me on the right track in dimension $16k + 12$. He also explained to me some properties of the free involutions on Stong manifolds.

2. SU and Spin-Manifolds

We begin with some examples. The quaternionic projective spaces HP^{2n} are Spin manifolds and have signature 1, hence $\sigma(\Omega^{Spin}_{8n}) = Z$ for all $n > 0$ by taking powers of $[HP^{2n}]$. In dimension 4, the complex surface V^4 in CP^3 defined by $\Sigma z_i^4 = 0$ has signature - 16 and $c_1 = 0$, hence it is an SU-manifold; thus V^4 is a Spin-manifold (and in fact a symplectic manifold as well since $Sp(1) = SU(2)$ and its tangent bundle reduces to an $SU(2)$-bundle).

The first divisibility theorem on the signature is

Rochlin's Theorem [16]: For a 4-dimensional Spin-manifold, $\sigma(M^4) \equiv 0\ (16)$.

Hence we've shown that this is best possible, in fact that

$$\sigma(\Omega_4^{Sp}) = \sigma(\Omega_4^{SU}) = \sigma(\Omega_4^{Spin}) = 16Z .$$

In 1973 Oshanin proved a generalization.

Oshanin's Theorem [13]: For an SU-manifold of dimension $8k + 4$, $\sigma(M^{8k+4}) \equiv 0(16)$.

Earlier Stong [19] showed that

$$\Omega_{8k+4}^{SU} \to \Omega_{8k+4}^{Spin}/\text{torsion}$$

is onto. Hence we obtain

Corollary: For a Spin-manifold of dimension $8k + 4$, $\sigma(M^{8k+4}) \equiv 0(16)$.

In fact, in this dimension $HP^{2k} \times V^4$ has signature -16, so we have completed the problem for Spin-manifolds, and shown that the periodicity is 8. In the same paper, Oshanin proved that the SU case is as stated, with period 16. More precisely, Conner and Floyd [3] showed that M^{16k+8} SU implies $\sigma(M)$ is even, and that there is a 16-dimensional SU-manifold with odd signature, hence one having signature 1 (we recall that the signature and Euler characteristic both reduce mod 2 to the top Stiefel-Whitney number $w_{4n}[M^{4n}]$, and that our problems are 2-primary since

$$\Omega_*^{Sp} \otimes Z[\tfrac{1}{2}] \to \Omega_*^{SO} \otimes Z[\tfrac{1}{2}]$$

is an isomorphism). Also Oshanin gave examples of SU-manifolds in dimensions 8 and 12 with signatures 2 and 16 respectively.

We end this section by stating Oshanin's theorem on Kervaire invariants. Assume given a homomorphism $K : \Omega_{8k+2}^{Spin} \to Z/2$ satisfying the two conditions

(a) N^{8k} Spin, $[\bar{S}^1] \in \Omega_1^{Spin}$ the nonzero class, implies $K(N^{8k} \times \bar{S}^1 \times \bar{S}^1) = \sigma(N^{8k}) \bmod 2$;

(b) V^{8k+2} Spin, $H^{4k+1}(V;Z/2) = 0$ implies $K(V) = 0$.

On the other hand, we can define an invariant on Ω_{8k+2}^{Spin} by the following route. For a Spin-manifold V^{8k+2}, characteristic numbers show easily that $[V^{8k+2}][\bar{S}^1] = 0$ in Ω^{Spin}_{8k+3}, so $V^{8k+2} \times \bar{S}^1 = \partial W^{8k+4}$ with W a Spin-manifold. Then ∂W has an automorphism φ reversing the Spin-structure (i.e., $id_V \times c$ where $c : S^1 \to S^1$ has degree -1). Then form $M^{8k+4} = W \cup W$, a closed Spin-manifold, hence $\sigma(M) \equiv 0\ (16)$ and $\sigma(W) \equiv 0\ (8)$. One obtains an invariant by putting

$$k(V) = \frac{1}{8}\,\mathrm{Sign}(W) \in Z/2 .$$

<u>Theorem [14]: If</u> V^{8k+2} <u>is a Spin-manifold whose unoriented bordism class in</u> $\mathfrak{N}_{8k+2}$ <u>contains a U-manifold</u> (Stiefel-Whitney numbers involving an odd w_i vanish), <u>then</u> $K(V) = k(V)$.

In particular, the theorem applies whenever V^{8k+2} is an SU-manifold.

3. <u>Symplectic Manifolds</u>

We give a brief account of the determination of $\mathrm{Sign}(\Omega^{Sp}_{4n})$ by Jones [8, 9]. The periodicity has dimension 32, which is reasonable since $[RP^2]^{16} \in \mathfrak{N}_{32}$ is known to contain an Sp-manifold M^{32}[18,17,6], hence $\sigma(M^{32})$ is odd and so there is a symplectic N^{32} with signature 1. Now Nigel Ray [15] has tabulated Ω^{Sp}_* for dimensions ≤ 16, and it is known that KO-characteristic numbers determine all divisibility relations in dimensions ≤ 30 [12]. This makes it possible to produce examples with the tabulated signatures in dimensions 4,8,...,28. Still one needed two <u>divisibility theorems:</u>

(1) M^{32k+16} symplectic implies $\sigma(M)$ even;

(2) M^{16k+8} sympectic implies $\sigma(M) \equiv 0(4)$.

The $16k + 8$ result is delicate; here the SU result implies that Sign(M) is even, so one needs to improve this by a factor of 2.

The $32k + 16$ result goes as follows. Let M^{32k+16} be symplectic; one wants to show that $w_{32k+16}[M] = 0$. Now Floyd [5] defined a subalgebra P_* of $\mathfrak{N}_*$, of the form

$$P_* = Z_2[x_2^2, x_4^2, x_5^2, x_6, \dots]$$

where the x_i $(i \neq 2^\ell - 1)$ are generators of η_*, and the generators of P_* are x_i^2, unless i is even and not a power of 2 in which case x_i is a generator. Floyd proved that

$$\mathrm{Im}(\Omega_*^{Sp} \to \eta_*) \subset P_*^8 ,$$

and recently Kochman [10] has announced that this is an equality. Hence $[M^{32k+16}] = [N^{4k+2}]^8$ with $[N^{4k+2}] \in P_{4k+2}$, and

$$w_{32k+16}[M] = w_{4k+2}[N] .$$

Then one proves that $w_{4k+2}[N] = 0$ on P_{4k+2} to finish the argument.

4. Self-conjugate bordism

In this section we shall determine the values of the signature on self-conjugate bordism Ω_*^{SC}. This is the bordism ring obtained from manifolds whose stable normal bundle has a complex structure together with an isomorphism to the conjugate complex vector bundle. Gozman [6] showed that $[RP^2]^8 \in \eta_{16}$ contains an SC-manifold, and Buchstaber [2] has proved that

$$\mathrm{Im}(\Omega_*^{SC} \to \eta_*) = P_*^4 .$$

In particular, there is a SC-manifold M^{16} with signature 1. Recently Gozman [7] has computed Ω_*^{SC} in dimensions ≤ 9, and shown that there is arbitrarily high 2-torsion in Ω_*^{SC}.

Our problem breaks into two parts: divisibility and examples.

Divisibility. It is useful to introduce two subrings in Ω_*^U which contain the image of $\Omega_*^{SC} \to \Omega_*^U$. Let

L = all $[M] \in \Omega_*^U$ such that every Chern number involving an odd Chern class vanishes, and

$Z(\psi^U)$ = all $[M] \in \Omega_*^U$ such that every Chern number involving c_1 vanishes.

These are both subrings of Ω_*^U, and

$$\mathrm{Im}(\Omega_*^{SC} \to \Omega_*^U) \subset L \subset Z(\mathscr{W}^U) .$$

So it will suffice to prove the desired divisibility on $Z(\mathscr{W}^U)$. Here $\mathscr{W}^U$ is the complex Wall subgroups of Ω_*^U [3], consisting of classes $[M]$ such that all Chern numbers $c_1^2 c_\omega[M] = 0$. There is a boundary operator $\partial : \mathscr{W}_{2n}^U \to \mathscr{W}_{2n-2}^U$ (dualize c_1), and $Z(\mathscr{W}^U)$ coincides with the cycles, $B(\mathscr{W}^U)$ denotes the boundaries and $H(\mathscr{W}^U)$ the homology. Despite the fact that $\mathscr{W}^U$ is not a subalgebra of Ω_*^U and ∂ is not a derivation, everything works well mod 2 and the cycles do form a subring.

From Conner and Floyd [3] we recall that

$$H(\mathscr{W}^U) = Z/2[c_4, c_{8k}(k \geq 2)]$$

and

$$\mathrm{Im}(\Omega_j^{SU} \to \Omega_j^U) = \begin{cases} Z_j(\mathscr{W}^U) & \text{if } j \neq 8k+4 \\ B_j(\mathscr{W}^U) & \text{if } j = 8k+4 . \end{cases}$$

Divisibility Theorem: $\sigma(Z_{16k+8}(\mathscr{W}^U)) \subset 2Z$, $\sigma(Z_{16k+4}(\mathscr{W}^U)) \subset 8Z$ and $\sigma(Z_{16k+12}(\mathscr{W}^U)) \subset 16Z$.

Proof: If $[M^{16k+8}]$ is a cycle, it comes from SU and so the SU result applies.

If $[M^{16k+4}]$ is a cycle, then $2[M]$ comes from SU, so the SU result gives only $\sigma([M]) \equiv 0(8)$.

Finally, let $[M^{16k+12}]$ be a cycle. Its class in $H_{16+12}(\mathscr{W}^U)$ must lie in the ideal generated by c_4^3 and $c_4 c_{16\ell+8}$ $(\ell \geq 1)$. Choose representatives $[C^4]$ and $[C^{8\ell}]$ for c_4 and $c_{8\ell}$ in $Z(\mathscr{W}^U)$. Hence in $Z(\mathscr{W}^U)$ we have

$$[M^{16k+12}] = [C^4]^3[N^{16k}] + \sum_{k=1}^{\ell} [C^4][C^{16\ell+8}][N^{16(k-\ell)}]$$

mod $B_{16k+12}(\mathscr{V}^U)$. Now we're done: the signature is $\equiv 0(16)$ on $B_{16k+12}(\mathscr{V})$ by the SU result. Since $2[C^4]$ is in $B_4(\mathscr{V})$, $\sigma[C^4] \equiv 0(8)$; and $[C^{16\ell+8}]$ in $Z_{16k+8}(\mathscr{V}^U)$ also comes from SU, so has even signature. Hence

$$\mathrm{Sign}[C^4]^3 \equiv 0(8^3)$$

and

$$\mathrm{Sign}[C^4][C^{16\ell+8}] \equiv 0(16) ,$$

so that we have the desired conclusion that also

$$\mathrm{Sign}[M^{16k+12}] \equiv 0(16) . \qquad \square$$

Thus we obtain the divisibility results tabulated in the introduction for Ω_*^{SC}; they also hold for the subalgebra L. Hence the examples we are going to produce show that the tabulated values hold for L and $Z(\mathscr{V}^U)$ as well as for Ω_*^{SC}.

Examples. To finish the SC problem, we need to produce self-conjugate manifolds M^4, M^8 and M^{12} for which $\sigma(M^4) = 8$, $\sigma(M^8) = 2$ and $\sigma(M^{12}) = 16$. In fact, we can find a symplectic M^{12} with signature 16, so we need only find suitable M^4 and M^8. Notice that symplectic manifolds won't do.

There are two ways to proceed. One is to take coefficients of power series arising in connection with 2-valued formal groups [1,2,6]. The more direct way is to use Stong manifolds, which are examples of symplectic manifolds, and to show that they carry free involutions whose orbit spaces are SC-manifolds — hence having half the signature of the original Stong manifolds.

In dimension 12, one of the generators produced by Nigel Ray [15] was a coefficient of an Adams operation; in fact Stong manifolds can't suffice to generate Ω_{12}^{Sp}, since if they gave a symplectic manifold M^{12} with signature 16, then we could get a self-conjugate M^{12}/T with signature 8.

<u>Stong Manifolds</u>. Stong [20] gave the following construction of symplectic manifolds. Let CP_*^{2n+1} denote CP^{2n+1} with stable tangent bundle viewed as $(n+1)(\xi \oplus \bar{\xi})$, where ξ is the canonical complex line bundle. Thus CP_*^{2n+1} is a symplectic manifold. Consider

$$M(m,n) \subset CP_*^{2n+1} \times CP_*^{2m+1}$$

dual to $\zeta = \xi_1 \otimes \xi_2 + \bar{\xi}_1 \otimes \bar{\xi}_2$, where ξ_1 and ξ_2 are the canonical complex line bundles over the two factors, so ζ is a symplectic line bundle. Then $M(m,n)$ is a symplectic manifold.

More generally, if $\omega = (k_1,\ldots,k_{2s})$ then the submanifold

$$M(\omega;q) \subset \prod_{\ell=1}^{2s} CP_*^{(2k_\ell+1)} = CP_*(\omega)$$

dual to $q\zeta$, for $\zeta = \xi_1 \otimes \ldots \otimes \xi_{2s} + \bar{\xi}_1 \otimes \ldots \otimes \bar{\xi}_{2s}$, is an Sp-manifold. Note that $M(m,n) = M(m,n;1)$.

As examples, we note that

$$M(0,0,0,0;1) \subset (CP_*^1)^4$$

is a symplectic 4-manifold with signature - 16, and that

$$M(1,1;1) \subset CP_*^3 \times CP_*^3$$

is a symplectic 8-manifold with signature 4.

Now Buchstaber [1] (also Nadiradze [11]) points out that <u>each</u> $M(\omega;q)$ <u>has a free involution</u> T <u>such that</u> $M(\omega;q)/T$ <u>is an</u> SC-manifold. (Hence Stong manifolds can always be divided by 2 in the subalgebra L of Ω_*^U). Since $M(\omega;q) \to M(\omega;q)/T$ is a double cover of oriented manifolds,

$$\mathrm{Sign}(M(\omega;q)) = 2\,\mathrm{Sign}(M(\omega;q)/T)\ .$$

So in dimensions 4 and 8 we get SC manifolds with signature 8 and 2 respectively, as desired. So the problem for SC is solved.

Since the details concerning these involutions have not been published, we examine them in the next section.

5. Free involutions on Stong manifolds

The involutions arise as follows. We have

$$CP^{2n+1} = CP(C^{2n+2}) = CP(H^{n+1})$$

and then multiplication by j yields a fixed point free involution T on CP_*^{2n+1}, hence also on $CP_*(\omega)$, and so on the $M(\omega;q)$ if they are chosen to be invariant under T. But it still remains to see that $M(\omega;q)$ is an SC-manifold.

Lemma (Buchstaber) (a) The stable tangent bundle of $CP_*(\omega)/T$ is an SC-bundle. (b) There is an SC-bundle $\zeta(T)$ on $CP_*(\omega)/T$ such that $\pi^*\zeta(T) = \zeta$, where $\pi : CP_*(\omega) \to CP_*(\omega)/T$ is the orbit projection.

From the lemma, one can dualize $q\zeta(T)$ first, and then take the inverse image under π to get a suitable invariant $M(\omega;q)$, such that $M(\omega;q)/T$ is the original dual of $q\zeta(T)$. In this way, $M(\omega;q)/T$ has an SC-structure.

It remains to prove the lemma. We begin with the second part, noting first the conjugate-linear isomorphism of bundles

$$\begin{array}{ccc} \xi & \xrightarrow{T} & \xi \\ \downarrow & & \downarrow \\ CP^{2n+1} & \longrightarrow & CP^{2n+1} \end{array}$$

given by $T(v) = jv$ over each point of CP^{2n+1}. It is conjugate-linear since

$$j(iv) = jiv = -ijv = -i(jv) .$$

More generally, if $\omega = (k_1,\ldots,k_t)$ where t need not be even, we get a similar conjugate-linear isomorphism T on the bundle $\xi_1 \otimes \ldots \otimes \xi_t \to CP_*(\omega)$. Recall that $\zeta \to CP_*(\omega)$ is $\xi_1 \otimes \ldots \otimes \xi_t + \overline{\xi_1 \otimes \ldots \otimes \xi_t}$. We then obtain a complex bundle map

$$\begin{array}{ccc} \zeta & \xrightarrow{T} & \zeta \\ \downarrow & & \downarrow \\ CP_*(\omega) & \longrightarrow & CP_*(\omega) \end{array}$$

by $T(v_1,v_2) = (-jv_2, jv_1)$ on each fibre. Note also the conjugate-linear isomorphism of ζ given by $\varphi(v_1,v_2) = (-v_2,v_1)$, and that $\varphi T = T\varphi$. (The reason for the minus sign will appear when we look at tangent bundles.)

From this, it follows that ζ/T is a complex vector bundle on $CP_*(\omega)$ and that φ determines a conjugate-linear isomorphism φ/T of ζ/T. This is the data which specifies the SC-bundle $\zeta(T)$ on $CP_*(\omega)/T$, as required in the second part of the lemma.

We now turn to the tangent bundle τ of $CP_*(\omega)$. Begin with CP^{2n+1}, noting that the differential dT of the involution T (given by j) is conjugate-linear (T is a *conjugation* of CP^{2n+1}). The usual isomorphism

$$\tau \cong \mathrm{Hom}_C(\xi,\xi^\perp)$$

respects the conjugate-linear isomorphisms. Note also that under the standard isomorphism $\mathrm{Hom}_C(\xi,\xi) \cong C$, the fact that $j(zv) = \bar{z}j(v)$ shows that we must view C as $R^{1,1} = R \oplus iR$, i.e., T acts as complex conjugation on C. Thus *as real bundles with involution*, we obtain

$$\begin{aligned} \tau \oplus R \oplus iR &\cong \mathrm{Hom}_C(\xi,\xi^\perp) \oplus \mathrm{Hom}_C(\xi,\xi) \\ &\cong \mathrm{Hom}_C(\xi,C^{2n+2}) = \mathrm{Hom}_C(\xi,H^{n+1}) \\ &\cong (n+1)\mathrm{Hom}_C(\xi,H) \cong (n+1)\bar{\xi} \otimes_C H \, . \end{aligned}$$

Now the involution on $\bar{\xi} \otimes_C H$ is

$$T(v \otimes q) = jv \otimes jq \, ,$$

so in particular

$$T(v \otimes 1) = jv \otimes j, \; T(v \otimes j) = -jv \otimes 1 \, .$$

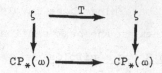

This shows that as bundles with involution, $\bar{\xi} \otimes_C H$ is isomorphic to $\xi \oplus \bar{\xi}$ with the same involution $T(v_1,v_2) = (-jv_2, jv_1)$ we used previously. The isomorphism is given by $(v_1,v_2) \mapsto v_1 \otimes 1 + v_2 \otimes j$.

We have found that for CP_*^{2n+1}, as bundles with involution we have

$$\tau \oplus R \oplus iR \cong (n+1)(\xi \oplus \bar{\xi}),$$

where the involution T on $\xi \oplus \bar{\xi}$ is complex linear and commutes with the conjugate-linear isomorphism $\varphi(v_1,v_2) = (-v_2,v_1)$.

Finally we pass to $CP_*(\omega)$ with $\omega = (k_1,\ldots,k_{2s})$, and observe that as bundles with involution

$$\tau \oplus 2sR \oplus 2siR = \bigoplus_{\ell=1}^{2s} (k_\ell+1)(\xi_\ell \oplus \bar{\xi}_\ell)$$

and so for $CP_*(\omega)/T$ we have

$$\tau \oplus 2sR \oplus 2s\tilde{R} \cong \bigoplus_{\ell=1}^{2s} (k_\ell+1)\zeta_\ell(T)$$

with $2sR$ trivial, $\tilde{R} \to CP_*(\omega)/T$ the twisted real line bundle associated to the double cover, and $\zeta_\ell(T)$ SC-bundles of complex dimension 2. Since $2s$ is even, $2s\tilde{R} = s(\tilde{R} \otimes_R C)$ can be viewed as an SC-bundle, so τ is stably an SC-bundle for $CP_*(\omega)/T$ as desired. Q.E.D.

REFERENCES

[1] V. M. Buchstaber, Two-valued formal groups in the apparatus of cobordism theory, Uspekhi Mat. Nauk 32 (194) (1977), 205-206 (Russian).

[2] ——————, Topological applications of the theory of two-valued formal groups, Izvestiya Akad. Nauk SSSR, Ser. Mat. 42 (1978), 130-184 (Russian).

[3] P. E. Conner and E. E. Floyd, Torsion in SU-bordism, Memoirs AMS 60 (1966).

[4] J. Daccach, The signature and the Kervaire invariant as cobordism invariants, Rutgers University Thesis, 1976.

[5] E. E. Floyd, Stiefel-Whitney numbers of quaternionic and related manifolds, Trans. AMS 155 (1971), 77-94.

[6] N. Ya. Gozman, On the image of the ring of self-conjugate cobordisms in the complex and unoriented cobordism rings, Dokl. Akad. Nauk SSSR 216 (1974), 1212-1214 (= Soviet Math. Dokl. 15 (1974), 953-956).

[7] ——————, Theory of self-conjugate cobordism, Mat. Zametki 22 (1977), 885-896 (= Math. Notes 22 (1977), 984-990).

[8] L. P. Jones, On the image of the signature homomorphism for symplectic cobordism, Rutgers University Thesis, 1975.

[9] ——————, The signature of symplectic manifolds, Trans. AMS 240 (1978), 253-262.

[10] S. Kochman, The symplectic cobordism ring I, Memoirs AMS (to appear).

[11] R. G. Nadiradze, Involutions on Stong manifolds and cobordism of self-conjugate manifolds, Bull. Acad. Sci., Georgian SSR 85 (1977), 301-303 (Russian, English summary).

[12] R. Okita, On the MSp Hattori-Stong problem, Osaka J. Math. 13 (1976), 547-566.

[13] S. D. Oshanin, The signature of SU-manifolds, Mat. Zametki 13 (1973), 97-102 (= Math. Notes 13 (1973), 57-60).

[14] S. Ochanine, Signature et invariants de Kervaire généralisés, C.R. Acad. Sci., Paris 285 (1977), Série A, 211-213.

[15] N. Ray, Realizing symplectic bordism classes, Proc. Cambridge Philos. Soc., 71 (1972), 301-305.

[16] V. A. Rochlin, New results in the theory of four-dimensional manifolds, Dokl. Akad. Nauk SSSR 84 (1952), 221-224 (Russian).

[17] F. Roush, On the image of symplectic cobordism in unoriented cobordism, Proc. AMS 38 (1973), 647-652.

[18] D. Segal, Divisibility conditions on characteristic numbers of stably symplectic manifolds, Proc. AMS 27 (1971), 411-415.

[19] R. E. Stong, Relations among characteristic numbers II, Topology 5 (1966), 133-148.

[20] ——————, Some remarks on symplectic cobordism, Annals of Math. 86 (1967), 425-433.

HOMOLOGY ISOMORPHISMS [1]

by

Denis Sjerve[2]

§1. Introduction: One of the main theorems in algebraic topology is the Whitehead theorem: if $f:X \to Y$ induces an isomorphism in homology and X,Y are simply connected CW complexes, then f is a homotopy equivalence. If we drop the hypothesis that X,Y are CW complexes, but still assume that they are simply connected, then f will in general only be a weak homotopy equivalence. On the other hand, without the hypothesis on simple connectivity, the homology isomorphism $f:X \to Y$ may fail to be even a weak homotopy equivalence. Thus the following problem naturally suggests itself:

(1.1) Problem: For a given CW complex Y study the set of all pairs (X,f) such that X is a CW complex and $f:X \to Y$ is a homology isomorphism.

It is convenient to examine this problem by converting it into one in homological algebra. To do this recall the Kan-Thurston theorem [2]:

Theorem (Kan-Thurston): For any path connected based space W there exists a fibration $p_W:TW \to W$ such that

(i) p_W is a homology isomorphism (even with respect to local coefficient systems on W).

---

(1) This paper is part of the text of a lecture given at the Algebraic Topology Conference held at the University of Waterloo during May and June 1978. I would like to thank the organizers of that conference for their efforts and hospitality.

(2) This research was partially supported by N.R.C. Contract A7218.

(ii) TW is an Eilenberg-MacLane space of type $K(\pi,1)$.

(iii) The fibration p_W is functorial in W.

Thus we can paraphrase our problem as follows

(1.2) Problem: For a given group π study the set of all pairs (G,α) such that $\alpha: G \to \pi$ is a homology isomorphism.

If we take π to be the trivial group then (1.2) is equivalent to studying all acyclic groups. To avoid this intractable case we shall assume that the commutator subgroup $[G,G]$ is free. In §2 we characterize those groups G satisfying

(1.3)
(i) $[G,G]$ is free of finite rank.
(ii) abelianization $G \to G_{ab}$ is a homology isomorphism.
(iii) G_{ab} is finite cyclic.

Then in §3 we relate such groups G to torsion in the general linear group $Gl_k(Z)$. The explicit problem we consider is the classification of congruence classes of extensions

$$1 \longrightarrow F \longrightarrow G \xrightarrow{\alpha} \pi \longrightarrow 1$$

where F is free of rank $k<\infty$, π is finite cyclic, and α is a homology isomorphism. The last section contains some applications. For example, if $k=2$ and $\pi \cong Z_6$ (the cyclic group of order 6) then the set of congruence classes of such extensions is in 1-1 correspondence with the cosets of $\langle T \rangle$ in $Gl_2(Z)$, where $\langle T \rangle$ is the subgroup of $Gl_2(Z)$ generated by the matrix

$$T = \begin{bmatrix} 0 & 1 \\ -1 & 1 \end{bmatrix}$$

Furthermore, it turns out that all such groups G are isomorphic to the projective special linear group $PSl_2(Z)$.

Finally I would like to acknowledge helpful conversations I had with S. Page, L. Roberts and R. Westwick.

§2. Homology Isomorphisms and Cyclic Groups: If π is a finite abelian group and $\varepsilon:G\to\pi$ is a homology isomorphism then it follows that $\ker\varepsilon=[G,G]$ and ε is, up to an isomorphism $G_{ab}\cong\pi$, the abelianization homomorphism $G\to G_{ab}$. Therefore, without loss of generality, we may assume that $\pi=G_{ab}$ and ε is abelianization. In fact we shall always use the letter ε to denote an abelianization homomorphism.

The notation $|G|$ is used for the order of a group G and (p,q) is used for the greatest common divisor of the integers p,q. Then we have

(2.1) Theorem: Let G be the free product A*B, where A and B are finite cyclic groups. Then

(i) $[G,G]$ is free of rank $(|A|-1)(|B|-1)$.

(ii) $\varepsilon:G\to G_{ab}$ is a homology isomorphism if, and only if, G_{ab} is cyclic.

(iii) G_{ab} is cyclic if, and only if, $(|A|,|B|) = 1$.

Proof: For any 2 abelian groups A,B the commutator subgroup of A*B is generated by the mixed commutators $[a,b]$, where $a\in A-1$ and $b\in B-1$. Using the normal form for words in a free product it follows that these commutators freely generate the commutator subgroup. Thus we have proved (i).

The inclusions $A\rightarrowtail G$, $B\rightarrowtail G$ induce Mayer-Vietoris isomorphisms

$$\tilde{H}_*(A)\oplus\tilde{H}_*(B)\xrightarrow{\cong}\tilde{H}_*(G)$$

On the other hand the homology of $G_{ab}=A\times B$ is computed by means of the Künneth formula

$$\tilde{H}_*(A)\oplus\tilde{H}_*(B)\oplus(\text{tensor and tor terms})\xrightarrow{\cong}\tilde{H}_*(G_{ab})$$

Restricting this isomorphism to the first 2 terms gives the commutative diagram

$$\begin{array}{ccc} & & \tilde{H}_*(G) \\ & \overset{\cong}{\nearrow} & \big\downarrow \varepsilon_* \\ \tilde{H}_*(A)\oplus\tilde{H}_*(B) & \searrow & \tilde{H}_*(G_{ab}) \end{array}$$

Therefore ε_* is an isomorphism if, and only if, the Künneth formula reduces to the isomorphism $\tilde{H}_*(A)\oplus\tilde{H}_*(B)\xrightarrow{\cong}\tilde{H}_*(G_{ab})$. But this happens if, and only if, the tensor and tor terms are zero, i.e., precisely when $(|A|,|B|)=1$.

Finally it is trivial that $G_{ab}=A\times B$ is cyclic if, and only if, A and B have relatively prime orders.

Q.E.D.

Next we consider amalgamated free products of finite cyclic groups. That is we put $G=A\underset{C}{*}B$, where A,B,C are all finite cyclic, and we want to find necessary and sufficient conditions in order that $\varepsilon:G\longrightarrow G_{ab}$ is a homology isomorphism.

The centre of G is the cyclic group C and the quotient group is just the free product ${}^{A}/_{C}*{}^{B}/_{C}$. Thus we have the central extension

$$1 \longrightarrow C \longrightarrow G \xrightarrow{\alpha} Q \longrightarrow 1, \quad Q=A/C*B/C$$

Since $H_2(Q)=0$ the Stallings-Stammbach [7], [9] exact sequence reduces to

(2.2) $$0 \longrightarrow C \longrightarrow G_{ab} \xrightarrow{\alpha_*} Q_{ab} \longrightarrow 0$$

These 2 exact sequences fit into the commutative diagram

(2.3) $$\begin{array}{ccccccccc} 1 & \longrightarrow & C & \longrightarrow & G & \xrightarrow{\alpha} & Q & \longrightarrow & 1 \\ & & \| & & \downarrow{\scriptstyle \varepsilon} & & \downarrow{\scriptstyle \varepsilon} & & \\ 0 & \longrightarrow & C & \longrightarrow & G_{ab} & \longrightarrow & Q_{ab} & \longrightarrow & 0 \end{array}$$

which in turn yields a commutative diagram of Stallings-Stammbach exact sequences

(2.4) $$\begin{array}{ccccccccccc} \cdots & \longrightarrow & H_2(G) & \longrightarrow & H_2(Q) & \longrightarrow & C & \longrightarrow & H_1(G) & \longrightarrow & H_1(Q) & \longrightarrow & 0 \\ & & \downarrow{\scriptstyle \varepsilon_*} & & \downarrow{\scriptstyle \varepsilon_*} & & \| & & \downarrow{\scriptstyle \varepsilon_*} & & \downarrow{\scriptstyle \varepsilon_*} & & \\ \cdots & \longrightarrow & H_2(G_{ab}) & \longrightarrow & H_2(Q_{ab}) & \longrightarrow & C & \longrightarrow & H_1(G_{ab}) & \longrightarrow & H_1(Q_{ab}) & \longrightarrow & 0 \end{array}$$

Now assume that $\varepsilon_*:H_*(G) \longrightarrow H_*(G_{ab})$ is an isomorphism. Then a simple diagram chase in (2.4) proves that $\varepsilon_*:H_2(Q) \longrightarrow H_2(Q_{ab})$ is an epimorphism. But $H_2(Q)=0$ and so $H_2(Q_{ab})=0$. Since Q_{ab} is the direct product $A/C \times B/C$ it follows that the orders of A/C, B/C must be relatively prime.

Conversely let us suppose that $(|A/C|,|B/C|)=1$. Then (2.1) shows that $\varepsilon:Q \longrightarrow Q_{ab}$ is a homology isomorphism. Applying a Hochschild-Serre spectral sequence argument to (2.3) now proves that $\varepsilon:G \longrightarrow G_{ab}$ is a homology isomorphism.

Putting these arguments together we see that we have proved part of the following theorem:

<u>(2.5) Theorem</u>: Suppose $G=A \underset{C}{*} B$ where A,B,C are finite cyclic groups. Then

(i) $[G,G]$ is free of rank $(|A/C|-1)(|B/C|-1)$.

(ii) $\varepsilon:G\longrightarrow G_{ab}$ is a homology isomorphism if, and only if, G_{ab} is cyclic.

(iii) G_{ab} is cyclic if, and only if, $(|A/C|,|B/C|)=1$.

Proof: Since $\alpha:G\twoheadrightarrow Q$ is an epimorphism it maps $[G,G]$ onto $[Q,Q]$. But from (2.2) we see that the composite

$$C\rightarrowtail G\twoheadrightarrow G_{ab}$$

is a monomorphism, and therefore $[G,G]\cap\ker\alpha = 1$. Hence $\alpha:G\twoheadrightarrow Q$ induces an isomorphism $[G,G]\cong[Q,Q]$. Part (i) now follows from (2.1). If $a_1=1, a_2,\cdots$(resp. $b_1=1, b_2,\cdots$) are coset representatives of C in A (resp. B) then it follows from the proof of (2.1) that the mixed commutators $[a_i,b_j]$, where $i,j\geq 2$, form a free basis for $[G,G]$.

The arguments immediately preceding (2.5) show that $\varepsilon:G\twoheadrightarrow G_{ab}$ is a homology isomorphism if, and only if, $(|A/C|,|B/C|)=1$. To conclude the proof of the theorem it suffices to show that $\varepsilon:G\twoheadrightarrow G_{ab}$ is a homology isomorphism if, and only if, G_{ab} is cyclic.

If G_{ab} is cyclic then so is Q_{ab} and hence, according to (2.1), $\varepsilon:Q\twoheadrightarrow Q_{ab}$ is a homology isomorphism. But a Hochschild-Serre spectral sequence argument applied to (2.3) shows that $\varepsilon:G\twoheadrightarrow G_{ab}$ is also a homology isomorphism.

Now assume that $\varepsilon:G\twoheadrightarrow G_{ab}$ is a homology isomorphism. Since G is the amalgamated free product $A\underset{C}{*}B$ we have the Mayer-Vietoris sequence

$$\cdots\longrightarrow H_2(A)\oplus H_2(B)\longrightarrow H_2(G)\longrightarrow C\longrightarrow A\oplus B\longrightarrow H_1(G)\longrightarrow 0$$

But $H_2(A)=H_2(B)=0$ and $C\longrightarrow A$, $C\longrightarrow B$ are monomorphisms, and therefore

$H_2(G)=0$. Thus we have $H_2(G_{ab})=0$ and from this it follows that G_{ab} is cyclic.

Q.E.D.

The 2 previous theorems provide us with numerous examples of groups satisfying (1.3). In order to be able to iterate these examples we need to know when an amalgamated free product $A\underset{C}{*}B$ satisfies (1.3), given that C is finite cyclic and A,B satisfy (1.3). The following theorem is quite helpful in this regard (see[3]).

Karass-Solitar Subgroup Theorem: Suppose H is a subgroup of the non-trivial amalgamated free product $G=A\underset{C}{*}B$ and suppose that $H\cap gCg^{-1}=1$ for all $g\varepsilon G$. Then there exist sets $\{g_\alpha\},\{g_\beta\}$ of double coset representatives for G mod (H,A), G mod (H,B) respectively and there exists a set $\{t_1,t_2,\cdots\}$, possibly empty, of elements of G such that

(i) $t_1,t_2,\cdots$ freely generate a free group F.

(ii) $H=F*(\underset{\alpha}{*}g_\alpha Ag_\alpha^{-1}\cap H)*(\underset{\beta}{*}g_\beta Bg_\beta^{-1}\cap H)$.

Remark: There is a more general subgroup theorem which holds when H meets some conjugates of C. However, in this case H is not a free product, but rather a free product with certain amalgamations.

(2.6) Theorem: Suppose A,B are groups with free commutator subgroups and that C is a finite abelian subgroup of both A and B. Then for the group $G=A\underset{C}{*}B$ we have

(i) $[G,G]$ is free

(ii) If $H_1(A)$, $H_1(B)$ are finite and $[A,A]$, $[B,B]$ have finite rank then $[G,G]$ also has finite rank.

Proof: First note that since C is finite and $[A,A]$, $[B,B]$ are free we must have $C\cap[A,A]=C\cap[B,B]=1$. That is, the natural morphisms $C\rightarrowtail H_1(A)$, $C\rightarrowtail H_1(B)$ are monomorphisms. Therefore we have the Mayer-Vietoris sequence

$$0\longrightarrow C\longrightarrow H_1(A)\oplus H_1(B)\longrightarrow H_1(G)\longrightarrow 0$$

But then it follows that $H_1(A)\longrightarrow H_1(G)$ is a monomorphism, for if $a\varepsilon H_1(A)$ is in the kernel then the element $(a,o)\varepsilon H_1(A)\oplus H_1(B)$ must come from C. However, $C\longrightarrow H_1(B)$ is a monomorphism and this forces a to be zero. Likewise $H_1(B)\longrightarrow H_1(G)$ is a monomorphism.

Thus we have proved that $A\cap[G,G]=[A,A]$ and $B\cap[G,G]=[B,B]$. It follows that $C\cap[G,G]=1$, and since $[G,G]$ is normal in G, this implies that $[G,G]$ satisfies the hypothesis for the Karass-Solitar subgroup theorem. Therefore there are elements $g_\alpha, g_\beta \varepsilon G$ such that

$$[G,G]=F*(\underset{\alpha}{*}g_\alpha A g_\alpha^{-1}\cap[G,G])*(\underset{\beta}{*}g_\beta B g_\beta^{-1}\cap[G,G])$$

for some free group F. Now

$$gAg^{-1}\cap[G,G]=gAg^{-1}\cap g[G,G]g^{-1}=g(A\cap[G,G])g^{-1}=g[A,A]g^{-1}$$

Likewise we have $gBg^{-1}\cap[G,G]=g[B,B]g^{-1}$. Hence

$$[G,G]=F*(\underset{\alpha}{*}g_\alpha[A,A]g_\alpha^{-1})*(\underset{\beta}{*}g_\beta[B,B]g_\beta^{-1})$$

All the factors in this product are free and so $[G,G]$ is also free.

To prove part (ii) note that $H_1(G)$ is finite if $H_1(A)$, $H_1(B)$ are finite. If in addition $[A,A]$, $[B,B]$ have finite rank then A,B are finitely generated, and hence G is finitely generated. Now $[G,G]$ is a subgroup of finite index in a finitely generated group and therefore $[G,G]$ is finitely generated. That is, $[G,G]$ is free of finite rank.

Q.E.D.

(2.7) Remark: The working ingredient in the above proof is the fact that $C \rightarrowtail H_1(A)$, $C \rightarrowtail H_1(B)$ are monomorphisms. It is because of this that $[G,G]$ has the form

$$[G,G]=F*(\ast_\alpha g_\alpha [A,A] g_\alpha^{-1})*(\ast_\beta g_\beta [B,B] g_\beta^{-1})$$

for some free group F. Also, it is important to realize that the sets $\{g_\alpha\}$, $\{g_\beta\}$ are not empty.

The next two theorems tell us that the group $G=A \ast_C B$, where C is finite cyclic, satisfies (1.3) if, and only if, A and B satisfy (1.3) and $H_1(G)$ is finite cyclic.

(2.8) Theorem: Suppose A,B satisfy the conditions of (1.3) and C is a finite cyclic subgroup of both A and B. Let G be the group $A \ast_C B$. Then

(i) $[G,G]$ is free of finite rank.

(ii) $\varepsilon: G \to G_{ab}$ is a homology isomorphism if, and only if, G_{ab} is cyclic.

(iii) G_{ab} is cyclic if, and only if, $(|H_1(A)/C|, |H_1(B)/C|)=1$.

Proof: Part (i) is a restatement of (2.6). Implicit in the statement of (iii) is that C is a subgroup of both $H_1(A)$ and $H_1(B)$. But this was noted in the proof of (2.6). In order to prove (ii) and (iii) let H be $H_1(A) \ast_C H_1(B)$. Then the abelianizations $\varepsilon: A \to A_{ab}$, $\varepsilon: B \to B_{ab}$ induce an epimorphism $\Theta: G \twoheadrightarrow H$, which in turn induces a morphism from the Mayer-Vietoris sequence of $G=A \ast_C B$ to that of $H=H_1(A) \ast_C H_1(B)$

$$\text{(2.9)}\qquad \begin{array}{ccccccccc} \cdots \to & H_i(C) & \longrightarrow & H_i(A)\oplus H_i(B) & \longrightarrow & H_i(G) & \longrightarrow & H_{i-1}(C) & \to \cdots \\ & \| & & \downarrow \varepsilon_* \oplus \varepsilon_* & & \downarrow \Theta_* & & \| & \\ \cdots \to & H_i(C) & \longrightarrow & H_i(A_{ab})\oplus H_i(B_{ab}) & \longrightarrow & H_i(H) & \longrightarrow & H_{i-1}(C) & \to \cdots \end{array}$$

Since $\varepsilon:A \to A_{ab}$, $\varepsilon:B \to B_{ab}$ are homology isomorphisms it follows from the 5 lemma that $\Theta:G \to H$ is also a homology isomorphism.

Now consider the commutative diagram

(2.10).
$$\begin{array}{ccc} G & \xrightarrow{\Theta} & H \\ \downarrow{\scriptstyle\varepsilon} & & \downarrow{\scriptstyle\varepsilon} \\ G_{ab} & \xrightarrow[\cong]{\Theta^*} & H_{ab} \end{array}$$

Thus $\varepsilon:G \to G_{ab}$ is a homology isomorphism if, and only if, $\varepsilon:H \to H_{ab}$ is one. However, from (2.5) we know that $\varepsilon:H \to H_{ab}$ is a homology isomorphism if, and only if, H_{ab} is cyclic, and this happens exactly when the orders of $H_1(A)/C$, $H_1(B)/C$ are relatively prime. This concludes the proof of (2.8).

Q.E.D.

(2.11) Theorem: Suppose $G = A \underset{C}{*} B$, where C is finite cyclic. If G satisfies (1.3) then so do A and B.

Proof: The commutator subgroups $[A,A]$, $[B,B]$ are free since they are subgroups of the free group $[G,G]$. Thus $C \rightarrowtail H_1(A)$ and $C \rightarrowtail H_1(B)$ are monomorphisms and we have the Mayer-Vietoris sequence

$$0 \to C \to H_1(A) \oplus H_1(B) \to H_1(G) \to 0$$

It follows, as in the proof of (2.6), that $H_1(A) \to H_1(G)$ and $H_1(B) \to H_1(G)$ are monomorphisms. Therefore $H_1(A)$, $H_1(B)$ are finite cyclic. From (2.7) we see that $[A,A]$, $[B,B]$ have finite rank.

To conclude the proof we put $H = H_1(A) \underset{C}{*} H_1(B)$. Then $H_{ab} \cong G_{ab}$ is finite cyclic and consequently (2.5) implies that $\varepsilon:H \to H_{ab}$ is a homology isomorphism. From the commutative diagram (2.10) it now follows that $\Theta:G \to H$ is a homology isomorphism. Finally, the 5 lemma applied to (2.9) shows that $\varepsilon:A \longrightarrow A_{ab}$ and $\varepsilon:B \longrightarrow B_{ab}$ are homology isomorphisms.

Q.E.D.

Theorem (2.8) allows us to build up a large collection of examples of groups satisfying (1.3). Specifically, let $\mathcal{C}$ be the smallest set of (isomorphism classes) of groups such that

(i) $\mathcal{C}$ contains all finite cyclic groups.

(ii) if $A,B \in \mathcal{C}$ and C is a finite cyclic subgroup of A,B then $A *_C B \in \mathcal{C}$.

Simple induction arguments allow one to establish some properties for groups in $\mathcal{C}$. To set up these arguments we define classes $\mathcal{C}_i$, $i \geq 1$, inductively as follows:

$\mathcal{C}_1$ = the class of all finite cyclic groups. Assuming $\mathcal{C}_1, \cdots, \mathcal{C}_n$ have already been defined we define $\mathcal{C}_{n+1}$ by the requirements:

$\mathcal{C}_{n+1} \supseteq \mathcal{C}_n$; if $A,B \in \mathcal{C}_n$ and C is a finite cyclic subgroup of A,B then $A *_C B \in \mathcal{C}_{n+1}$.

The first requirement is redundant since we are not disallowing non-proper free products (i.e., $A *_C C \cong A$) in the second requirement. In any case we have

$$\mathcal{C}_1 \subseteq \mathcal{C}_2 \subseteq \cdots \quad \text{and} \quad \mathcal{C} = \bigcup_{n=1}^{\infty} \mathcal{C}_n$$

Such a representation for $\mathcal{C}$ makes induction on n possible. For example, it is easy to see that if $G \in \mathcal{C}$ then G is finitely presented, [G,G] is free of finite rank, and $H_1(G)$ is finite. In a similar vein we have

(2.12) Theorem: Let G be a group in $\mathcal{C}$. Then

(i) for all $i\geq 1$ $H_{2i}(G)=0$ and $H_{2i-1}(G)\cong G_{ab}$.

(ii) if F is a finite subgroup of G then F is cyclic and the inclusion $\iota: F\subseteq G$ induces a monomorphism $\iota_*: H_*(F)\rightarrowtail H_*(G)$.

(iii) rank $[G,G]=1$ if, and only if, $G\cong Z_{2m} \underset{Z_m}{*} Z_{2m}$ for some $m\geq 1$.

Proof: Parts (i) and (ii) are true for cyclic groups. Assume that we have proved (i) and (ii) for all groups in $\mathcal{C}_n$. If $G\varepsilon\mathcal{C}_{n+1}-\mathcal{C}_n$ then G is a non-trivial free product.

$G=A\underset{C}{*}B$, where C is finite cyclic and $A,B\varepsilon\mathcal{C}_n$.

By induction $H_*(C)\rightarrowtail H_*(A)$, $H_*(C)\rightarrowtail H_*(B)$ are monomorphisms and therefore the Mayer-Vietoris sequence for G reduces to short exact sequences

$$0\longrightarrow H_j(C)\longrightarrow H_j(A)\oplus H_j(B)\longrightarrow H_j(G)\longrightarrow 0,\ j\geq 1$$

For $j=2i$ we have $H_{2i}(A)=H_{2i}(B)=0$ and so $H_{2i}(G)=0$. On the other hand if $j=2i-1$ then there are isomorphisms $H_{2i-1}(A)\cong A_{ab}$, $H_{2i-1}(B)\cong B_{ab}$; and we can choose these isomorphisms so that

$$\begin{array}{ccc} H_{2i-1}(C) & \rightarrowtail & H_{2i-1}(A)\oplus H_{2i-1}(B) \\ \wr\Vert & & \wr\Vert \\ C & \rightarrowtail & A_{ab}\oplus B_{ab} \end{array}$$

is a commutative diagram, where the horizontal morphisms come from Mayer-Vietoris sequences. Thus $H_{2i-1}(G)\cong G_{ab}$, and this proves (i).

To substantiate the above claim we need only show that for any 2 generators c_1, c_2 of a subgroup C of the cyclic group $D=\{x\mid x^p=1\}$

there exists an automorphism $f \in Aut(D)$ such that $f(c_1)=c_2$. Thus suppose C has order r. Then we can find an integer s such that

$$C \text{ is generated by } x^s \text{ and } s|p$$

Therefore $r=p/s$ and any other generator of C has the form x^{st} for some t relatively prime to r. To prove the existence of the automorphism f we must show that there exists an integer n satisfying

$$(n,p)=1 \text{ and } x^{sn}=x^{st}$$

But $x^{sn}=x^{st}$ if, and only if $n \equiv t\ (r)$. Thus we must find an integer m so that $(t+mr,p)=1$. Since $(t,r)=1$ we may appeal to the Dirichlet theorem on primes in an arithmetic progression to establish the existence of such an integer m.

Now suppose F is a finite subgroup of G. Then F is conjugate to a subgroup of either A or B and therefore is cyclic. For arguments sake suppose there is an inner automorphism $c:G \to G$ such that

$$c(F)=F_1=\text{a subgroup of } A.$$

Thus we have the commutative diagram

$$\begin{array}{ccc} H_*(F) & \xrightarrow{\iota_*} & H_*(G) \\ \cong \downarrow c_* & & \cong \downarrow c_* \\ H_*(F_1) & \xrightarrow{\iota_*} & H_*(G) \end{array}$$

where $\iota_*:H_*(F_1) \to H_*(G)$ is the composite $H_*(F_1) \rightarrowtail H_*(A) \to H_*(G)$. By induction $H_*(C) \rightarrowtail H_*(A)$, $H_*(C) \rightarrowtail H_*(B)$ are monomorphisms and so we have the short exact sequence

$$0 \to H_*(C) \to H_*(A) \oplus H_*(B) \to H_*(G) \to 0$$

and it follows, as in the proof of (2.6), that $H_*(A) \to H_*(G)$ is a monomorphism. Hence so is $\iota_*:H_*(F) \to H_*(G)$. This proves (ii).

To prove (iii) observe that $Z_{2m} \underset{Z_m}{*} Z_{2m}$ is in $\mathcal{C}_2$ and, according to (2.5), the commutator subgroup is free of rank 1. For the converse suppose $G \varepsilon \mathcal{C}$ is such that rank $[G,G] = 1$. Then G is not cyclic and so $G = A \underset{C}{*} B$ in a non-trivial way, where C is finite cyclic. Since $H_1(G)$ is finite we have

number of ends of G = number of ends of Z = 2

But the number of ends of an amalgamated product $A \underset{C}{*} B$ is infinite if either $[A:C] > 2$ or $[B:C] > 2$ (see [8]). Therefore $[A:C] = [B:C] = 2$. Since the only finite groups in $\mathcal{C}$ are cyclic this forces A, B to be cyclic of order 2m, where $m = |C|$.

Q.E.D.

<u>(2.13) Theorem:</u> A group G satisfies (1.3) if, and only if, $G \varepsilon \mathcal{C}$ and $H_1(G)$ is cyclic.

<u>Proof</u>: First we'll show that G satisfies (1.3) if $G \varepsilon \mathcal{C}$ and $H_1(G)$ is cyclic. This is obviously true if $G \varepsilon \mathcal{C}_1$. Thus suppose we have proved it for all groups in $\mathcal{C}_n$. If $G \varepsilon \mathcal{C}_{n+1}$ then $G = A \underset{C}{*} B$ where C is finite cyclic and $A, B \varepsilon \mathcal{C}_n$. The usual Mayer-Vietoris argument proves that $H_1(A)$, $H_1(B)$ are finite cyclic. Induction now implies that A,B satisfy (1.3) and therefore so does G according to (2.8).

Conversely, suppose G satisfies (1.3). Then we have the extension

$$1 \longrightarrow [G,G] \longrightarrow G \longrightarrow G_{ab} \longrightarrow 1$$

where $[G,G]$ is free of finite rank.

If rank $[G,G] = 1$ then G has exactly 2 ends and so there is a finite normal subgroup N of G such that

either $G/N \cong Z$ or $G/N \cong Z_2 * Z_2$ (see [8])

However, both possibilities contradict the hypothesis that $H_1(G)$ is finite cyclic. Therefore rank $[G,G]\neq 1$.

If rank $[G,G]=0$ then G is cyclic and we are done. Thus suppose rank $[G,G]\geq 2$. Then number of ends of G = number of ends of $[G,G] = \infty$. By the Stallings structure theorem for groups with infinitely many ends [8] it follows that either G is an H.N.N. construction or $G\cong A\underset{C}{*}B$, where C is a proper finite subgroup of A,B and of index at least 3 in one of them.

If G is an H.N.N. construction then $H_1(G)\cong Z\oplus *$ is not finite and so $G\cong A\underset{C}{*}B$. Since C is finite and $[G,G]$ is free we have $C\cap[G,G]=1$. Thus C is a subgroup of $H_1(G)$ and therefore cyclic. From (2.11) it now follows that A,B satisfy (1.3), and therein lies the basis for an induction.

The induction will be on the rank of $[G,G]$. That is, we assume that if H is a group satisfying (1.3) and also rank $[H,H]<$rank $[G,G]$ then $H\in\mathcal{C}$. Notice that the induction starts with the cyclic groups.

From the Karrass-Solitar subgroup theorem there is a free group F and non-empty sets $\{g_\alpha\}$, $\{g_\beta\}$ such that

$$[G,G]=F*(\underset{\alpha}{*}g_\alpha[A,A]g_\alpha^{-1})*(\underset{\beta}{*}g_\beta[B,B]g_\beta^{-1})$$

Thus we certainly have rank $[A,A]$, rank $[B,B]\leq$rank $[G,G]$. If rank $[A,A]$ = rank $[G,G]$ then $[A,A] = [G,G]$, i.e., every commutator in G can be expressed in terms of commutators in A. Now consider a mixed commutator $[a,b]$, where $a\in A-C$ and $b\in B-C$. This element can be chosen to be in normal form and so is not in $[A,A]$. We therefore have a contradiction; hence rank $[A,A]$ < rank $[G,G]$. Likewise rank $[B,B]$ <

rank $[G,G]$. By induction $A,B\in\mathcal{C}$, and by the definition of $\mathcal{C}$ we finally have that $G=A\underset{C}{*}B$ belongs to $\mathcal{C}$.

Q.E.D.

We conclude this section by giving an example of (2.13). Suppose $A_1,\cdots,A_{n+1}, C_1,\cdots,C_n$ are finite cyclic groups of respective orders $a_1,\cdots,a_{n+1}, c_1,\cdots,c_n$. Moreover, suppose that C_i is a subgroup of both A_i and A_{i+1} for $1\le i\le n$. Then the iterated product $G=A_1\underset{C_1}{*}\cdots\underset{C_n}{*}A_{n+1}$ makes sense and we have

<u>(2.14) Corollary</u>: G satisfies (1.3) if, and only if,

$$\left(\frac{a_1\ldots a_i}{c_1\ldots c_i}\ ,\ \frac{a_{i+1}}{c_i}\right) = 1 \text{ for } 1\le i\le n.$$

<u>Proof</u>: The group G is in $\mathcal{C}$ and therefore we automatically have that $[G,G]$ is free of finite rank. By induction on n we can prove that the order of G_{ab} (for any such group G) is given by

$$|G_{ab}| = \frac{a_1\ldots a_{n+1}}{c_1\ldots c_n}$$

Now we express G as follows

$$G=A\underset{C}{*}B,\text{ where } A=A_1\underset{C_1}{*}\cdots\underset{C_{n-1}}{*}A_n,\ B=A_{n+1},\ C=C_n.$$

According to (2.8) and (2.11) G satisfies (1.3) if, and only if, A satisfies (1.3) and $(|H_1(A)/C|,|B/C|)=1$. But we now have by induction on n

(i) A satisfies (1.3) if, and only if,

$$\left(\frac{a_1\ldots a_i}{c_1\ldots c_i}\ ,\ \frac{a_{i+1}}{c_i}\right)= 1 \text{ for } 1\le i\le n-1.$$

(ii) $(|H_1(A)/C|,|B/C|) = 1$ if, and only if,

$$\left(\frac{a_1\ldots a_n}{c_1\ldots c_n},\ \frac{a_{n+1}}{c_n}\right) = 1.$$

Finally, notice that (2.5) starts the induction.

Q.E.D.

§3 Torsion Automorphisms: In the last section we completely determined all those groups G such that $[G,G]$ is free of finite rank, $H_1(G)$ is finite cyclic, and $\varepsilon:G\twoheadrightarrow G_{ab}$ is a homology equivalence. The purpose of this section is to study the structure of all such pairs (G,ε). Thus, suppose we are given an extension

(3.1) $$1\longrightarrow F\longrightarrow G\overset{\alpha}{\longrightarrow}\pi\longrightarrow 1$$

where $F\cong F_k$ is a free group of rank $k<\infty$ and π is an arbitrary group. If $k\neq 1$ then the congruence class of (3.1) is determined completely by the associated abstract kernel $\rho:\pi\rightarrow\mathcal{O}(F)$, where $\mathcal{O}(F)$ is the group of outer automorphisms of F. The reason for this is that F has trivial centre when $k\neq 1$.

We shall always assume that a fixed basis $x_1,\ldots,x_k$ has been chosen for F. This then gives a basis for F_{ab} and it is a classical result that the natural homomorphism $Aut(F)\longrightarrow Aut(F_{ab})=Gl_k(Z)$ is an epimorphism. Clearly all inner automorphisms of F map to the identity matrix and therefore there is the induced epimorphism

$$\kappa:\mathcal{O}(F)\twoheadrightarrow Gl_k(Z)$$

To the extension (3.1) we now associate the representation $\kappa\rho:\pi\rightarrow Gl_k(Z)$. This gives the usual π module structure on F_{ab}. Then we have

(3.2) Theorem: $\alpha_*:H_*(G)\longrightarrow H_*(\pi)$ is an isomorphism if, and only if, $H_*(\pi;F_{ab})=0$.

Proof: We prove this by means of the Hochschild-Serre spectral sequence

$$E^2_{s,t}\cong H_s(\pi;H_t(F))\Longrightarrow H_{s+t}(G)$$

As usual there is a filtration on $H_*(G)$ so that the composition factors are the E^∞ terms, i.e.,

$$H_n(G)=A_{n,o}\supseteq A_{n-1,1}\supseteq\ldots\supseteq A_{o,n}\supseteq 0$$

$$\text{where } E^\infty_{s,t}\cong A_{s,t}/A_{s-1,t+1}$$

Since F is free we have for all r, $2\le r\le\infty$, that $E^r_{s,t}=0$ if $t\neq o,1$. In particular it follows that the only possible non-zero differential is $d^2:E^2_{s,o}\longrightarrow E^2_{s-2,1}$ and hence $E^3=E^4=\ldots=E^\infty$. The edge homomorphism $\alpha_*:H_n(G)\longrightarrow H_n(\pi)$ is the composite

$$H_n(G)=A_{n,o}\twoheadrightarrow A_{n,o}/A_{n-1,1}\cong E^\infty_{n,o}\rightarrowtail E^2_{n-o}\cong H_n(\pi)$$

Therefore $\alpha_*:H_n(G)\longrightarrow H_n(\pi)$ is an isomorphism if, and only if,

$$E^\infty_{n-1,1}=0 \text{ and } E^\infty_{n,o}=E^2_{n,o}$$

But this happens if, and only if,

$$d^2:E^2_{n+1,o}\longrightarrow E^2_{n-1,1} \text{ is an epimorphism, and}$$

$$d^2:E^2_{n,o}\longrightarrow E^2_{n-2,1} \text{ is the zero homomorphism.}$$

Thus $\alpha_*:H_n(G)\longrightarrow H_n(\pi)$ is an isomorphism for all n if, and only if, $E^2_{*,1}\cong H_*(\pi;F_{ab})=0$.

Q.E.D.

Now we specialize to the situation $\pi\cong Z_n$. If $t\varepsilon\pi$ is a fixed generator then the extension (3.1) determines the automorphisms

$$\tau=\rho(t)\varepsilon\mathcal{O}(F) \text{ and } T=\kappa(\tau)\varepsilon Gl_k(Z)$$

For the remainder of this section t, τ, T will always be understood in this context.

The group ring $Z[\pi]$ is the truncated polynomial ring $Z[t]/(1-t^n)$ and the augmentation ideal I is the principal ideal (1-t). That is, every element in I has a representative of the form (1-t)p(t). For any π module M there is the norm homomorphism

$$N: M \longrightarrow M, \quad m \mapsto \sum_{x\epsilon\pi} x.m$$

Some standard notation is

M^{π} = the π invariant part of M

NM = the image of the norm homomorphism

${}_N M$ = the kernel of the norm homomorphism

Then the homology of π is given as follows

$$H_o(\pi;M) \cong M/IM, \; H_1(\pi;M) \cong M^{\pi}/NM, \; H_2(\pi;M) \cong {}_N M/IM$$

Since $H_*(\pi;M)$ is periodic of period 2 this gives $H_*(\pi;M)=0$ if, and only if,

$$M=IM={}_N M \text{ and } M^{\pi} = NM$$

Now suppose that M is the π module F_{ab} coming from the extension (3.1). Thus the action of the generator $t\epsilon\pi$ is given by the matrix $T\epsilon Gl_k(Z)$. If $H_*(\pi;M)=0$ then M=IM, which is equivalent to $I-T:M \longrightarrow M$ being an epimorphism. Hence, if $H_*(\pi;M)=0$ then $I-T\epsilon Gl_k(Z)$.

Conversely, suppose that $I-T\epsilon Gl_k(Z)$. Then M=IM and +1 is not an eigenvalue of T. Therefore $M^{\pi}=0$. Since $NM \subseteq M^{\pi}$ it follows that the norm homomorphism $N:M \longrightarrow M$ is the zero homomorphism. Thus it follows that I-T being invertible over Z implies that $M=IM={}_N M$ and $M^{\pi}=NM=0$, i.e., $H_*(\pi;M)=0$.

This proves the following corollary.

<u>(3.3) Corollary:</u> $H_*(\pi;F_{ab})=0$ if, and only if, $\det(I-T)=\pm 1$.

Putting (3.2) and (3.3) together then gives

<u>(3.4) Theorem:</u> Suppose given an extension $1 \longrightarrow F \longrightarrow G \xrightarrow{\alpha} \pi \longrightarrow 1$ where F is

a free group of rank $k<\infty$ and π is a finite cyclic group. Let $T\varepsilon Gl_k(Z)$ be the linear automorphism determined by a fixed generator $t\varepsilon\pi$. Then α is a homology isomorphism if, and only if, $I-T\varepsilon Gl_k(Z)$.

If G is a group satisfying (1.3) then we know that rank $[G,G]\neq 1$ (see the proof of (2.13)). Therefore, if $1\longrightarrow F\longrightarrow G\xrightarrow{\alpha}\pi\longrightarrow 1$ is an extension as in (3.4) and if α is a homology isomorphism then rank $F\neq 1$. This implies that the congruence class of the extension is determined by the element $\tau\varepsilon\mathcal{O}(F)$. Indeed the extension is congruent to the induced extension (see chapter 4 of [5]).

$$(E_\tau)\qquad \begin{array}{ccccccccc} 1 & \longrightarrow & F & \longrightarrow & G_\tau & \longrightarrow & \pi & \longrightarrow & 1 \\ & & \| & & \downarrow & & \downarrow\rho & & \\ 1 & \longrightarrow & F & \longrightarrow & \mathrm{Aut}F & \xrightarrow{\omega} & \mathcal{O}(F) & \longrightarrow & 1 \end{array}$$

$$G_\tau=\{(t^a,\sigma)\varepsilon\pi\times\mathrm{Aut}F\,|\,\omega(\sigma)=\tau^a\}$$

<u>(3.5) Theorem:</u> Suppose F is a free group of rank $k<\infty$ and π is a cyclic group of order n. Then there exists a 1-1 correspondence between congruence classes of extensions $1\longrightarrow F\longrightarrow G\xrightarrow{\alpha}\pi\longrightarrow 1$ such that $\alpha_*:H_*(G)\longrightarrow H_*(\pi)$ is an isomorphism, and outer automorphisms $\tau\varepsilon\mathcal{O}(F)$ such that $\tau^n=1$ and $I-T\varepsilon Gl_k(Z)$.

Therefore the study of all such pairs (G,α) is equivalent to the following

(3.6)
(i) find all matrices $T\varepsilon Gl_k(Z)$ such that $T^n=I$ and $\det(I-T)=\pm 1$.
(ii) for any such matrix T determine all $\tau\varepsilon\mathcal{O}(F)$ satisfying $\kappa(\tau)=T$ and $\tau^n=1$.

The first thing to be noticed about (i) is that the 2 properties listed for T depend only on its similarity class over the integers, i.e., if $U\varepsilon Gl_k(Z)$ and $S=UTU^{-1}$ then $S^n=I$ (resp. $I-S\varepsilon Gl_k(Z)$) if, and only if,

$T^n=I$ (resp. $I-T\varepsilon Gl_k(Z)$). The second thing to be noticed is that a matrix $T\varepsilon Gl_k(Z)$ will satisfy these properties if, and only if, T considered as a matrix over the rational number field Q satisfies them.

A classical result is that if $A\varepsilon Gl_k(Q)$ has order n then A is similar over Q to a matrix $T\varepsilon Gl_k(Z)$, which necessarily has order n. That is, the canonical inclusion $Gl_k(Z)\subseteq Gl_k(Q)$ induces an epimorphism from the similarity classes of torsion in $Gl_k(Z)$ to those in $Gl_k(Q)$. Unfortunately it is possible to have matrices $S,T\varepsilon Gl_k(Z)$ of finite order which are similar over Q but not over Z. A particularly easy example for order 2 is given by

$$S = \begin{bmatrix} -1 & 1 \\ 0 & 1 \end{bmatrix} \qquad T = \begin{bmatrix} -1 & 0 \\ 0 & 1 \end{bmatrix}$$

In fact the similarity classification over Z of integral matrices is still an open problem. For example, the general linear group $Gl_{p-1}(Z)$, where p is a prime, contains p torsion. If $A\varepsilon Gl_{p-1}(Z)$ has order p then the minimal polynomial of A is the cyclotomic polynomial $1+x+\ldots+x^{p-1}$ and A is similar over Q to the $(p-1)x(p-1)$ companion matrix

$$\begin{bmatrix} 0 & 1 & 0 & \ldots\ldots & 0 \\ & & \ddots & & \vdots \\ & & & \ddots & 0 \\ & & & & 1 \\ -1 & -1 & \ldots & \ldots & -1 \end{bmatrix}$$

Therefore there is only one canonical form over Q for p torsion in $Gl_{p-1}(Z)$. However, it follows from a result of Latimer and MacDuffee [4] that the number of similarity classes of p torsion in $Gl_{p-1}(Z)$ is the ideal class number of the cyclotomic extension $Q(\zeta)$, where $\zeta=e^{2\pi i/p}$.

To solve (3.6) we must first find explicit canonical forms for torsion in $Gl_k(Q)$. These forms do not seem to be in the literature, but rather than derive the results here we shall simply state them and publish the details elsewhere.

Similarity over Z or Q will be denoted by $\underset{Z}{\sim}$, $\underset{Q}{\sim}$ respectively, and for a square matrix A we shall use the notation $\mu_A(x)$, $\gamma_A(x)$ for the minimal and characteristic polynomials. To any monic polynomial $p(x)=a_o+a_1x+\ldots+x^k$ we associate the kxk companion matrix

$$C(p(x)) = \begin{bmatrix} 0 & 1 & 0 \ldots 0 \\ & & 0 \\ & & 1 \\ -a_o & \ldots & -a_{k-1} \end{bmatrix}$$

where this is to be interpreted as $[-a_o]$ when k=1. Since $\det C(p(x))=\pm a_o$ we see that $C(p(x))\varepsilon Gl_k(Z)$ if the a_i are integers and $a_o=\pm 1$.

For any integer $d\geq 1$ let $\Phi_d(x)$ be the cyclotomic polynomial associated to the primitive d^{th} roots of unity. If C(d) denotes the companion matrix of $\Phi_d(x)$ then

$$C(d)\varepsilon Gl_{\phi(d)}(Z) \text{ has order exactly } d,$$

where $\phi(d)$ is the Euler totient function. It follows that the matrix $e_1C(d_1)\oplus\ldots\oplus e_rC(d_r)$ has size kxk and order n, where

$$k=e_1\phi(d_1)+\ldots+e_r\phi(d_r), \text{ and}$$

$$n=\text{the least common multiple of the } d_i.$$

Conversely, from the theory of canonical forms over Q, we have

<u>(3.7) Theorem</u>: Suppose $A\varepsilon Gl_k(Q)$ has order n. Then there are unique distinct divisors $d_1,\ldots,d_r$ of n and unique positive integers $e_1,\ldots,e_r$ such that

(i) $A \underset{Q}{\sim} e_1C(d_1) \oplus \ldots \oplus e_rC(d_r)$.

(ii) n=the least common multiple of $d_1, \ldots, d_r$.

(iii) $k=e_1\phi(d_1)+\ldots+e_r\phi(d_r)$.

(3.8) Remark: Since $Gl_k(Q)$ has torsion of order n if, and only if, $Gl_k(Z)$ does, it follows that (ii) and (iii) of (3.7) are also valid for $Gl_k(Z)$. However, (i) does not give the similarity classification over Z.

If all the e_i in (3.7) are equal to 1 then

$$\mu_A(x)=\gamma_A(x)=\Phi_{d_1}(x)\ldots\Phi_{d_r}(x) \qquad (3.9)$$

Conversely, if A is a torsion matrix whose minimal and characteristic polynomials are equal then the rational canonical form is given by (3.7) with all e_i=1. In accordance with terminology in linear algebra we refer to such matrices as non-derogatory torsion (recall that a matrix A is said to be non-derogatory if $\mu_A(x)=\gamma_A(x)$).

The theorem of Latimer and MacDuffee [4] referred to above, is concerned with the similarity classification over Z of non-derogatory matrices with a fixed minimal polynomial. To be precise suppose $\mu(x)=a_0+a_1x+\ldots+x^k$ is a monic polynomial over the integers such that $a_0 \neq 0$ and $\mu(x)$ has no repeated factors in its factorization into primes. Then

Theorem (Latimer and MacDuffee): There is a 1-1 correspondence between similarity classes of kxk integral matrices A satisfying $\mu_A(x)=\mu(x)$ and classes of non-singular ideals in the truncated polynomial ring $Z[x]/(\mu(x))$.

We can use this theorem to get a hold on non-derogatory torsion in $Gl_k(Z)$, because if A is such a matrix then (3.9) states that the hypothesis for the Latimer-MacDuffee result are satisfied.

<u>(3.10) Theorem</u>: Suppose $A\varepsilon Gl_k(Z)$ is a non-derogatory matrix of order n. Then there exist unique distinct divisors $d_1,\ldots,d_r$ of n such that

(i) $A \underset{Q}{\sim} C(d_1)\oplus\ldots\oplus C(d_r)$.

(ii) n=the least common multiple of $d_1,\ldots,d_r$.

(iii) $k=\phi(d_1)+\ldots+\phi(d_r)$.

Moreover, the Z similarity classes of such A having the rational canonical form $C(d_1)\oplus\ldots\oplus C(d_r)$ are in 1-1 correspondence with the classes of non-singular ideals in the ring $Z[x]/\Phi_{d_1}(x)\ldots\Phi_{d_r}(x))$.

Given an integer n we would like to determine the least k for which $Gl_k(Z)$ has n torsion. If $A\varepsilon Gl_k(Z)$ has order n and k is minimal then A must be non-derogatory. Thus we must minimize the sum $k=\phi(d_1)+\ldots+\phi(d_r)$ subject to the constraint that $d_1,\ldots,d_r$ are distinct divisors of n with least common multiple equal to n.

<u>(3.11) Theorem</u>: Suppose $n=p_1^{s_1}\ldots p_t^{s_t}$ is the prime factorization of n. Then the least k for which $Gl_k(Z)$ has n torsion is given by

$$k = \begin{cases} 1 \text{ if } n=2 \\ \phi(p_1^{s_1})+\ldots+\phi(p_t^{s_t}) \text{ if } n\not\equiv 2(4) \\ \phi(p_1^{s_1})+\ldots+\phi(p_t^{s_t})-1 \text{ if } n\equiv 2(4),\ n>2. \end{cases}$$

The rational canonical forms for non-derogatory torsion are given as follows:

<u>(3.12) Theorem</u>: Let $n=p_1^{s_1}\ldots p_t^{s_t}$ be the prime factorization of n, where $2\le p_1<\ldots<p_t$. If k is the least integer for which there exists $A\varepsilon Gl_k(Z)$ having order n then

(i) either $A \underset{Q}{\sim} C(4)\oplus C(3)\oplus C(p_3^{s_3})\oplus\ldots\oplus C(p_t^{s_t})$ or

$A \underset{Q}{\sim} C(12)\oplus C(p_3^{s_3})\oplus\ldots\oplus C(p_t^{s_t})$ if $n=4\cdot 3\cdot p_3^{s_3}\ldots p_t^{s_t}$.

(ii) $A \underset{Q}{\sim} C(p_1^{s_1}) \oplus \ldots \oplus C(p_t^{s_t})$ if $n \neq 4\cdot 3\cdot p_3^{s_3}\ldots p_t^{s_t}$ and $n \not\equiv 2$ (4).

(iii) $A \underset{Q}{\sim} C(2^{\varepsilon_2} p_2^{s_2}) \oplus \ldots \oplus C(2^{\varepsilon_t} p_t^{s_t})$ if $n \equiv 2$ (4), where the ε_j are either o or 1 and $\varepsilon_j = 1$ for at least one j.

To solve the first part of (3.6) we must still determine for which torsion matrices $T \in Gl_k(Z)$ we also have $I - T \in Gl_k(Z)$. From (3.7) we have

$$T \underset{Q}{\sim} e_1 C(d_1) \oplus \ldots \oplus e_r C(d_r)$$

It follows that $I - T \in Gl_k(Z)$ if, and only if, $I - C(d_i)$ has determinant ± 1 for $1 \leq i \leq r$.

For any integer d the characteristic polynomial of C(d) is by definition $\det(xI - C(d))$. Thus $I - C(d)$ is invertible over Z if, and only if, $\Phi_d(1) = \pm 1$.

<u>(3.13) Lemma</u>: $\Phi_d(1) \geq 1$ for all $d > 1$, with equality if, and only if, d is not a prime power.

<u>Proof</u>: Let $n \geq 2$ be any integer. Since $\Phi_1(x) = x - 1$ we have the factorization

$$1 + x + \ldots + x^{n-1} = \prod_{\substack{d|n \\ d \neq 1}} \Phi_d(x)$$

Putting x = 1 gives $n = \prod_{d|n, d\neq 1} \Phi_d(1)$. Because $\Phi_2(1) = 2$ we can use induction on n to prove that $\Phi_d(1) \geq 1$ for all $d > 1$.

Now consider the identity

$$1 + x + \ldots + x^{n-1} = \prod_{p^s|n} \Phi_{p^s}(x) \prod_{d|n} \Phi_d(x)$$

where the first product is over all non-trivial prime power divisors and the second product is over all divisors which are not prime powers. But

$$\Phi_{p^s}(x) = x^{p^{s-1}(p-1)} + x^{p^{s-1}(p-2)} + \ldots + x^{p^{s-1}} + 1$$

and therefore $\Phi_{p^s}(1) = p$. Hence $\prod_{p^s|n} \Phi_{p^s}(1) = n$ and it follows that $\Phi_d(1) = 1$ for all divisors d in the second product. Since n was arbitrary this concludes the proof.

Q.E.D.

The following theorem is our solution to the first part of (3.6).

<u>(3.14) Theorem</u>: Let k,n be fixed integers. Then there exist matrices $T \varepsilon Gl_k(Z)$ such that n=order T and $I-T \varepsilon Gl_k(Z)$ if, and only if, there are distinct divisors $d_1,\ldots,d_r$ of n and positive integers $e_1,\ldots,e_r$ such that

(i) n=the least common multiple of the d_i.

(ii) the d_i are not prime powers.

(iii) $k=e_1\phi(d_1)+\ldots+e_r\phi(d_r)$.

Moreover $T \underset{Q}{\sim} e_1C(d_1) \oplus \ldots \oplus e_rC(d_r)$.

Since the similarity classification over Z for torsion in $Gl_k(Z)$ is still an open problem it is not possible to give a complete solution to part (ii) of (3.6). However, we do have partial results.

Recall that we are assuming that the free group F has a preferred basis $x_1,\ldots,x_k$. Then to any monic polynomial $p(x)=a_o+a_1x+\ldots+x^k \varepsilon Z[x]$ we associate the word $W=x_1^{-a_o}x_2^{-a_1}\ldots x_k^{-a_{k-1}}$ and the corresponding endomorphism

$$f_W : F \longrightarrow F \qquad \begin{array}{l} x_1 \longrightarrow x_2 \\ \vdots \qquad \vdots \\ x_{k-1} \longrightarrow x_k \\ \\ x_k \longrightarrow W \end{array}$$

The induced endomorphism on F_{ab} is given by the companion matrix $C(p(x))$.

If $a_o = \pm 1$ then f_W is an automorphism. In particular, for the cyclotomic polynomials $\Phi_d(x)$ we have automorphisms $f_d \varepsilon Aut(F_{\phi(d)})$ such that the associated matrix is $C(d) \varepsilon Gl_{\phi(d)}(Z)$. Thus the f_d are at least candidates for automorphisms of finite order. It is a result of Baumslag and Taylor [1] that the kernels of the epimorphisms

$$Aut(F) \twoheadrightarrow Gl_k(Z), \mathcal{O}(F) \twoheadrightarrow Gl_k(Z)$$

are torsion free. Therefore, if f_d has finite order in either $Aut(F)$ or $\mathcal{O}(F)$ it must have order d.

The following lemma will prove useful.

<u>(3.15) Lemma</u>: Let W be a word in F such that $x_2, x_3, \ldots, x_{k-1}$, W is a basis for F. Then

(i) the associated endomorphism f_W is an automorphism.

(ii) the order of f_W is the least positive integer n such that $f_W^n(x_k) = x_k$.

<u>Proof</u>: The first part is a consequence of the fact that an epimorphism of a free group is an automorphism. For the second part assume there is such an n. Because f_W is an automorphism we have $f_W^{n-i}(x_k) = x_{k-i}$ for $o \leq i < k$. But then

$$f_W^n(x_{k-i}) = f_W^{n-i} f_W^i(x_{k-i}) = f_W^{n-i}(x_k) = x_{k-i}, \quad o \leq i < k$$

Therefore f_W has order n.

Q.E.D.

Now we consider the automorphism f of F_k, $k = \phi(p^s) = p^{s-1}(p-1)$, associated to the cyclotomic polynomial

$$\Phi_{p^s}(x) = x^{p^{s-1}(p-1)} + x^{p^{s-1}(p-2)} + \ldots + x^{p^{s-1}} + 1$$

Thus $f \varepsilon Aut(F)$ is defined by

$$\begin{array}{l} x_1 \longrightarrow x_2 \\ \vdots \qquad \vdots \\ x_{k-1} \longrightarrow x_k \\ \\ x_k \longrightarrow W \end{array} \qquad W = \prod_{i=o}^{p-2} (x_{ip^{s-1}+1})^{-1}$$

Note that the order of the terms in the product for W is from i=o to i=p-2.

(3.16) Theorem: Let p be any prime. Then the automorphism $f \varepsilon \mathrm{Aut}(F_k)$, where $k=p^{s-1}(p-1)$, associated to the cyclotomic polynomial $\Phi_{p^s}(x)$ has order p^s.

Proof: From the definition of f we see that

$$f^j(x_k) = \prod_{i=o}^{p-2} (x_{ip^{s-1}+j})^{-1} \text{ if } 1 \leq j \leq p^{s-1}$$

In particular

$$f^{p^{s-1}}(x_k) = \prod_{i=o}^{p-2} (x_{ip^{s-1}+p^{s-1}})^{-1} = \prod_{i=1}^{p-1} (x_{ip^{s-1}})^{-1}$$

Applying f one more time gives

$$f^{p^{s-1}+1}(x_k) = \prod_{i=1}^{p-2} (x_{ip^{s-1}+1})^{-1} W^{-1} = x_1$$

But then we have

$$f^{p^s}(x_k) = f^{p^{s-1}+1+p^{s-1}(p-1)-1}(x_k) = f^{p^{s-1}(p-1)-1}(x_1) = x_k$$

According to (3.15) this implies that f has order p^s.

Q.E.D.

Now suppose that p is an odd prime. Then $Gl_{\phi(2p^s)}(Z)$ has torsion of order $2p^s$, namely the companion matrix of the cyclotomic polynomial

$$\Phi_{2p^s}(x) = x^{p^{s-1}(p-1)} - x^{p^{s-1}(p-2)} + \ldots - x^{p^{s-1}} + 1$$

The associated automorphism $f\in Aut(F_k)$, where $k=\text{rank } F=\phi(2p^s)=p^{s-1}(p-1)$, is by definition

$$\begin{array}{l} x_1 \longrightarrow x_2 \\ \vdots \qquad \vdots \\ x_{k-1} \longrightarrow x_k \\ x_k \longrightarrow W \end{array} \qquad W = \prod_{i=o}^{p-2} (x_{ip^{s-1}+1})^{(-1)^{i+1}}$$

Again we must be careful about the order of terms in a product. The notation $\prod_{i=o}^{p-2} y_i$ denotes the product $y_o y_1 \ldots y_{p-2}$, whereas, $\prod_{i=p-2}^{o} y_i$ denotes $y_{p-2} \ldots y_1 y_o$.

If $f\in Aut(F_k)$ has finite order this order must be $2p^s$. Thus we calculate $f^{2p^s}(x_k)$. From the definition of f it follows that

$$f^j(x_k) = \prod_{i=o}^{p-2} (x_{ip^{s-1}+j})^{(-1)^{i+1}} \quad \text{for } 1\le j\le p^{s-1}$$

Therefore

$$f^{p^{s-1}}(x_k) = \prod_{i=o}^{p-2} (x_{ip^{s-1}+p^{s-1}})^{(-1)^{i+1}} = \prod_{i=1}^{p-1} (x_{ip^{s-1}})^{(-1)^i}$$

The last term in this product is x_k and thus

$$f^{p^{s-1}+1}(x_k) = \prod_{i=1}^{p-2} (x_{ip^{s-1}+1})^{(-1)^i} \prod_{i=o}^{p-2} (x_{ip^{s-1}+1})^{(-1)^{i+1}} = x_1^{-1}X$$

$$\text{where } X = \prod_{i=o}^{p-2} (x_{ip^{s-1}+1})^{(-1)^i} \prod_{i=o}^{p-2} (x_{ip^{s-1}+1})^{(-1)^{i+1}}$$

Then we have

$$f^{p^{s-1}+1+j}(x_k) = x_{1+j}^{-1} f^j(X), \quad o\le j\le p^{s-1}(p-1)-1$$

The following calculation shows that $f^{p^{s-1}}(X) = X$:

$$f^{p^{s-1}}(X) = \prod_{i=o}^{p-2} f^{p^{s-1}}(x_{ip^{s-1}+1})^{(-1)^i} \prod_{i=o}^{p-2} f^{p^{s-1}}(x_{ip^{s-1}+1})^{(-1)^{i+1}}$$

$$= \prod_{i=o}^{p-3} f^{p^{s-1}}(y_i)^{(-1)^i} f^{p^{s-1}}(y_{p-2})^{-1} \prod_{i=o}^{p-3} f^{p^{s-1}}(y_i)^{(-1)^{i+1}} f^{p^{s-1}}(y_{p-2})$$

where $y_i = x_{ip^{s-1}+1}$.

Since $f^{p^{s-1}}(y_i) = y_{i+1}$ if $i \leq p-3$ and $f^{p^{s-1}}(y_{p-2}) = f(x_k)$ we have

$$f^{p^{s-1}}(X) = \prod_{i=1}^{p-2} y_i^{(-1)^{i+1}} f(x_k)^{-1} \prod_{i=1}^{p-2} y_i^{(-1)^i} f(x_k)$$

$$= \prod_{i=1}^{p-2} y_i^{(-1)^{i+1}} \prod_{i=p-2}^{o} y_i^{(-1)^i} \prod_{i=1}^{p-2} y_i^{(-1)^i} \prod_{i=o}^{p-2} y_i^{(-1)^{i+1}}$$

$$= y_o \prod_{i=1}^{p-2} y_i^{(-1)^i} \prod_{i=o}^{p-2} y_i^{(-1)^{i+1}} = X \text{(because of cancellation)}$$

Taking $j=p^{s-1}(p-1)-1$ then gives

$$f^{p^s}(x_k) = x_k^{-1} f^{-1}(x)$$

This then implies that $f^{p^s+1}(x_k) = W^{-1}X$, which in turn yields

$$f^{p^s+1+p^{s-1}}(x_k) = f^{p^{s-1}}(W)^{-1}X$$

But $f^{p^{s-1}}(W) = \prod_{i=o}^{p-3} (x_{ip^{s-1}+1+p^{s-1}})^{(-1)^{i+1}} f^{p^{s-1}}(x_{(p-2)p^{s-1}+1})$

$$= \prod_{i=1}^{p-2} (x_{ip^{s-1}+1})^{(-1)^i} f(x_k)$$

$$= \prod_{i=1}^{p-2} (x_{ip^{s-1}+1})^{(-1)^i} \prod_{i=o}^{p-2} (x_{ip^{s-1}+1})^{(-1)^{i+1}}$$

Therefore $f^{p^s+1+p^{s-1}}(x_k)$ is given by the expression

$$\prod_{i=p-2}^{o} y_i^{(-1)^i} \prod_{i=p-2}^{1} y_i^{(-1)^{i+1}} \prod_{i=o}^{p-2} y_i^{(-1)^i} \prod_{i=o}^{p-2} y_i^{(-1)^{i+1}}$$

$$= \prod_{i=p-2}^{o} y_i^{(-1)^i} \prod_{i=p-2}^{o} y_i^{(-1)^{i+1}} \cdot y_o \cdot \prod_{i=o}^{p-2} y_i^{(-1)^i} \prod_{i=o}^{p-2} y_i^{(-1)^{i+1}} = X^{-1}x_1X$$

Applying f $k-1=p^{s-1}(p-1)-1$ more times then gives

$$f^{2p^s}(x_k) = f^{-1}(X^{-1})x_kf^{-1}(X)$$

Thus we have

$$f^{2p^s+k-j}(x_j) = f^{-1}(X^{-1})f^{k-j}(x_j)f^{-1}(X),\ 1\le j\le k,$$

which finally yields

$$f^{2p^s}(x_j) = f^{j-k-1}(X)^{-1}x_jf^{j-k-1}(X),\ 1\le j\le k.$$

Hence f^{2p^s} is an inner automorphism if, and only if, $f(X)=X$. But this happens if, and only if $s=1$.

We have proved the following theorem.

<u>(3.17) Theorem</u>: Let p be an odd prime and let $f\epsilon \mathrm{Aut}(F_k)$, $k=p^{s-1}(p-1)$, be the automorphism associated to the cyclotomic polynomial $\Phi_{2p^s}(x)$. Then

$$f^{2p^s}(x_j)=f^{j-k-1}(X)^{-1}x_jf^{j-k-1}(X),\ 1\le j\le k,$$

where $X = \prod_{i=o}^{p-2}(x_{ip^{s-1}+1})^{(-1)^i}\prod_{i=o}^{p-2}(x_{ip^{s-1}+1})^{(-1)^{i+1}}$. Moreover, f determines an element of finite order in $\mathcal{O}(F)$ if, and only if, $s=1$.

In theorem (3.11) we gave, for a fixed n, the least integer k such that $Gl_k(Z)$ has n torsion. Now we would like to do the same for $\mathrm{Aut}(F_k)$ and $\mathcal{O}(F_k)$. If $n=p_1^{s_1}p_2^{s_2}\cdots p_t^{s_t}$, $2\le p_1<p_2<\ldots p_t$, is the prime factorization of n then the matrix

$$A = C(p_1^{s_1}) \oplus \ldots \oplus C(p_t^{s_t}) \varepsilon Gl_k(Z)$$
$$k = \phi(p_1^{s_1}) + \ldots + \phi(p_t^{s_t})$$

has order n. Furthermore, this value of k is the least such value so that $Gl_k(Z)$ has n torsion; except when $p_1^{s_1}=2$ and $n \neq 2$, in which case the least value of k is $\phi(p_1^{s_1})+\ldots+\phi(p_t^{s_t})-1$.

Let $f_i \varepsilon Aut(F_{\phi(p_i^{s_i})})$ be the automorphism associated to the cyclotomic polynomial $\Phi_{p_i^{s_i}}(x)$. According to (3.19) the automorphism f_i has order $p_i^{s_i}$. It follows that $f_1 * \ldots * f_t \varepsilon Aut(F_k)$ has order n. This proves

<u>(3.18) Theorem</u>: Let the prime factorization of n be $n=p_1^{s_1}\ldots p_t^{s_t}$. If $n \not\equiv 2(4)$ then the least value of k such that $Aut(F_k)$ has torsion of order n is

$$k = \phi(p_1^{s_1}) + \ldots + \phi(p_t^{s_t})$$

If $n \equiv 2(4)$ then this least value is either $k=\phi(p_1^{s_1})+\ldots+\phi(p_t^{s_t})$ or $k=\phi(p_1^{s_1})+\ldots+\phi(p_t^{s_t})-1$.

Meskin [6] has shown that $Aut(F_2)$ does not have an element of order 6. This supports the following conjecture:

<u>(3.19) Conjecture</u>: The least value of k for which $Aut(F_k)$ has torsion of order $n=p_1^{s_1}\ldots p_t^{s_t}$ is

$$k = \phi(p_1^{s_1}) + \ldots + \phi(p_t^{s_t})$$

As far as outer automorphisms are concerned we have

<u>(3.20) Theorem</u>: Let the prime factorization of n be $n=p_1^{s_1}\ldots p_t^{s_t}$, $2 \le p_1 < p_2 < \ldots < p_t$. Then the least value of k so that $\mathcal{O}(F_k)$ has an element

of order n is:

(i) $k=\phi(p_1^{s_1})+\ldots+\phi(p_t^{s_t})$ if $p_1^{s_1}\neq 2$.

(ii) $k=\phi(p)=p-1$ if $n=2p$, where p is an odd prime.

(iii) either $k=\phi(p_1^{s_1})+\ldots+\phi(p_t^{s_t})$ or $k=\phi(p_1^{s_1})+\ldots+\phi(p_t^{s_t})-1$ in all other cases.

Proof: Parts (i) and (iii) follow from (3.18), whereas part (ii) follows from (3.17).

Q.E.D.

(3.21) Conjecture: The least value of k for which $\mathcal{O}(F_k)$ has torsion of order $n=p_1^{s_1}\ldots p_t^{s_t}$, where $2\leq p_1<\ldots<p_t$, is

(i) $k=p-1$ if $n=2p$, where p is an odd prime.

(ii) $k=\phi(p_1^{s_1})+\ldots+\phi(p_t^{s_t})$ otherwise

We conclude this section with a corollary

(3.22) Corollary: Suppose G is a non-cyclic group satisfying (1.3). Then there exist distinct positive integers $d_1,\ldots,d_r$ and positive integers $e_1,\ldots,e_r$ such that

(i) $|G_{ab}|$ is a multiple of the least common multiple of the d_i.

(ii) rank $[G,G] = e_1\phi(d_1)+\ldots+e_r\phi(d_r)$.

(iii) the d_i are not prime powers.

In particular it follows that $[G,G]$ has even rank and that the order of G_{ab} is not a prime power.

Proof: The extension $1\longrightarrow[G,G]\longrightarrow G\longrightarrow G_{ab}\longrightarrow 1$ corresponds to an element $\tau\in\mathcal{O}(F)$ such that $\tau^m=1$, where $m=|G_{ab}|$ and $F=[G,G]$. Let the actual order of τ be n. Then $n|m$ and the result follows from (3.14).

Q.E.D.

§4 Applications:

As a first application we consider the case where $[G,G]$ is free of rank 2 and $H_1(G)$ is cyclic of order 6. According to (3.22) this is the first non-cyclic possibility. We shall determine all congruence classes of extensions

(4.1) $$1 \longrightarrow F_2 \longrightarrow G \xrightarrow{\alpha} Z_6 \longrightarrow 1$$

such that $\alpha_*: H_*(G) \longrightarrow H_*(Z_6)$ is an isomorphism.

By the theory developed in §3 this amounts to finding all $\tau \in \mathcal{O}(F_2) = Gl_2(Z)$ such that

$$\tau^6 = 1 \text{ and } I-T \in Gl_2(Z)$$

Thus we use (3.7) to determine the rational similarity classes of torsion in $Gl_2(Z)$. The following table gives the cyclotomic polynomials of degree ≤2 together with their companion matrices

d = order C(d)	$\Phi_d(x)$	C(d)
1	$x-1$	$[1]$
2	$x+1$	$[-1]$
3	x^2+x+1	$\begin{bmatrix} 0 & 1 \\ -1 & -1 \end{bmatrix}$
4	x^2+1	$\begin{bmatrix} 0 & 1 \\ -1 & 0 \end{bmatrix}$
6	x^2-x+1	$\begin{bmatrix} 0 & 1 \\ -1 & 1 \end{bmatrix}$

Therefore we have the following rational canonical forms for torsion in $Gl_2(Z)$:

form	$\begin{bmatrix}1&0\\0&1\end{bmatrix}$	$\begin{bmatrix}-1&0\\0&-1\end{bmatrix}$	$\begin{bmatrix}1&0\\0&-1\end{bmatrix}$	$\begin{bmatrix}0&1\\-1&-1\end{bmatrix}$	$\begin{bmatrix}0&1\\-1&0\end{bmatrix}$	$\begin{bmatrix}0&1\\-1&1\end{bmatrix}$
order	1	2	2	3	4	6

Only $T = \begin{bmatrix}0&1\\-1&1\end{bmatrix} = C(6)$ satisfies the condition $I-T\varepsilon Gl_2(Z)$. Since

$$\mu_T(x) = \gamma_T(x) = x^2-x+1$$

the conditions of the Latimer-MacDuffee theorem are met. Thus the similarity classes of matrices $A\varepsilon Gl_2(Z)$ such that $\mu_A(x) = x^2-x+1$ are in 1-1 correspondence with the classes of ideals in $Z[x]/(x^2-x+1)$. But this ring is just the ring of integers in the cyclotomic extension $Q(\zeta),\zeta = e^{\pi i/3}$, and therefore the class number is 1. This means that every matrix $A\varepsilon Gl_2(Z)$ of order 6 is similar to T as integral matrices.

Hence, there is a 1-1 correspondence between congruence classes of extensions (4.1) such that $\alpha_*:H_*(G)\longrightarrow H_*(Z_6)$ is an isomorphism and distinct elements of the form UTU^{-1}, $U\varepsilon Gl_2(Z)$. Now $UTU^{-1} = VTV^{-1}$ if, and only if, $V^{-1}U\varepsilon \mathfrak{Z}(T)$ = the centralizer of T in $Gl_2(Z)$. By a direct calculation we see that

$$\mathfrak{Z}(T) = \langle T\rangle = \text{the cyclic group generated by } T.$$

Therefore the congruence classes of such extensions are in 1-1 correspondence with the cosets of $\langle T\rangle$ in $Gl_2(Z)$. A matrix $U\varepsilon Gl_2(Z)$ determines the congruence class

$$1\longrightarrow F_2\longrightarrow G_U\longrightarrow Z_6\longrightarrow 1$$

(E_U)

$$G_U = \{(t^a,\sigma)\varepsilon Z_6 x Aut(F_2)\,|\,\omega(\sigma) = UT^aU^{-1}\}$$

and $(E_U) = (E_V)$ if, and only if, $V^{-1}U\varepsilon\langle T\rangle$.

On the other hand we always have $G_U \cong G_V$. To see this merely observe that the automorphism

$$Z_6 x\mathrm{Aut}(F_2) \rightarrowtail Z_6 x\mathrm{Aut}(F_2),\ (t^a,\sigma) \longrightarrow (t^a, \phi\upsilon^{-1}\sigma\upsilon\phi^{-1})$$

maps G_U isomorphically onto G_V, where $\upsilon,\phi\varepsilon\mathrm{Aut}(F_2)$ are automorphisms such that $\omega(\upsilon) = U$, $\omega(\phi) = V$. Thus there exists a unique isomorphism class of groups G such that $[G,G]$ is free of rank 2, $H_1(G)\cong Z_6$, and $\varepsilon : G \longrightarrow H_1(G)$ is a homology isomorphism. However, by (2.1), the group $G = Z_2 * Z_3$ satisfies these conditions and therefore every such group G_U is isomorphic to $Z_2 * Z_3$. Since $PSl_2(Z) \cong Z_2 * Z_3$ we have proved the following result.

<u>(4.2) Theorem</u>: The set of congruence classes of extensions $1 \longrightarrow F_2 \longrightarrow G \xrightarrow{\alpha} Z_6 \longrightarrow 1$ such that $\alpha_* : H_*(G) \longrightarrow H_*(Z_6)$ is an isomorphism is in 1-1 correspondence with the coset space $Gl_2(Z)/\langle T\rangle$, where $\langle T\rangle$ is the cyclic subgroup of order 6 generated by the matrix $T = \begin{bmatrix} 0 & 1 \\ -1 & 1 \end{bmatrix}$. Moreover all such groups G are isomorphic to $PSl_2(Z)$.

For the second application suppose G is a group such that $[G,G]$ is free of rank 2, G_{ab} is finite cyclic, and $\varepsilon : G \longrightarrow G_{ab}$ is a homology isomorphism. It then follows from (3.25) that $|G_{ab}| = 6m$ for some integer $m \geq 1$. If T is the matrix as in (4.2) we have

<u>(4.3) Theorem</u>: The set of congruence classes of extensions

$$1 \longrightarrow F_2 \longrightarrow G \xrightarrow{\alpha} Z_{6m} \longrightarrow 1$$

such that $\alpha_* : H_*(G) \longrightarrow H_*(Z_{6m})$ is an isomorphism is in 1-1 correspondence with $Gl_2(Z)/\langle T\rangle$. If $U \varepsilon Gl_2(Z)$ then the corresponding extension is the one induced from $1 \longrightarrow F_2 \longrightarrow \mathrm{Aut}(F_2) \longrightarrow Gl_2(Z) \longrightarrow 1$ by the representation

$$\rho : Z_{6m} \longrightarrow Gl_2(Z),\ t \longrightarrow UTU^{-1}$$

Moreover every such group G is isomorphic to $Z_{2m} \underset{Z_m}{*} Z_{3m}$.

<u>Proof</u>: As in the proof of (4.2) we see that the congruence classes of such extensions are in 1-1 correspondence with elements $\tau \varepsilon \mathcal{O}(F_2) = Gl_2(Z)$ such that

$$\tau^{6m} = 1 \text{ and } 1-\tau \varepsilon Gl_2(Z)$$

However the only such τ are the conjugates UTU^{-1}, where $U \varepsilon Gl_2(Z)$, and consequently we can classify the extensions just as in (4.2).

The proof of (4.2) also shows that all such groups G are isomorphic. According to (2.5) one of the possible groups is $Z_{2m} \underset{Z_m}{*} Z_{3m}$.

Q.E.D.

Since $Z_4 \underset{Z_2}{*} Z_6 \cong Sl_2(Z)$ we have the interesting corollary:

<u>(4.4) Corollary</u>: If G is a group such that $[G,G]$ is free of rank 2, $G_{ab} \cong Z_{12}$, and $\varepsilon : G \longrightarrow G_{ab}$ is a homology isomorphism, then $G \cong Sl_2(Z)$.

For the last application we shall assume that G is a group satisfying (1.3) and that $G_{ab} \cong Z_{2p}$, where p is an odd prime. If G is not cyclic it follows from (3.22) that rank $[G,G] = e(p-1)$ from some $e \geq 1$. Now we choose e=1 and ask for the congruence classification of all extensions

(4.5) $$1 \longrightarrow F_k \longrightarrow G \xrightarrow{\alpha} Z_{2p} \longrightarrow 1, \; k = p-1$$

such that $\alpha_* : H_*(G) \longrightarrow H_*(Z_{2p})$ is an isomorphism.

If $T \varepsilon Gl_k(Z)$ is the linear automorphism associated to this extension then

$$T^{2p} = I \text{ and } I-T \varepsilon Gl_k(Z)$$

From (3.14) it follows that the order of T is 2p and that $T \underset{Q}{\sim} C(2p)$.

In fact if T is any element of $Gl_k(Z)$ having order 2p then we automatically have $T \underset{Q}{\sim} C(2p)$ and $I-T \varepsilon Gl_k(Z)$. Therefore the congruence classes of extensions (4.5) are in 1-1 correspondence with the elements of order 2p in $\mathcal{O}(F_k)$.

Now let T be the companion matrix C(2p). If $\tau \varepsilon \mathcal{O}(F_k)$ is the outer automorphism associated to T then τ has order 2p (see (3.17)). Therefore $\upsilon\tau\upsilon^{-1}$ has order 2p for each $\upsilon \varepsilon \mathcal{O}(F_k)$. Hence, to each $\upsilon \varepsilon \mathcal{O}(F_k)$ we have an extension

$$E_\upsilon \qquad 1 \longrightarrow F_k \longrightarrow G_\upsilon \xrightarrow{\alpha} Z_{2p} \longrightarrow 1$$

such that $\alpha_*: H_*(G_\upsilon) \longrightarrow H_*(Z_{2p})$ is an isomorphism. Moreover 2 such extensions E_ϕ, E_υ are congruent if, and only if, $\phi\tau\phi^{-1} = \upsilon\tau\upsilon^{-1}$. Finally, the same argument used in (4.2) shows that for any $\phi, \upsilon \varepsilon \mathcal{O}(F_k)$ we have $G_\phi \cong G_\upsilon$.

We have proved the following theorem

<u>(4.6) Theorem</u>: Let p be an odd prime and k=p-1. Let $\tau \varepsilon \mathcal{O}(F_k)$ be the outer automorphism associated to C(2p). Then

(i) There is a 1-1 correspondence between congruence classes of extensions (4.5) and elements of order 2p in $\mathcal{O}(F_k)$.

(ii) If E_υ is the above extension, $\upsilon \varepsilon \mathcal{O}(F_k)$, then E_ϕ, E_υ are congruent if, and only if, $\phi\tau\phi^{-1} = \upsilon\tau\upsilon^{-1}$.

(iii) $G_\phi \cong G_\upsilon$ for all ϕ, $\upsilon \varepsilon \mathcal{O}(F_k)$.

References

1. Baumslag, G. and Taylor, T.: The Centre of Groups with one defining Relator, Math. Ann., 175, 315-319 (1968).

2. Kan, D. M. and Thurston, W. P.: Every connected Space has the Homology of a $K(\pi,1)$, Topology, 15, 253-258 (1976).

3. Karass, A. and Solitar, D.: The Subgroups of a free Product of two Groups with an amalgamated Subgroup, T. A. M. S., 150, 227-255 (1970).

4. Latimer, C. G. and MacDuffee, C. C.: A Correspondence between Classes of Ideals and Classes of Matrices, Ann. of Math., 34, 313-316(1933).

5. MacLane, S.: Homology, volume 114, Die Grundlehren der mathematischen Wissenschaften, Springer-Verlag (1963).

6. Meskin, S.: Periodic Automorphisms of the two-generator free Group, L. N. M. 372, 494-498 (1974).

7. Stallings, J.: Homology and central Series of Groups, J. of Algebra, 170-181 (1965).

8. Stallings, J.: Group Theory and 3 dimensional Manifolds, volume 4 Yale Univ. Press (1971).

9. Stammbach, U.: Anwendungen der Homologietheorie der Gruppen auf Zentralreihen und auf Invarianten von Präsentierungen, Math. Z., 94, 157-177 (1966).

AN ISOMORPHISM BETWEEN PRODUCTS OF ABELIAN GROUPS

Richard Steiner

(supported by the National Research Council)

1. Introduction

This paper is concerned with a curious isomorphism arising when a product of abelian groups is given an exotic abelian group structure by "homogeneous multilinear maps". Consider for instance a commutative (not anticommutative) graded ring $A = \bigoplus_{q=0}^{\infty} A_q$. Multiplication gives $\prod_{q=1}^{\infty} A_q = \{1\} \times \prod_{q=1}^{\infty} A_q$ an exotic abelian group structure; call the resulting group UA. Let r be a positive integer and let R_r be the subring of the complex numbers generated by the rth roots of unity. Then $R_r \otimes A = \bigoplus_{q=0}^{\infty} R_r \otimes A_q$ is a graded ring, so we have an abelian group $U(R_r \otimes A)$. In this case the theorem of this paper gives an isomorphism

$$R_r \otimes UA \cong U(R_r \otimes A),$$

provided each group A_q with $q \geq 1$ is a $Z[r^{-1}]$-module ($Z[r^{-1}]$ denotes the ring obtained from the integers Z by inverting r).

Note that some restriction on A is necessary. For instance, if $r = 4$, so that $R_r = Z \oplus Zi$ ($i^2 = -1$), and $A = (Z/2)[t]/(t^3)$ (deg $t = 1$), then $UA \cong Z/4$, so

$$R_r \otimes UA \cong Z/4 \oplus Z/4,$$

while $U(R_r \otimes A)$ has sixteen elements, exactly two of which (1 and $1+t^2$) are squares, so

$$U(R_r \otimes A) \cong Z/4 \oplus Z/2 \oplus Z/2 .$$

To state the theorem in general we need a precise definition of homogeneous multilinear maps. To this end, define a category Mult as follows. The objects are families

$$A = (A_\alpha : \alpha \in I)$$

of abelian groups A_α indexed by arbitrary sets I, each group A_α having a positive integral degree $|\alpha|$. The morphisms from $(A_\alpha: \alpha \in I)$ to $(B_\beta: \beta \in J)$ are functions $f: \prod_\alpha A_\alpha \to \prod_\beta B_\beta$ which can be put in the following form: for $x = (x_\alpha) \in \prod_\alpha A_\alpha$

$$f(x) = (f_\beta(x)) = (\sum_\phi f_{\beta,\phi}(x^\phi)),$$

where ϕ runs through functions from I to $\{0,1,2,\ldots\}$ with $\phi(\alpha) = 0$ for all but finitely many α and

$$\sum_\alpha \phi(\alpha)|\alpha| = |\beta|$$

(this is the homogeneity condition). $x^\phi \in \prod_\alpha A_\alpha^{\phi(\alpha)}$ is the point with each A_α coordinate equal to x_α, and the

$$f_{\beta,\phi}: \prod_\alpha A_\alpha^{\phi(\alpha)} \to B_\beta$$

are multilinear maps, only finitely many of them non-zero for each β. Note that f does not determine the $f_{\beta,\phi}$ uniquely; for instance a quadratic form can be the restriction of various bilinear forms.

It is easy to see that Mult has products given by disjoint union of families, and that the obvious functor P from Mult to sets sending $A = (A_\alpha: \alpha \in I)$ to $PA = \prod_\alpha A_\alpha$ is product-preserving. So if A is an abelian group object in Mult, then PA is an abelian group. Also

$$R_r \otimes A = (R_r \otimes A_\alpha: \alpha \in I)$$

(recall that R_r is the ring generated by the complex rth roots of unity) is an abelian group object of Mult in an obvious way, so $P(R_r \otimes A)$ is an abelian group.

<u>Theorem</u>. Let r be a positive integer. For $A = (A_\alpha: \alpha \in I)$ an abelian group object of Mult with each A_α a $Z[r^{-1}]$-module there is a natural isomorphism

$$R_r \otimes PA \cong P(R_r \otimes A).$$

The case of the multiplicative group of a graded ring occurs in my thesis ([3], I.7). It is related to certain cohomology theories constructed by Segal in [2]: Let A be an anticommutative graded ring; then cup-product gives an exotic multiplication on the product of Eilenberg-MacLane spaces

$$K(A) = \{1\} \times \prod_{q=1}^{\infty} K(A_q, q),$$

and Segal shows that $K(A)$ is an infinite loop space. In [3] I constructed a natural isomorphism

$$R_r \otimes gA^*(X) \cong g(R_r \otimes A)^*(X)$$

for gA the cohomology theory associated to $K(A)$ and X any spectrum, provided A_q is a $Z[r^{-1}]$-module for $q \geq 1$. I hope to construct cohomology theories G delooping in exotic ways other products $\prod_{\alpha \in I} K(A_\alpha, |\alpha|)$ of Eilenberg-MacLane spaces; if $|\alpha| \geq 1$ and A_α is a $Z[r^{-1}]$-module for all α, then there should be a natural isomorphism

$$R_r \otimes G^*(X) \cong G_r{}^*(X)$$

for G_r an appropriate delooping of $\prod_{\alpha \in I} K(R_r \otimes A_\alpha, |\alpha|)$.

2. Proof of the Theorem.

Step 1. This consists of general remarks about the abelian group object A. First, it is convenient to truncate A above, obtaining abelian group objects

$$A[1,q] = (A_\alpha : 1 \leq |\alpha| \leq q)$$

with homomorphisms

$$A \to A[1,q] \to A[1,q-1],$$

and to truncate the $A[1,q]$ below obtaining abelian group objects

$$A[q] = (A_\alpha : |\alpha| = q)$$

with homomorphisms

$$A[q] \to A[1,q].$$

We observe that

(1) $$PA = \varprojlim_q PA[1,q],$$

the sequences

(2) $$0 \to PA[q] \to PA[1,q] \to PA[1,q-1] \to 0$$

are exact, and

(3) $$PA[q] \cong \prod_{|\alpha|=q} A_\alpha \quad \text{as abelian groups;}$$

the last is an easy deduction from the fact that the functions giving the product on $PA[q]$ are linear, by homogeneity.

We shall use (1), (2) and (3) for inductive proofs; for instance they show

(4) $$PA \text{ is a } Z[r^{-1}]\text{-module.}$$

For the $PA[q]$ are $Z[r^{-1}]$-modules by (3), hence the $PA[1,q]$ are so by induction using (2), hence PA is so by (1).

We shall work inside $R_\infty \otimes A$, where R_∞ is the subring of the complex numbers generated by all the nth roots of unity with n dividing a power of r (that is, n a unit in R_r, notation $n|r^\infty$). For $\zeta \in R_\infty$ we have an "Adams operation"

$$\psi^\zeta : P(R_\infty \otimes A) \to P(R_\infty \otimes A)$$

given by multiplying the $R_\infty \otimes A_\alpha$ co-ordinate by $\zeta^{|\alpha|}$. It is easy to see that the ψ^ζ are homomorphisms, and that

(5) $$\psi^1 = 1, \ \psi^\zeta\psi^\omega = \psi^{\zeta\omega} \quad \text{for} \quad \zeta,\omega \in R_\infty.$$

By the analogue of (4) for $P(R_\infty \otimes A)$ we can define an endomorphism e_n of $P(R_\infty \otimes A)$ for $n|r^\infty$ by

(6) $$e_n = n^{-1} \sum_{\zeta^n=1} \psi^\zeta;$$

in fact e_n gives endomorphisms of PA and $P(R_r \otimes A)$ by a symmetry argument. We see from (5) that the e_n are idempotent and commute with each other.

Step 2. Imitating the work of Bergman in [1, 26.3E] we use the commuting idempotents e_n to decompose PA (and similarly $P(R_r \otimes A)$, $P(R_\infty \otimes A)$). To be precise,

$$PA = \prod_{k|r^\infty} P_k A,$$

where

(7) $$P_k A = \{x \in PA: e_n x = x \text{ for } n|r^\infty \text{ and } n|k,$$
$$e_n x = 0 \text{ for } n|r^\infty \text{ and } n\nmid k\}.$$

This is not quite immediate, because there are infinitely many idempotents (unless $r = 1$); also other distributions of eigenvalues are conceivable. However, it is immediate from (3) that

(8) $PA[q] = P_k A[q]$ for k the largest common divisor of r^∞ and q,

since

$$n^{-1} \sum_{\zeta^n=1} \zeta^q = 1 \quad \text{if } n|q, \quad 0 \text{ if } n\nmid q.$$

It follows in particular that $e_n = 0$ on $PA[q]$ for $n > q$; the exact sequence

$$0 \to e_n PA[q] \to e_n PA[1,q] \to e_n PA[1,q-1] \to 0$$

derived from (2) shows by induction on q that $e_n = 0$ on $PA[1,q]$ for $n > q$. So there are really only finitely many idempotents on $PA[1,q]$, and there is really a finite decomposition

$$PA[1,q] = \prod_{k|r^\infty} P_k A[1,q];$$

only the factors shown can be non-trivial, for if E is the projection operator corresponding to any other distribution of eigenvalues, then $EPA[q] = 0$ by (8), so $EPA[1,q] = 0$ just as $e_n PA[1,q] = 0$ for $n > q$. The decomposition of PA follows by inverse limits, using (1).

Step 3. We construct a homomorphism

$$\theta: R_r \otimes P_k A \to P_k(R_r \otimes A)$$

for each $k \mid r^\infty$. Since we have the inclusion $P_k A \to P_k(R_r \otimes A)$, it suffices to put an R_r-module structure on $P_k(R_r \otimes A)$. Write

(9) $$\zeta(s) = e^{2\pi i/s}$$

for $s = 1,2,\ldots$; then $R_r = Z[\zeta(r)]/(\Phi_r(\zeta(r)))$, where Φ_r is the rth cyclotomic polynomial. So it suffices to find an endomorphism ξ of $P_k(R_r \otimes A)$ satisfying $\Phi_r(\xi) = 0$. We use the following Lemma.

Lemma. The endomorphism $\psi^{\zeta(kr)}$ of $P(R_\infty \otimes A)$ maps $P_k(R_r \otimes A)$ to itself and

$$\Phi_r(\psi^{\zeta(kr)}) \mid P_k(R_r \otimes A) = 0.$$

The proof will be given later.

Step 4. We assume the Lemma and prove that $\theta: R_r \otimes P_k A \to P_k(R_r \otimes A)$ constructed as in Step 3 is an isomorphism; this of course completes the proof of the Theorem. By (2) and (1), it suffices to prove that

$$\theta\colon R_r \otimes P_k\ A[q] \to P_k(R_r \otimes A[q])$$

is an isomorphism for each q. By (8) we may assume that k is the largest common divisor of r^∞ and q; we are then considering the whole of $PA[q]$. Using (3), a typical element of $R_r \otimes PA[q]$ has the form

$$f(\zeta(r)) \otimes (x_\alpha)$$

with $f(\zeta(r))$ an integral polynomial in $\zeta(r)$. Chasing the definitions and using (5) we see that

$$\begin{aligned}\theta(f(\zeta(r)) \otimes (x_\alpha)) &= (f(\zeta(rk)^q) \otimes x_\alpha) \\ &= (f(\zeta(r)^{q/k}) \otimes x_\alpha).\end{aligned}$$

Now k is the largest common divisor of q and r^∞, so q/k is an integer prime to r. It follows that θ is an isomorphism, as required.

Proof of the Lemma. Since $\psi^{\zeta(kr)}$ commutes with the e_n, by (5), $\psi^{\zeta(kr)}$ certainly maps $P_k(R_r \otimes A)$ into $P_k(R_\infty \otimes A)$. To show that $\psi^{\zeta(kr)}$ maps $P_k(R_r \otimes A)$ to itself, we shall show that

$$\alpha\psi^{\zeta(kr)}x = \psi^{\zeta(kr)}x$$

for $x \in P_k(R_r \otimes A)$ and α a member of the Galois group of R_∞/R_r acting on $P(R_\infty \otimes A)$ in the obvious way. Indeed $\alpha\zeta(kr) = \zeta(kr)\zeta(k)^s$ for some s, as α fixes the kth power of $\zeta(kr)$, and

$$\psi^{\zeta(k)}x = x \quad \text{for } x \in P_k(R_r \otimes A), \tag{10}$$

as $\psi^{\zeta(k)}e_k = e_k$ by (5) and (6) and $e_k x = x$ by (7). Therefore

$$\alpha\psi^{\zeta(kr)}x = \psi^{\alpha\zeta(kr)}\alpha x = \psi^{\zeta(kr)}(\psi^{\zeta(k)})^s x = \psi^{\zeta(kr)}x,$$

as required.

To show that $\Phi_r(\psi^{\zeta(kr)})$ vanishes on $P_k(R_r \otimes A)$ we recall that

$$\Phi_r(X) = \prod_{d|r}(X^d-1)^{m(d)}$$

for certain positive or negative integers $m(d)$ with $m(r) = 1$. So it suffices to show that $(\psi^{\zeta(kr)})^r-1$ acts trivially on $P_k(R_r \otimes A)$ and $(\psi^{\zeta(kr)})^d-1$ acts isomorphically on $P_k(R_r \otimes A)$ for d a proper divisor of r. Now $(\psi^{\zeta(kr)})^r-1 = \psi^{\zeta(k)}-1$ acts trivially by (10). If d is a proper divisor of r, then

$$(\psi^{\zeta(kr)})^d = \psi^{\zeta(n)} \quad \text{with } n|r^\infty,\ n\nmid k.$$

Let

$$\phi = n^{-1}[1+2\psi^{\zeta(n)}+3(\psi^{\zeta(n)})^2+\ldots+n(\psi^{\zeta(n)})^{n-1}]$$

(the division by n is legitimate by (4)). By (5) and (6),

$$(\psi^{\zeta(n)}-1)\phi = \phi(\psi^{\zeta(n)}-1) = -n^{-1}[1+\psi^{\zeta(n)}+(\psi^{\zeta(n)})^2+\ldots+(\psi^{\zeta(n)})^{n-1}]+\psi^1$$

$$= -e_n+1.$$

As e_n acts trivially on $P_k(R_r \otimes A)$ by (7), $\psi^{\zeta(n)}-1$ acts isomorphically. This completes the proof.

3. Remarks.

The proof just given seems unduly computational and does not really explain the isomorphism. It might help understanding to consider other rings of algebraic integers. The statement of the Theorem may be too pretentious: R_r is a free abelian group and $R_r \otimes PA$ should perhaps be replaced by a direct sum of copies of PA.

REFERENCES

1. D. Mumford, with a contribution by G.M. Bergman, Lectures on curves on an algebraic surface, Annals of mathematics studies 59, Princeton University Press, Princeton, N.J., 1966.

2. G. Segal, The multiplicative group of classical cohomology, Quart. J. Math. Oxford (2) 26 (1975), 289-293.

3. R.J. Steiner, Infinite loop spaces and products in cohomology theories, Ph.D. dissertation, Cambridge, 1977.

Axiomatic Homotopy Theory*

by D. W. Anderson

1. Introduction

The axioms given by Eilenberg and Steenrod [AT] for homology theory have been remarkably successful. These axioms characterized singular homology on finite complexes (and, with slight strengthening, on all CW-complexes). They also provided a guide to generalized homology and cohomology theories. Much of their success, I believe, is due to the feature that while verifying the axioms is rather difficult to do from first principles, once one has the axioms most of the desirable properties of homology theory follow from fairly straightforward arguments whose most subtle parts are algebraic rather than topological. A second key to their success has been that they can be altered individually in small ways (particularly the excision axiom can be altered) to suit circumstances without changing the nature of the arguments based on them.

I have long been interested in the question of whether or not homotopy theory could be axiomatized in such a way that would cover a broad sample of situations which are "obviously" homotopy theory. One requirement of such an axiomatization would be that the representable functors should satisfy some version of Ed Brown's generalization of the Eilenberg-Steenrod axioms for homology theory (see Brown [CT]). If possible, it would be desirable to be able to add a simple axiom which would characterize the homotopy theory of CW-complexes the way that the Eilenberg-Steenrod axioms characterize singular homology. Further, such an axiomatization should include the homotopy theory of "spaces with special structures", whether

*Research partially supported by NSF grant no. MCS77-01593.

they are loop spaces, E_∞-spaces, spaces with operads acting on them, spaces with groups acting on them, spectra, special Γ-spaces, or whatever. In this way one would have a good setting for equivalence statement--rather than the statement "every A_∞-space is the loop space of a CW-complex" we would have the statement "the homotopy category of A_∞-spaces is equivalent to the homotopy category of connected pointed CW-complexes".

There is an axiomatization of homotopy theory given by Quillen [HA] for what he calls model categories. These were modified in his paper [RH] to cover his "closed" model categories, and in this form these axioms have been shown by various people to be satisfied in a remarkably wide collection of situations, and lead to a beautiful and general theory of homotopy.

Unfortunately, when I began to investigate whether or not spaces with various "up to homotopy" structures (such as special Γ-spaces, A_∞-spaces, and so on) satisfied Quillen's axioms, I discovered that Quillen's axioms did not apply in a number of cases which I found interesting. In order of increasing complexity, these include the categories of finite CW-complexes, equivariant homotopy theory, simplicial sheaves, and special Γ-spaces. That the axioms proposed here are satisfied by such categories will be proved in my [HFS].

The key to the success of Quillen's axioms is the close connection between fibrations and cofibrations. Indeed, the finite CW-complexes primarily fail to satisfy Quillen's axioms because it takes infinite complexes to produce path spaces and other types of fibrations, so that this category is deficient in fibrations. Similarly, as was noticed in Ken Brown's M.I.T. thesis [AHT], simplicial sheaves are deficient in cofibrations. However, in each case there is a homotopy theory, and Brown showed how to "denature" Quillen's axioms so that they would apply in this case. However, Brown's variants on Quillen's axioms were still not satisfactory for equivariant homotopy theory or special Γ-spaces, in the sense that if one began with a category satisfying these axioms, and looked at the objects with a fixed group acting on them or looked at the special Γ-objects, there seemed to be no way to show that this new category satisfied Brown's axioms.

My approach was to get rid of both the cofibrations and the fibrations (as well as the colimits and limits) and concentrate on what homotopy theorists tend to use in practice--the homotopy theoretic fiber product and cofiber sum. Due to the fact that higher order information was prone to being lost by concentrating on so specialized a construction, I threw in the homotopy Kan extensions, and at this point produced a rather abstract, difficult to verify set of axioms which had the virtue of being small in number and of allowing one the freedom of always working "up to homotopy". Further, and more importantly, these axioms allowed one to make all sorts of standard constructions and to prove that representable functors satisfy Ed Brown's axioms for cohomology. Also, these axioms allow one to construct standard spectral sequences, including the Eilenberg-Moore spectral sequence for the homology of the homotopy theoretic orbit space of an object with a group action. Finally, these axioms are stable under the construction of diagram categories.

I do not mean to imply that my axioms are entirely satisfactory. They resist attempts to construct comma categories which also satisfy the axioms, though since the arrow (derived) categories satisfy the axioms, relative homotopy theory exists in a global setting. The absence of constructions with given spaces and maps is unnerving, though the desired construction can generally be made "up to diagram homotopy type".

Our general approach is the following. Homotopy theory is characterized by two categories, the category of spaces and the homotopy category, and the homotopy category is a localization of the category of spaces. The study of homotopy theory could be defined as the study of the relationship between these two categories. We then define completeness, cocompleteness, regularity, and coregularity for a homotopy theory--indeed, we do not give numbered axioms, but rather study the consequences of these properties. Ordinary categories will determine "discrete" homotopy theories in the way that sets determine discrete topological spaces; discrete homotopy theories are about as interesting as discrete topologies. There are also "indiscrete" homotopy theories associated to categories; these are slightly more interesting to study than indiscrete topologies.

There are variations on the properties mentioned in the way of finiteness restrictions. With such restriction, finite CW-complexes determine a finitely coregular homotopy theory. In general, some care needs to be taken with respect to the "size" of things even if infinite constructions are allowed. We shall assume that we have some fixed Grothendieck universe, that a "small" category has as classes of objects and morphisms objects in that universe, and that "category" without the term "small" modifying it means a category whose individual objects and morphisms lie in the given universe.

One of the most pleasant consequences of our approach is that the homotopy theory of CW-complexes plays a central role in homotopy theory. In fact, a coregular or regular homotopy theory has the feature that for any pair of "spaces" T, S, there is a CW-complex HOM(T,S), well defined up to homotopy, such that π_0 (HOM(T,S)) is the set of homotopy classes of maps of T to S. Indeed, a second approach to homotopy theory is suggested by this result. For each object S there is a homotopy continuous functor Y(S)(T) = HOM(T,S) which is a Yoneda type of functor. Are the regular and coregular homotopy theories characterized by this functor the way topos are characterized by the Yoneda functor? Are the topos simply special cases of homotopy theories in this sense? A related question is whether a coregular homotopy theory gives rise to a coregular homotopy theory when localized with respect to a homology theory (that is, does Bousfield's localization theorem hold for coregular homotopy theories or does it only hold for model categories)?

Finally, there is the problem of verifying that a model category is regular or coregular. This is ignored here, and represents a good bit of the harder slog in my forthcoming paper [HFS]. I hope that this talk will convince people that once coregularity (or regularity) have been established, one can safely forget cofibrations and fibrations, except up to homotopy. More ambitiously, I hope that the axiomatization of homotopy theory given here will bring some order to the expansion of homotopy theory into functional analysis and algebraic geometry, as well as throwing some light on more classical situations of equivariant homotopy theory and CW-complexes.

2. Axioms for homotopy

At the absolute minimum, to do homotopy theory one needs a family of objects, which we call spaces, and two sets $Hom(X,Y)$ and $[X,Y]$ for each pair of spaces X,Y, together with a map $\tau(X,Y)\colon Hom(X,Y) \to [X,Y]$ and a composition law for Hom and $[\ ,\ ]$ respectively, so that each of these defines a category with the spaces, and such that τ defines a functor. We refer to $Hom(X,Y)$ as the set of maps from X to Y and to $[X,Y]$ as the set of homotopy classes of maps from X to Y. If $f \in Hom(X,Y)$, we refer to $\tau(f)$ as the homotopy class of f.

By a homotopy theory $\mathcal{J}$, we shall mean two categories $\underline{Sp}(\mathcal{J})$, $\underline{Ho}(\mathcal{J})$, together with a functor $\tau\colon \underline{Sp}(\mathcal{J}) \to \underline{Ho}(\mathcal{J})$, such that τ is a localization. We call a map $f \in Hom(X,Y)$ a weak equivalence if $\tau(f)$ is an isomorphism. We call $\underline{Sp}(\mathcal{J})$ the category of spaces of the theory $\mathcal{J}$, and $\underline{Ho}(\mathcal{J})$ the homotopy category of $\mathcal{J}$. If $\mathcal{J}', \mathcal{J}''$ are two homotopy theories, by a map $\varphi\colon \mathcal{J}' \to \mathcal{J}''$ we mean a pair of functors $\underline{Sp}(\varphi)\colon \underline{Sp}(\mathcal{J}') \to \underline{Sp}(\mathcal{J}'')$, $\underline{Ho}(\varphi)\colon \underline{Ho}(\mathcal{J}') \to \underline{Ho}(\mathcal{J}'')$ such that

$\underline{Ho}(\varphi)\tau' = \tau'' \underline{Sp}(\varphi)$. Since τ' is a localization, such maps φ are in a bijective correspondence with functors $\psi: \underline{Sp}(\mathfrak{J}') \to \underline{Sp}(\mathfrak{J}'')$ which carry weak equivalences to weak equivalences, under the correspondence $\psi = \underline{Sp}(\varphi)$. Notice that every category C is the category of spaces of two homotopy theories, the discrete homotopy theory $\mathfrak{D}(C)$ where $\underline{Ho}(\mathfrak{D}(C)) = C$ and the localization is the identity, and the indiscrete homotopy theory $\mathfrak{I}(C)$, where $\underline{Ho}(\mathfrak{I}(C))$ is the localization of C with respect to all of its morphisms. Just as in topology, the discrete and the indiscrete structures tend to be uninteresting.

If $\varphi_0, \varphi_1 : \mathfrak{S} \to \mathfrak{J}$ are two morphisms of homotopy theories, by a natural transformation from φ_0 to φ_1 we mean a natural transformation from $\underline{Sp}\ \varphi_0$ to $\underline{Sp}\ \varphi_1$. Recall (Segal [CSSS]) that a natural transformation between two ordinary functors $\gamma_0, \gamma_1 : C \to \mathfrak{D}$ can be thought of as a functor $\gamma: C \times \mathfrak{I} \to \mathfrak{D}$, where $\mathfrak{I}$ is the ordered set $\{0,1\}$ thought of as a category, $\gamma|C \times \{i\} = \gamma_i$ for $i = 0,1$. Notice that since the category of topological theories has finite products, and since $\underline{Sp}$ and $\underline{Ho}$ preserve finite products, a natural transformation $\eta: \varphi_0 \to \varphi_1$ determines natural transformations $\underline{Sp}\ \eta: \underline{Sp}\ \varphi_0 \to \underline{Sp}\ \varphi_1$, $\underline{Ho}(\eta): \underline{Ho}(\varphi_0) \to \underline{Ho}(\varphi_1)$, and is equivalent to a functor $\mathfrak{S} \times \mathfrak{D}(\mathfrak{I}) \to \mathfrak{J}$ which restricts to φ_0 and φ_1 at the "ends" of $\mathfrak{D}(\mathfrak{I})$.

For an ordinary category $\mathfrak{D}$ and a homotopy theory $\mathfrak{J}$, define the "homotopy theory of functors from $\mathfrak{D}$ to $\mathfrak{J}$" as follows. Let $\mathfrak{F}(\mathfrak{D},\mathfrak{J})$ be the homotopy theory whose category of spaces is the category $\mathfrak{F}(\mathfrak{D},\underline{Sp}\mathfrak{J})$ of functors from $\mathfrak{D}$ to $\underline{Sp}\mathfrak{J}$, and whose weak equivalences are the inverse image of the isomorphisms of $\mathfrak{F}(\mathfrak{D},\underline{Ho}(\mathfrak{J}))$. We write $\underline{Ho}(\mathfrak{D},\mathfrak{J})$ for $\underline{Ho}(\mathfrak{F}(\mathfrak{D},\mathfrak{J}))$, the localization of $\mathfrak{F}(\mathfrak{D},\underline{Sp}\mathfrak{J})$ with respect to its weak equivalences. If $\varphi: \mathfrak{D}' \to \mathfrak{D}''$, we have an induced functor $\varphi^*: \mathfrak{F}(\mathfrak{D}'',\mathfrak{J}) \to \mathfrak{F}(\mathfrak{D}',\mathfrak{J})$. We denote $\underline{Ho}(\varphi^*)$ by $\varphi^\#$. Notice that if $\varphi\psi$ is defined, $(\varphi\psi)^\# = \psi^\#\varphi^\#$.

Definition 2.1. If $\mathcal{J}$ is a homotopy theory, we say that $\mathcal{J}$ is complete (resp. cocomplete) if for all $\varphi: \mathcal{D}' \to \mathcal{D}''$ with $\mathcal{D}', \mathcal{D}''$ small, there is a right adjoint $\varphi_{\#}$ (resp. a left adjoint φ_{+}) to $\varphi^{\#}$.

Every complete (resp. cocomplete) model category determines a complete (resp. cocomplete) homotopy theory (see my [HFS]).

Remark 2.2 All model categories are finitely complete and cocomplete. If one replaces the phrase "for all $\varphi: \mathcal{D}' \to \mathcal{D}''$ with $\mathcal{D}', \mathcal{D}''$ small" by the phrase "for all $\varphi: \mathcal{D}' \to \mathcal{D}''$ with the nerves of $\mathcal{D}', \mathcal{D}''$ finite simplicial sets" we get a proper definition for "finitely complete" or "finitely cocomplete" respectively.

Definition 2.3. If $\varphi: \mathcal{D}' \to \mathcal{D}''$ is such that $\varphi_{\#}$ (resp. φ_{+}) exists, $\varphi_{\#}$ (resp. φ_{+}) is called the homotopy right (resp. left) Kan extension along φ. If $\mathcal{D}''$ is the terminal category, $\varphi_{\#}$ (resp. φ_{+}) is called homotopy limit (resp. colimit).

Given a finitely cocomplete homotopy theory $\mathcal{J}$ for which $\underline{Ho}(\mathcal{J})$ is base-pointed (i.e., the initial object is also a terminal object), it is possible to define suspensions, Toda brackets, and other common constructions. In order to show that these have any desirable properties, we need to know that there is some coherence for homotopy Kan extensions. Since for any φ, ψ which can be composed we have $(\psi\varphi)^{\#} = \varphi^{\#}\psi^{\#}$, automatically $(\psi\varphi)_{\#}$ is naturally isomorphic to $\psi_{\#}\varphi_{\#}$ and $(\psi\varphi)_{+}$ is naturally isomorphic to $\psi_{+}\varphi_{+}$, since adjoints are unique up to isomorphism. The sort of coherence desired has to do with restriction along fiber products.

The algebraic geometers call a functor $\pi: \mathcal{E} \to \mathcal{B}$ a prefibration if for all B in $\mathcal{B}$, the inclusion $\pi^{-1}(B) \to \pi/B$ has a right adjoint. Here, $\pi^{-1}(B)$ is the subcategory of $\mathcal{E}$ whose objects are those objects carried to B by π, and whose morphisms are those morphisms of $\mathcal{E}$ carried to the identity morphism of B, and

π/B is the category of diagrams of the form $\beta\colon \pi(E) \to B$, where a map from (β,E) to (β',E') is a map $\alpha\colon E \to E'$ with $\beta'\pi(\alpha) = \beta$. Projections of products onto factors are prefibrations, and if $\pi\colon \mathcal{E} \to \mathcal{B}$ is a prefibration, $\mathcal{B}' \to \mathcal{B}$ is any functor, and $\pi'\colon \mathcal{E}' \to \mathcal{B}'$ is the pullback (=base extension = fiber product) of π along $\mathcal{B}' \to \mathcal{B}$, then π' is also a prefibration. If for all B, $\pi^{-1}(B) \to B/\mathcal{B}$ has a left adjoint, π is called a preopfibration; π is a preopfibration if and only if its opposite functor is a prefibration.

Definition 2.4. A homotopy theory $\mathcal{J}$ is said to be regular (resp. coregular) if for all prefibrations (resp. preopfibrations) $\pi\colon \mathcal{E} \to \mathcal{B}$ and for all $\beta\colon \mathcal{B}' \to \mathcal{B}$, where $\mathcal{B}'$ and $\mathcal{E}$ are small, if $\pi^{\#}\colon \underline{\mathrm{Ho}}(\mathcal{B},\mathcal{J}) \to \underline{\mathrm{Ho}}(\mathcal{E},\mathcal{J})$ has a right (resp. left) adjoint $\pi_{\#}$ (resp. π_{+}), then so does $(\pi')^{\#}$, and $(\pi')_{\#}(\beta')^{\#} = \beta^{\#}\pi_{\#}$ (resp. $(\pi')_{+}\beta^{\#} = \beta^{\#}\pi_{+}$), where $\pi'\colon \mathcal{E}' \to \mathcal{B}'$ is the pullback of π along β, $\beta'\colon \mathcal{E}' \to \mathcal{E}$ is the pullback of β along π.

Every complete (resp. cocomplete) model category determines a regular (resp. coregular) homotopy theory. This is proved in my paper [HFS]. If $\mathbb{C}$ is a complete (resp. cocomplete) category, the discrete theory $\mathcal{D}(\mathbb{C})$ is regular (resp. coregular).

For some functors homotopy Kan extensions will exist for a special reason. If $\varphi\colon \mathcal{S} \to \mathcal{J}$, $\psi\colon \mathcal{J} \to \mathcal{S}$ are maps of homotopy theories, we say that φ is left adjoint to ψ if $\underline{\mathrm{Sp}}\,\varphi$ is left adjoint to $\underline{\mathrm{Sp}}\,\psi$. Notice that for ψ to have a left adjoint, it is necessary and sufficient that $\underline{\mathrm{Sp}}\,\psi$ have a left adjoint which preserves weak equivalences. It is easy to check that φ is left adjoint to ψ if and only if there are natural transformations $\eta\colon 1 \to \psi\varphi$, $\varepsilon\colon \varphi\psi \to 1$ such that the resulting compositions $\varphi \to \varphi\psi\varphi \to \varphi$, $\psi \to \psi\varphi\psi \to \psi$ are the identity. From this observation, we obtain the following lemma immediately.

Lemma 2.5. If $\varphi: \mathcal{D}' \to \mathcal{D}''$ is a functor with a left (resp. right) adjoint $\psi: \mathcal{D}'' \to \mathcal{D}'$, then for all homotopy theories $\mathcal{J}$, φ^* has a right (resp. left) adjoint given by ψ^*. Thus $\varphi^{\#}$ has a right (resp. left) adjoint given by $\psi^{\#}$, so $\varphi_{\#} = \psi^{\#}$ (resp. $\varphi_+ = \psi^{\#}$).

One of the useful features of regular complete (or coregular cocomplete) homotopy theories is that certain diagrams in the homotopy category can be shown to be, up to isomorphism, the images of diagrams of spaces. Another useful feature is that homotopy Kan extensions behave much like ordinary Kan extensions. This is reflected in the following lemma, whose proof is a straightforward verification which is left to the reader.

Lemma 2.6. If $\varphi: \mathcal{D}' \to \mathcal{D}''$, let $\mathcal{E} = \mathcal{D}''/\Phi$ (resp. $\Phi/\mathcal{D}''$) be the category whose objects are triples (d,D',D'') where D' is in $\mathcal{D}'$, D'' is in $\mathcal{D}''$, and $d: D'' \to \Phi(D')$ (resp. $d: \Phi(D') \to D''$), and whose morphisms are pairs of maps which produce commutative squares in the obvious way. Let $\pi': \mathcal{E} \to \mathcal{D}'$, $\pi'': \mathcal{E} \to \mathcal{D}''$ be the two projection maps, and let $i: \mathcal{D}' \to \mathcal{E}$ be given by $i(D') = (1,D',\varphi(D'))$. Then $\varphi = \pi''i$, i is right (resp. left) adjoint to π', and π'' is a prefibration (resp. preopfibration). Thus if $\mathcal{J}$ is any complete (resp. cocomplete) homotopy theory, $\varphi_{\#} = (\pi'')_{\#}i^{\#}$ (resp. $\varphi_+ = (\pi'')_+i^{\#}$). Further, if for D'' in $\mathcal{D}''$ we let D''/φ (resp. φ/D'') be $(\pi'')^{-1}(D'')$, then if $\mathcal{J}$ is also regular (resp. coregular), then for all Φ in $\underline{\text{Ho}}(\mathcal{D}',\mathcal{J})$, the adjunction map $\varphi_{\#}(\Phi)(D'') \to \text{Holim}(i^{\#}(\Phi)|(D''/\varphi))$ (resp. $\text{Hocolim}(i^{\#}(\Phi)|(\varphi/D'')) \to \varphi_+(\Phi)(D'')$) is an isomorphism.

Notice that the fact that π'' is a prefibration (resp. preopfibration) shows that regularity (resp. coregularity) implies completeness (resp. cocompleteness). Indeed, we could restrict the requirements for regularity so that only fibrations need to be considered without losing the elementary parts of homotopy theory. However, model categories satisfy the more general requirements, and all arguments based on subdivisions of categories (of which my [HFS] has many) require the form of completeness given here.

For those readers unfamiliar with fibrations and opfibrations of categories, we shall show how these tend to arise (indeed, up to suitable equivalence, this is the general case). Suppose that C is a category, Cat is the category of small categories, and $\Phi: C \to \mathrm{Cat}$ is any functor. Then the "total category", "wreath product", or "Grothendieck construction" on Φ is a category $\mathrm{Wr}(\Phi)$, whose objects are pairs (C,X) for C in C, X in $\Phi(C)$, and whose morphisms $(C_0,X_0) \to (C_1,X_1)$ are pairs of morphisms $\gamma: C_0 \to C_1$, $\mu: \Phi(\gamma)(X_0) \to X_1$. Composition of morphisms is defined in the obvious manner. Then $\pi(\Phi): \mathrm{Wr}(\Phi) \to C$ given by $\pi(C,X) = C$ is an opfibration. Notice that coregularity implies that for any $\Delta: \mathcal{D} \to C$ with $\mathcal{D},C$ small, $\pi(\Phi\Delta)_+\mathrm{Wr}(\Delta)^\# = \Delta^\#\pi(\Phi)_+$. The situation in (2.6) is described by letting $\Phi: \mathcal{D}'' \to \mathrm{Cat}$ by $\Phi(D'') = \varphi/D''$.

Lemma 2.7. If $\varphi: \mathcal{D}' \to \mathcal{D}''$ is full and faithful, then for every regular (resp. coregular) homotopy theory $\mathcal{J}$, and for all $\Psi: \mathcal{D}' \to \underline{\mathrm{Ho}}(\mathcal{J})$, the canonical map $\Psi \to \varphi^\#\varphi_\#(\Psi)$ (resp. $\varphi_+\varphi^\#(\Psi) \to \Psi$) in $\underline{\mathrm{Ho}}(\mathcal{D}',\mathcal{J})$ is an isomorphism.

<u>Proof</u>. If φ is full and faithful, the projection $\mathcal{D}''/\varphi \to \mathcal{D}''$ (resp. $\varphi/\mathcal{D}'' \to \mathcal{D}''$) has the property that its base extension along φ has a left (resp. right) adjoint given by the identity morphism. Thus right (resp. left) homotopy Kan extension along the base extension of the projection is given by composing with this section.

Notice that if $\mathcal{D}$ is a category with an initial object ϕ, then for all $\Phi: \mathcal{D} \to \underline{\mathrm{Ho}}(\mathcal{J})$, the homotopy limit of Φ is naturally isomorphic to $\Phi(\phi)$. More surprising is the following fact. If $\mathcal{D}$ is a category with a terminal object $*$, and $\Phi\colon \mathcal{D} \to \underline{\text{Ho}}(\mathcal{J})$ is a constant functor, then if $\mathcal{J}$ is regular, the homotopy limit of Φ will be isomorphic to $\Phi(*)$. To see this, let $\xi\colon \{*\} \to \mathcal{D}$ be the inclusion of the terminal object. For all D in $\mathcal{D}$, observe that D/ξ has a single object, the terminal map $D \to *$. Thus $\xi_{\#}\xi^{\#}(\Phi)$ is naturally isomorphic to the constant functor with value $\Phi(*)$; thus if Φ is constant, $\Phi \to \xi_{\#}\xi^{\#}(\Phi)$ is an isomorphism. If $\lambda\colon \mathcal{D} \to \{*\}$ is the terminal functor, $\lambda\xi$ is the identity. Since homotopy limit is given by $\lambda_{\#}$, $\underline{\text{Ho}}\lim(\Phi) = \lambda_{\#}\xi_{\#}\xi^{\#}(\Phi) = (\lambda\xi)_{\#}\xi^{\#}(\Phi) = \xi^{\#}(\Phi) = \Phi(*)$. This gives us the following result.

Lemma 2.8. If $\mathcal{J}$ is regular (resp. coregular), and if $\pi\colon \mathcal{E} \to \mathcal{B}$ is a prefibration (resp. preopfibration) with the property that for all B in $\mathcal{B}$ the category $\pi^{-1}(B)$ has a terminal (resp. initial) object, then for all $\Phi\colon \mathcal{B} \to \underline{\text{Sp}}(\mathcal{J})$, the natural map $\Phi \to \pi_{\#}\pi^{\#}(\Phi)$ (resp. $\pi_{+}\pi^{\#}(\Phi) \to \Phi$) is an isomorphism in $\underline{\text{Ho}}(\mathcal{B},\mathcal{J})$.

Corollary 2.9. If $\mathcal{J}$ is a regular (resp. coregular) homotopy theory, and if $\pi\colon \mathcal{E} \to \mathcal{B}$ is any prefibration (resp. preopfibration) of small categories in which every fiber has a terminal (resp. initial) object, then $\pi^{\#}\colon \underline{\text{Ho}}(\mathcal{B},\mathcal{J}) \to \underline{\text{Ho}}(\mathcal{E},\mathcal{J})$ is full and faithful and has a right (resp. left) adjoint left inverse given by $\pi_{\#}$ (resp. π_{+}).

This result is central to the use of subdivisions of categories to study homotopy limits and colimits. In my [HFS], (2.9) is the starting point for showing that complete model categories are regular.

If $\Gamma\colon \mathcal{J} \to \mathcal{S}$ is a map of homotopy theories, Γ is said to be continuous (resp. cocontinuous) if for all $\varphi\colon \mathcal{D}' \to \mathcal{D}''$ with $\mathcal{D}',\mathcal{D}''$ small, the adjunction map $\underline{\text{Ho}}(\Gamma)\varphi_{\#} \to \varphi_{\#}\underline{\text{Ho}}(\Gamma)$ (resp. $\varphi_{+}\underline{\text{Ho}}(\Gamma) \to \underline{\text{Ho}}(\Gamma)\varphi_{+}$) is always an isomorphism. Less generally, we will say that Γ preserves right (resp. left) Kan extensions along φ if, for the given functor φ, the adjunctions above are isomorphisms. The following follows immediately from (2.6).

Lemma 2.10. If $\mathcal{S}$ and $\mathcal{T}$ are regular (resp. coregular) homotopy theories, $\Gamma: \mathcal{T} \to \mathcal{S}$ is continuous (resp. cocontinuous) if and only if Γ preserves homotopy limits (resp. colimits).

The following is proved in a more general form in my [HFS]. We shall omit the proof here as it requires an exposition of the subdivision construction for categories which is behond us here.

Theorem 2.11. If $\mathcal{S}$ and $\mathcal{T}$ are regular (resp. coregular), $\Gamma: \mathcal{S} \to \mathcal{T}$ is continuous (resp. cocontinuous) if and only if Γ preserves homotopy product (resp. coproducts) and either homotopy pullbacks or homotopy equalizers (resp. homotopy pushouts or homotopy coequalizers).

Similarly, Γ preserves finite homotopy right (resp. left) Kan extensions if and only if it preserves finite products (resp. coproducts) and homotopy equalizers (resp. coequalizers).

Our conditions of regularity, completeness, and their duals, are all stated in terms of functor homotopy theories. This fact gives the following result immediately, which we note here. The analog of this for model categories is not nearly so simple to state.

Lemma 2.12. If $\mathcal{T}$ is a homotopy theory, then for all small categories $\mathcal{D}$, if $\mathcal{T}$ is complete, cocomplete, regular, or coregular, respectively, so is $\mathcal{F}(\mathcal{D},\mathcal{T})$.

3. Realizing diagrams

A homotopy theory which is regular or coregular has the property that certain types of diagrams which exist in the homotopy category exist, up to isomorphism, in the category of spaces. The diagrams which can be so realized are those for which no "higher homotopies" are required.

Recall that an ordered graph $\mathcal{G}$ is a set O and a set M together with maps $\sigma,\tau\colon M \to O$. Associated to a graph $\mathcal{G}$ is a category $\mathcal{P}(\mathcal{G})$ with objects O and morphisms all paths $(a_0,a_1,\dots,a_n)$ while $a_i \in M$, $\sigma(a_i) = \tau(a_{i-1})$. We call $\mathcal{P}(\mathcal{G})$ the category of paths generated by $\mathcal{G}$.

Theorem 3.1. If $\mathcal{J}$ is a coregular or regular homotopy theory, then for any graph $\mathcal{G}$, every functor $\Phi\colon \mathcal{P}(\mathcal{G}) \to \underline{Ho}(\mathcal{J})$ is isomorphic to some $\Psi\colon \mathcal{P}(\mathcal{G}) \to \underline{Sp}(\mathcal{J})$.

Theorem 3.2. If $\mathcal{J}$ is a coregular homotopy theory, and if $\alpha_i = (A_i \to B_i)$ are objects in $\underline{Ho}(\cdot\!\to\!\cdot,\mathcal{J})$ for $i = 0,1$ such that A_0 is isomorphic to A_1, then there is a $\beta = (B_0' \leftarrow A' \to B_1')$ in $\underline{Ho}(\cdot\!\leftarrow\!\cdot\!\to\!\cdot,\mathcal{J})$ so that the restrictions of β to the two arrows are isomorphic to the two α_i by isomorphisms which are compatible with the given isomorphisms when restricted to source objects.

The most important application of (3.2) is the following.

Theorem 3.3. If $\mathcal{J}$ is coregular, D the homotopy colimit of a diagram $B \leftarrow A \to C$, then for all E, given $\beta\colon B \to D$, $\gamma\colon C \to D$ in $\underline{Ho}(\mathcal{J})$ with $\beta|A = \gamma|A$, then there is $\delta\colon D \to E$ in $\underline{Ho}(\mathcal{J})$ with $\delta|B = \beta$, $\delta|C = \gamma$. The dual statement holds for $\mathcal{J}$ regular.

The final result on realizing diagrams in the category of spaces has to do with the following question. Suppose $\varphi\colon \mathcal{D}' \to \mathcal{D}''$ is a functor between small categories, $\Phi\colon \mathcal{D}' \to \underline{Sp}(\mathcal{J})$, $\Psi\colon \mathcal{D}'' \to \underline{Sp}(\mathcal{J})$, and $\xi\colon \Phi \to \Psi\varphi$ in $\underline{Ho}(\mathcal{D}',\mathcal{J})$. Now (3.1) tells us that Φ is isomorphic to some Φ' and $\Psi\varphi$ to some Φ'' so that there is a map $\xi'\colon \Phi' \to \Phi''$ in $(\mathcal{D}',\underline{Sp}(\mathcal{J}))$ which is conjugate to ξ under the isomorphisms given in $\underline{Ho}(\mathcal{D}',\mathcal{J})$. The question then arises: can we find Ψ' in $\underline{Ho}(\mathcal{D}'',\mathcal{J})$, isomorphic to Ψ, so that for a suitable choice of Φ' there is a map $\xi''\colon \Phi' \to \Psi'$ which represents ξ in the homotopy category? The answer is yes, as we shall see. From now on, let $\mathcal{J}$ be a coregular homotopy theory.

Lemma 3.4. If $A_0 \leftarrow B_0 \rightarrow A_1 \leftarrow \dots \rightarrow A_n$ is a diagram with homotopy colimit C such that each $A_i \leftarrow B_i$ is a weak equivalence, then the natural map $A_n \rightarrow C$ is an isomorphism in $\underline{Ho}(\mathfrak{J})$.

Proof. The diagram $B_0 \rightarrow A_1 \leftarrow \dots \rightarrow A_n$, when homotopy left Kan extended to a diagram of the type above, is isomorphic to the diagram above in the homotopy category of such diagrams. Since the homotopy colimit of a homotopy left Kan extension is a homotopy colimit, we see that the homotopy colimit of $A_0 \leftarrow \dots \rightarrow A_n$ is the same as that of $B_0 \rightarrow A_1 \leftarrow \dots \rightarrow A_n$. However, in this last diagram, the inclusion of the subcategory $A_1 \leftarrow \dots \rightarrow A_n$ is right adjoint to the functor which collapses the arrow $B_0 \rightarrow A_n$; thus homotopy colimits do not change if taken over this smaller category. We now induct on n.

If $\varphi\colon \mathfrak{D}' \rightarrow \mathfrak{D}''$ is any functor, we define the comapping cylinder $M_+(\varphi)$ to be the following category. The objects of $M_+(\varphi)$ are the disjoint union of the objects of $\mathfrak{D}'$ and $\mathfrak{D}''$, and both $\mathfrak{D}'$, $\mathfrak{D}''$ are full subcategories of $M_+(\varphi)$. There are no morphisms from an object of $\mathfrak{D}''$ to an object of $\mathfrak{D}'$. If D'' is in $\mathfrak{D}''$, D' is in $\mathfrak{D}'$, then $\mathrm{Hom}(D',D'') = \mathrm{Hom}(\varphi(D'),D'')$. Notice that if $\mathfrak{D}''$ is the terminal object, $M_+(\varphi)$ is the right cone on $\mathfrak{D}'$. Further, we have obvious functors $\zeta_0\colon \mathfrak{D}' \rightarrow M_+(\varphi)$, $\zeta_1\colon \mathfrak{D}'' \rightarrow M_+(\varphi)$, $\xi\colon M_+(\varphi) \rightarrow \mathfrak{D}''$ such that ζ_0, ζ_1 are full and faithful, $\xi\zeta_1$ is the identity. Also, ζ_1 is right adjoint to ξ.

Lemma 3.5. Given $\varphi\colon \mathfrak{D}' \rightarrow \mathfrak{D}''$ as above, for all $\Phi\colon \mathfrak{D}' \rightarrow \underline{Sp}(\mathfrak{J})$, $\varphi_+(\Phi) = \zeta_1^{\#}(\zeta_0)_+(\Phi)$, $\zeta_0^{\#}(\zeta_0)_+(\Phi) = \Phi$. If $\varepsilon\colon \mathfrak{D}' \times \{0{\rightarrow}1\} \rightarrow M_+(\varphi)$ by $\varepsilon|\mathfrak{D}' \times \{0\} = \zeta_0$, $\varepsilon|\mathfrak{D}' \times \{1\} = \zeta_1\varphi$, then $\varepsilon^{\#}(\zeta_0)_+(\Phi)$ defines a map $\Phi \rightarrow \varphi^{\#}\varphi_+(\Phi)$ which is the usual adjunction.

We now turn to the proofs of (3.2) and (3.3). To do this, we need a definition.

Definition 3.6. By a quasilinear partially ordered set we shall mean any poset $\{x_0,x_1,\ldots,x_n\}$ such that each pair (x_i,x_{i+1}) is comparable, and if a pair (x_i,x_{i+j}) is comparable, then $\{x_i,x_{i+1},\ldots,x_{i+j}\}$ is a totally ordered set. By a path in $\underline{Sp}(\mathcal{J})$ from an object A to an object B, we mean a quasilinear poset $\mathcal{P} = \{x_0,\ldots,x_n\}$, together with a functor $\Pi: \mathcal{P} \to \underline{Sp}(\mathcal{J})$ such that $\Pi(x_0) = A$, $\Pi(x_n) = B$, and if for all i, if $x_i > x_{i+1}$, then $\Pi(x_{i+1}) \to \Pi(x_i)$ is a weak equivalence; $(\Pi,\mathcal{P})$ is a constant path if $A = B$ and all maps are the identity; $(\Pi,\mathcal{P})$ is quasiconstant if all maps are weak equivalences.

Lemma 3.7. If $\Pi: \mathcal{P} \to \underline{Sp}(\mathcal{J})$ is a path from A to B, then if $\mathcal{J}$ is regular (resp. coregular), the natural map $\mathrm{Holim}(\Pi) \to A$ (resp. $B \to \mathrm{Hocolim}(\Pi)$) is an isomorphism in $\underline{Ho}(\mathcal{J})$.

__Proof__. The proof is given by induction on the "length" of $\mathcal{P}$. Let $\mathcal{P}' = \{x_0,\ldots,x_{n-1}\}$, $\Pi' = \Pi|\mathcal{P}'$. It suffices to show that $\mathrm{Hocolim}(\Pi) \to \mathrm{Hocolim}(\Pi')$ is an isomorphism. If $x_{n-1} < x_n$, the inclusion $i: \mathcal{P}' \to \mathcal{P}$ has a right adjoint, so the result follows from (2.5) and the fact that homotopy limits are homotopy right Kan extensions along terminal functors. If $x_n < x_{n-1}$, then the map $\Pi \to i_\# i^\#(\Pi)$ is a weak equivalence, since $\Pi(x_n) \to \Pi(x_{n-1})$ is. Thus, since $\mathrm{Holim}(\Pi') = \mathrm{Holim}(i_\#(\Pi'))$, we have the desired result.

Theorem 3.2 now can be proved as follows. Choose an isomorphism $A_0 \to A_1$ in $\underline{Ho}(\mathcal{J})$. Then, since $\underline{Sp}(\mathcal{J}) \to \underline{Ho}(\mathcal{J})$ is a localization, there is a path from A_0 to A_1 which represents the given isomorphism. Let $\Pi: \mathcal{P} \to \underline{Sp}(\mathcal{J})$ represent this path, and let $\mathcal{D} = \mathcal{P} \cup \{y_0,y_1\}$, $y_0 > x_0$, $y_1 > x_n$, $\Xi: \mathcal{D} \to \mathrm{Sp}(\mathcal{J})$ be given by $\Xi|\mathcal{P} = \Pi$, $\Xi(y_0) = B_0$ $\Xi(y_1) = B_1$. Let $\mathcal{L} = \{y_0,z,y_1\}$ with $y_0 > z < y_1$, and let $\varphi: Q \to \mathcal{L}$ be given by $\varphi(y_1) = y_i$ for $i = 0,1$, $\varphi(\mathcal{D}) = \{z\}$. Notice that $\varphi_+(\Xi)$ as the restriction to $\mathcal{L}$ of $i_+(\Xi)$ for $i: \mathcal{D} \to M_+(\varphi)$, we see that $\varphi_+(\Xi)$ is the desired diagram.

To prove (3.3), we need to consider "homotopies" between paths. If A,B are two objects in $\underline{Sp}(\mathcal{J})$, and if $(\Pi_0,\mathcal{P}_0)$, $(\Pi_1,\mathcal{P}_1)$ are two paths from A to B, by a homotopy $(\Gamma,\mathcal{P})$ from $(\Pi_0,\mathcal{P}_0)$ to $(\Pi_1,\mathcal{P}_1)$ we mean a partially ordered set $Q = (q_0,\ldots,q_r)$ whose partial order is generated by the fact that each pair (q_i,q_{i+1}) is comparable, and a functor Γ from $\mathcal{Q}$ to the category of paths from A to B, where a map of paths $\Lambda: (\Pi_0,\mathcal{P}_0) \to (\Pi_1,\mathcal{P}_1)$ is a pair $\Lambda: \mathcal{P}_0 \to \mathcal{P}_1$ which preserves endpoints, together with a natural transformation $\lambda: \Pi_0 \to \Pi_1\Lambda$ which is the identity on the endpoints.

Recall from Gabriel-Zisman [CFHT] that one can associate to any localization $\gamma: \mathcal{C} \to \mathcal{L}$ a category $\mathcal{P}(\mathcal{C})$ with the same objects as $\mathcal{C}$, but whose morphisms are the paths in the sense described above. Composition of paths is described in the obvious manner, and composition of the corresponding maps in a path in $\mathcal{L}$ gives a map $\pi: \mathcal{P}(\mathcal{C}) \to \mathcal{L}$. Gabriel and Zisman show that π is epimorphic and give generators for the equivalence relation on morphisms given by π.

Lemma 3.8. If A,B are objects of $\mathcal{C}$ as above, the relation $\mathcal{P}(\mathcal{C})(A,B) \xrightarrow{\pi} \mathcal{L}(A,B)$ is the relation of homotopy described above.

Proof. First we observe that if a commutative square has both top and bottom arrows isomorphisms, the square remains commutative if the top and bottom arrows are reversed and the inverses of the given isomorphisms are used. From this it follows immediately that homotopic paths are carried by π to the same morphism. Thus we need only show that if two paths are carried to the same path by π, then they are homotopic.

If $(\Pi_0,\mathcal{P}_0)$, $(\Pi_1,\mathcal{P}_1)$ are two paths, we say that $(\Pi_1,\mathcal{P}_1)$ is a consolidation of $(\Pi_0,\mathcal{P}_0)$ if they have the same endpoints and there is a map $\Lambda: (\Pi_0,\mathcal{P}_0) \to (\Pi_1,\mathcal{P}_1)$ such that for all but at most one $x_i \in \mathcal{P}_1$, $\Lambda^{-1}(x_i)$ is a single point, and such that for the remaining x_i, $\Lambda^{-1}(x_i)$ is one of the following categories:

(a) $\{x_i \le x_{i+1}\}$, (b) $\{x_i \le x_{i+1} \ge x_{i+2}\}$, (c) $\{x_i \le x_{i+1} \ge x_{i+2}\}$. We also require that $j > j'$, $\Lambda(x_i) = x_j$, $\Lambda(x_{i'}) = x_{j'}$ implies $i > i'$. Finally, we require the following conditions on Π_0, Π_1, λ in each of the cases above. In all cases, $\lambda: \Pi_0(x_i) \to \Pi_1(\Lambda(x_i))$ is the identity map if $\Lambda^{-1}\Lambda(x_i) = \{x_i\}$. In case (a), we require either that $\Pi_0(x_i) = \Pi_0(x_{i+1}) = \Pi_1(x_i)$, and all associated maps are the identity, or we require $\Pi_0(x_i) = \Pi_1(x_i)$, $\Pi_0(x_{i+2}) = \Pi_1(x_{i+1})$, $\Pi_1(x_i) \to \Pi_1(x_{i+1})$ is the composition $\Pi_0(x_i) \to \Pi_0(x_{i+1}) \to \Pi_0(x_{i+2})$, and the $\Pi_0(x_i) \to \Pi_1(x_i)$, $\Pi_0(x_{i+2}) \to \Pi_1(x_{i+1})$ are the identity. In case (b), we require that $\Pi_1(x_i) = \Pi_0(x_i) = \Pi_0(x_{i+2})$ with corresponding natural transformations the identity, and the two maps $\Pi_0(x_i) \to \Pi_0(x_{i+1})$ and $\Pi_0(x_{i+2}) \to \Pi_0(x_{i+1})$ are the same (weak equivalence). Part (c) is the reverse of part (b).

It is pointed out in the beginning of Gabriel-Zisman that the relation determined by π is generated by the relation defined by consolidation above. Notice that every consolidation is a homotopy of paths. Since homotopies of paths can be composed, we see that homotopy of paths is an equivalence relation, and that therefore they give all of the relations defined by π.

Notice that associated to a homotopy $(\Gamma,\mathcal{D})$ we have associated a diagram (with the obvious definition) $Wr(\Gamma): Wr(\mathcal{D}) \to \underline{Sp}(J)$. Let $\xi: Wr(\mathcal{D}) \to \mathcal{D}$ be the projection. This is an opfibration, and thus a preopfibration. Thus ξ_+ can be computed by taking homotopy colimits on the fibers $\xi^{-1}(q_i) = \Gamma(q_i)$. Thus $\xi_+(Wr(\Gamma)$ will be a constant path at the common target of the paths $\Gamma(q_i)$. Consequently, $\mathrm{Hocolim}(Wr(\Gamma)) = \mathrm{Hocolim}(\xi_+(Wr(\Gamma)))$ is naturally isomorphic to the common target of the paths.

Consider the situation now in (3.3). We have maps $B \leftarrow A \to C$ in $\underline{Sp}(J)$, together with paths from B and C to E such that the resulting paths from A to E agree in $\underline{Ho}(J)$. Since the resulting paths agree in $\underline{Ho}(J)$, there is a homotopy $(\Gamma,\mathcal{D})$ from the first to the second. We shall modify this homotopy slightly so that it will give us a map of the diagram $B \leftarrow A \rightarrow C$ to the constant diagram $E \leftarrow E \rightarrow E$ in $\underline{Ho}(\{\cdot\leftarrow\cdot\rightarrow\cdot\},\mathcal{J})$.

If (Π,ρ) is a path from A to B, by the subdivision of this path we mean a path (Π',ρ') defined as follows. The objects of ρ' are all pairs (x_i,x_j) with $|i-j| \leq 1$, and $\xi\colon \rho' \rightarrow \rho$ by $\xi(x_i,x_j) = x_k$ for $x_k = \max(x_i,x_j)$, and $\Pi' = \Pi\xi$. Notice that every comma category ξ/x_i is a quasilinear partially ordered set. It is not difficult to show that $\xi_+\xi^{\#}(\Pi) \rightarrow \Pi$ is always an isomorphism in $\underline{Ho}(\rho,\mathcal{J})$.

Observe that subdivision of paths is functorial. Thus, if we have a homotopy $(\Gamma,\mathcal{D})$ between two paths, $(\Gamma',\mathcal{D})$ is a homotopy between their subdivision, where $\Gamma'(y_i)$ is $\Gamma(y_i)'$ for all y_i in $\mathcal{D}$. We now define a quasilinear poset $\mathcal{D}_0$ as follows. If $\mathcal{D} = \{y_0,\ldots,y_m\}$, $\mathcal{D}_0 = \{y_{-2},y_{-1},y_0,\ldots,y_{m+2}\}$, where $y_{-2} \leq y_{-1} \geq y_0$, $y_m \leq y_{m+1} \geq y_{m+2}$. Let Γ_0 be defined on $\mathcal{D}_0$ by $\Gamma_0|\mathcal{D} = \Gamma'$, $\Gamma_0(y_{-2}) = \Gamma_0(y_{-1}) = \Gamma(y_0)$, $\Gamma_0(y_{m+2}) = \Gamma_0(y_{m+1}) = \Gamma(y_m)$, $\Gamma_0(y_0) \rightarrow \Gamma_0(y_{-1})$ and $\Gamma_0(y_m) \rightarrow \Gamma_0(y_{m+1})$ are the maps associated to the subdivision, $\Gamma_0(y_{-2}) \rightarrow \Gamma_0(y_{-1})$, $\Gamma_0(y_{m+2}) \rightarrow \Gamma_0(y_{m+1})$ are the identity. Notice that each $\Gamma_0(y_i)$ begins with a path of the form $A \rightarrow A'$; $\Gamma_0(y_{-2})$ begins as $A \rightarrow B$ and $\Gamma_0(y_{m+2})$ begins as $A \rightarrow C$.

We now construct a map $\zeta\colon \mathrm{Wr}(\Gamma_0) \rightarrow \mathcal{S}$, where $\mathcal{S}$ is the commutative square diagram, as follows. The inverse image under ζ of the initial object is the path of initial objects, the inverse images under ζ of the two intermediate objects are the two pairs (y_{-2},B) and (y_{m+2},C). The inverse image of the terminal object is everything else. The comma categories ζ/x for the three non-terminal objects x are just paths, and the comma category over the terminal object is the underlying category of $\mathrm{Wr}(\Gamma_0)$. Consequently, we see that $\zeta_+(\mathrm{Wr}(\Gamma_0))$ is a commutative square; consequently it determines a map of a diagram isomorphic to $B \leftarrow A \rightarrow C$ to a constant diagram $E' \leftarrow E' \rightarrow E'$ for E' isomorphic to E. This establishes (3.3).

4. Enriching homotopy theories

In this section, $\mathcal{J}$ will be a coregular homotopy theory. We shall show how to "enrich" the homotopy category of $\mathcal{J}$ by providing it with a bifunctor HOM with values in the homotopy category of simplicial sets ($\cong$ the homotopy category of CW-complexes). By simplicial set we shall mean the simplicial sets whose objects lie in some fixed universe so that all required coproducts can be constructed (set theoretic paradoxes arise, of course, if inside a category $\mathcal{C}$ one attmpts to construct the coproduct of all of the objects of $\mathcal{C}$). Notice that $\underline{Ho}(\mathcal{J})$ will not be a small category with respect to this universe, in general.

If X is a simplicial set, associated to X we have the covering category $Cov(X)$ of X whose objects are maps $\Delta^n \to X$ of a standard simplex to X, and whose morphisms are the commutative triangles arising from compatible maps of standard simplices. Clearly Cov is a functor. Let $Norm(X) \subset Cov(X)$ be the full subcategory of $Cov(X)$ whose objects are nondegenerate--i.e., do not factor through a nontrivial epimorphism of standard simplices. The inclusion $Norm(X) \to Cov(X)$ has a left inverse left adjoint which assigns to $\Delta^n \to X$ the map $\Delta^n \to X$ arising in its unique factorization $\Delta^n \to \Delta^m \to X$, where $\Delta^n \to \Delta^m$ is an epimorphism and $\Delta^m \to X$ is nondegenerate. It is easy to verify that Norm is also a functor. The following is immediate from the observation above.

Lemma 4.1. If X is a simplicial set, $\Phi\colon Cov(X) \to \underline{Sp}(\mathcal{J})$, then the map $Hocolim(\Phi|Norm(X)) \to Hocolim(\Phi)$ is an isomorphism in $\underline{Ho}(\mathcal{J})$.

We can now define a bifunctor $(T,X) \mapsto T \otimes X$ from $\underline{Sp}(\mathcal{J}) \times s\mathcal{E}$ to $\underline{Ho}(\mathcal{J})$, where $s\mathcal{E}$ is the category of simplicial sets (in our universe) by letting $T \otimes X$ be the homotopy colimit over $Norm(X)$ of the constant functor with value T.

The functoriality of this tensor product can be improved in a certain sense. If $\mathcal{D}$ is a small category, $\phi\colon \mathcal{D} \to \underline{Sp}(\mathcal{J}) \times s\mathcal{E}$, we can factor $\phi\colon \mathcal{D} \to \underline{Ho}(\mathcal{J})$

through $\underline{Sp}(J)$ as follows. Let $\mathcal{D}'$ have as objects pairs (D,σ) where D is in $\mathcal{D}$ and σ is in $Cov(\pi_2\Phi(D))$, where $\pi_2\colon \underline{Sp}(J) \times s\mathcal{E} \to s\mathcal{E}$ is projection onto the second factor. By a morphism from (D_0,σ_0) to (D_1,σ_1) we mean a morphism $\delta\colon D_0 \to D_1$ and a morphism $\varepsilon\colon Cov(\pi_2\Phi(\delta))(\sigma_0) \to \sigma_1$. Then projection onto the first factor gies us a functor $\xi\colon \mathcal{D}' \to \mathcal{D}$. Let $\Phi'\colon \mathcal{D}' \to \underline{Sp}(J)$ assign to (D,σ) the value $\pi_1\Phi(D)$, where $\pi_1\colon \underline{Sp}(J) \times s\mathcal{E} \to \underline{Sp}(J)$ is projection onto the first factor. The following is now straightforward to verify.

Lemma 4.2. ξ is a preopfibration, and for all D in $\mathcal{D}$, $\xi_+(\Phi')(D)$ is naturally isomorphic to $\otimes\, \Phi(D)$.

Notice that $\xi_+(\Phi')$ is in $\underline{Ho}(\mathcal{D},J)$. In particular, if Φ is the identity functor, if $\xi_+(\Phi)$ exists, it will give us a representative for $\otimes$ in $\underline{Ho}(\underline{Sp}(J) \times s\mathcal{E},J)$. The problem here is that $\underline{Sp}(J) \times s\mathcal{E}$ is not a small category. There are consequently no problems associated to realizing $T \otimes X$ as a functor in $\underline{Sp}(J)$ so long as it is required to be a functor only on a small category over $\underline{Sp}(J) \times s\mathcal{E}$.

Theorem 4.3. The functor $\otimes$ factors through $Ho(J \times s\mathcal{E}) = \underline{Ho}(J) \times \underline{Ho}(s\mathcal{E})$ and is copontinuous in both variables. There is a functor $HOM\colon Ho(J^0 \times J) \to \underline{Ho}(s\mathcal{E})$ together with a natural transformation $[X,HOM(T,S)] \to [T \otimes X,S]$ which is an isomorphism for X a finite simplicial set. If $\underline{Sp}(J)$ is pointed, the same theorem holds for pointed simplicial sets, and the isomorphism $[X,HOM(T,S)] \cong [T \otimes X,S]$ for X pointed connected. In either case, there is an associative composition $HOM(T,S) \times HOM(R,T) \to HOM(R,S)$ and the usual associativity laws for tensored enriched categories, hold.

We will prove the unpointed version of 4.3 and sketch the proof in the pointed case. For all T,S there will be natural isomorphisms $[T,S] \cong \pi_0 HOM(T,S)$, $T \otimes \Delta^0 \cong T$ for Δ^0 the usual 0-simplex, $T \otimes (X \times Y) \cong (T \otimes X) \otimes Y$ for all simplicial sets X,Y. As will be described at greater length in [HFS], HOM will be a homotopy continuous functor.

Remark 4.4. There is one point which has been smoothed over here. The functor $T \otimes X$ will be defined for X a *small* simplicial set; however, HOM(S,T) will be a large simplicial set--that is, it will be a simplicial set over the universe for Sp(J). If J is finitely regular and coregular, the tensor and cotensor structure can be defined with respect to finite simplicial sets (those whose geometric realization is a finite complex), but HOM(T,S) will generally be infinite.

Lemma 4.5. For all T,X,Y, the natural map $T \otimes (X \times Y) \to (T \otimes X) \otimes Y$ is an isomorphism.

Proof. $Cov(X \times Y) = Cov(X) \times Cov(Y)$, colimits over $Cov(X \times Y)$ agree with left Kan extensions along the preopfibration $Cov(X \times Y) \to Cov(Y)$ followed by colimit.

Since, for all X, $Norm(X) \to Cov(X)$ has a left adjoint, by (2.5) we see that for any $\Phi: Cov(X) \to \underline{Sp}(J)$, $Hocolim(\Phi|Norm(X)) \to Hocolim(\Phi)$ is an isomorphism. From this observation we can deduce the following result.

Lemma 4.6. If X is a simplicial set such that Norm(X) is isomorphic to Norm(C(Y)) for C(Y) the cone on some simplicial set Y, then for all T in Sp(J), the natural map $T \otimes X \to T = T \otimes \Delta^0$ arising from the constant map $X \to \Delta^0$ is an isomorphism in Ho(J).

Proof. Since $T \otimes X$ depends, up to isomorphism, only on the isomorphism classes of T and Norm(X), we may as well assume $X = CY$. Let $p: \Delta^0 \to X$ be the cone point of X. Then the nondegenerate simplices of X consist of p, the nondegenerate simplices of Y, and pairs consisting of p and a nondegenerate simplex of Y. Let C be the full subcategory of Norm(X) which has all the objects except those of Norm(Y). Then the inclusion $C \to$ Norm(X) has a left adjoint which is the identity of C and which assigns to an object σ of Norm(Y) the object (p,σ) in Norm(X). Thus $T \otimes X$ is the homotopy colimit over C of the constant T-valued functor. Notice that p is the initial object of C. Thus, the homotopy left Kan extension of any functor from $\{p\}$ to C is isomorphic to the corresponding constant functor on C. Thus, since the homotopy colimit of a homotopy left Kan extension is the homotopy colimit, Top: $T \otimes p \to T \otimes X$ is an isomorphism.

The reason for stating (4.6) as we did is that the horns $\Lambda^n_k \subset \Delta^n$ of simplicial topology are not simplicially cones for all k, but Λ^n_n is a cone and all Norm(Λ^n_k) are isomorphic as categories. Since $- \otimes -$ clearly preserves homotopy coproducts, we have the following result.

Corollary 4.7. For all T and all families $f_i: \Lambda^{n_i}_{k_i} \to \Delta^{n_i}$ of inclusions, $T \otimes \coprod \left\{\Lambda^{n_i}_{k_i}\right\} \to T \otimes \coprod \left\{\Lambda^{n_i}\right\}$ is an isomorphism in $\underline{\mathrm{Ho}}(\mathcal{J})$.

Lemma 4.8. For all T and all diagrams $X \leftarrow A \to B$ of simplicial sets with $A \to X$ a monomorphism, if Y is the colimit of $X \leftarrow A \to B$, then $T \otimes Y$ is the homotopy colimit of $T \otimes X \leftarrow T \otimes A \to T \otimes B$.

Proof. Let $\mathcal{J}$ be the category $\cdot \leftarrow \cdot \to \cdot$. Then we have $\Xi: \mathcal{J} \to$ Cat given by Norm(X) $\leftarrow$ Norm(A) $\to$ Norm(B). Let Ψ be the total category associated to Ξ. Then $\Psi \to \mathcal{J}$ is a preopfibration, so the homotopy left Kan extension along this functor of the constant T-valued functor is the diagram $T \otimes X \leftarrow T \otimes A \to T \otimes B$. Thus, it suffices to show that the homotopy left Kan extension of the constant T-valued functor along $\Psi \to$ Norm(Y) is isomorphic to the constant T-valued functor. However, this can be done by verifying that the natural map is an equivalence for all objects of Norm(Y). For those nondegenerate simplices which do not lie in A, the relevent comma categories have terminal objects, so there is nothing to prove. Since $A \to X$ is a monomorphism, any simplex of Y which comes from A comes from a unique simplex in B. Thus, if $f: \Delta^n \to Y$ is a nondegenerate simplex, f factors uniquely through a nondegenerate simplex $f': \Delta^n \to B$. Further, the comma category ψ/f has as objects the three following types of things. First, nondegenerate simplices over f'. Second, if $A_f = A \times_B \Delta^n$ is the inverse image in A of $f'(\Delta^n)$, the nondegenerate simplices of A_f. Third, if $X_f = X \times_Y \Delta^n$, the nondegenerate simplices of X_f. Now $A_f \to X_f$ is an isomorphism, since $A \to X$ is a monomorphism. Thus, if ψ_f is the full subcategory of ψ/f whose objects are all except the nondegenerate simplices of X_f, the homotopy left Kan extension from ψ_f to ψ/f of a constant functor is constant. Thus the homotopy colimit of a constant functor over ψ/f agrees with its homotopy colimit over ψ_f. However, ψ_f has a terminal object, so homotopy colimits over ψ_f are computed by evaluation on the terminal objects, so that the homotopy colimit over ψ/f of a constant functor is the common value of that constant functor.

Corollary 4.9. If $X \to Y$ is a map of simplicial sets arising from attaching simplices to X along horns, then for all T, $T \otimes X \to T \otimes Y$ is an isomorphism.

The cobase extensions of the inclusions of horns in simplices generate, under retraction, composition, and sequential infinite composition, a collection of monomorphisms of simplicial sets which were called the "anodyne extensions" by Gabriel-Zisman [CFHT]. They prove that any functor defined on simplicial sets which takes the anodyne extensions to isomorphisms has a unique extension to the homotopy category of simplicial sets ($\cong$ the homotopy category of CW-complexes, cf. Quillen [HA]). Notice that a retract of an isomorphism is an isomorphism, so any functor which takes a class of maps to isomorphisms also takes all retracts of those maps to isomorphisms.

Lemma 4.10. If $X_0 \subset X_1 \subset \ldots$ is an increasing sequence of inclusions of simplicial sets with union X, then if for some T every $T \otimes X_i \to T \otimes X_{i+1}$ is a weak equivalence, then so is $T \otimes X_0 \to T \otimes X$.

Proof. Let $\mathcal{J}$ be the ordered set of the natural numbers, considered as a set, and let $\mathcal{I}$ be the ~~wreath~~ of the functor on $\mathcal{J}$ given by $i \mapsto \mathrm{Cov}(X_i)$. Let $\pi: \mathcal{I} \to \mathcal{J}$, $\psi: \mathcal{I} \to \mathrm{Cov}(X)$ be the obvious maps. Then if $\xi: \mathcal{I} \to \underline{\mathrm{Sp}}(\mathcal{J})$ is the constant functor at T, $\pi_+(\xi)(i) = T \otimes X_i$, so $\mathrm{Hocolim}(\xi) = \mathrm{Hocolim}(\pi_+(\xi)) = T \otimes X_0$, since $\pi_+(\xi)$ is the homotopy left Kan extension from $\{0\}$ to $\mathcal{J}$ of the constant functor at $T \otimes X_0$. Thus $T \otimes X_0 \cong \mathrm{Hocolim}(\psi_+(\xi))$. It suffices to show that $\psi_+(\xi)$ is constant at T.

We now investigate the comma category ψ/σ for $\sigma: \Delta^n \to X$. There is some minimal i such that σ factors uniquely through X_i. Clearly, ψ/σ consists of pairs (j,τ) where $\tau: \Delta^m \to \Delta^n$, $j \geq i$, and thus is isomorphic to $\mathcal{J} \times \mathrm{Cov}(\Delta^n)$. Since $\psi_+(\xi)(\sigma)$ is the homotopy colimit over $\mathcal{J} \times \mathrm{Cov}(\Delta^n)$ of a constant functor and $\mathcal{J}$ has an initial object, $\psi_+(\xi)(\sigma) = \xi(i,1) = T$. Thus $\psi_+(\xi)$ is the constant T-valued functor on $\mathrm{Cov}(X)$.

We now have proved the following result. Notice that cocontinuity must be understood in light of (4.4); the interpretation to be used is the obvious one.

Lemma 4.11. The functor $(T,X) \mapsto T \otimes X$ factors through $\underline{\mathrm{Ho}}(\mathcal{J}) \times \underline{\mathrm{Ho}}(s\mathcal{C})$, and is cocontinuous in both variables.

At this point, we have proved the unpointed version of (4.3), since if $H(X) = [T \otimes X, S]$, H will be a cohomology functor depending functorially on T, S. We let $\mathrm{HOM}(T,S)$ be the CW-complex (or, equivalently, in the homotopy category, simplicial set) which represents this cohomology functor on finite simplicial sets (see the Appendix for details).

Whether $[X,\mathrm{HOM}(T,S)] \to [T \otimes X,S]$ is an isomorphism for all CW-complexes depends upon whether or not the following variation on J.H.C. Whitehead's theorem is true or not: "A natural transformation $H \to K$ of cohomology functors defined on (unbased) CW-complex such that $H(P) \to K(P)$ is an isomorphism for all finite polyhedra P is an isomorphism for all CW-complexes".

If $\underline{Sp}(J)$ has an initial (resp. terminal) object, it is not difficult to show that this object is also an initial (resp. terminal) object in $\underline{Ho}(J)$. In this case, for T in $\underline{Sp}(J)$ and X a pointed simplicial set, let $T^*: \mathrm{Cov}(X) \to \underline{Sp}(J)$ be the functor which is constant at X except at the basepoint $\Delta^0 \to X$, where it is the basepoint. This defines $T \otimes X = \mathrm{Hocolim}(T^*)$. Everything now proceeds exactly as in the unpointed case, except that we now have the Whitehead theorem for all connected pointed CW-complexes.

Appendix: Brown's representability theorem without basepoints

Singular homology has the following feature, whose proof can be safely left to the reader.

Lemma A.1. If $f: X \to Y$ is a map of spaces such that for all finite polyhedra P, $[P,X] \to [P,Y]$ is an isomorphism on the homotopy classes of (unpointed) maps from P to the two spaces, then $f_*: H_*(X) \to H_*(Y)$ is an isomorphism.

The converse to A.1 is, of course, not true. However, notice that if $f: X \to Y$ has the property that $f_*: H_*(X) \to H_*(Y)$ is an isomorphism, then the Z-localization $f_Z: X_Z \to Y_Z$ is a homotopy equivalence (for CW-complexes) (see Bousfield [LSH]).

Theorem A.2. If H is an additive set valued cohomology functor defined on the category of CW-complexes, there is a space X and an element $i \in H(X)$ such that

(a) If $\Phi: [-,X] \to H$ is given by $\Phi(f) = f^*(i)$, then Φ induces an isomorphism for all finite polyhedra.

(b) If K is a second additive cohomology functor with a natural transformation $\Psi: H(P) \to K(P)$ for P a finite polyhedron, then there is a class $j \in K(X)$ such that $\Psi\Phi(f) = f^*(j)$ for all $f: P \to X$ with P a finite polyhedron.

Proof. Notice that the class of homotopy types of finite polyhedra is small--indeed, it is countable, as it agrees with the classes of homotopy types of finite simplicial sets. This observation will assure that our constructions stay in the universe of CW complexes where our cohomology functors are defined. For each homotopy type of a finite polyhedron, choose a representative.

Let X_0 be the disjoint union of one copy $P_x = P$ for each P a chosen representative of a finite polyhedron homotopy class and each $x \in H(P)$. Then let $i_0 \in H(X_0) = \prod\{H(P_x)\}$ have as components in $H(P_{x_0})$ the class x. Let Φ_0: $[-,X_0] \to H(-)$ by $\Phi_0(f) = f^*(i_0)$. Next, we let X_1 be the homotopy colimit of $X_0 \leftarrow \coprod (P_{y_0} \cup P_{y_1}) \to \coprod P_{y_0,y_1}$, where y_0,y_1 run over all homotopy classes of maps $y_0,y_1: P \to X_0$ for which $\Phi_0(y_0) = \Phi_0(y_1)$, $P_{y_0,y_1} = P_{\Phi_0(y_0)}$. Then y_0 and y_1 represent the homotopy class in X_1 given by the inclusion $P_{y_0,y_1} \to X_1$, and thus represent the same homotopy class. Observe that $H(X_1)$ maps epimorphically to the fiber product of $H(X_0)$ with $\prod H(P_{y_0,y_1})$ over $\prod (H(P_{y_0}) \times H(P_{y_1}))$. Notice that since $y_0^*(i_0) = y_1^*(i_0) = \Phi_0(y_0) = \Phi_0(y_1)$, if we take the class $\prod\{\Phi_0(y_0)\} \in \prod H(P_{y_0,y_1})$, this defines, together with i_0, a class j_1 in the fiber product. Let $i_1 \in H(X_1)$ map to j_1. Continue inductively to construct a chain $X_0 \to X_1 \to X_2 \to \dots$ of cofibrations such that we have $i_n \in H(X_n)$, the image of i_n in $H(X_{n-1})$ is i_{n-1}, and if $f,g: P \to X_n$ are such that $f^*(i_n) = g^*(i_n)$, then the compositions of f and g with $X_n \to X_{n+1}$ are homotopic. Let

X be the union of the X_n in the weak topology. Then for all finite polyhedra P, $[P,X]$ is the direct limit of the $[P,X_n]$, since each P is compact. Since the maps are all cofibrations, it is a standard argument that X is the homotopy colimit of the X_n, so there is $i \in H(X)$ which restricts to i_n in each $H(X_n)$. Since $\Phi_0: [P,X_0] \to H(P)$ is epimorphic, so is $\Phi: [P,X] \to H(P)$ given by $\Phi(f) = f^*(i)$. Further, the relations on Φ are trivial, again by compactness.

Part (b) follows immediately from the construction of the class i above. Notice that no classes of uniqueness are made. Further, notice that we do not claim that $[-,X]$ agrees with $H(-)$ on all spaces--only on finite polyhedra.

Notice that if $H(-) = [-,Y]$, and if X is the representing space which we have constructed for H, then there is a map $X \to Y$ such that $[P,X] \to [P,Y]$ is an isomorphism for all finite polyhedra P. In particular, this implies that $X \to Y$ induces an isomorphism on singular homology groups. Using bouquets of spheres, one can make arguments (which must be known, but not to me) which indicate that $X \to Y$ induces an isomorphism on the nilpotent completions of the homotopy groups and has acyclic homotopy theoretic fibers on each component.

Bibliography

Anderson, D. W. [HFS] Homotopy Functors and Sheaves (in preparation).

Brown, E. H. [CT] Cohomology Theories, Annals of Math (2) v. 75 (1962) 467-484.

Brown, K. [AHT] Abstract Homotopy Theory and Generalized Sheaf Cohomology, Trans. AMS v. 186 (1973) 419-485.

Eilenberg, S. and Steenrod, N. [AT] Foundations of Algebraic Topology (1952) Princeton University Press.

Gebriel, P. and Zisman, M. [CFHT] Calculus of Fractions and Homotopy Theory, Ergebnisse der Math, Band 35 (1967) Springer Verlag.

Segal, Graeme [CSSS] Classifying Spaces and Spectral Sequences, Publications IHES No. 34 (1968) 105-112.

Quillen, D. [HA] Homotopical Algebra, Lecture Notes in Math 43 (1967), Springer Verlag.

______ [RH] Rational Homotopy Theory, Annals of Math (2) 90 (1969) 205-295.

Construction of mod p H-spaces II.

John R. Harper*

This paper surveys some recent techniques used to construct finite mod p H-spaces. Paper $[H_1]$ with this title discussed algebraic methods. The emphasis here is on geometric ideas. Most of the work discussed here will appear in [CHZ]. Some remarks concerning non-simply connected examples are also made.

The general question in this area is

Q1. Find all p-local spaces Y with $H^*(Y:Z/pZ)$ that of a finite complex such that Y is an H-space.

Hopf's theorem provides the first necessary condition. A space X is an <u>H_0-space</u> if $H^*(X:\mathbb{Q})$ is an exterior algebra on odd dimensional generators $\{x_i\}$. Let $n_i = \dim x_i$. The <u>type</u> of X is the sequence $(n_1, \ldots, n_r)$ arranged in non-decreasing order. The <u>rank</u> of X is the number of generators r.

Another useful necessary condition is supplied by the fact that the λ-th power map on an H-space induces multiplication by λ on homotopy groups. In what follows p is a prime (odd by nature of the arguments) and $\lambda \varepsilon (Z/pZ)^*$ is usually a generator.

Definition $[Z_1]$. A space X is a <u>π_λ-space</u> if there

*Research supported by NSF grant MCS76-0715

exists $\psi: X_{(p)} \to X_{(p)}$ inducing multiplication by λ on $\pi_*(X_{(p)}, p)$. The pair (X,ψ) is called a $\underline{\pi_\lambda\text{-pair}}$.

The point in using homotopy with coefficients is to permit iteration of ψ without changing its algebraic property. In view of the intent to iterate, we have the following from $[Z_1]$.

Definition. A map $f:(X,\psi) \to (Y,\phi)$ of π_λ-pairs is a $\underline{\pi_\lambda\text{-map}}$ provided for some k, $\phi^{p^k} f_{(p)} \simeq f_{(p)} \psi^{p^k}$.

A useful step toward answering Q1 would be to give a practical answer to Q2. Suppose X is a finite CW complex which is an H_0-space and $X_{(p)}$ is a π_λ-space. When is $X_{(p)}$ an H-space?

An answer for low rank is given in [CHZ].

Theorem. Let X be a 1-connected finite CW complex which is an H_0-space, and $H_*(X:Z)$ is p-torsion free. Let λ generate $(Z/pZ)^*$. If X is a π_λ-space and rank $X < p/2$. then $X_{(p)}$ is an H-space. More precisely write the π_λ-pair as (X,ψ). If rank $X < p/3$ then $X_{(p)}$ has a homotopy associative and homotopy commutative multiplication. Regardless, $X_{(p)}$ has a multiplication m which is a π_λ-map

$$m:(X \times X, \psi \times \psi) \to (X,\psi) \ \mathrm{mod}\ p.$$

The proof is by means of fairly standard obstruction theory. The main point is to show that in the presence of ψ the obstructions to an H-structure are a-priori k-fold products in $H^*(X \times X:Z/pZ)$ where $k \equiv 1 \mod(p-1)$ and $k > 1$. The conditions, no p-torsion and rank $< p/2$ imply these obstructions are 0. The main steps are the following.

There exists a set of elements $\{Z_i\}$ $1 \leq i \leq r$, $Z_i \in H^{n_i}(X;Z_{(p)})$ *and a self-map* $\tilde{\psi}$ *such that*

(a) $\{Z_i\}$ *project to a basis for* $QH^*(X;Z_{(p)})$

(b) $\tilde{\psi}^*(Z_i) = \lambda^{p^k} Z_i$ *for some* k.

This is proved in $[Z_3]$ lemma 4.2.1.

A basic tool in the evaluation of obstructions is the Lifting Theorem of $[Z_1],[Z_3]$. We describe a special case used here. Let (X,ψ), (B,ϕ), (B_0,ϕ_0) be π_λ-pairs with $B_0 = K(G,n)$ and assume G is a Z/pZ-module. Suppose $f:X \to B$ and $g:B \to B_0$ are π_λ-maps and $\phi_0 g \simeq g\phi$, $\phi f \simeq f\psi$. Let E be the homotopy fibre of g and $\overline{\phi}:E \to E$ the self-map induced on E. Iterating if necessary, we obtain a diagram of π_λ-pairs and π_λ-maps.

$$\begin{array}{ccccc} & & (E,\overline{\phi}) & \to & (PB_0,P\phi_0) \\ & & \downarrow & & \downarrow \\ (X,\psi) & \xrightarrow{f} & (B,\phi) & \xrightarrow{g} & (B_0,\phi_0) \end{array}$$

We denote the composition $[X,\Omega B_0] \times [X,E] \to [X,E]$ induced by the principal action $\Omega B_0 \times E \to E$ by $*$. It is straight forward to check that the principal action is a π_λ-map.

Lifting Theorem LT. *If* f *lifts, then a lifting* $h:X \to E$ *exists which is a* π_λ*-map. Furthermore if* h_1, h_2 *are two lifts of* f *which are* π_λ*-maps, then there exists a* π_λ*-map* $d:X \to \Omega B_0$ *such that* $h_1 \simeq d * h_2$.

Proof of Theorem

Write type $X = (n_1, \ldots, n_r)$. Assume X is localized at p and define

$$K(X) = K(Z_{(p)}, n_1) \times \ldots \times K(Z_{(p)}, n_r).$$

Let $\{Z_i\}$, $Z_i \in H^{n_i}(X; Z_{(p)})$ and $\psi: X \to X$ be the elements and self-map given by lemma 2.6. Define $f: X \to K(X)$ by $f^*(\iota_{n_i}) = Z_i$. Then $f: (X, \psi) \to (K(X), \phi)$ is a π_λ-map where ϕ is defined by multiplication by λ^{p^k} on fundamental classes. By construction $F =$ fibre f has $\pi_*(F)$ a graded finitely generated p-torsion group. Using theorem LT the map $f: X \to K(X)$ has a Moore-Postnikov factorization consisting of π_λ-pairs (X_n, ϕ_n) and each k-invariant k_n a Z/pZ-characteristic class with characteristic value λ,

$$\begin{array}{ccccc} X \to \ldots \to X_{n+1} & \xrightarrow{p_{n+1}} & X_n & \to \ldots \to K(X) = X_0 \\ & & \downarrow k_n & \\ & & K(G_n, n+1) & \end{array}$$

Let m_0 be the product multiplication on $X_0 = K(X)$. Then m_0 is a π_λ-map. Assume inductively that X_n has a multiplication m_n covering m_0 such that m_n is a π_λ-map. We have $\phi_n^*(k_n) = \lambda k_n$. Hence $(\phi_n \times \phi_n)^* \bar{m}_n^* (k_n) = \lambda \bar{m}_n^*(k_n)$ is a λ-characteristic class of $H^{n+1}(X_n \wedge X_n; Z/pZ)$. Since $H^{n+1}(X_n \wedge X_n) \cong H^{n+1}(X \wedge X)$ and rank $X < p/2$, 0 is the only such λ-characteristic class. Thus k_n is primitive and X_{n+1} has an induced H-structure, denote it by m'_{n+1}. The obstruction to

m'_{n+1} being a π_λ-map may be interpreted as the H-deviation of the self-map ϕ_{n+1}, and hence factors through $X_{n+1} \wedge X_{n+1}$. Since $m_n \circ p_n \times p_n$ is a π_λ-map, the lifting theorem implies that a multiplication m_{n+1} on X_{n+1} exists which is a π_λ-map. The theorem follows by induction on n. Regarding homotopy commutativity and homotopy associativity for $r < p/3$,

$$\begin{array}{ccc} X_{n+1} \times X_{n+1} & \xrightarrow{m_{n+1}} & X_{n+1} \\ \big\downarrow T & \nearrow & \\ X_{n+1} \times X_{n+1} & m_{n+1} & \end{array}$$

the two routes may be regarded as π_λ-liftings of the same π_λ -map m_n. The second part of the lifting theorem provides that the obstruction to commutativity or associativity is a λ-characteristic class in the 2 or 3 fold smash of X_{n+1}. As before this class is 0.

The following construction and the theorem can be used to answer Q1 for simply connected spaces with no p-torsion and rank $< p/2$. The basic idea is well known. The novelty is in the use of a condition somewhat weaker than homotopy associativity or commutativity.

<u>Construction</u>. Let X be a mod p H-space and Λ the λ-th power map. Suppose some multiplication can be found such that

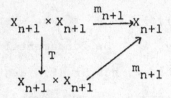

$$m:(X\times X,\Lambda\times\Lambda)\to(X,\Lambda)$$

is a π_λ-map mod p. In particular if m is hty. assoc. and hty. comm. and Λ is the λ-th power map w.r.t. m then Λ is an H-map. Iterating if necessary we can assume Λ is an H-map. Let $\alpha:S^{2q}\to X$ be any homotopy class and ρ the degree λ map on S^{2q}. Then $\Lambda\alpha\simeq\alpha\rho$. Since Λ is an H-map, the Hopf construction on m yields a commutative diagram

$$\begin{array}{ccccc} X\to X* X & \xrightarrow{H(m)} & \Sigma X \\ \Lambda\downarrow\quad\ \downarrow\Lambda*\Lambda & & \downarrow\Sigma\Lambda \\ X\to X* X & \longrightarrow & \Sigma X \end{array}\ .$$

Let $(X,\Lambda)\to(Y^\alpha,\tilde\Lambda)\to(S^{2q+1},\Sigma\rho)$ be the fibration of spaces with self-maps induced from the Hopf fibration by $\Sigma\alpha:S^{2q+1}\to\Sigma X$

$$\begin{array}{ccc} (Y^\alpha,\tilde\Lambda) & \to & (X* X,\Lambda*\Lambda) \\ \downarrow & & \downarrow H(m) \\ (S^{2q+1},\Sigma\rho) & \xrightarrow{\Sigma\alpha} & (\Sigma X,\Sigma\Lambda). \end{array}$$

The mod p H-structure on S^{2q+1} implies that $\Sigma\rho$ induces multiplication by λ on $\pi_*(S^{2q+1},p)$. A straightforward argument with the homotopy sequence yields that $(Y^\alpha,\tilde\Lambda^p)$ is a π_λ-pair. Thus if X has p-torsion free homology and rank(r-1) with $r<p/2$, the theorem implies Y^α is a mod p H-space (with p-torsion free homology) of rank r. The theorem supplies multiplication which are π_λ-maps, making itera-

tion of the construction effective.

The analysis in [CHZ], also [CH] shows that all mod p H-spaces within the scope of the theorem can be obtained this way. This is not exactly a classification since the same space can arise out of different sequences of the construction.

Many useful rank 2 examples are available. Let $p \geq 5$ Then S^n (n odd) admits a homotopy associative and homotopy commutative multiplication. Let $\alpha \in \pi_{q-1}(S^n)$ (q odd). The mod p fibration $S^n \to Y^\alpha \to S^q$ obtained by the construction is a mod p H-space. Its mod p cell decomposition is $S^n \cup_\alpha e^q \cup e^{n+q}$. If $p = 3$, Y^α is still a π_λ-space.

The construction of mod p H-spaces with torsion is much more obscure. So far the only known simply connected non-Lie examples arise from the following $[H_2]$. Recall that $B_1(p)$ is a mod p group [C-E] [W]. Let $g: S^{2p+1} \to B_1(p)$ represent a generator of π_{2p+1}. A g can be chosen which is an A_{p-1}-map mod p. Let K(p) be the mod p $B_1(p)$-bundle over the (p-1)-st reduced product of James $S^{2p+2}_{(p-1)}$ as in the diagram

$$\begin{array}{ccc} & B_1(p) & \\ \swarrow & & \searrow \\ K(p) & \longrightarrow & EB_1(p) \\ \downarrow & & \downarrow \\ S^{2p+2}_{(p-1)} & \xrightarrow{\tilde{g}} & BB_1(p) \end{array}$$

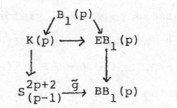

Obviously this construction can be attempted with other mod p groups. In case $p=2$ using $SU(3)$ and S^5 produces the Lie group G_2. It is open what happens in general but in $[H_2]$ it is shown that $K(p)$ is a mod p H-space. The argument is algebraic, outlined in $[H_1]$.

The non-simply connected situation appears to be interesting. In $[H_3]$ it is shown that the generalized Lens space $L_p = S^{2p-1}/Zp$ is a mod p H-space. One proof is to show that L_p is a mod p factor of $PSU(p)$. Two questions are

Q3. What non-simply connected mod p H-spaces can be obtained from mod p decomposition of compact Lie groups?

Q4. If G is a finite abelian group acting on S^{2n+1} and $H^*(S^{2n+1}/G;Z/pZ)$ satisfies the Borel structure theorem for the cohomology of H-spaces, is S^{2n+1}/G a mod p H-space?

Additional Remarks, Nov. 1978

1. Zabrodsky has discovered further restrictions on coproducts which yield that a π_λ-space X with p-torsion free homology is a mod p H-space provided rank $< p-1$. The construction then gives all examples up to this rank.

2. I have proved that $B_n(3)$ is a mod 3 H-space for all $n \equiv 2(3)$.

References

[C-E] A. Clark and J. Ewing, The realization of polynomials algebras as cohomology rings. Pacific J. Math. 40 (1974).

[CHZ] G. Cooke, J. Harper and A. Zabrodsky, Torsion free mod p H-spaces of low rank, to appear.

[CH] G. Cooke and J. Harper, Torsion free mod p H-spaces, Geometric Applications of Homotopy Theory I, Springer Lecture Notes 657, (1978).

[H_1] J. Harper, On the construction of mod p H-spaces. Proc. Symposia Pure Math. 77 Amer. Math. Soc.

[H_2] ________, H-spaces with torsion, to appear.

[H_3] ________, Regularity of finite H-spaces, Ill. J. Math. to appear.

[M] M. Mimura, On the mod p H-structures of spherical fibrations. Manifolds, Tokyo (1973).

[W] C. Wilkerson, Bull. Amer. Math. Soc. 79 (1973).

[Z_1] A. Zabrodsky, Power spaces. Inst. Adv. Study mimeograph.

[Z_2] ___________, On relations in mod 3 cohomology rings of H-spaces, to appear.

[Z_3] ___________, Endomorphisms in the homotopy category, to appear.

The Mod 3 Cohomology of the Exceptional Lie Group E_8 .

Richard Kane

§1. Introduction

Let $\mathbb{Z}/p$ be the integers reduced mod p . Given a (mod p) finite H-space then $H^*(X;\mathbb{Z}/p)$ is a Hopf algebra over the Steenrod algebra $A^*(p)$. In recent years a number of general structure theorems have been obtained for the mod p cohomology of finite H-spaces (For the algebra structure and the Steenrod module structure see [14], [5], and [9]. For the coalgebra structure see [6]). The basic tool used to obtain all of these results has been $\mathbb{Z}/p$ secondary cohomology operations (see [13], and [21]). The main purpose of this paper is to demonstrate that BP operations provide another tool which can be effectively used to understand the (mod p) cohomology of finite H-spaces. Moreover, BP operations complement $\mathbb{Z}/p$ operations in the sense that our use of BP operations depends upon and extends the results obtained by $\mathbb{Z}/p$ operations.

Let E_8 be the exceptional Lie group. In this paper we will use BP operations to study the structure of $H^*(E_8;\mathbb{Z}/3)$ as a Hopf algebra over $A^*(3)$. This study serves two related purposes. First of all, the structure of $H^*(E_8;\mathbb{Z}/3)$ is already known (see [12]). However, the arguments used in [12] to determine $H^*(E_8;\mathbb{Z}/3)$ depend on specific geometrical properties of E_8 . Our arguments, on the other hand, are purely

(supported by NRC Grant A-3140)

homotopy theoretic in nature. Secondly, our arguments are prototypes for more general BP arguments. These arguments should produce general structure theorems for all (mod p) finite H-spaces. What some of these theorems should be is indicated at the end of the paper.

Our method of procedure will be to study how the algebra structure of $H^*(E_8;\mathbb{Z}/3)$ affects its coalgebra structure and its Steenrod module structure. We pick a (mod 3) finite H-space and assume that $H^*(X;\mathbb{Z}/3) \cong H^*(E_8;\mathbb{Z}/3)$ as an algebra then (see[12]).

(1:1) $$H^*(X;\mathbb{Z}/3) \cong E(x_3,x_7,x_{15},x_{19},x_{27},x_{35},x_{39},x_{40}) \otimes T(x_8,x_{20})$$

where E denotes an exterior algebra and T denotes a polynomial algebra with all elements truncated at height 3. We will proceed by studying the consequences for $H^*(X;\mathbb{Z}/3)$ of the presence of the even dimensional generators x_8 and x_{20} in 1:1. In §2 we outline some basic facts about BP theory which will be used in our arguments. In §3 and §4 we study the interrelation between the algebra structure of $H^*(X;\mathbb{Z}/3)$ and the action of $A^*(3)$ on $H^*(X;\mathbb{Z}/3)$. In particular we can almost completely determine the action of $A^*(3)$ on the module of indecomposables $Q(H^*(X;\mathbb{Z}/3))$. In §5 and §6 we study the interrelation between the algebra structure of $H^*(X;\mathbb{Z}/3)$ and its coalgebra structure. In §5 we show that the generators x_{35} and x_{47} from 1:1 can never be chosen to be primitive. In §6 we impose the hypothesis that $H^*(X;\mathbb{Z}/3)$ is coassociative and show that $H^*(X;\mathbb{Z}/3) \cong H^*(E_8;\mathbb{Z}/3)$ as a Hopf algebra over $A^*(3)$. In §7 we discuss some general theorems suggested by our study of $H^*(E_8;\mathbb{Z}/3)$.

It is conjectured that, for any 1-connected compact Lie group G and any prime p, the algebra structure of $H^*(G;\mathbb{Z}/p)$ determines the coalgebra and Steenrod module structure of $H^*(G;\mathbb{Z}/p)$. So in particular our study of $H^*(E_8;\mathbb{Z}/3)$ verifies this conjecture for $p = 3$ and $X = E_8$.

We will assume a knowledge of general homology theory and, in particular of BP theory (see [2]); a knowledge of Hopf algebra (see [18]); and a knowledge of the Steenrod algebra (see [17]). However, we should remark that the action of $A^*(3)$ on $H_*(X;\mathbb{Z}/3)$ which we will use is the left one. It is related to the usual left action of $A^*(3)$ on $H^*(X;\mathbb{Z}/3)$ by the formula

$$\langle \chi(\theta)x,y\rangle = (-1)^{|x||\theta|}\langle x,\theta y\rangle$$

for any $x \in H^*(X;\mathbb{Z}/3)$, $y \in H_*(X;\mathbb{Z}/3)$ and $\theta \in A^*(3)$. ($\chi : A^*(3) \to A^*(3)$ is the canonical anti-automorphism). Regarding notation we will use the same symbol to denote elements of a module and their images in any quotient module. Likewise for maps between modules.

As a final comment on the techniques in this paper we observe that we are studying finite H-spaces which have integral 3 torsion by using general cohomology operations. On the other hand, Zabrodsky in [22] studies finite H-spaces without integral 3 torsion by using $\mathbb{Z}/3$ secondary and tertiary operations. This is a reversal of the usual state of affairs in finite H-space theory. Up to now $\mathbb{Z}/p$ operations have been used to study torsion while general cohomology operations (in particular K-theory operations) have been used to study torsion free spaces.

§2. BP-Theory

In this section we recall a few basic facts about BP homology and cohomology. We will state the results for an arbitrary prime p even though we will only use them for the prime $p = 3$. Given the prime p let $\mathbb{Z}_{(p)}$ be the integers localized at the prime p. For BP at that prime $BP_*(X)$ is a module over $BP_*(pt) = \mathbb{Z}_{(p)}[v_1,v_2,\ldots]$ (dim $v_s = 2p^s - 2$) while $BP^*(X)$ is a module over $BP_*(pt) = \mathbb{Z}_{(p)}[v_1,v_2\ldots]$ (dim $v_s = -2p^s + 2$). Using the identity $BP^i(pt) = BP_{-i}(pt)$ there is a pairing between $BP^*(X)$ and $BP_*(X)$. The algebra of BP operations, $BP^*(BP)$, acts from the left on both $BP^*(X)$ and $BP_*(X)$. The duality between $BP^*(X)$ and $BP_*(X)$ also produces a duality between these actions. The duality is established by using the coaction map $\lambda : BP_*(X) \to BP_*(BP) \otimes BP_*(X)$ (see Lecture III of [1] for all of the above). For any sequence $E = (e_1,e_2,\ldots)$ of non-negative integers with only finitely many non-zero terms we have the Quillen element $r_E \in BP^*(BP)$. The action of $BP^*(BP)$ on $BP^*(pt)$ (and, hence, on $BP_*(pt)$) is non trivial. It can be determined from a knowledge of the action of $BP^*(BP)$ on $BP^*(pt) \otimes Q$ (see [23]) plus a knowledge of the imbedding $BP^*(pt) \subset BP^*(pt) \otimes Q$ (see [4]). We will use the following facts. Let $v_0 = p$ and let Δ_i be the sequence $(0,\ldots,0,1,0,\ldots)$ with 1 in the i-th position.

(2:1) $$r_E v_n \equiv \begin{cases} v_{n-i} \bmod (v_0,v_1,\ldots)^2 & \text{if } E = p^{n-i}\Delta_i \\ 0 \bmod (v_0,v_1,\ldots)^2 & \text{otherwise} \end{cases}$$

(2:2) the ideals $(v_0,v_1,\ldots)^n$ are invariant

The action of r_E on $BP^*(X)$ is related to the action of the Milnor element $\mathcal{P}^E \in A^*(p)$ on $H^*(X;\mathbb{Z}/p)$. Let $T : BP^*(X) \to H^*(X;\mathbb{Z}/p)$ be the Thom map. Then we have the following commutative diagram

(2:3)
$$\begin{array}{ccc} BP^*(X) & \xrightarrow{\;r_E\;} & BP^*(X) \\ {\scriptstyle T}\downarrow & & \downarrow{\scriptstyle T} \\ H^*X;\mathbb{Z}/p) & \xrightarrow{\;\chi(\mathcal{P}^E)\;} & H^*(X;\mathbb{Z}/p) \end{array}$$

where $\chi : A^*(p) \to A^*(p)$ is the canonical anti-automorphism. There is a similar diagram for homology. Because of 2:3 we can use BP operations to analyze the action of $A^*(p)$ on $H^*(X;\mathbb{Z}/p)$ and on $H_*(X;\mathbb{Z}/p)$. Moreover, the operations $\{r_E\}$ satisfy a series of relations analogous to the composition law proven by Milnor in [17]. Namely

(2:4)
$$r_E r_F \equiv \sum_{\substack{R(X) = F \\ S(X) = E}} b(X)\, r_{T(X)} \mod (v_1, v_2, \ldots)$$

where X , (X) , R(X) , S(X) , and T(X) are defined as in Theorem 4 of [17]. For proofs of 2:3 and 2:4 see (10). We should also remark that we will only be using some of the simpler cases of 2:4 . These cases can also be deduced directly from the coalgebra structure of $BP_*(BP)$ (see Table I of [23] for an indication as to how this is done).

We will also need the following fact about the Thom map for finite H-spaces (see [11]). Let $\{Q_s\}_{s \geq 0}$ be the element of $A^*(p)$ defined in [17].

If p is odd and (X,μ) a (mod p) finite H-space then

(2:5) $x \in H^*(X;\mathbb{Z}/p)$ lies in the image of $T : BP^*(X) \to H^*(X;\mathbb{Z}/p)$ if, and only if, $Q_s(x) = 0$ for $s \geq 0$.

§3. The Steenrod Module Structure of $Q(H^*(X;\mathbb{Z}/p))$ Part I

We assume that $p = 3$ and that (X,μ) is a mod 3 finite H-space satisfying 1:1 . By the Cartan formula, the action of $A^*(3)$ on $H^*(X;\mathbb{Z}/3)$ induces an action of $A^*(3)$ on $Q(H^*(X;\mathbb{Z}/3))$, the module of indecomposables. The elements $\{x_s\}$ of 1:1 project to a basis of $Q(H^*(X;\mathbb{Z}/3))$. In this section we will state what can be deduced about the action of $A^*(3)$ on $Q(H^*(X;\mathbb{Z}/3))$ by using secondary $\mathbb{Z}/p$ cohomology operations. These facts will be used in the BP arguments of the next few sections.

(3:1)
(a) $\mathcal{P}^1(x_3) = x_7 \qquad \mathcal{P}^1(x_{15}) = x_{19}$

(b) $\mathcal{P}^3(x_7) = x_{19} \qquad \mathcal{P}^3(x_8) = x_{20}$

(c) $\beta_p(x_7) = x_8 \qquad \beta_p(x_{19}) = x_{20}$

These results can be deduced from the fact that the following relations must hold: $x_8 = \beta_p\mathcal{P}^1(x_3)$, $x_{20} = \beta_p\mathcal{P}^3(x_7)$, and $x_{20} = Q_1(x_{15})$ (see [14]. In particular we see that the even dimensional generators x_8 and x_{20} force the presence of generators in dimension 3, 7, 15, and 19.

§4. The Steenrod Module Structure of $Q(H^*(X;\mathbb{Z}/3))$ Part II

We continue the study of $Q(H^*(X;\mathbb{Z}/3))$ begun in §3. We will prove

$$\text{(4:1)}\quad \begin{aligned} &\text{(a)}\quad \mathcal{P}^3(x_{27}) = x_{39} \\ &\text{(b)}\quad \mathcal{P}^3(x_{35}) = x_{47} \\ &\text{(c)}\quad \mathcal{P}^1(x_{35}) = x_{39} \end{aligned}$$

This leaves open the question as to whether $\mathcal{P}^3(x_{15}) = x_{27}$ and, hence, whether all the generators of 1:1 are tied together by Steenrod powers. In §6, when we consider the coalgebra structure of $H^*(\mathbb{Z}/3)$, we will show that $\mathcal{P}^3(x_{15}) = x_{27}$, if $H^*(X;\mathbb{Z}/3)$ is coassociative. However, we should point out that the arguments of this section do show that (X,μ) is indecomposable even if $\mathcal{P}^3(x_{15}) = 0$. For we prove 4:1 by using the relation $\mathcal{P}^3(x_8) = x_{20}$ to force the relations in 4:1. Thus, combining 3:1 and 4:1, it follows that all of the odd dimensional generators of 1:1 are related to the even dimensional ones.

We deduce 4:1 by passing to the loop space ΩX. It follows from 1:1 and 3:1, via an Eilenberg-Moore spectral sequence argument, that

$$\text{(4:2)}\quad H_*(\Omega X;\mathbb{Z}/3) = \mathbb{Z}/3\,[a_2,a_6,a_{14},a_{18},a_{26},a_{34},a_{38},a_{46}]\big/ I \otimes \mathbb{Z}/3\,[b_{22},b_{58}]$$

as an algebra where I is the ideal (a_2^3,a_6^3) (see [7] for the type of argument used). If we pass to the modules of indecomposables $Q(H_*(\Omega X;\mathbb{Z}/3))$ then the elements $\{a_i\} \cup \{b_j\}$ project to a basis. There is a duality between $Q(H_*(\Omega X;\mathbb{Z}/3))$ and the primitives $P(H^*(\Omega X;\mathbb{Z}/p))$ (see [18]).

The elements $\{a_i\}$ are duals of elements from the image of the loop map $\Omega^*: Q(H^*(X;\mathbb{Z}/3)) \to P(H^*(\Omega X;\mathbb{Z}/3))$. In fact

(4:3) $\langle \Omega^*(x_{i+1}), a_j \rangle = \sigma_{ij}$ (Kronecker delta)

The elements b_{22} and b_{58} are duals of transpotence elements. Since Image Ω^* is invariant under the action of $A^*(3)$ it follows that the $\mathbb{Z}/3$ module generated by x_{22} and x_{58} is invariant under the action of $A^*(3)$. In particular $\mathcal{P}^i(x_{58}) = 0$ for $0 < i < 9$. The elements b_{22} and b_{58} are related by the conjugate of $\mathcal{P}^9$. That is

(4:4) $\chi(\mathcal{P}^9)\, b_{58} = b_{22}$

To see this observe that b_{22} is dual to the transpotence element $t(x_8)$ produced by x_8 while b_{58} is dual to the transpotence element $t(x_{20})$ produced by x_{20}. If we take the Eilenberg-Moore spectral sequence $\{E_r\}$ converging to $H^*(X;\mathbb{Z}/3)$ and define $E_2 = \mathrm{Tor}_{H^*(X;\mathbb{Z}/3)}(\mathbb{Z}/3;\mathbb{Z}/3)$ via the bar construction (see [16]) then $t(x_8)$ and $t(x_{20})$ are defined, respectively, by the elements $x_8^2 \otimes x_8$ and $x_{20}^2 \otimes x_{20}$ in $H^*(X;\mathbb{Z}/3) \otimes H^*(X;\mathbb{Z}/3)$. It follows from 2:1 that $\mathcal{P}^9(x_8^2 \otimes x_8) = x_{20}^2 \otimes x_{20}$. Since the spectral sequence $\{E_r\}$ is a spectral sequence of Steenrod modules (see [19]) it follows that $\chi(\mathcal{P}^9)\, b_{58} = b_{22}$ (see the duality formula at the end of §1).

The rest of this section will be spent in showing that 4:4 implies the following results about the Steenrod module structure of $Q(H_*(\Omega X;\mathbb{Z}/3))$ Up to units in $\mathbb{Z}/3$, we have the following relations

(4:5) (a) $\mathcal{P}^3(a_{46}) = a_{34}$

(b) $\mathcal{P}^3(a_{34}) = b_{22}$

(c) $\mathcal{P}^1(a_{38}) = a_{34}$

(d) $\mathcal{P}^1(a_{26}) = b_{22}$

(e) $\mathcal{P}^3(a_{38}) = a_{26}$

In particular, by the duality established in 4:3, these results imply that 4:1 holds. For example, $\mathcal{P}^3(a_{38}) = a_{26}$ implies $\chi(\mathcal{P}^3)\, x_{27} = x_{39}$. Since $\chi(\mathcal{P}^3) = -\mathcal{P}^3$ (see [17] it follows that $\mathcal{P}^3(x_{27}) = x_{39}$ (up to a unit in $\mathbb{Z}/3$).

We prove 4:5 by using BP operations to analyze the action of $A^*(3)$ on $Q(H_*(\Omega X;\mathbb{Z}/3))$. For if we pass to the modules of indecomposables, $Q(BP_*(\Omega X))$ and $Q(H_*(\Omega X;\mathbb{Z}/3))$, then 2:3 is still valid.

Since $H_*(\Omega X) = H_*(\Omega X;\mathbb{Z}) \otimes_{\mathbb{Z}} \mathbb{Z}_{(3)}$ is torsion free (see [14]) it follows that $BP_*(\Omega X)$ is a free $BP_*(pt)$ module and the Thom map $T : BP_*(\Omega X) \to H_*(\Omega X;\mathbb{Z}/3)$ is surjective. It follows from 4:2 that

(4:6) $$BP_*(\Omega X) = BP_*(pt)[A_2,A_6,A_{14},A_{18},A_{26},A_{34},A_{38},A_{46},B_{22},B_{58}]/J$$

as an algebra where J is the ideal generated by elements

$$R_6 = A_2^3 + \alpha_6 A_6 + \alpha_2 A_2 + d_6$$

$$R_{18} = A_6^3 + \beta_{18} A_{18} + \beta_{14} A_{14} + \beta_6 A_6 + \beta_2 A_2 + d_{18}$$

(here α_i, $\beta_i \in BP_*(pt)$, d_6, d_{18} are decomposables, A_i and B_i are representatives in $BP_*(\Omega X)$ for a_i and b_i) Let $M = Q(BP_*(\Omega X))/K$ where K is the ideal generated by A_2, A_6, A_{14}, and A_{18} . Then M is the free $BP_*(pt)$ module generated by the elements $B_{22}, B_{58}, A_{26}, A_{34}, A_{38}$, and A_{46} . Further the action of $BP^*(BP)$ on $BP_*(\Omega X)$ induces an action on M .

We now proceed to the proof of 4:5 . We will work in M throughout our proof. We will prove 4:5 by using divisibility arguments. Since we are only concerned with the power of 3 dividing given integers we will simply ignore units in $\mathbb{Z}_{(3)}$ when convenient.

<u>PROOF OF 4:5 (a) and (b)</u>

By 2:3 and 4:4 it follows that $r_9(B_{58}) \not\equiv 0 \mod (3, v_1, v_2, \ldots)$. Therefore

(1) $3r_9(B_{58}) \not\equiv 0 \mod (3^2, v_1, v_2, \ldots)$

We also have, up to units in $\mathbb{Z}_{(3)}$, the identity

(2) $3r_9(B_{58}) \equiv r_6r_3(B_{58}) + r_5r_3r_1(B_{58}) + r_5r_1r_3(B_{58}) \mod (v_1, v_2, \ldots)$

This follows from the relation $84r_9 \equiv r_6r_3 - r_5r_{01} \mod (v_1, v_2 \ldots)$ (see 2:4) plus the relation $r_{01} = r_pr_1 - r_1r_p$ (see [23]). It follows from (1) and (2) that

(3) $r_6r_3(B_{58}) \not\equiv 0 \mod (3^2, v_1, v_2 \ldots)$

To prove (3) we need only show that $r_5r_3r_1(B_{58}) \equiv r_5r_1r_3(B_{58}) \equiv 0$ mod $(3^2, v_1, v_2, \ldots)$. Consider the term $r_5r_3r_1(B_{58})$. Write

$$r_1(B_{58}) = \gamma_{46}A_{46} + \gamma_{38}A_{38} + \gamma_{34}A_{34} + \gamma_{26}A_{26} + \gamma_{22}B_{22}$$

where $\gamma_s \in BP_*(pt)$. For dimension reasons each coefficient γ_s is divisible by either v_1^2 or by v_i for $i > 1$. Thus it follows from 2:1 and 2:2 that

$$r_5r_3r_1(B_{58}) \equiv \gamma_{46}r_5r_3A_{46} + \gamma_{38}r_5r_3(A_{38}) + \gamma_{34}r_5r_3(A_{38}) + \gamma_{26}r_5r_3(A_{26})$$
$$+ \gamma_{22}r_5r_3(B_{22}) \mod (3,v_1,v_2,\ldots)^2$$
$$\equiv 0 \mod (3^2,v_1,v_2,\ldots)$$

Next consider the term $r_5r_1r_3(B_{58})$. Write
$r_3(B_{58}) = \sigma_{46}A_{46} + \sigma_{38}A_{38} + \sigma_{34}A_{34} + \sigma_{26}A_{26} + \sigma_{22}B_{22}$

where $\sigma_s \in BP_*(pt)$. The coefficient σ_{46} belongs to $\mathbb{Z}_{(3)}$. It is divisible by 3 . For, by 2:3 , $\chi(\mathcal{P}^3)(b_{58}) = 0$ implies $r_3(B_{58})$ 0 mod $(3,v_1,v_2,\ldots)$. For dimension reasons the coefficient σ_s for $s < 48$ are divisible by either v_1^2 or by v_i for $i > 1$. It follows from 2:1 and 2:2 that

$$r_5r_1r_3(B_{58}) \equiv \sigma_{46}r\ r\ (A_{46}) + \sigma_{38}r_5r_1(A_{38}) + \sigma_{34}r_5r_1(A_{34}) + \sigma_{26}r_5r_1(A_{26})$$
$$+ \sigma_{22}r_5r_1(B_{22}) \mod (3,v_1,v_2,\ldots)^2$$
$$\equiv \sigma_{48}r_5r_1(A_{46}) \mod (3^2,v_1,v_2,\ldots)$$

Since $\chi(\mathcal{P}^5)\chi(\mathcal{P}^1)(a_{46}) = 0$ it follows from 2:3 that $r_5r_1(A_{46}) \equiv 0$ mod $(3,v_1,v_2,\ldots)$. Furthermore, as noted above, σ_{46} is divisible by 3 . Thus

$$\sigma_{46}r_5r_1(A_{46}) \equiv 0 \mod (3^2,v_1,v_2,\ldots)$$

This establishes (3). It follows from (3) that

(4) $r_6(A_{46}) \not\equiv 0 \mod (3,v_1,v_2,\ldots)$

For the expansion of $r_3(B_{58})$ in the proof of (3) plus the argument following implies that

$$r_6 r_3(B_{58}) \equiv \sigma_{46} r_6(A_{46}) \mod (3^2, v_1, v_2, \ldots)$$

where σ_{46} is divisible by 3. Comparing this identity with (3) we conclude that (4) holds.

We can now deduce 3:5 (a) and (b) from (4). By (4) and (2:3) it follows that $\chi(\mathcal{P}^6)(a_{46}) \neq 0$ in $Q(H_*(X;\mathbb{Z}/3))$. By [17]. $\chi(\mathcal{P}^6) = 2\mathcal{P}^6 + 2\mathcal{P}^{2,1}$ and $2\mathcal{P}^6 = \mathcal{P}^3\mathcal{P}^3 + 2\mathcal{P}^{2,1}$. But $\mathcal{P}^{2,1}(a_{46}) = \mathcal{P}^2\mathcal{P}^{01}(a_{46}) = 0$ ($Q(H_{30}(X;\mathbb{Z}/3)) = 0$ implies $\mathcal{P}^{01}(a_{46}) = 0$). Thus $\mathcal{P}^3\mathcal{P}^3(a_{46}) = 2\mathcal{P}^6(a_{46}) = \chi(\mathcal{P}^6)(a_{46}) \neq 0$. The only possibilities are $\mathcal{P}^3(a_{46}) = a_{34}$ and $\mathcal{P}^3(a_{34}) = b_{22}$ (up to units in $\mathbb{Z}/3$).

PROOF OF 4:5 (c) and (d)

We deduce 4:5 (c) from 4:5 (a) and 4:5 (d) from 4:5 (b). The proofs are analogous so we will only prove (d). By 4:5 (b) $\chi(\mathcal{P}^3)(a_{34}) = -\mathcal{P}^3(a_{34}) \neq 0$. Thus, by 2:3, $r_3(A_{34}) \not\equiv 0 \mod (3, v_1, v_2, \ldots)$. Hence

(5) $3r_3(A_{34}) \not\equiv 0 \mod (3^2, v_1, v_2, \ldots)$

By 2:4 we have the identity

(6) $3r_3(A_{34}) \equiv r_2 r_1(A_{34}) \mod (v_1, v_2, \ldots)$.

We can write

$$r_1(A_{34}) = \psi_{26}A_{26} + \psi_{22}B_{22}$$

for some coefficients $\psi_s \in BP_*(pt)$. For dimension reasons ψ_{26} is divisible by v_1 while ψ_{22} is divisible by v_1^2. It follows from

2:1 and 2:2 that

(7) $r_2r_1(A_{34}) \equiv r_1(\psi_{26})r_1(A_{26}) + \psi_{26}r_2(A_{26}) + \psi_{22}r_2(B_{22}) \mod (3,v_1,v_2,\ldots)^2$

$$\equiv r_1(\psi_{26})r_1(A_{26}) \mod (3^2,v_1,v_2,\ldots) .$$

It follows from (5), (6), and (7) that $r_1(\psi_{26})r_1(A_{26}) \not\equiv 0 \mod (3^2,v_1,v_2,\ldots)$. Also, by 2:2 , $r_1(\psi_{26}) \equiv 0 \mod (3,v_1,v_2,\ldots)$. Hence $r_1(A_{26}) \not\equiv 0 \mod (3,v_1,v_2,\ldots)$. It follows from 2:3 that $\chi(\mathcal{P}^1)(a_{26}) = -\mathcal{P}^1(a_{26}) \neq 0$. The only possibility is $\mathcal{P}^1(a_{26}) = b_{22}$ (up to an unit in $\mathbb{Z}/3$) .

PROOF OF 4:5 (e)

It follows from 2:4 that, up to an unit in $\mathbb{Z}_{(3)}$ we have the identity

(8) $3r_3(A_{38}) \equiv r_1r_1r_1(A_{38}) \mod (v_1,v_2,\ldots)$

Write

$$r_1(A_{38}) = \phi_{34}A_{34} + \phi_{26}A_{26} + \phi_{22}B_{22}$$

where $\phi_s \in BP_*(pt)$. It follows from 4:5 (c) and 2:3 that ϕ_{34} is an unit in $\mathbb{Z}(3)$. Assume $\phi_{34} = 1$. For dimension reasons ϕ_{26} and ϕ_{22} belong to $(3,v_1,v_2,\ldots)^2$. Thus, by 2:2 ,

(9) $r_1r_1r_1(A_{38}) \equiv r_1r_1(A_{34}) \mod (3,v_1,v_2,\ldots)^2$

As we demonstrated in the proof of 3:5 (d)

$$r_1(A_{34}) \equiv \psi_{26}A_{26} \mod (3,v_1,\ldots)^2$$

where ψ_{26} is divisible by v_1 . Using 2:2 , it follows that

(10) $r_1r_1(A_{34}) \equiv r_1(\psi_{26})A_{26} + \psi_{26}r_1(A_{26}) \mod (3,v_1,v_2,\ldots)^2$

$$\equiv r_1(\psi_{26})A_{26} \mod (3^2,v_1,v_2,\ldots) .$$

But $r_1(\psi_{26}) \not\equiv 0 \mod (3^2, v_1, v_2, \ldots)$. For statements (5), (6) and (7) in the proof of 3:5 (d) imply that $r_1(\psi_{26}) r_1(A_{26}) \not\equiv 0 \mod (3^2, v_1, v_2, \ldots)$. It follows that

(11) $r_1(\psi_{26}) A_{26} \not\equiv 0 \mod (3^2, v_1, \ldots)$

Putting together statements (8), (9), (10) and (11) we conclude that $r_3(A_{38}) \not\equiv 0 \mod (3, v_1, v_2, \ldots)$. By 2:4 $\chi(\mathcal{P}^3)(a_{38}) = -\mathcal{P}^3(a_{38}) \neq 0$ The only possibility is $\mathcal{P}^3(a_{38}) = a_{26}$ (up to an unit in $\mathbb{Z}/3$).

<u>Remark</u> For any choice of elements $b_{22}, a_{26}, a_{34}, a_{38}$, and a_{46} the identities in 4:5 are only valid up to units in $\mathbb{Z}/3$. There is no choice for which all of the identities (a), (b), (c), (d), (e) are simultaneously valid. The proof of 4:5 (e) can be extended to show that if we pick the generators such that (a), (b), (c), and (d) hold then we must have $\mathcal{P}^3(a_{38}) = 2a_{26}$.

§5. The Coalgebra Structure of $H^*(X;\mathbb{Z}/3)$ Part I

We now begin our study of the coalgebra structure of $H^*(X;\mathbb{Z}/3)$. In this section we work with no assumptions on the coalgebra structure. In the next section we will assume that $H^*(X;\mathbb{Z}/3)$ is coassociative and extend the results obtained here. Observe that by 3:1 the generators x_3, x_7, x_8, x_{19}, and x_{20} can all be chosen to be primitive. In the next section we will show that, when $H^*(X;\mathbb{Z}/3)$ is coassociative, none of the remaining generators $x_{15}, x_{27}, x_{35}, x_{39}$, or x_{47}, can be chosen to be primitive. In this section, working without the coassociativity hypothesis, we will show that neither x_{35} nor x_{47} can be chosen to be primitive. This result is best possible. For it follows from the work of Zabrodsky that $X = E_8$ has a multiplication such that all generators but x_{35} and x_{47}, can be chosen to be primitive.

To show that x_{35} and x_{47} cannot be chosen to be primitive it suffices, by 3:1 (b), to show that x_{47} is never primitive. Our proof of this fact will be by contradiction. Suppose x_{47} can be chosen to be primitive. It follows that

(5:1) all elements of positive dimension from $A^*(3)$ act trivially on x_{47}.

For, given $\Theta \in A^*(3)$, $\Theta(x_{47})$ is primitive. By [18] any non-zero primitive in $H^*(X;\mathbb{Z}/3)$ is either indecomposable or a pth power. Thus, by 1:1, $H^*(X;\mathbb{Z}/3)$ has no non-zero primitives in dimension > 47. Consequently $\Theta(x_{47}) = 0$ if $|\Theta| > 0$. We now proceed to show that it is impossible for x_{47} to satisfy 4:1. First of all it follows from 5:1 and 2:5 that x_{47} lifts to $X_{47} \in BP^*(X)$ under the Thom map

$T : BP^*(X) \to H^*(X;\mathbb{Z}/3)$. Next pass to $BP^*(\Omega X)$ via the loop map.

$$\Omega^* : Q(BP^*(X))\big/\text{Torsion} \to P(BP^*(\Omega X))$$

It is induced by the map $\Omega : \sum \Omega x \to x$. The resulting map $\Omega^* : BP^*(X) \to BP^*(\sum \Omega X) = BP^*(\Omega X)$ annihilates decomposables since it factors through $BP^*(\sum \Omega X)$. It annihilates torsion since $BP^*(\Omega X)$ is a free $BP^*(pt)$ module. By the commutative diagram

$$\begin{array}{ccc} BP^*(X) & \xrightarrow{\Omega^*} & BP^*(\Omega X) \\ \downarrow & & \downarrow \\ BP^*(X) \otimes Q & \xrightarrow{\Omega^*} & BP^*(\Omega X) \otimes Q \\ & & \\ H^*(X;Q) \otimes (BP^*(pt) \otimes Q) & \xrightarrow{\Omega^*} & H^*(\Omega X;Q) \otimes (BP^*(pt) \otimes Q) \end{array}$$

(see [8] for the vertical isomorphism) it follows that Image $\{\Omega^* : BP^*(X) \to BP^*(\Omega X)\}$ lies in $P(BP^*(\Omega X))$. For Image $\{\Omega^* : H^*(X;Q) \to H^*(\Omega X;Q)\}$ is known to lie in $P(H^*(\Omega X;Q))$.

We will prove the following conflicting statements about X_{47} or, more precisely, about $\Omega^*(X_{47})$.

(5:2) $\quad r_3\Omega^*(X_{47}) \equiv 0 \mod (3^2,v_1,v_2,\ldots)$

(5:3) $\quad r_3\Omega^*(X_{47}) \not\equiv 0 \mod (3^2,v_1,v_2,\ldots)$

This produces the required contradiction.

PROOF OF 5:2

First of all we prove a result about $Q(BP^*(X))\big/\text{Torsion}$. Pick $x_{19} \in H^*(X;\mathbb{Z}/3)$ to be primitive and let $x_{20} = \beta_p(x_{19})$. By an argument analogous to that for 4:1 we can show $Q_s(x_{19}) = 0$ for $s \geq 1$ and $Q_s(x_{20}) = 0$ for $s \geq 0$. Therefore. $Q_s(x_{19}x_{20}^2) = 0$ for $s \geq 0$ (observe $Q_0(x_{19}x_{20}^2) = x_{20}^3 = 0$. It follows from 2:5 that $x_{19}x_{20}^2$ lifts to $X_{59} \in BP^*(X)$. We will show

$$(5:4) \qquad Q^s(BP^*(X))/\text{Torsion} = \begin{cases} \mathbb{Z}_{(3)} \text{ with generator } X_{59} \text{ if } s = 59 \\ 0 \text{ if } s > 59 \end{cases}$$

To prove 5:4 let $H^*(X) = H^*(X;\mathbb{Z}) \otimes \mathbb{Z}_{(3)}$ and consider $H^*(X)/\text{Torsion}$

By calculating the cohomology Bockstein spectral sequence (see [3]) using 3:1 (c) we can deduce that

$$B_\infty = B_2 = E(\{x_3\},\{x_{15}\},\{x_7x_8^2\},\{x_{27}\},\{x_{35}\},\{x_{39}\},\{x_{47}\},\{x_{19}x_{20}^2\})$$

Since $B_\infty = H^*(X)/\text{Torsion} \otimes \mathbb{Z}/3$ it follows that

$$H^*(X)/\text{Torsion} = E(Y_3,Y_{15},Y_{23},Y_{27},Y_{35},Y_{39},Y_{47},Y_{59})$$

where the algebra generators of $H^*(X)/\text{Torsion}$ map into those of B_∞ in the obvious fashion. In particular, since the Thom map factors as $BP^*(X) \to H^*(X) \to H^*(X;\mathbb{Z}/3)$ we can choose Y_{59} to be the image of X_{59} under $BP^*(X) \to H^*(X)/\text{Torsion}$. Since $BP^*(X) \otimes Q = (H^*(X)/\text{Torsion} \otimes Q) \otimes (BP^*(pt) \otimes Q)$ it follows that

$$BP^*(X) \otimes Q = E(X_3,X_{15},X_{23},X_{27},X_{35},X_{39},X_{47},X_{59})$$

where X_{59} is the element chosen previously. It follows that $Q(BP^*(X) \otimes Q)$ is a free $BP^*(pt) \otimes Q$ module with generators $X_3,X_{15},\dots,X_{59}$

We can now deduce 4:4 . For if we pick a set of $BP_*(pt)$ module generators of $Q(BP^*(X)/_{\text{Torsion}} \subset Q(BP^*(X)) \otimes Q \subset Q(BP^*(X) \otimes Q)$ then, for each generator x , $3^s x$ (for some $s \geq 0$) can be written (over $BP^*(pt)$) in terms of the elements $X_3, X_{15}, \ldots, X_{59}$. Since $BP^*(pt)$ is negatively graded it follows that $Q^s(BP^*(X))/_{\text{Torsion}} = 0$ for $s > 59$. Since $X_{59} \in Q(BP^*(X))/_{\text{Torsion}}$ is not divisible by 3 (it maps to Y_{59} in $Q(H^* X)/_{\text{Torsion}}$ it follows that $Q^{59}(BP^*(X))/_{\text{Torsion}} = \mathbb{Z}_{(3)}$ generated by X_{59} . This proves 5:4 .

It follows from 5:4 that, in $Q(BP^*(X))/_{\text{Torsion}}$, $r_3(X_{47}) = \alpha X_{59}$ for some $\alpha \in \mathbb{Z}_{(3)}$. It follows from 5:1 and 2:4 that α is divisible by 3 . Also, by using the commutative diagram

$$\begin{array}{ccc} Q(BP^*(X)) & \xrightarrow{\ \Omega^*\ } & P(BP^*(\Omega X)) \\ \downarrow T & & \downarrow T \\ Q(H^*(X;\mathbb{Z}/3)) & \xrightarrow{\ \Omega^*\ } & P(H^*(\Omega X;\mathbb{Z}/3)) \end{array}$$

and the fact that $T(X_{59}) = x_{19}x_{20}^2$ it follows that $\Omega^*(X_{59}) \equiv 0 \mod (3, v_1, v_2, \ldots)$. Thus $r_3\Omega^*(X_{47}) = \Omega^*(r_3 X_{47}) = \alpha r_*(X_{59}) \equiv 0 \mod (3^2, v_1, v_2, \ldots)$.

PROOF OF 5:3

In proving 5:3 we will use the following facts about the action of $BP^*(BP)$ on $Q(BP_*(\Omega X)$.

(5:5) $\quad r_3(B_{58}) \equiv 3A_{46} \mod (3^2, v_1, v_2, \ldots)$

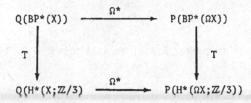

(5:6) $\quad r_2(B_{58}) \equiv 0 \mod (3^2, v_1^2, v_2, v_3, \ldots)$

Statement (5:5) follows from the proof of 4:5 (a) and (b). For in that proof the following statements occur.

$$r_3(B_{58}) \equiv \delta_{48} A_{48} \mod (3^2, v_1, v_2, \ldots)$$

$$\delta_{48} r_6(A_{48}) \equiv r_6 r_3(B_{58}) \not\equiv 0 \mod (3^2, v_1, v_2, \ldots)$$

Statement (5:6) then follows from (5:5) by an argument involving the relation $r_1 r_2 \equiv 3 r_3 \mod (v_1, v_2, \ldots)$.

Now the pairing between $BP^*(\Omega X)$ and $BP_*(\Omega X)$ induces a pairing between $P(BP^*(\Omega X)$ and $Q(BP_*(\Omega X)$. Furthermore there is a duality between the action of $BP^*(BP)$ on $Q(BP_*(\Omega X)$ and $P(BP^*(\Omega X)$ which is achieved via the coaction map

$$\lambda : Q(BP_*(\Omega X)) \to BP_*(BP) \otimes Q(BP_*(\Omega X))$$

(see Lecture III of [1]).

We will use these dualities to prove 5:3. We will prove 5:3 by showing

(5:7) $\quad \langle r_3 \Omega^*(X_{47}), B_{58} \rangle \not\equiv 0 \mod (3^2, v_1, v_2, \ldots)$

To simplify our work we will pass from $Q(BP_*(\Omega X))$ to the quotient module $N = Q(BP_*(\Omega X))/L$ where L is the ideal generated by $\{A_2, A_6, A_{14}, A_{18}, B_{22}, A_{26}, A_{34}, A_{38}\}$. Thus N is the free $BP_*(pt)$ module generated by the elements A_{46} and B_{58}. The action of $BP^*(BP)$ and the coaction of $BP_*(BP)$ on $Q(BP_*(\Omega X))$ induce actions and coactions on N. Also, the pairing of $\Omega^*(X_{47})$ with the elements of $Q(BP_*(\Omega X))$

induces a pairing with the elements of N (for $\Omega^*(X_{47})$ annihilates L) . Write

$$\lambda(B_{58}) = 1 \otimes B_{58} + C_{12} \otimes A_{46}$$

where $C_{12} \in BP_*(BP)$. We have the identities

$$(5{:}8)\qquad \langle r_3\Omega^*(X_{47}),B_{58}\rangle = \langle r_3, 1\ \ \Omega^*(X_{47},B_{58})\rangle + \langle r_3,C_{12}\langle\Omega^*(X_{47})A_{46}\rangle\rangle$$
$$= \langle r_3,C_{12}\rangle$$

The first identity is Proposition 3 of Lecture III of [1]. The second identity follows from the fact that we can choose A_{46} and B_{58} such that $\Omega^*(X_{47})$ is dual to A_{46} .

Next, letting $c : BP_*(BP) \to BP_*(BP)$ be the conjugate map, we have the identity

$$(5{:}9)\qquad \langle r_3,C_{12}\rangle \equiv -\langle r_3,c(C_{12})\rangle \bmod (3^2,v_1,v_2,\ldots)$$

First of all, since $BP_*(BP) = BP_*(pt)\ [t_1,t_2,\ldots]$ (deg $t_s = 2\ 3^s-2$) it follows that

$$C_{12} = \alpha t_1^3 + \beta v_1 t_1^2 + \gamma v_1^2 t_1 + \delta v_1^3$$

where $\alpha,\beta,\gamma,\delta \in \mathbb{Z}_{(3)}$. Thus $\langle r_3,C_{12}\rangle = \alpha$ and to prove 5:8 we need only show that $\langle r_3,c(C_{12})\rangle \equiv -\alpha \bmod (3^2,v_1,v_2,\ldots)$. Since $c(t_1) = -t_1$ and $c(v_1) = v_1 + 3t_1$ it follows that

$$c(C_{12}) \equiv -\alpha t_1^3 + \beta v_1 t_1^2 + \beta t_1^3 \bmod (3^2,v_1^2,v_2,v_3,\ldots)$$

Furthermore, β is divisible by 3 . For $\beta_1 v_1 \equiv \langle r_2,c(C_{12})\rangle$ mod $(3^2,v_1^2,v_2,v_3,\ldots)$. And, by Proposition 2 of Lecture III of [1], $r_2(B_{58}) = \langle r_2,c(C_{12})\rangle A_{46}$. Thus it follows from 5:6 that

$\langle r_2, c(C_{12})\rangle \equiv 0 \mod (3^2, v_1^2, v_2, v_3, \ldots)$. Thus $\beta v_1 \equiv 0 \mod (3^2, v_1^2, v_2, v_3, \ldots)$ and it follows that β is divisible by 3 . Therefore

$$c(C_{12}) \equiv -\alpha t_1^3 \mod (3^2, v_1, v_2, \ldots)$$

and $\langle r_3, c(C_{12})\rangle \equiv -\alpha \mod (3^2, v_1, v_2, \ldots)$ as required.

Our last relation is

(5:10) $\langle r_3, c(C_{12})\rangle \not\equiv 0 \mod (3^2, v_1, v_2, \ldots)$

For, by Proposition 2 of Lecture III of [1], $r_3(B_{58}) = \langle r_3, c(C_{12})\rangle A_{48}$. Thus, by 5:5, $\langle r_3, c(C_{12})\rangle \not\equiv 0 \mod (3^2, v_1, v_2, \ldots)$.

It now follows from statements 5:8, 5:9 and 5:10 that $\langle r_3 \Omega^*(X_{47}) , B_{58}\rangle \not\equiv 0 \mod (3^2, v_1, v_2, \ldots)$.

<u>Remark</u> Our arguments in this section show not only that x_{47} , is never primitive but that x_{47} cannot satisfy 5:1 . It follows that $H^*(X;\mathbb{Z}/3)$ can never be a $\cup(M)$ module in the sense of [15].

§6. The Coalgebra Structure of $H^*(X;\mathbb{Z}/3)$ Part II

In this section we will assume that $H^*(X;\mathbb{Z}/3)$ is coassociative and show that the coalgebra structure of $H^*(X;\mathbb{Z}/3)$ as well as its Steenrod module structure are uniquely determined from 1:1. Actually we will dualize and study $H_*(X;\mathbb{Z}/3)$. In particular we can assume that $H_*(X;\mathbb{Z}/3)$ is associative. It is also true that $H_*(X;\mathbb{Z}/3)$ is primitively generated. This follows from 1:1 using 4:23 of [18]. Let $P = P(H_*(X;\mathbb{Z}/3))$. Then P has a restricted Lie algebra structure given by the Frobenius power map

$$\xi_3 : P \to P$$

$$\xi_3(x) = x^3$$

and the Lie bracket product

$$[\ ,\ ] : P \otimes P \to P$$

$$[x,y] = xy - (-1)^{|x||y|}yx$$

Furthermore the action of $A^*(3)$ on $H_*(X;\mathbb{Z}/3)$ restricts to give an action of $A^*(3)$ on P. This action respects the restricted Lie algebra structure of P. In particular $\mathcal{P}^n$ acts on $[x,y]$ by the rule $\mathcal{P}^n[x,y] = \sum_{i+j=n} [\mathcal{P}^i(x),\mathcal{P}^j(y)]$. Since $H_*(X;\mathbb{Z}/3)$ is associative and primitively generated it follows that the algebra structure of $H_*(X;\mathbb{Z}/3)$ over $A^*(3)$ is determined by the Lie algebra structure of P over $A^*(3)$ (by §5 of [18] $H_*(X;\mathbb{Z}/3)$ is the universal envelopping Hopf algebra of P). Thus we must show that the Lie algebra structure of P is uniquely determined.

First of all, the Frobenius map is trivial (see 5:4:1 of [14]). So we need only consider the Lie bracket product and the action of $A^*(3)$. Now P is dual, as a $\mathbb{Z}/3$ module, to $Q(H^*(X;\mathbb{Z}/3))$. Thus P has a $\mathbb{Z}/3$ basis consisting of elements $\alpha_3,\alpha_7,\alpha_8,\alpha_{15},\alpha_{19},\alpha_{20},\alpha_{27}$, $\alpha_{35},\alpha_{39},\alpha_{47}$ (dim $\alpha_s = s$). By dualizing 3:1 and 4:1 it follows that

(6:1) (a) $\mathcal{P}^3(\alpha_{19}) = \alpha_7$, $\mathcal{P}^3(\alpha_{20}) = \alpha_8$, $\mathcal{P}^3(\alpha_{39}) = \alpha_{27}$, $\mathcal{P}^3(\alpha_{47}) = \alpha_{35}$

(b) $\mathcal{P}^1(\alpha_7) = \alpha_3$, $\mathcal{P}^1(\alpha_{19}) = \alpha_{15}$, $\mathcal{P}^1(\alpha_{39}) = \alpha_{35}$

(c) $\beta_3(\alpha_8) = \alpha_7$, $\beta_3(\alpha_{20}) = \alpha_{19}$

The only other possible non trivial Steenrod power is $\mathcal{P}^3(\alpha_{27})$. All the remaining Steenrod powers can be eliminated either by using the fact that $P_i = 0$ in most dimensions or by dualizing and using the fact that $A^*(3)$ acts unstably on $H^*(X;\mathbb{Z}/3)$ and hence on $Q(H^*(X;\mathbb{Z}/3))$.

Regarding Lie brackets, since $P_i = 0$ in most dimensions, the possible non-zero brackets in P are

dimension 15 $[\alpha_7,\alpha_8]$

dimension 27 $[\alpha_7,\alpha_{20}]$, $[\alpha_8,\alpha_{19}]$

dimension 35 $[\alpha_{15},\alpha_{20}]$, $[\alpha_8,\alpha_{27}]$

dimension 39 $[\alpha_{19},\alpha_{20}]$

dimension 47 $[\alpha_8,\alpha_{39}]$, $[\alpha_{20},\alpha_{27}]$.

We must determine which brackets are non-zero and what is the relation between different brackets in the same dimension. In particular, observe that $[\alpha,\beta] = -[\beta,\alpha]$ for all of the above Lie brackets. We will be using this fact.

Our analysis of the Lie algebra structure of P over $A^*(3)$ will be done on a dimension by dimension case.

(a) DIMENSION 47

Since x_{47} is not primitive (see §5) it follows that at least one of the brackets $[\alpha_8,\alpha_{39}]$ or $[\alpha_{20},\alpha_{27}]$ is non-zero. (By the arguments in [16], lack of primitivity in $H^*(X;\mathbb{Z}/3)$ is equivalent to non trivial Lie brackets in P) . Since $[\alpha_{20},\alpha_{39}] = 0$ it follows that

$$\begin{aligned} 0 &= \mathcal{P}^3[\alpha_{20},\alpha_{39}] \\ &= [\mathcal{P}^3\alpha_{20},\alpha_{39}] + [\alpha_{20},\mathcal{P}^3\alpha_{39}] \\ &= [\alpha_8,\alpha_{39}] + [\alpha_{20},\alpha_{27}] \ . \end{aligned}$$

Thus

$$(6{:}2) \qquad [\alpha_8,\alpha_{39}] = [\alpha_{27},\alpha_{20}] = \alpha_{47} \ .$$

(b) DIMENSION 35

By 6:2 we can write $\alpha_{47} = [\alpha_8,\alpha_{39}]$. By 6:1 we can write $\alpha_{35} = \mathcal{P}^3\alpha_{47}$. Thus

$$\begin{aligned} \alpha_{35} &= \mathcal{P}^3[\alpha_8,\alpha_{39}] \\ &= [\alpha_8,\mathcal{P}^3\alpha_{39}] \\ &= [\alpha_8,\alpha_{27}] \ . \quad \text{(by 6:1)} \end{aligned}$$

Next, since $[\alpha_{20},\alpha_{39}] = 0$, it follows that

$$0 = \mathcal{P}^6 [\alpha_{20}, \alpha_{39}]$$
$$= [\mathcal{P}^3 \alpha_{20}, \mathcal{P}^3 \alpha_{39}] + [\alpha_{20}, \mathcal{P}^6 \alpha_{39}]$$
$$= [\alpha_8, \alpha_{27}] + [\alpha_{20}, \mathcal{P}^6 \alpha_{39}] \quad \text{(by 6:1)}$$

From the above two identities it follows that $[\alpha_{20}, \mathcal{P}^6 \alpha_{39}] = -[\alpha_8, \alpha_{27}] = -\alpha_{35} \neq 0$. Hence $\mathcal{P}^6 \alpha_{39} = 0$. Factoring we have

$$\mathcal{P}^6 (\alpha_{39}) = 2\mathcal{P}^3 \mathcal{P}^3 (\alpha_{39}) + \mathcal{P}^{01} \mathcal{P}^2 (\alpha_{39}) \quad \text{(by [17])}$$
$$= -\mathcal{P}^3 (\alpha_{27}) \quad \text{(by 6:1)}$$

Thus $\mathcal{P}^3 (\alpha_{27}) \neq 0$ and the only possibility is $\mathcal{P}^3 (\alpha_{27}) = \alpha_{15}$ (up to an unit in $\mathbb{Z}/3$) . Also $[\alpha_8, \alpha_{27}] = -[\alpha_{20}, \mathcal{P}^6 \alpha_{39}] = -[\alpha_{20}, -\alpha_{15}] = [\alpha_{20}, \alpha_{15}]$. We conclude

(6:3) $\quad \mathcal{P}^3 (\alpha_{39}) = \alpha_{15}$ and $[\alpha_8, \alpha_{27}] = [\alpha_{20}, \alpha_{15}] = \alpha_{35}$

(c) DIMENSION 39

Since $\mathcal{P}^1 [\alpha_{20}, \alpha_{19}] = [\alpha_{20}, \mathcal{P}^1 \alpha_{19}]$

$$= [\alpha_{20}, \alpha_{15}] \quad \text{(by 6:1)}$$

if follows from 6:3 that

(6:4) $\quad [\alpha_{20}, \alpha_{19}] = \alpha_{39}$

(d) DIMENSION 15

By 6:4 we can write $\alpha_{39} = [\alpha_{20}, \alpha_{19}]$. By 6:3 $\mathcal{P}^6 (\alpha_{39}) = -\alpha_{15}$. Thus

$$\alpha_{15} = -\mathcal{P}^6 [\alpha_{20},\alpha_{19}]$$
$$= -[\mathcal{P}^3 \alpha_{20},\mathcal{P}^3 \alpha_{19}]$$
$$= -[\alpha_8,\alpha_7]$$

Therefore

(6:5) $[\alpha_8,\alpha_7] = \alpha_{15}$

(e) DIMENSION 27

Since $[\alpha_8,\alpha_{20}] = 0$ it follows that

$$0 = \beta_3[\alpha_8,\alpha_{20}]$$
$$= [\beta_3\alpha_7,\alpha_{20}] + [\alpha_8,\beta_3\alpha_{20}]$$
$$= [\alpha_7,\alpha_{20}] + [\alpha_8,\alpha_{19}] \quad \text{(by 6:1)}$$

Hence $[\alpha_7,\alpha_{20}] = [\alpha_{19},\alpha_8]$. Also,

$$\mathcal{P}^3 [\alpha_7,\alpha_{20}] = [\alpha_7,\mathcal{P}^3 \alpha_{20}]$$
$$= [\alpha_7,\alpha_8] \quad \text{(by 6:1)}$$

Thus, by 6:5 , $[\alpha_7,\alpha_{20}] \neq 0$. We conclude

(6:6) $[\alpha_7,\alpha_{20}] = [\alpha_{19},\alpha_8] = \alpha_{27}$.

§7. Possible Extensions

In this section we will discuss some possible extensions of the results of this paper. Our method in this paper has been to study the consequences for $H^*(X;\mathbb{Z}/3)$ of the presence of the even dimensional generators x_8 and x_{20}. More exactly we used the fact that $\mathcal{P}^3(x_8) = x_{20}$ to force our consequences. We can ask the obvious question as to whether these results are a particular case of general structure theorems which hold for all finite H-spaces.

First of all consider §3 and §4. There we studied the algebra structure and the Steenrod module structure of $H^*(X;\mathbb{Z}/3)$. In §3 we observed that the generators x_8 and x_{20} forced the presence of generators in dimension 3, 7, 15, and 19 with $A^*(3)$ acting as in 3:1. We used the structure theorems of [14]. Our arguments in §4 suggest that there should be another type of structure theorem relating even dimensional generators to odd ones. If two even dimensional indecomposables in $H^*(X;\mathbb{Z}/p)$ are related by a Steenrod power then their transpotence elements in $H^*(\Omega X;\mathbb{Z}/p)$ are also related by a Steenrod power. And we must then be able to fill in the "gap" between these transpotence elements with elements from Image $\Omega^* : Q^{odd}(H^*(X;\mathbb{Z}/p)) \to P^{even}(H^*(\Omega X;\mathbb{Z}/p))$. For example, in our case, we showed that because $\mathcal{P}^3(x_8) = x_{20}$ it follows that none of the generators x_{27}, x_{35}, x_{39}, or x_{47} can be removed from 1:1. Unfortunately our argument does not show that $\mathcal{P}^3(x_8) = x_{20}$ forces generators in exactly dimensions 27, 35, 39, and 47. It can only be used to show that there are at least four generators between dimensions 23 and 59. So more preciseness in our results seem desirable before general structure theorems are formulated.

Next consider §5 and §6. There we studied the coalgebra structure of $H^*(X;\mathbb{Z}/3)$. In particular we showed in §5 that if $\mathcal{P}^3(x_8) = x_{20}$ it follows that the generators x_{35} and x_{47} can never be primitive. In a future paper we will show that if p is odd and (X,μ) a 1-connected mod p finite H-space where $H^*(X;\mathbb{Z}/p)$ is primitively generated then $A^*(p)$ acts trivially on $Q^{even}(H^*(X;\mathbb{Z}/p))$. It follows that $Q^{2n}(H^*(X;\mathbb{Z}/p)) = 0$ except for $n = p^s + \ldots + p + 1$ $(s \geq 1)$. Thus the coalgebra structure of $H^*(X;\mathbb{Z}/p)$ affects the algebra structure and vice-versa. And as this paper illustrates detailed information about this relationship is obtainable.

References

(1) Adams, J.F., Lectures on Generalized Cohomology, Lecture Notes in Mathematics, 99, Springer-Verlag (1969).

(2) Adams, J.F., Stable Homotopy and Generalized Homology, University of Chicago Press (1974).

(3) Browder, W., Torsion in H-Spaces, Annals of Math. 74(1961), 24-51.

(4) Hazewinkel, M., A Universal Formal Group and Complex Cobordism, Bull. Amer. Math. Soc. 81(1975), 930-933.

(5) Kane, R., The Module of Indecomposables for Finite H-Spaces, Trans. Amer. Math. Soc. 222(1976), 303-318.

(6) Kane, R., Torsion in Homotopy Associative H-Spaces, Illinois J. Math. 20(1976), 476-485.

(7) Kane, R., The BP Homology of H-Spaces, Trans. Amer. Math. Soc. 241(1978), 99-120.

(8) Kane, R., Rational BP Operations and the Chern Character. Math. Proc. Camb. Phil. Soc. 84(1978), 65-72.

(9) Kane, R., The Module of Indecomposables for Mod 2 Finite H-Spaces, Trans. Amer. Math. Soc. (to appear).

(10) Kane, R., BP Operations and Steenrod Modules, Quart. J. Math Oxford (to appear).

(11) Kane, R., BP Torsion in Finite H-Spaces (to appear).

(12) Kono, A., and Mimura, M., Cohomology Operations and the Hopf Algebra Structures of the Compact Exceptional Lie Groups E_7 and E_8, Proc. London Math. Soc. (3) 35(1977), 345-358.

(13) Lin, J., Torsion in H-Spaces I, Annals of Math. 103(1976), 457-487.

(14) Lin, J., Torsion in H-Spaces II, Annals of Math. 107(1978), 41-88.

(15) Massey, W.S., and Peterson, F.P., The Cohomology Structure of Certain Fibre Spaces - I, Topology 4 (1965), 47-65.

(16) May, J.P., The Cohomology of Restricted Lie Algebra and of Hopf Algebras, J. Algebra (1966), 123-146.

(17) Milnor, J., The Steenrod Algebra and its Dual, Annals of Math. 67(1958), 150-171.

(18) Milnor, J., and Moore, J.C., On the Structure of Hopf Algebras, Annals of Math. 81(1965), 211-264.

(19) Rector, D.L., Steenrod Operations in the Eilenberg-Moore Spectral Sequence, Comm. Math. Helv. 45(1970), 540-552.

(20) Quillen, D.G., On the Formal Group Laws of Unoriented and Complex Cobordism Theory, Bull. Amer. Math. Soc. 75(1969), 1293-1298.

(21) Zabrodsky, A., Secondary Cohomology Operations in the Module of Indecomposables, Algebraic Topology Conference, Aarhus, 1970.

(22) Zabrodsky, A., Some Relations in the Mod 3 Cohomology of H-Spaces, (to appear).

(23) Zahler, R., The Adams-Novikov Spectral Sequence for the Spheres, Annals of Math. 96(1972), 480-504

University of Alberta
Edmonton, Alberta

A Counterexample to the Transfer Conjecture

by

David Kraines and Thomas Lada

In this paper we discuss the following conjecture, which has been attributed to D. Quillen, and present a counterexample to it.

Conjecture: If a representable homotopy functor admits a transfer, then it extends to a cohomology theory.

This conjecture appeared in a preliminary version (circa 1970) of [S2], at that time entitled "Homotopy everything H spaces." In that preprint, G. Segal presented his permutative category approach to the study of infinite loop spaces. As an application of his methods, he outlined a supposed proof of this conjecture. However, as work on transfer and infinite loop spaces continued in the early 1970's, many doubts were raised about the truth of this conjecture.

If $X = \Omega Y$ then X has an associative H space structure and the functor $h(\ ) = [\ ,X]$ takes values in the category of groups. Conversely if $h(\ ) = [\ ,X]$ takes values in the category of groups, then X has a homotopy associative H structure, but is not necessarily of the homotopy type of a loop space. Stasheff [S4] introduced the notion of A_k structures and showed that a connected space X is homotopy associative if and only if it has an A_3 structure, while X has the homotopy type of a loop space if and only if it has an A_∞ structure.

Assume that there is a sequence $\{X_n\}$ with $\Omega X_n = X_{n-1}$. Then we say that X_0 is a perfect infinite loop space. In this case X has a Q algebra structure (Definition 1.1 and Theorem 1.2), and $h(\ ) = [\ ,X]$ admits a transfer. Conversely if $h(\ ) = [\ ,X]$ admits a transfer, then we say X is a transfer space. In sections 2 and 3 we introduce the notion of Q_k structures. A connected space X is a transfer space if and only if it has a Q_2 structure while X is equivalent to an infinite loop space if and only if X has a Q_∞ structure (Theorems 2.4 and 2.5).

If $X = \Omega Y$ then the Eilenberg-Moore spectral sequence arising from the Milnor construction converges to $H^*(Y;\Lambda)$ from a functor of the coalgebra $H^*(X;\Lambda)$. Stasheff [S4] found a relationship between k cycles and A_{k+1} structures and was able to use this to construct an A_k space with no A_{k+1} structure.

If X is a perfect infinite loop space, then Haynes Miller has constructed an infinite delooping spectral sequence [M8]. We prove that k cycles in this spectral sequence correspond to Q_k maps and that these induce Q_k spaces (Theorems 3.3 and 5.2). Thus to construct a counterexample to the transfer conjecture, we need to find a 2 cycle which is not a 3 cycle in the Miller spectral sequence.

In an earlier draft we were able to construct a 2 stage Postnikov system P and a map $f: P \to K(Z/p,n)$ which represented a 2 cycle, but not a 3 cycle. The induced fiber space E turns out to be a 3 stage Postnikov system such that $h(\ ) = [\ ,E]$ is a counterexample to the transfer conjecture. The proof required a great deal of ad hoc technical constructions and proofs.

Recently work of Madsen and Snaith ([MST],[M4]) was brought to our attention. They show that if s_k is the primitive generator of $H^{2k}(BU;Z_{(2)})$, then $2s_k$ is represented by a transfer commuting map, which we also write $2s_k: BU \to K(Z_{(2)},2k)$. Furthermore, they showed that the fiber of $2s_7$ was not an infinite loop space. Unfortunately, this fiber was not a transfer space so no counterexample was obtained.

Our techniques apply directly to this type of construction. We show that the fiber induced by $4s_k$ is a transfer space for all k, but that the fiber E of $4s_{15}$ does not have an infinite loop structure.

Thus, [,E] does give a counterexample to the transfer conjecture. The proof of this result is far less technical than our proof for the Postnikov system counterexample.

In section 1 we review infinite loop space theory and Q algebras. If (X,ρ) is a Q algebra, then following Beck [B] and May [M6] we construct a simplicial spectrum $B(\Sigma^{\infty},Q,X)_*$ whose realization gives an infinite delooping of X. In sections 2 and 3 we introduce Q_k structures and establish its relationship with transfer. In sections 4 and 5 the spectral sequence of the simplicial spectrum of section 1 is constructed and the relationship between its cycles and Q_k maps is proven. In sections 6 and 7 we prove that under suitable conditions $E_r^{s,t}(X,Z_{(p)})$ is a Z/p module for $s \geq 1$ and $r \geq 2$. This fact allows us to conclude that $4s_k$ is a Q_2 map and so its induced fiber E is a transfer space. In section 8 we prove that although E has a 2 fold loop sturcture it does not have a 3 fold loop structure. This implies that h( ) = [,E] is a counterexample to the transfer conjecture.

In our constructions we rely heavily on the theory of infinite loop spaces built up by P. May. The volumes [M6] and [M7] serve as background references for much of this paper.

We appreciate the interest and encouragement of J. Stasheff and P. May. We are grateful to I. Madsen for his discovery of an error in a previous manuscript. It was in the understanding and correction of that error that we were led to our present spectral sequence approach. We feel that the Miller delooping spectral sequence will have many interesting applications.

§1. Infinite loop spaces and Beck's Theorem

We will work throughout in the category of pointed compactly generated spaces with $H_*(X)$ of finite type. By a representable homotopy functor h we mean $h(Y) = [Y,X]$ = based homotopy classes of maps where X is determined up to weak homotopy type. We say that h extends to a cohomology theory if there is a sequence $\{h^k\}$ with $h^k(\Sigma Y)$ naturally equivalent to $h^{k-1}(Y)$ and h^0 naturally equivalent to h.

By an Ω spectrum we will mean a sequence of spaces $\{X_k\}$ which are connected for $k > 0$ and such that there are weak homotopy equivalences $\Omega X_k \simeq X_{k-1}$ for $k > 0$. If $\{X_k\}$ is an Ω spectrum, let $\Omega^\infty\{X_k\} = X_0$ be the 0th space. If $X \simeq \Omega^\infty\{X_k\}$ for some Ω spectrum then we say that X is an infinite loop space. The following classical result follows immediately using the adjointness of Σ and Ω.

<u>Proposition 1.1</u>. The functor $h(Y) = [Y,X]$ extends to a cohomology theory if and only if X is an infinite loop space.

We say that the Ω spectrum $\{X_k\}$ is perfect if $\Omega X_k = X_{k-1}$ for $k > 0$. For any space Y let $\Sigma^\infty Y$ be the perfect Ω spectrum $\{Q\Sigma^k Y\}$ where $Q = \varinjlim \Omega^N\Sigma^N$. The adjunctions between Σ and Ω induce the adjunctions

$$\eta: 1 \to \Omega^\infty\Sigma^\infty = Q$$

$$\text{and} \qquad \varepsilon: \Sigma^\infty\Omega^\infty \to 1 \quad \text{(see [M5])}.$$

Let $\mu = \Omega^\infty\varepsilon\Sigma^\infty: Q^2 \to Q$. Then for functorial reasons there are identities $\mu Q\eta = 1 = \mu\eta Q$ and $\mu\mu Q = \mu Q\mu$. We call (Q,μ,η) a monad.

<u>Definition 1.2</u>. If there is a map $\rho: QX \to X$ satisfying $\rho\eta = 1_X$ and $\rho\mu = \rho Q\rho: Q^2X \to X$ then we say that (X,ρ) is a Q algebra and that X has a Q structure.

A more complete description of these concepts and of the important application below can be found in [B] and §2 [M6]. See also the more general treatment in Chapter VI [M2].

Theorem 1.3. (Beck). If X is a perfect infinite loop space, then X has a Q structure. If (X,ρ) is a connected Q algebra, then X is an infinite loop space.

Proof. Assume $\{X_k\}$ satisfies $\Omega X_k = X_{k-1}$ and $X_0 = X$. Define $\rho\colon QX \to X$ to be the limit of $\rho_N\colon \Omega^N \Sigma^N X_0 = \Omega^N \Sigma^N \Omega^N X_N \xrightarrow{\Omega^N \varepsilon_N} \Omega^N X_N = X_0$. The verification that (X,ρ) is a Q algebra is standard [M6].

Conversely assume that (X,ρ) is a Q algebra. Define the simplicial Ω spectrum $B_* = B(\Sigma^\infty, Q, X)_*$ by

$$B_q = \Sigma^\infty Q^q X \quad \text{and}$$

$$(1.4) \qquad \partial_i = \begin{cases} \varepsilon \Sigma^\infty Q^{q-1} & \text{if } i = 0 \\ \Sigma^\infty Q^{i-1} \mu Q^j & \text{if } 0 < i < q \text{ and } i + j = q - 1 \\ \Sigma^\infty Q^{q-1} \rho & \text{if } i = q \end{cases}$$

$$s_i = \Sigma^\infty Q^i \eta Q^j \text{ for } i + j = q.$$

The realization $||B_*||$ is the Ω spectrum $\{X_k\}$ with X_k defined by

$$\coprod_{q \geq 0} \Delta^q \times Q\, \Sigma^k Q^q X/\sim$$

where $\sim$ is the standard equivalence relation $(u,\partial_i x) \sim (\delta_i u, x)$ and $(u, s_i x) \sim (\sigma_i u, x)$. By §12 [M6], this is indeed an Ω spectrum. Furthermore if X is connected then the inclusion of $X = \Delta^\circ \times Q^\circ X$ into $\coprod_q \Delta^q \times Q^{q+1}X/\sim = X_o = \Omega^\infty ||B^*||$ is a strong deformation retraction by Theorem 9.10 [M6].

The fact that many infinite loop spaces do not have a (strict) Q algebra structure has necessitated the introduction of various infinite loop space machines, such as those by Boardman and Vogt, May and Segal. Further generalizations by the second author are discussed in the next section. On the other hand, if $\{X_n\}$ is an Ω spectrum, then $\{\lim\limits_{\rightarrow} \Omega^N X_{n+N}\}$ is a perfect Ω spectrum equivalent to $\{X_n\}$ up to weak homotopy [M5]. Thus we may replace infinite loop spaces by Q algebras which contain the same homotopy theoretical information.

§2. Transfer and Q_k Spaces

We now consider the definition of transfer for a representable homotopy functor and discuss its relationship with infinite loop space structures. See also [L1] and [R].

Definition 2.1. We say that the functor $h(\_) = [\_,X]$ admits a transfer for finite coverings if given a covering $p: \tilde{Y} \to Y$, there is a map of pointed sets $\tau_p: [\tilde{Y},X] \to [Y,X]$ such that

1) τ is natural with respect to pullbacks
2) If $id: Y \to Y$ is the identity covering, then $\tau_{id} = id$
3) Given a composition of coverings $\tilde{\tilde{Y}} \xrightarrow{p_2} \tilde{Y} \xrightarrow{p_1} Y$, then $\tau_{p_1 \circ p_2} = \tau_{p_1} \circ \tau_{p_2}$
4) Given the covering $p = id \coprod id: X \coprod X \to X$, then $\tau_p[id \coprod *] = [id] = \tau_p[* \coprod id]$, where $\coprod$ means disjoint union and $*$ denotes the constant map $X \to *$, the basepoint of X.

With this definition, one can immediately deduce

Proposition 2.2. If the functor $h(\_) = [\_,X]$ admits a transfer, then h takes on values in the category of abelian monoids. [E, pp. 12-13], [L1 , pp. 54-62], [M4].

Remark: This proposition implies that X is a homotopy associative, homotopy commutative H space.

A generalization up to homotopy of Q structures on spaces has been developed by Lada in [L2] and may be summarized by the following definition and theorem.

Definition 2.3. A space X is a Q_k space if there is a family of homotopies

$$h_q: I^q \times Q^{q+1}X \to X \quad \text{for } q < k$$

such that

$$h_q(t_1,\dots,t_q,z) = h_{q-1} \circ (1 \times Q^{j-1}\mu Q^{q-j})(t_1,\dots,\hat{t}_j,\dots,t_q,z) \text{ if } t_j=0,$$

and $h_q(t_1,\dots,t_q,z) = h_{j-1} \circ (1 \times Q^j h_{q-j})(t_1,\dots,\hat{t}_j,\dots,t_q,z)$ if $t_j=1$,

and $h_o \circ \eta = \mathrm{id}: X \to X$.

Note that $h_o: QX \to X$ is a retraction and that the homotopy $h_1: I \times Q^2X \to X$ requires only that $\rho\mu$ be homotopic to $\rho Q\rho$ where $h_o = \rho$.

Theorem 2.4. A connected space X is an infinite loop space if and only if X has a Q_∞ structure, i.e., a Q_k structure for all k.

With the above definitions in hand, we are now able to discuss the relationship between transfer and infinite loop space structures. The following theorem has been proven by a number of authors, [E], [KP], [L1], [M4], [R].

Theorem 2.5. The functor h(_) = [_,X] admits a transfer if and only if X is a Q_2 space.

For this reason Madsen calls a Q_2 space a transfer space. Thus the transfer conjecture can be reformulated as follows.

Conjecture′ Every Q_2 structure on X extends to a Q_∞ structure.

It is when the conjecture is stated in this form that it appears unlikely to be true. To find a counterexample, all that one needs is a space X that is not an infinite loop space and yet possesses a Q_2 structure. The remainder of this section is occupied with a sketch of the main ideas in the proof of Theorem 2.5.

Let $W\Sigma_n$ be the normalized Milnor construction for Σ_n, the symmetric group on n symbols. We may regard QX as $\coprod(W\Sigma_n \times X^n)/\sim$ by the results of the preprint version of [DL] and [M7, §4]. If $[\_,X]$ admits a transfer, a Q_2 structure for X may be defined by the following argument; see [L1] for details. Consider the n-fold covering $p_n: W\Sigma_n \times X^n \times F_n \to W\Sigma_n \times X^n$ where $F_n = \{1,\ldots,n\}$, and $p_n = \text{id}$ on each part of the union. Define maps $f_n: W\Sigma_n \times X^n \times F_n \to X$ by projection of a tuple indexed by $i \in F_n$ onto the ith coordinate in X^n. One may then carefully choose equivalent elements of $\tau_{p_n}[f_n]$ to serve as building blocks of $h_o: QX \to X$.

To construct the homotopy $h_1: I \times Q^2X \to X$, consider the composition

$$W\Sigma_k \times W\Sigma_{j_1} \times X^{j_1} \times \ldots \times W\Sigma_{j_k} \times X^{j_k} \times F_j \xrightarrow{\coprod_{i=1}^{k} 1 \times p_{j_i}}$$

$$W\Sigma_k \times W\Sigma_{j_1} \times X^{j_1} \times \ldots \times W\Sigma_{j_k} \times X^{j_k} \times F_k \xrightarrow{p_k}$$

$$W\Sigma_k \times W\Sigma_{j_1} \times X^{j_1} \times \ldots \times W\Sigma_{j_k} \times X^{j_k} \quad \text{where } j = \sum_{i=1}^{k} j_i.$$ One then computes transfer of f_j through the composition and compares the answer with the result from applying the naturality property of transfer to the pullback diagram

$$\begin{array}{ccc} W\Sigma_k \times W\Sigma_{j_1} \times \ldots \times W\Sigma_{j_k} \times X^j \times F_j & \xrightarrow{\overline{\gamma}} & W\Sigma_j \times X^j \times F_j \\ \downarrow & & \downarrow \\ W\Sigma_k \times W\Sigma_{j_1} \times \ldots \times W\Sigma_{j_k} \times X^j & \xrightarrow{\gamma} & W\Sigma_j \times X^j \end{array} \quad ;$$

here γ is induced by a generalized wreath product. The composition property of transfer will then yield the requisite homotopy.

To see that a Q_2 structure implies the existence of a transfer map, we sketch Kahn and Priddy's approach. Let $p: \tilde{Y} \to Y$ be an n-fold covering and $P(\tilde{Y})$ the associated principal Σ_n covering. One then composes the obvious map $Y \to P(\tilde{Y}) \times_{\Sigma_n} \tilde{Y}^n$ with the classifying map $P(\tilde{Y}) \to W\Sigma_n$ to obtain a map $\bar{p}:Y \to W\Sigma_n \times_{\Sigma_n} \tilde{Y}^n$. If $f:\tilde{Y} \to X$ is a map, the transfer of f may be represented by the composition

$$Y \longrightarrow W\Sigma_n \times_{\Sigma_n} \tilde{Y}^n \xrightarrow{1 \times f^n} W\Sigma_n \times_{\Sigma_n} X^n \xrightarrow{h_o} X$$

where h_o is the Dyer Lashof map induced by $h_o:QX \to X$ [DL]. Properties 1, 2 and 4 of transfer may be readily deduced from this construction. To verify the composition property, property 3, let $q:\tilde{\tilde{Y}} \to Y$ be an m-fold covering. One then shows that $\overline{p \circ q}$ is the compostion

$$Y \xrightarrow{\bar{p}} W\Sigma_n \times_{\Sigma_n} \tilde{Y}^n \xrightarrow{1 \times \bar{q}^n} W\Sigma_n \times_{\Sigma_n} (W\Sigma_m \times_{\Sigma_m} \tilde{\tilde{Y}}^m)^n \quad \text{and that}$$

$$W\Sigma_n \times_{\Sigma_n} (W\Sigma_m \times_{\Sigma_m} \tilde{\tilde{Y}}^m)^n = W(\Sigma_n \int \Sigma_m) \times_{\Sigma_n \int \Sigma_m} \tilde{\tilde{Y}}^{nm}$$

where $\Sigma_n \int \Sigma_m \subset \Sigma_{mn}$ is the wreath product.

§3. Q_k maps induce Q_k spaces

Assume that (X,ρ) and (K,ϕ) are Q algebras. If $f:X \to K$ commutes with the algebra structure, i.e. if $\phi Qf = f\rho : QX \to K$, then it is easy to see that f extends to a map of simplicial spectra $f_*:B(\sum^\infty,Q,X)_* \to B(\sum^\infty,Q,K)_*$, and thus to their realizations. In this case f is a stable, or infinite loop map. Also in this case it is easy to verify that the fiber in $E \xrightarrow{\pi} X \xrightarrow{f} K$ is a Q algebra.

__Definition 3.1.__ A map $f:(X,\rho) \to (K,\phi)$ between Q algebras is called a Q_k map if there is a collection of maps

$$f_q:\Delta^q \times Q^qX \to K \quad \text{for } q \leq k$$

with $f_o = f$ and satisfying

$$f_q(t_o,\ldots,t_q,z) = \begin{cases} \phi Qf_{q-1}(t_1,\ldots,t_q,z) & \text{if } t_o=0 \\ f_{q-1}(t_o,\ldots,\hat{t}_i,\ldots,t_q,Q^{i-1}\mu Q^j z) & \text{if } t_i=0,\ i+j=q-1 \\ f_{q-1}(t_o,\ldots,t_{q-1},Q^{q-1}\rho z) & \text{if } t_q=0 \end{cases}$$

where $Q^{i-1}\mu Q^j:Q^{i-1}Q^2Q^jX \quad Q^{i-1}QQ^jX$.

If f is a Q_k map for all k, then we say that f is a strong homotopy Q (or shQ) map. Such maps can be lifted to infinite loop maps [L2]. If f is a Q_1 map, then we only require that $\phi Qf \simeq f\rho$. Such maps have been called transfer commuting maps [MST], [M4]. We will be most interested in Q_2 maps in this paper.

Let $\underline{F_k} = \coprod_{q \leq k} \Delta^q \times \Sigma^\infty Q^qX/\sim$ be a filtration of $||B_*||$. Let $\underline{K}$ be an Ω spectrum and let (K,ϕ) be the Q algebra constructed in Theorem 1.3. By a map $G:\underline{F_k} \to \underline{K}$ we mean a sequence of compositions for $q \leq k$ $\Delta^q \times \Sigma^\infty Q^qX \xrightarrow{i} \Sigma^\infty(\Delta^q \times Q^qX) \xrightarrow{g_q} \underline{K}$ where g_q is a map of spectra satisfying the coherence relations induced by (1.4).

The adjoint of $g_q : \Sigma^\infty(\Delta^q \times Q^q X) \longrightarrow \underline{K}$ is the map f_q given by the composition $\Delta^q \times Q^q X \xrightarrow{\eta} \Omega^\infty\Sigma^\infty(\Delta^q \times Q^q X) \xrightarrow{\Omega^\infty g_q} \Omega^\infty \underline{K} = K$. The compatibility conditions of (1.4) for g_q translate to the compatibility relations (3.1) which imply that f is a Q_k map. Conversely if $f:(X,\rho) \longrightarrow (K,\phi)$ is a Q_k map, then we may construct a map $G:\underline{F_k} \to \underline{K}$ by defining g_q to be the composition

$$\Sigma^\infty(\Delta^q \times Q^q X) \xrightarrow{\Sigma^\infty f_q} \Sigma^\infty K = \Sigma^\infty \Omega^\infty \underline{K} \xrightarrow{\varepsilon} K.$$

Compare Chapter IV [M2]. Thus we have proven the following.

<u>Theorem 3.2</u>. A map $f:(X,\rho) \longrightarrow (K,\phi)$ is a Q_k map if and only if there is a map $G:\underline{F_k} \longrightarrow \underline{K}$ such that f is the adjoint of $g_o : \Sigma^\infty X \longrightarrow \underline{K}$.

The Q_k spaces and maps are, of course, analogues of Stasheff's higher homotopy associative (A_k) spaces and maps. We now prove the analogue of his Theorem 6.1 [S4].

<u>Theorem 3.3</u>. Let $f:(X,\rho) \to (K,\phi)$ be a Q_k map between connected Q algebras. Let E be the fiber space over X induced by f from the path loop fibration over K. Then E is a Q_k space.

Proof: The Q_k structure on a space E involves maps $I^q \times Q^{q+1}E \to E$. It will be helpful in the proof if we translate the definition of Q_k maps from the simplicial to the cubical theory. Indeed one can check that f is a Q_k map if there are maps $f_q : I^q \times Q^q X \to K$ such that $f_o = f$ and

$$f_q(t_1,\ldots,t_q,z) = \begin{cases} \phi Q f_{q-1}(t_2,\ldots,t_q,z) & \text{if } t_1=0 \\ f_{q-1}(t_1,\ldots,\hat{t}_i,\ldots,t_q,Q^{i-1}\mu Q^j z) & \text{if } t_i=0,\ i+j=q \\ f_{i-1}(t_1,\ldots,t_{i-1},Q^{i-1}\rho^j z) & \text{if } t_i=1,\ i+j=q \end{cases}$$

where $\rho^i = \rho Q \rho \ldots Q^i \rho$.

We present the details of the proof of this theorem only for the case when $k=2$ as that is all that is required in subsequent sections. The existence of a Q_2 structure on f implies that we have homotopies

$$f_1: I \times QX \to K \quad \text{and} \quad f_2: I^2 \times Q^2X \to K$$

such that $f_1(0,z) = \phi Qf(z)$, $f_1(1,z) = f\rho(z)$, $f_2(0,t,z) = \phi Qf_1(t,z)$, $f_2(t,0,z) = f_1(t,\mu z)$, $f_2(1,t,z) = f\rho Q\rho(z)$, $f_2(t,1,z) = f_1(t,Q\rho z)$.

We regard $E \subset X \times PK$ to be defined by $\{(x,\lambda) \mid f(x) = \lambda(1)\}$. Note that by [M6, p. 6] we have $\rho_{PK}: QPK \to PK$ defined by $\rho_{PK}(z,\lambda_1,\ldots,\lambda_n)(s) = \rho_K(z,\lambda_1(s),\ldots,\lambda_n(s))$ where $(z,\lambda_1,\ldots,\lambda_n) \in W\Sigma_n \times (PK)^n$.

The Q structure map for E, $\rho_E: QE \to E$, may now be defined by

$$\rho_E(z,x_1,\lambda_1,\ldots,x_n,\lambda_n) =$$

$$(\rho(z,x_1,\ldots,x_n),\ \rho_{PK}(z,\lambda_1,\ldots,\lambda_n)(\cdot) + f_1(\cdot,z,x_1,\ldots,x_n)).$$

To see that the path addition in the definition is well defined, note that $\rho_{PK}(z,\lambda_1,\ldots,\lambda_n)(1) = \phi(z,\lambda_1(1),\ldots,\lambda_n(1)) = \phi(z,f(x_1),\ldots,f(x_n)) = \phi Qf(z,x_1,\ldots,x_n) = f_1(0,z,x_1,\ldots,x_n)$. The remaining step is to verify that $\rho_E(z,x_1,\lambda_1,\ldots,x_n,\lambda_n) \in E$. We have $[\rho_{PK}(z,\lambda_1,\ldots,\lambda_n) + f_1(z,x_1,\ldots,x_n)](1) = f_1(1,z,x_1,\ldots,x_n) = f \circ \rho(z,x_1,\ldots,x_n)$ and we are done.

At this point we have shown that a Q_1 structure on f yields a Q_1 structure on E. It will be useful to observe that the second coordinate of ρ_E may be described by

$$g(t) = \begin{cases} \rho_{PK}(2t) & t \leq 1/2 \\ f_1(2t-1) & t \geq 1/2 \end{cases}$$

We now turn our attention to the Q_2 structure on E. A point in Q^2E lies in some subspace of the form $W\sum_k \times (\coprod_n W\sum_n \times E^n/\sim)^k$. Denote such a point by $(w,z_1,\ldots,z_k,\overline{x}_{\alpha_1},\overline{\lambda}_{\alpha_1},\ldots,\overline{x}_{\alpha_k},\overline{\lambda}_{\alpha_k})$ where $\overline{x}_{\alpha_j}$ and $\overline{\lambda}_{\alpha_j}$ are i_j tuples indexed by the integers $\sum_{n=1}^{j-1} i_n + \ell$, $\ell=1,\ldots,i_j$ and the pth component of each $\overline{x}_{\alpha_j}$ and $\overline{\lambda}_{\alpha_j}$ satisfy $f(\overline{x}_{\alpha_j})_p = (\overline{\lambda}_{\alpha_j})_p(1)$. We also have $z_j \in W\sum_{i_j}$. Let us now define $\rho_E^1 : I \times Q^2E \to E$ by

$$\rho_E^1(t_2,w,z_1,\ldots,z_k,\overline{x}_{\alpha_1},\overline{\lambda}_{\alpha_1},\ldots,\overline{x}_{\alpha_k},\overline{\lambda}_{\alpha_k})(t_1) = (\rho_x^1(w,z_1,\ldots,z_k,$$

$$\overline{x}_{\alpha_1},\ldots,\overline{x}_{\alpha_k}),\ g_1(t_1,t_2,w,z_1,\ldots,z_k,\overline{x}_{\alpha_1},\overline{\lambda}_{\alpha_1},\ldots,\overline{x}_{\alpha_k},\overline{\lambda}_{\alpha_k}))$$

where

$$g_1(t_1,t_2) = \begin{cases} \rho_{PK}^1\left(\dfrac{4t_1}{2-t_2}\right) & 0 \leq t_1 \leq \dfrac{2-t_2}{4} \\ \rho_{PK} \circ Qf_1(4t_1 - (2-t_2)) & \dfrac{2-t_2}{4} \leq t_1 \leq \dfrac{1}{2} \\ f_2(2t_1-1,t_2) & t_1 \geq \dfrac{1}{2} \end{cases}$$

We regard the first coordinate, t_1, as the path parameter. The following diagram should clarify the construction of g_1.

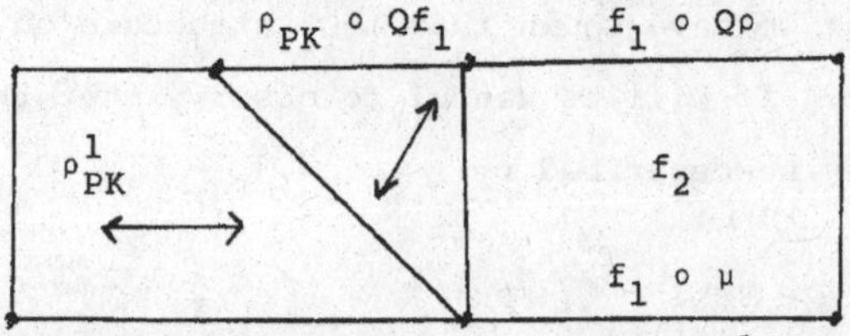

Note that it is easy to see that the image of ρ^1_E lies in E. To see that ρ^1_E does indeed define a Q_2 structure for E_1 we have to verify the compatibility requirements set out in Definition 2.3. We need only check the second coordinate of ρ^1_E, g_1, as compatibility is obviously satisfied in the first.

Suppose $t_2 = 0$. Then

$$g_1(t_1,0) = \begin{cases} \rho^1_{PK}(2t_1) & 0 \le t_1 \le \frac{1}{2} \\ \rho_{PK} \circ Qf_1(4t_1-1) & t_1 = \frac{1}{2} \\ f_2(2t_1-1,0) & t_1 \ge \frac{1}{2} \end{cases}$$

$$= \begin{cases} \rho_{PK} \circ \mu(2t_1) & 0 \le t_1 \le \frac{1}{2} \\ f_1(2t_1-1) \circ \mu & \frac{1}{2} \le t_1 \le 1 \end{cases}$$

$= g \circ \mu$. Thus $\rho^1_E|_0 = \rho_E \circ \mu$.

Now let $t_2 = 1$. Then

$$g_1(t_1,1) = \begin{cases} \rho^1_{PK}(4t_1) & 0 \le t_1 \le \frac{1}{4} \\ \rho_{PK} \circ Qf_1(4t_1-1) & \frac{1}{4} \le t_1 \le \frac{1}{2} \\ f_2(2t_1-1,1) & t_1 \ge \frac{1}{2} \end{cases}$$

$= (\rho^1_{PK} + \rho_{PK} \circ Qf_1) + f_1 \circ Q\rho_X$. A straightforward calculation shows that this path is indeed the second coordinate of $\rho_E \circ Q\rho_E$.

Remark: In general assume that $f_q: I^q \times Q^q X \to K$, $q \le k$ is a Q_k map. It is possible to define a Q_k structure on E, $\rho_E^q: I^q \times Q^{q+1}E \to E$ by $\rho_E^q = (\rho_X^q, g_q)$ where

$$g_q(t_1,\ldots,t_{q+1}) = \begin{cases} \rho_{PK}^{j-1} \circ Q^j f_{q-j+1}\left(\dfrac{2^{j+1}t_1 - \prod_{i=2}^{j+1}(2-t_i)}{\prod_{i=2}^{j}(2-t_i)}, t_{j+2},\ldots,t_{q+1}\right) & \text{if } \dfrac{\prod_{i=2}^{j+1}(2-t_i)}{2^{j+1}} \le t_1 \le \dfrac{\prod_{i=2}^{j}(2-t_i)}{2^j} \\ \rho_{PK}^q\left(\dfrac{2^{q+1}t_1}{\prod_{i=2}^{q+1}(2-t_i)}\right) & \text{if } t_1 \le \dfrac{\prod_{i=2}^{q+1}(2-t_i)}{2^{q+1}} \end{cases}$$

We let $j=0,1,\ldots,q$ and define $\prod_{i=2}^{1}(2-t_i) = 1 = \prod_{i=2}^{0}(2-t_i)$. The verification of the coherence formulas is quite tedious and in the interests of good taste will be omitted.

§4. Stable homology and the delooping spectral sequence

Let h_* be a connected homology theory, i.e., $h_q(X) = 0$ for $q \leq 0$. If $\underline{X} = \{X_k\}$ is an Ω spectrum, then we define the stable homology by

$$h_q^S(\underline{X}) = \lim_{\to} h_{q+k}(X_k)$$
$$= h_{q+k}(X_k) \quad \text{for } k > q.$$

If (X,ρ) is a Q algebra, then we define $h_q^S(X,\rho) = h_q^S(||B_*||)$.

Proposition 4.1. For a connected space X there is a suspension isomorphism

$$\Sigma: h_q(X) \approx h_q{}^S(\Sigma^\infty X).$$

Proof: Note that $\eta: \Sigma^n X \to Q\Sigma^n X$ is a $2n-1$ equivalence. Thus

$$h_q^S(\Sigma^\infty X) = h_{q+n}(Q\Sigma^n X) \quad \text{for } n > q$$
$$= h_{q+n}(\Sigma^n X) = h_q(X).$$

In an analogous way we may define the stable cohomology functor h_S^*. The usual duality and universal coefficient theorems hold.

Example 4.2. Let $h_q(\ ) = H_q(\ ;Z/p)$. If X is connected, then $H_q(QX; Z/p) = A\ T\ H_*(X;Z/p)$, the free commutative algebra on the free admissible Dyer Lashof module on $\hat{H}_*(X,Z/p)$ (see p. 42. [M7] and [DL]).

$$H_*^S(QX;Z/p) = H_*^S(\Sigma^\infty X;Z/p)$$
$$= \tilde{H}_*(X;Z/p).$$

Example 4.3. Let $\underline{K}(Z/p)$ be the perfect Ω spectrum $\{K(Z/p,n)\}$. Then $H_S^*(\underline{K}(Z/p),Z/p) = \lim H^{*+n}(K(Z/p,n),Z/p) \cong A(p)$, the mod p Steenrod algebra.

On the other hand $H^q(K(Z/p,0);Z/p) = Z/p[Z/p]$ if $q=0$ and 0 otherwise.

Example 4.4. Let $bu = \{BU[2n,\dots,\infty]\}$ be the Ω spectrum for connected K theory. Then $H^*(BU;Z/2) = Z/2[c_1,c_2,\dots]$ where c_k is the Chern class of degree 2k. Adams [A1] [AP] has computed that $H_S^*(bu;Z/2) \approx \Sigma^2 A/A(Sq^1,Sq^3)$. Results on the localized unstable and stable theories $[BU_{(p)},BU_{(p)}]$ and $[bu_{(p)},bu_{(p)}]$ have been obtained in [MST].

Let Y_* be a simplicial space and let $\partial \subset Y_q$ generically denote the subspace of degeneracies $\partial = \bigcup_i \operatorname{Im} s_i(Y_{q-1})$. We will assume that the inclusion is a cofibration. For a homology theory h_*, there is an associated spectral sequence with $E^1_{s,t} \approx h_s(Y_t,\partial)$ which converges to $h_*(||Y_*||)$ [S1]. This construction can be generalized to simplicial spectra.

Theorem 4.5. Let (X,ρ) be a Q algebra and h_* a connective homology theory. Then there is a first quadrant spectral sequence with $E^1_{s,t} = E^1_{s,t}(X,h_*) \approx h_t(Q^sX,\partial)$ which converges to $h^S_*(||B_*||) = h^S(X,\rho)$. Moreover the differential d^1 is induced by $\sum_{i=1}^{s}(-1)^i\zeta_{i*}$ where $\zeta: Q^sX \to Q^{s-1}X$ is given by

$$\zeta_i = \begin{cases} Q^{i-1}\mu Q^j & \text{for } 1 \le i < s,\ i+j = s-1 \\ Q^{s-1}\rho & \text{for } i=s. \end{cases}$$

Proof: Recall from section 3 that $\underline{F}_s = \coprod_{q \le s} \Delta^q \times \Sigma^\infty Q^q X/\sim$ is a filtration of $||B_*||$. Thus there is an exact couple

$$(4.6)\qquad \begin{array}{ccc} h_*^S(\underline{F}_{s-1}) & \xrightarrow{\ i_*\ } & h_*^S(\underline{F}_s) \\ & \nwarrow \quad \swarrow & \\ & h_*^S(\underline{F}_s,\underline{F}_{s-1}) & \end{array}$$

with an associated spectral sequence having $E^1_{s,t} = h^S_{s+t}(\underline{F}_s,\underline{F}_{s-1})$ and converging to $h^S_*(||B_*||)$. Moreover

$$\begin{aligned} h^S_{s+t}(\underline{F}_s,\underline{F}_{s-1}) &= \lim h_{s+t+n}(\textstyle\coprod_{q \le s} \Delta^q \times Q\Sigma^n Q^q X, \_) \\ &= h_{s+t}(\textstyle\coprod_{q \le s} \Delta^q \times Q^q X, \_) \\ &= h_t(Q^s X, \partial). \end{aligned}$$

(compare [S1] and [M1]).

The differential $d^1 : h^S_{s+t}(\underline{F}_s,\underline{F}_{s-1}) \to h^S_{s+t}(\underline{F}_{s-1},\underline{F}_{s-2})$ is induced from the alternating sum of the maps $\partial_i : \Sigma^\infty Q^s X \to \Sigma^\infty Q^{s-1} X$ defined in (1.4). If $i > 0$ then $\partial_i = \Sigma^\infty \zeta_i$. If $i=0$ then $\partial_0 = \varepsilon\Sigma^\infty Q^{s-1} : \Sigma^\infty Q^s X \to \Sigma^\infty Q^{s-1} X$ can be seen to induce the 0 map on $h^S_*(\Sigma^\infty Q^s X)/(\eta Q^{s-1})_* h^S_*(\Sigma^\infty Q^{s-1} X)$, and thus on $E^1_{s,t} = h^S_{s+t}(\underline{F}_s,\underline{F}_{s-1}) \cong h_t(Q^s X,\partial)$ (compare p. 110-112 [M6]). Thus there is no ζ_o needed in the formula.

We will write $E^r_{s,t}(X,\Lambda)$ for $E^r_{s,t}(X,H_*(\ ;\Lambda))$. For $\Lambda = Z/p$ we note that $d^1 : E^1_{t,1} \to E^1_{t,o}$ corresponds to $\rho_* : H_t(QX,\eta X;Z/p) \to H_t(X;Z/p)$. An element on the left is a formal polynomial in formal Dyer Lashof operations on classes of $H_*(X;Z/p)$. The differential is evaluation. For example if $x,y \in H_1(X;Z/2)$ and $Q^2 z \in H_3(X;Z/2)$, then $d^1(Q^3[xy]\cdot Q^6[Q^2 z]) = (\Sigma Q^{3-i} x Q^i y)(Q^6 Q^2 z) = (x^2 Q^2 y + Q^2 x y^2) Q^5 Q^3 z$ using the Cartan formula, unstability and the Adem relation $Q^6 Q^2 = Q^5 Q^3$.

The cohomology spectral sequence $E_1^{s,t}(X,h^*) = h^t(Q^s X,\partial)$ converges to $h_S^*(X)$. This is related to the homology spectral sequence by the usual duality theorem and the universal coefficient theorems if $h^* = H^*(;\Lambda)$.

The existence of such a spectral sequence was noted by P. May ([M6], p. 155) and D. W. Anderson [A2]. Using completely different methods, Haynes Miller defined a delooping spectral sequence and computed $E^2_{s,t}(X;Z/2)$ [M8]. Miller was able to compute the spectral sequence in certain cases and give some applications. We had discovered this spectral sequence independently after realizing that a certain computation was a d^3 in some spectral sequence and then identifying the spectral sequence.

In a forthcoming paper we will show that the spectral sequences above are equivalent. We will also describe $E^2(X;Z/p)$ and do a number of computations. We expect that there will be many applications of this spectral sequence to infinite loop space theory and stable homology theory.

§5. Cycles and Q_k maps

Let X be a Q algebra. The Eilenberg Moore spectral sequence has $E_2^{*,*}$ equal to a functor of $H^*(X;\Lambda)$ as a Λ coalgebra and converges to $H^*(BX;\Lambda)$. If $x \in H^t(X;\Lambda)$ is represented by a map $f: X \to K(\Lambda,t)$, then f is a loop map, i.e., $f \simeq \Omega g$ for $g: BX \to K(\Lambda,t+1)$, if and only if x is an infinite cycle, i.e., it represents a class in $H^*(BX;\Lambda)$. Moreover, x is a d_k cycle if and only if f is an A_k map [S4].

Similarly, f is an infinite loop map if and only if x is an infinite cycle in the Miller delooping spectral sequence. For in that case, x survives to a stable class $y \in H_S^t(X;\Lambda)$. We prove that x is a d_k cycle if and only if f is a Q_k map.

<u>Theorem 5.1</u>. Let $x \in h^t(X) = [X,K_t] = E_1^{o,t}$ be represented by a map $f: X \to K_t$. Then f is a Q_k map if and only if x is a k-cycle.

Proof: The class x is a k-cycle if and only if there is a class $x_k \in h_S^t(\underline{F_k})$ such that $(i^k)^* x_k = x \in h_S^t(\underline{F_o}) = h^t(X)$ (see 4.6 and Ch. XI [Ml]). Thus x is a k cycle if and only if there is a map $G_k: \underline{F_k} \to \underline{K_t}$ such that $i^k G_k g$ is adjoint to f. This means that f is a Q_k map by Theorem 3.2.

<u>Corollary 5.2</u>. Let $f: X \to K_t$ be a map between Q algebras. Then f is homotopic to an infinite loop map if and only if $[f] \in E_1^{o,t} = [X,K_t]$ is an infinite cycle. Furthermore [f] survives to $E_2^{o,t}$ if and only if f is transfer commuting, i.e. a Q_1 map.

If X is an infinite loop space then the Eilenberg Moore spectral sequence $E_r^{*,*} \Longrightarrow H^*(BX)$ is in the category of abelian Hopf algebras. This implies that if $d_r(x) \neq 0$ then $r = p^{k+1}-1$ or $2p^k-1$ (compare Theorem 2.4 [K]). It can be shown that if $d_r(x) \neq 0$ for r as above in the Eilenberg Moore spectral sequence, then $d_k(x) \neq 0$ in the Miller spectral sequence. Thus if there is an obstruction to $f: X \to K(Z/p,n)$ being an $A_p k$ map, there is also an obstruction to f being a Q_k map.

In Theorem A [MST] it was shown that every transfer commuting endomorphism of BSO was homotopic to a stable, i.e., infinite loop, map. This result implies that $E_2^{o,t} = E_\infty^{o,t}$ in the spectral sequence converging to $[bso_{(p)}, bso_{(p)}]$.

Assume now that $[f] \in h^t(X) = [X,K_t]$ is a 2 cycle and let $E \xrightarrow{\Pi} X \xrightarrow{f} K_t$ be the induced fibration. Then by Theorems 3.3 and 2.5, E is a transfer space. Thus to construct a transfer counterexample, we need to be able to compute differentials in the spectral sequence.

Let X be the stable 2 stage Postnikov system with k invariant $P^{p^2n}P^{pn}P^{n}\iota: K(Z/p,2n+1) \to K(Z/p,2np^3+1)$ for $n \not\equiv -1(p)$. Using techniques developed in [K], we can show that there is a class $\psi \in H^{2(n+1)p^3-2}(X;Z/p)$ which does indeed represent a 2 cycle but not a 3 cycle. Moreover an elementary Postnikov system argument proves that the fiber induced by ψ is not an infinite loop space. Indeed ψ is the p^3 transpotence of the fundamental class of BX. Thus this fiber is a counterexample to the transfer conjecture.

Unfortunately, the proof of this fact requires a fairly complete description of $E^2_{s,t}(X)$ and $E^\infty_{s,t}(X)$. Results of Madsen and Snaith [M4] concerning the transfer conjecture were recently brought to our attention. We extend their results to find a more "real life" counterexample. To establish this counterexample we need much less information about the E^2 term and no explicit knowledge of E^∞. Instead we need p torsion results on $E^r_{s,t}(X;Z_{(p)})$.

§6. Partial computation of E^2

Assume that X is a connected Q algebra of finite type. Since X is an infinite loop space, $H_*(X;Z/p)$ is an abelian Hopf algebra. Borel's theorem implies that $H_*(X;Z/p)$ is a free commutative algebra modulo relations of the form y^{p^k}. To simplify the following arguments we will make the strong assumption that $H_*(X;Z/p)$ is the free algebra on a Z/p module M with basis $\{y_j\}$. The examples we use to construct the counterexample are X = BU and X = BSU, which satisfy this hypothesis.

Definition 6.1. For a graded connected Z/p module N with basis $\{x_j\}$, let AN be the underlying module of the free commutative Z/p algebra on N. Let TN denote the Z/p module with basis $Q^I x_j$, where Q^I is an admissible Dyer Lashof operation of excess greater than the degree of x_j.

Theorem 6.2. If $H_*(X;Z/p)$ is the free commutative algebra on a module M, then

$$H_*(Q^sX;Z/p) = (AT)^s AM$$
$$= ATA\ldots TAM.$$

Proof: This follows easily from p. 42 [M7].

Thus $H_t(Q^sX;Z/p)$ is generated by monomials in Dyer Lashof operations on monomials in Dyer Lashof operations on ... monomials in basis elements $\{x_j\}$ of $M \approx QH_*(X;Z/p)$. For $u \in H_t(Q^{s-1}X;Z/p)$ we write $1(u) = Q^{\emptyset}(u)$ for $\eta_*(u)$ in $H_t(Q^sX;Z/p)$.

Example 6.3. $\alpha = Q^6\{Q^2(x)1(y)\}Q^7Q^4\{Q^1(z)\}$ is an element of $H_{23}(Q^2X;Z/2)$ where $x,y,z \in H_1(X;Z/2)$.

The $E^1_{s,t}$ term of the Miller spectral sequence is the quotient of $(AT)^s AM \approx H_*(Q^sX;Z/p)$ by the image of the degeneracies $s_i: Q^{s-1}X \to Q^sX$ of (1.4). The differential of Theorem 4.5 extends to $d = \sum_{i=0}^{s} (-1)^i d_i$:

$(AT)^sAM \to (AT)^{s-1}AM$ for $s > 0$. Note that the map d_o is given by the composite

$$H_t(Q^sX) \underset{\sim}{\overset{\Sigma}{\to}} H_t^s(\Sigma^\infty Q^sX) \xrightarrow{\sigma_*} H_t^s(\Sigma^\infty Q^{s-1}X) \underset{\sim}{\overset{\Sigma^{-1}}{\to}} H_t(Q^{s-1}X).$$

Thus $(AT)^*AM$ is the unnormalized version of $E^1_{*,*}$ and so the two are chain equivalent.

The complex $(AT)^*AM$ is much too large and complicated to use effectively. Consider the subcomplex $T^*M = \{T^sM\}$. We may write a generator of T^sM as $Q^{I_1}|\ldots|Q^{I_s}|x$ where x is a generator of M and I_j is an admissible sequence of excess greater than $\deg I_{j+1}+\ldots+\deg I_s + \dim x$. The face map d_j removes the jth bar. If $i = 0$ we get 0 unless $Q^{I_1} = Q^{\emptyset} = 1$, in which case the first term is omitted. Thus T^*M is an unstable unnormalized bar construction for the R module $M \approx QH_*(X;Z/p)$.

<u>Theorem 6.4</u>. The inclusion of the subcomplex T^*M into $(AT)^*AM$ is a chain equivalence.

<u>Proof</u>. Our method is to successively contract out the algebra structure of $(AT)^sAM$. First define a filtration of $(AT)^*AM$, $T^sM \approx F^s_o \subset F^s_1 \subset \ldots \subset F^s_s \subset F^s_{s+1} = (AT)^sAM$, by setting $F^s_k = \operatorname{Im}[(AT)^kT^{s-k}M \to (AT)^sAM]$ for $0 \le k \le s$. Recall that $d_i = (AT)^{i-1}\mu(AT)^{s-i}$ where $\mu: ATAT \longrightarrow AT$ is evaluation. Thus $d_i(F^s_k) \subset F^{s-1}_{k-1}$ if $i < k$ and $d_i(F^s_k) \subset F^{s-1}_k$ if $i \ge k$.

We will show that the quotient complex F^*_k/F^*_{k-1} is acyclic for $k \ge 1$. Thus $F^*_{k-1} \longrightarrow F^*_k$ is a chain equivalence and the theorem follows by iteration. Alternatively the filtration gives rise to a spectral sequence converging to $H_*((AT)^*AM)$. We show that $E^2_{*,t} \approx H_*(F^*_t, F^*_{t-1}) = 0$ for $t \ne 0$ and thus $H_*(T^*M) \approx E^2_{*,o} = E^\infty_{*,o} \approx H_*((AT)^*AM)$.

By Definition 6.1. AN is the module with generators of the form $x_1 \ldots x_n$ where the x_i's are generators of N. Also there are generators $l(x_i) = Q^{\emptyset}(x_i)$ in TN. Define a homomorphism $c: AN \longrightarrow ATN$ by $c(x_1 \ldots x_n) = l(x_1) \ldots l(x_n)$. Letting $N = T^{s-k+1}M$ we have an extension

$$c_k = (AT)^{k-1}c: F_k^s \longrightarrow F_k^{s+1}$$

for $k > 0$. Since $\mu(l(x_1) \ldots l(x_n)) = x_1 \ldots x_n$, $d_k c_k(x_1 \ldots x_n) = x_1 \ldots x_n$. M reover $d_i c_k = c_k d_{i-1}$ if $i > k$. Since $d_i(F_k^{s+1}) \subset F_{k-1}^{s+1}$ for $i < k$, c_k extends to a homomorphism

$$c': F_k^s / F_{k-1}^s \longrightarrow F_k^{s+1} / F_{k-1}^{s+1}$$

such that $dc' - c'd = 1$. Thus c' is a contradiction and so F_k^* / F_{k-1}^* is acyclic for $k > 0$ as required.

In an attempt to make the formula $dc' - c'd = 1$ more comprehensible, we carry out the necessary computations for the element $\alpha \in \mathrm{Im}(ATATM) = F_2^2$ of Example 6.3.

$d_o\alpha = 0$

$d_1\alpha = \sum_t (Q^{6-t}Q^2(x)Q^t(y))(Q^7Q^4Q^1(z))$

$= 0$ (Here we use the Cartan formula, excess, and the Adem relations $Q^7Q^4Q^1 = Q^7Q^3Q^2 = 0$.)

$d_2\alpha = Q^6\{(Q^2x)(y)\}Q^7Q^4\{z^2\}$ (Here we use $Q^1z = z^2$.)

$c(\alpha) = Q^6\{1[Q^2(x)]1[1(y)]\}\ Q^7Q^4\{1[Q^1(z)]\}$

$d_o c(\alpha) = 0$

$d_1 c(\alpha) = (\sum_t Q^{6-t}[Q^2(x)]Q^t[1(y)])\ Q^7Q^4[Q^1(z)] \in F_1^2$

$d_2 c(\alpha) = \alpha$

$d_3 c(\alpha) = Q^6\{1[Q^2x]1[y]\}\ Q^7Q^4\{1[z^2]\} = c(d_2\alpha)$.

A standard argument will show that T^sM is chain equivalent to $\tilde{T}^sM$ where $\tilde{T}M = TM/(1M)$ is the normalized version. Thus a generator of $\tilde{T}^sM$ may be written $Q^{I_1}|\ldots|Q^{I_s}|x_j$ where x_j is a generator of M, I_j is admissible and nontrivial ($I_j \neq \emptyset$), and excess of I_j is greater than the degree of $Q^{I_{j+1}}\ldots Q^{I_s}x_j$. This theorem implies that $E^1_{s,t}$ is chain equivalent to $\tilde{T}^sM$. In order to form $E^2_{s,t}$ we must still take into account the Adem relations.

Miller [M9] started with essentially the complex $\tilde{T}^sM$. He described $E^2_{s,t}$ as an unstable Tor functor on the Dyer Lashof algebra and computed $E^2_{s,t}(X,Z/2)$ when the Dyer Lashof action on $H_*(X;Z/2)$ is trivial. Thus the Miller spectral sequence is analogous to an unstable Adams spectral sequence.

Remark 6.5. If $H_*(X;Z/p)$ is not a free algebra, but instead there are relations $y^{p^{k+1}} = 0$ for dim $y = 2n$, then we must add the generators $Q^{I_1}|\dots|Q^{I_{s-1}}|Q_o^k|y$ to the above collection where $Q_o^k = Q^{p^k n}\dots Q^{pn}Q^n$.

This situation will be fully discussed in our forthcoming paper.

§7. p torsion in $E^r_{s,t}$

If X is an infinite loop space, then it is possible to define homology Pontrjagin-Thomas p^{th} power operations

$$\mathcal{P}: H_{2n}(X;Z/p^{k-1}) \to H_{2np}(X;Z/p^k) \to ([M3],[M7])$$

Let β_k be the Bockstein operator associated with

$0 \to Z/p \to Z/p^{k+1} \xrightarrow{r} Z/p^k \to 0$. Then

$$\text{(7.1)} \quad \begin{aligned} &\text{a)} \quad r_*\mathcal{P}(x) = x^p \\ &\text{b)} \quad \beta_k\mathcal{P}(x) = x^{p-1}\beta_{k-1}x \quad \text{if } p > 2 \text{ or } k > 1 \\ &\qquad\qquad\quad = x\beta x + Q^{2n}\beta x \text{ if } p = 2 \text{ and } k = 1. \end{aligned}$$

If all of the higher p torsion of X arises from Pontrjagin products, then X is called Henselian. More precisely let $\tilde{\beta}_k$ be the Bockstein operator associated with $0 \to Z_{(p)} \xrightarrow{p^k} Z_{(p)} \xrightarrow{r} Z/p^k \to 0$. Note that $r\tilde{\beta}_k = \beta_k$ if r is the reduction $Z_{(p)} \to Z/p$. Then X is Henselian if the p^k torsion of $H_*(X;Z_{(p)})$ for $k > 1$ is generated by elements of the form $\beta_k\mathcal{P}\ldots\mathcal{P}(x)$ for $x \in H_*(X;Z/p)$ (compare definition 1.7 [M3]).

<u>Theorem 7.2</u>. If $H_*(X;Z_{(p)})$ has no p^2 torsion, then QX is Henselian at p.

Proof: See p. 63 [M7].

Note that if QX is Henselian, then the mod p reduction of higher torsion is decomposable, unless p=2 and k=2. Since $E^2_{1*}(X;Z_{(p)})$ is a subquotient of $H_*(QX;Z_{(p)})$ in which decomposables have been contracted out, by Theorem 6.4, we may expect that there is no higher torsion in that group. Indeed more is true.

Theorem 7.3. Assume that $H_*(X;Z/p)$ is a polynomial algebra and that $H_*(X;Z_{(p)})$ contains no p^2 torsion. Then $E^r_{s,t}(X;Z_{(p)})$ and $E_r^{s,t}(X;Z_{(p)})$ are Z/p modules for $r \geq 2$ and $s \geq 1$.

Proof: By the universal coefficient theorem and the fact that the homology of a Z/p module is a Z/p module, it suffices to prove that $E^2_{s,t}$ is a Z/p module for $s \geq 1$. From the Bockstein spectral sequence for $H_*(QX)$ (p. 48 [M7]), the infinite factors of $H_*(QX;Z_{(p)})$ are in the image of $\eta: X \to QX$ or arise from formal products of generators of infinite factors of $H_*(X;Z_{(p)})$ such as $1(x)1(y)$. But such elements are either in the image of the degeneracies $Q^i\eta Q^j: Q^{s-1}X \to Q^sX$ or they are decomposables. Moreover if there were an infinite factor in $E^2_{s,t}(X;Z_{(p)})$, then it would reduce non trivially to $E^2_{s,t}(X;Z/p)$. Since $E^2_{s,t}(X;Z/p)$ is the homology of $\tilde{T}^*M$, in which degeneracies and decomposables are divided out, there is a contradiction.

Similarly if $z \in E^2_{s,t}(X;Z_{(p)})$ generated a p^k factor for $k \geq 2$, then it must be degenerate or of the form $\beta_k \mathcal{P}(y)$. If $p > 2$ or $k > 2$ then by (7.1), z is represented by a degenerate or a decomposable in $E^2(X;Z/p)$ and so we reach a contradiction again. Finally if $p=2$ and $k=2$, then the 4 torsion element is represented by $Q^{2n}\beta x$ in $E^2_{s,t}(X;Z/2)$. But this element is in the d^1 image of $Q^{2n}[\beta x]$ modulo terms of lower filtration. Thus $Q^{2n}\beta x$ cannot represent a nonzero element in $E^2_{s,t}(X;Z/2)$, and the proof is complete.

Note that while $E_r^{s,t}(X;Z_{(p)})$ is a Z/p module for $s > 0$, the edge homomorphism

$$E_r^{o,t} \to H^t(X;Z_{(p)})$$

is a monomorphism. Thus the edge term will often have infinite factors.

Corollary 7.4. Assume that $H_*(X;Z/p)$ is a polynomial algebra and that $H_*(X;Z_{(p)})$ has no p^2 torsion. Let $c \in E_2^{o,t}(X;Z_{(p)})$. Then $p^r c$ is an r+1 cycle for all $r \geq 1$.

Proof: Assume that $p^{r-1}c$ is an r cycle and that $d^{r+1}(p^{r-1}c) = y$. Then $d^{r+1}(p^r c) = py = o$.

§8. The counterexample

In this section all spaces will be localized at 2 and $H^*(\ )$ will mean $H^*(\ ;Z_{(2)})$. We will write a class $\zeta \in H^n(X)$ and a representing map $\zeta: X \to K(Z_{(2)},n)$ interchangeably.

Recall that $H_*(BU) \approx Z_{(2)}[a_1,a_2,\ldots]$ as algebras [L3] and so the hypotheses of Corollary 7.4 are satisfied. Also $H^*(BU) \approx Z_{(2)}[c_1,c_2,\ldots$ as algebras where c_k is the Chern class of dimension 2k. Let $S_k \in PH^{2k}(BU) \approx Z_{(2)}$ be the primitive class dual to a_k in the basis of monomials. We may express S_k as the Newton polynomial kc_k + decomposables (Chapter IV [L3]).

<u>Theorem 8.1</u>. For each k, $4S_k: BU \to K(Z_{(2)},2k)$ is a Q_2 map and its induced homotopy fiber E_k is a transfer space.

Proof: Madsen [M4] has shown that $2S_k$ is a transfer commuting or Q_1 map. By Corollary 5.2 $[2S_k]$ represents a nonzero class in $E_2^{o,2k}$. By Corollary 7.4, the class $4S_k$ is a d_2 cycle and so by Theorem 5.1, the map $4S_k$ is a Q_2 map. Finally, Theorem 3.3 implies that E_k is a Q_2 or transfer space.

In summary, $[\ ,E_k]$ is a representable homotopy functor which admits a transfer. Let $\alpha(k)$ be the number of 1's in the diadic expansion of k. It follows from work of Adams [Al] (see also [M4]) that 2^nS_k is a stable class if and only if $n \geq \alpha(k)-1$. If $\alpha(k) \leq 3$ then $4S_k: BU \to K(Z_{(2)},2k)$ may be taken to be an infinite loop map and so $[\ ,E_k]$ extends to a cohomology theory. However if $\alpha(k) \geq 4$, we then will get a counterexample to the transfer conjecture.

<u>Theorem 8.2</u>. The fiber $E = E_{15}$ of $4S_{15}$ has a Q_2 structure which does not extend to a Q_∞ structure. Thus $[\ ,E]$ is a counterexample to the transfer conjecture.

By Proposition 2.2, the Q_2 structure on E determines its H structure. Thus it suffices to show that there is no infinite loop space which is H equivalent to E. In fact, we will show that there is no H space F such that $H_*(\Omega^2 F) \cong H_*(E)$ as algebras.

We first outline the proof. Assume to the contrary that such an F existed. Then we will show that a Postnikov approximation of F fibers over a Postnikov approximation of BSU. Moreover, it will be induced by a map τ: BSU $\to K(Z_{(2)},32)$ with $\Omega^2\tau \simeq 4S_{15}$. However, if t_{16} is the generator of QH^{32}(BSU), then we will show that $\tau = 4t_{16}$ modulo decomposables whereas PH^{32}(BSU) is generated by $8t_{16}$ modulo decomposables. This will imply that τ cannot be chosen to be primitive and that F will not in fact have an H structure.

We first record some classical facts about BSU.

<u>Lemma 8.3</u>. $H^*(BSU) \cong Z_{(2)}[t_2,t_3,\ldots]$ as algebras. If $\sigma^2: QH^{2k+2}(BSU) \to PH^{2k}(BU)$ is the 2-fold loop suspension followed by the identification $\Omega^2 BSU \simeq BU$, then $\sigma^2 t_{k+1} = S_k$. Finally PH^{32}(BSU) is generated by a class γ which equals $8t_{16}$ modulo decomposables.

Proof: These results can be found in the literature (e.g. [L3], [S3]). The first is classical. The second follows from the collapse of the Eilenberg-Moore spectral sequences ${}_1E_2^{**} = \mathrm{Tor}_{H^*(BSU)}(Z_{(2)},Z_{(2)}) \Longrightarrow H^*(\Omega BSU) \cong H^*(SU)$ and ${}_2E_2^{**} = \mathrm{Tor}_{H^*(SU)}(Z_{(2)},Z_{(2)}) \Longrightarrow H^*(\Omega SU) \cong H^*(BU)$. To see the last result recall [L3] that the Newton polynomial $S_{16} = 16c_{16} + 2D + 2c_2^8 + c_1^{16}$ generates PH^{32}(BU) for D a decomposable. If we replace c_k by t_k for $k > 1$ and 0 for $k = 1$ in the above polynomial, then we get a primitive $2\gamma = 16t_{16} + \ldots + 2t_2^8$ in H^{32}(BSU). But γ is primitive and not divisible by 2 and so generates PH^{32}(BSU).

For a simply connected H space X, there is a Postnikov decomposition $X \to \ldots \to X^n \xrightarrow{\pi} X^{n-1} \to \ldots \to X_1 = *$. In particular $\pi: X^n \to X^{n-1}$ is a principle fibration induced by an H map $k^n: X^{n-1} \to K(\pi_n(X), n+1)$. We will assume knowledge of the Postnikov systems for BU and BSU (see §2 [AP]).

Lemma 8.4. Assume that F is an H space such that $H^*(\Omega^2 F) \cong H^*(E)$ as Hopf algebras. Then $F^n = BSU^n$ for $n < 32$.

Proof: Note that $\pi_k(F) \cong \pi_{k-2}(E) \cong \pi_{k-2}(BU) \cong \pi_k(BSU)$ for $k < 32$. Thus $F^4 = F^5 = K(Z_{(2)}, 4)$. Also $F^6 = F^7$ is the 2-stage Postnikov system with k invariant $k: K(Z_{(2)}, 4) \to K(Z_{(2)}, 7)$. But $H^7(K(Z_{(2)}, 4); Z_{(2)}) \cong Z_{(2)}$ is generated by $\tilde{\beta} Sq^2 \iota_4$ where $\tilde{\beta}$ is the Bockstein associated with $0 \to Z_{(2)} \to Z_{(2)} \to Z/2 \to 0$.

Thus the k invariant for F^6 is $v\tilde{\beta}Sq^2\iota$ for $v \in Z_{(2)}$ and so the k invariant for ΩF^6 is $v(\iota_3)^2$. Moreover the k invariant for $\Omega^2 F^6 = E^4$ is 0 and so $\Omega^2 F^6 \simeq K(Z_{(2)}, 2) \times K(Z_{(2)}, 4)$. The H structure on $\Omega^2 F^6$ depends on v. More precisely

$$\Delta \iota_4 = \iota_4 \otimes 1 + v(\iota_2 \otimes \iota_2) + 1 \otimes \iota_4.$$

Since $\Delta c_2 = c_2 \otimes 1 + c_1 \otimes c_1 + 1 \otimes c_2$ in $H^*(BU)$ and so in $H^*(E^4)$, we may assume that $v = 1$. Since the first k invariant for BSU is $\tilde{\beta}Sq^2\iota$ ([AP], [S3]), we have $F^6 = BSU^6$.

Inductively assume that $F^{2n-2} = F^{2n-1} = BSU^{2n-1}$ for $2n < 32$. Then the k invariant k^{2n} for F^{2n} is in $PH^{2n+1}(BSU^{2n-1})$. It is not hard to compute that this group is $Z_{(2)}$ and is generated by a class x with $j^*x = \tilde{\beta}Sq^2\iota_{2n-2} \in H^{2n+1}(K(Z_{(2)}, 2n-2), Z_{(2)})$. Since $\sigma^2(k^{2n})$ is the k invariant for $E^{2n-2} = BU^{2n-2}$, knowledge of the k invariants for BU and BSU implies that $F^{2n} = BSU^{2n}$.

Using these results it follows that F^{32} appears in the following diagram of induced fibrations.

$$\begin{array}{ccccc} & & F^{32} & & \\ & & \downarrow & & \\ K(Z_{(2)},32) & \xrightarrow{j} & BSU^{32} & \xrightarrow{\tau} & K(Z_{(2)},32) \\ & & \downarrow \pi & & \\ K(Z_{(2)},30) & \xrightarrow{j} & BSU^{30} & \xrightarrow{k} & K(Z_{(2)},33) \end{array}$$

where j denotes the fiber inclusion, $j^*(k) = \tilde{\beta}Sq^2\iota$ and $\sigma^2(\tau) = 4S_{15}$.

This is no longer a Postnikov tower since dim k > dim τ. Let λ be determined by $j^*(\tau) = \lambda\iota \in H^{32}(K(Z_{(2)},32),Z_{(2)}) \approx Z_{(2)}$. (It can be shown that $\lambda = 4\cdot 15!$). The final stage of the Postnikov system is thus

$$\begin{array}{ccccc} K(Z/\lambda,31) & \xrightarrow{j} & F^{32} & & \\ & & \downarrow & & \\ K(Z_{(2)},30) & \xrightarrow{j} & BSU^{30} & \xrightarrow{k'} & K(Z/\lambda,32). \end{array}$$

Moreover $r_2j^*(k') = Sq^2\iota$ in $H^{32}(K(Z_{(2)},30);Z/2)$ and $\pi^*(k') = r_\lambda(\tau) \in$ $\pi^*(k') = r_\lambda(\tau) \in H^{32}(BSU^{32};Z/\lambda)$ where r is the appropriate reduction homomorphism and π: $BSU^{32} \to BSU^{30}$. Q.E.D.

To finish the proof of Theorem 8.2, it suffices to show that k' cannot be chosen to be a primitive, for then F^{32} and thus F would not be an H space. Since $\sigma^2(\tau) = 4S_{15}$, Lemma 8.3 implies that τ cannot be chosen to be primitive. In fact $\Delta\tau$ contains the term $t_2^8 \otimes t_2^8$. Thus $\Delta r_\lambda(\tau)$ contains a nonzero middle term. Since π is an H map, k' cannot be primitive and the proof is complete.

Several remarks are in order about this example. First we are not claiming that E has no infinite loop structure. Indeed it is possible that the ⊗ infinite loop structure on BU may induce some Q_∞ structure on E. If, however, we consider the fiber E' of $4S_{30}$: BSU $\to K(Z_{(2)},60)$, then using the uniqueness of the infinite loop structure on BSU [AP] it is possible to prove that E' has no infinite loop structure.

Corollary 5.2 implies that $[4S_{15}] \in E_r^{0,30}$ is not an infinite cycle. The class $\alpha = \tilde{\beta}(Q^{16}|Q^8|Q^4|a) \in H_{27}(Q^3BU,\partial)$ represents an element $[\alpha] \in E^3_{3,27}$. It can be shown that $< d_3(4S_{15}),\alpha > \neq 0$ under the pairing of the torsion submodules of $H^{28}(Q^3BU)$ and $H_{27}(Q^3BU)$. On the other hand, for dimension reasons $8S_{15}$ will be an infinite cycle. Its mod 2 reduction can be shown to represent $Sq^{16}Sq^8Sq^4\iota$ in $H_S^{30}(bu;Z/2) \cong \Sigma^2 A/A(Sq^1,Sq^3)$. The Miller spectral sequence for $H_*(BU;Z/p)$ and $H_*(BO;Z/p)$ will be completely analyzed in our forthcoming paper.

BIBLIOGRAPHY

[A1] J. F. Adams, Chern characters and the structure of the unitary group. Proc. Camb. Phil. Soc. 57(1961), pp. 189-199.

[AP] J. F. Adams and S. B. Priddy, Uniqueness of BSU. Math. Proc. Camb. Phil. Soc. 80(1976), pp. 475-509.

[A2] D. W. Anderson, Chain functors and homology theories. Lecture Notes in Mathematics 249, Springer-Verlag (1971), pp. 1-12.

[B] J. Beck, On H-spaces and infinite loop spaces. Lecture Notes in Mathematics 99, Springer-Verlag (1969), pp. 139-153.

[DL] E. Dyer and R. K. Lashof, Homology of iterated loop spaces. Amer. J. Math. 84(1962), pp. 35-88.

[E] P. Eccles, Does transfer characterize cohomology theories? mimeographed, Manchester (1974).

[KP] D. Kahn and S. Priddy, The transfer and stable homotopy theory, Math. Proc. Camb. Phil. Soc. 83(1978), pp. 103-111.

[K] D. Kraines, The kernel of the loop suspension map. Ill. J. Math. 21(1977), pp. 99-108.

[L1] T. Lada, Strong Homotopy Monads, Iterated Loop Spaces and Transfer. Notre Dame thesis (1974).

[L2] ------, Strong homotopy algebras over monads. Lecture Notes in Mathematics 533, Springer-Verlag (1976), pp. 399-479.

[L3] A. Liulevicius, On Characteristic Classes. Lecture notes, Aarhus Universitet (1968).

[M1] S. MacLane, Homology. Academic Press, New York (1963).

[M2] ------, Categories for the Working Mathematician. Springer-Verlag, New York, Berlin (1971).

[M3] I. Madsen, Higher torsion in SG and BSG. Math. Z. 143(1975), pp. 55-80.

[M4] ------, Remarks on normal invariants from the infinite loop space point of view. AMS Summer Institute, Stanford (1976).

[MST] I. Madsen, V. Snaith and J. Tornehave, Infinite loop maps in geometric topology. Math. Proc. Camb. Phil. Soc. 81(1977), pp. 399-430.

[M5] J. P. May, Categories of spectra and infinite loop spaces. Lecture Notes in Mathematics 99, Springer-Verlag (1969), pp. 448-479.

[M6] ------, The Geometry of Iterated Loop Spaces. Lecture Notes in Mathematics 271, Springer-Verlag (1972).

[M7] ------, Homology of E_∞ spaces. Lecture Notes in Mathematics 533, Springer-Verlag (1976), pp. 1-68.

[M8] H. Miller, A spectral sequence for infinite delooping. (to appear).

[R] F. W. Rousch, Transfer in Generalized Cohomology Theories. Princeton Thesis (1971).

[S1] G. Segal, Classifying spaces and spectral sequences. IHES 34(1968), pp. 105-112.

[S2] -----, Categories and cohomology theories. Topology 13(1974), pp. 293-312.

[S3] W. M. Singer, Connective fiberings over BU and U. Topology 7(1968), pp. 271-304.

[S4] J. Stasheff, Homotopy associativity of H spaces, II. Trans. Amer. Math. Soc. 108(1963), pp. 293-312.

Duke University
Durham, NC 27706

and

Institute for Advanced Study
Princeton, NJ 08540

North Carolina State University
Raleigh, NC 27650

INFINITE LOOP SPACE THEORY REVISITED

by J. P. May

Just over two years ago I wrote a summary of infinite loop space theory [37]. At the time, there seemed to be a lull in activity, with little immediately promising work in progress. As it turns out, there has been so much done in the interim that an update of the summary may be useful.

The initial survey was divided into four chapters, dealing with additive infinite loop space theory, multiplicative infinite loop space theory, descriptive analysis of infinite loop spaces, and homological analysis of infinite loop spaces. We shall devote a section to developments in each of these general areas and shall also devote a section to the newly evolving equivariant infinite loop space theory.

Two of the biggest developments will hardly be touched on here however. I ended the old survey with the hope that "much new information will come when we learn how the rich space level structures described here can effectively be exploited for calculations in stable homotopy theory." This hope is being realized by work in two quite different directions.

As discussed in [37, §4], the approximation theorem to the effect that $\Omega^n\Sigma^n X$ is a group completion of the simple combinatorial space C_nX plays a central role in the general theory. I stated there that "homotopical exploitation of the approximation theorem has barely begun." This is no longer the case. Such exploitation is now one of the more active areas of homotopy theory, recent contributions having been made by Mahowald, Brown and Peterson, R. Cohen, Sanderson and Koschorke, Caruso and Waner, and F. Cohen, Taylor, and myself. I plan to summarize the present state of the art in [42], and will content myself here with a remark in section two and a brief discussion of the equivariant approximation theorem in section five.

Second, the notion of E_∞ ring spectrum discussed in [37, §11] led to a simpler homotopical notion of H_∞ ring spectrum. This concept is really part of stable homotopy theory as understood classically, rather than part of infinite loop space theory, and seems to be basic to that subject. An introduction and partial summary of results based on this concept are given in [39]. A complete treatment will appear in the not too distant future [5]; meanwhile, the main results are available in the theses of Bruner [4], Steinberger [60], Lewis [28], and McClure [44].

I must end the introduction on a less sanguine note. Even in this short report, I shall have to mention a disconcertingly large number of published errors, both theoretical and calculational, both mine and those of many others. I do not know whether to ascribe this to carelessness, the complexity of the subject, or simple human blindness. Certainly the lesson is that an attitude of extreme skepticism is warranted towards any really difficult piece of work not supported by total detail. This pertains particularly to some of the embryonic theories discussed in sections two and five.

§1. *Additive infinite loop space theory*

The first change to be celebrated is in the state of the art of exposition. In an attempt to make the subject accessible to beginners, Frank Adams has written a truly delightful tract [1]. Anyone wishing a painless introduction, in particular to the various approaches to the recognition principle, is urged to read it.

In Adams' survey, there is a little of the flavor of competition between these approaches, and I was perhaps the worst offender in spreading this atmosphere. The point is that the black boxes for constructing spectra out of space level data looked so drastically different that it was far from obvious to me that they would produce equivalent spectra from the same data.

A major advance in the last two years is that we now have such a uniqueness theorem. There is only one infinite loop space machine, but there are various ways to construct it.

The first uniqueness theorem of this sort is due to Fiedorowicz [12], who axiomatized the passage from rings to the spectra of algebraic K-theory. (Actually, there are $\lim^1$ problems associated with getting the pairing he needs on the Gersten-Wagoner spectra; the argument in [13] is wrong, for the silly but substantive reason that η on page 165 fails to be a natural transformation.) Fiedorowicz' idea is based on the following simple, but extremely fruitful, observation which is at the heart of all the spectrum level uniqueness theorems discussed below. Let X be a bispectrum, namely a sequence of spectra $X_i = \{X_{i,j}\}$ and equivalences of spectra $X_i \rightarrow \Omega X_{i+1}$. Then the $0^{\underline{th}}$ spectrum $X_0 = \{X_{0,j}\}$ is equivalent to the spectrum $\{X_{i,0}\}$. Here spectra are at least Ω-spectra; one has variants depending on what category one is working in [43, App. A].

Thomason and I used this idea to axiomatize infinite loop space machines [43], and I want to say just enough about our work to explain precisely what such a gadget is.

Consider topological categories with objects the based sets $\underline{n} = \{0,1,\ldots,n\}$. Let F be the category of finite based sets; its objects are the $\underline{n}$ and its morphisms are all functions which take 0 to 0. Inside F, we have the subcategory Π consisting of the injections and projections, namely those morphisms $f:\underline{m} \to \underline{n}$ such that $f^{-1}(j)$ has at most one element for $1 \leq j \leq n$. We say that G is a category of operators if it contains Π and maps to F; we say that G is an E_∞ category if the map to F is an equivalence. Let T be the category of based spaces. A G-space is a functor $G \to T$, written $\underline{n} \to X_n$ on objects, such that the n projections $\underline{n} \to \underline{1}$ induce an equivalence $X_n \to X_1^n$ for $n \geq 0$ (and a technical cofibration condition is satisfied).

An infinite loop space machine E is an E_∞ category G and a functor E from G-spaces to spectra together with a natural group completion $\iota:X_1 \to E_0X$. Thus π_0E_0X is the universal group associated to the monoid π_0X_1 and, for any commutative coefficient ring, H_*E_0X is obtained from the Pontryagin ring H_*X_1 by localizing at $\pi_0X_1 \subset H_0X_1$.

With just this one axiom, we prove that any two infinite loop space machines defined on G-spaces are naturally equivalent. Actually, we prove the uniqueness theorem for F-spaces and deduce it for G-spaces by use of a functor from G-spaces to F-spaces suitably inverse to the pullback functor the other way. The proof for F-spaces proceeds by comparing any given machine to Segal's original machine [50]. An E_∞ operad (as in [37, §2]) gives rise to an E_∞ category G. May's original machine [35,36] was only defined on those G-spaces with X_n actually equal to X_1^n. We generalize its domain of definition to all G-spaces and so conclude that the May and Segal machines are equivalent. Any other machine which really is a machine must be equivalent to these.

I have also given an addendum [40] asserting the uniqueness of infinite loop space machines defined on permutative categories, the point being that there are several quite different ways of passing from such categories to the domain data (G-spaces) of infinite loop space theory.

Due to work of Thomason [64], we now have a much better understanding of this passage, together with a more general class of morphisms to which it can be applied. (On objects, restriction to permutative categories is harmless; see [37, §8].) Some discussion may be worthwhile, since I for one find the ideas illuminating. Given a permutative category $(A,\Box,*,c)$, so that $\Box:A\times A \to A$ is an associative product with unit $*$ and natural commutativity isomorphism c, one's first attempt to get into the domain of an infinite loop space machine is to try to write down a functor $F \to Cat$ with $n^{\underline{th}}$ category precisely A^n. In detail, for a morphism $f:\underline{m} \to \underline{n}$ in F, one defines a functor $f_*:A^m \to A^n$ by

$$f_*(A_1,\ldots,A_m) = (B_1,\ldots,B_n),\ B_k = \underset{f(j)=k}{\square} A_j,$$

on objects and morphisms. Due to permutations, these functors fail to define a functor $F \to Cat$, but it is a simple matter to use c to write down natural transformations $c(f,g):(fg)_* \to f_*g_*$. Upon writing out the formal properties satisfied by these data, one sees that one has a sort of system category theorists have known about for years, and have called a lax functor (up to opposite conventions on the c(f,g), hence the term op-lax in [64]). Ross Street [63] provides not just one but two ways of constructing an associated functor $F \to Cat$. Either way, the $n^{\underline{th}}$ category is equivalent to A_1^n and we obtain an F-space upon application of the classifying space functor B. A third way of getting such a functor is due to Segal [50] and explained in detail in [40]. Street [63] developed a notion of lax natural transformation between lax functors and showed that such things induce actual natural transformations under either of his constructions. Upon application of B, we deduce that lax natural transformations induce maps of F-spaces. This allows morphisms $F:A \to B$ with coherent natural transformations $FA\square FB \to F(A\square B)$ which need not be isomorphisms; neither Segal's construction nor my passage from permutative categories to E_∞ spaces is functorial with respect to such lax morphisms.

I should add that these observations are not the main thrust of Thomason's work in [64], his primary purpose being to show that B converts homotopy colimits of categories, suitably defined, to homotopy colimits of spaces. (A detailed categorical study of this comparison has since been given by Gray [18].) Thomason [65] later used this result, or rather its spectrum level version, to deduce some very interesting spectral sequences involving the algebraic K-theory of permutative categories.

Before leaving the additive theory, I want to say a bit about two more uniqueness theorems. The first reconciles two natural ways of looking at the stable classifying spaces of geometric topology. Consider Top for definiteness; needless to say, the argument is general. One can form $BTop = \varinjlim BTop(n)$. This is an L-space, where L is the linear isometries operad; see [37, §7]. On the other hand, one can regard $\amalg\, Top(n)$ as a permutative category. There result two spectra, and I proved in [41] that the first is in fact the connected cover of the second. While this may seem plausible enough, the lack of obvious technical relationship between the linear isometries data and the permutative data makes the proof one of the more difficult in the subject. With this result, the foundations seem to be complete; any two machine-built spectra which ought to be equivalent are equivalent.

The last uniqueness theorem I want to mention concerns A_∞ spaces (see [35, §3]) rather than E_∞ spaces and is due to Thomason [66]. In [35, p. 134], I gave two machines for constructing a classifying space, or delooping, functor on C-spaces X, where C is an A_∞ operad. One can either form a bar construction $B(S^1, C\times C_1, X)$ directly or replace X by an equivalent monoid $B(M,C,X)$ and take the classical classifying space of the latter. The second approach is more or less obviously equivalent to the delooping machines for A_∞ spaces of Boardman and Vogt [3] and Segal [50]. When X is an E_∞ space regarded as an A_∞ space by neglect of structure, one is looking at first deloopings in the May and Segal machines respectively, hence the two are equivalent by the spectrum level uniqueness theorem. In general, the total lack of commutativity in the situation, with the concomitant lack of the simple group completion notion, makes the consistency much harder. Thomason has given a quite ingenious proof that these two deloopings are always equivalent. The result gains interest from work to be mentioned in the next section.

§2. *Multiplicative infinite loop space theory*

Here the most significant development has been that mentioned in the introduction, the invention and exploitation of H_∞ ring spectra. As discussed in [37, §11], E_∞ ring spectra are defined in terms of actions by an E_∞ operad G on spectra. H_∞ ring spectra are defined in the stable category, without reference to operads, but are really given in terms of actions up to homotopy by E_∞ operads. While H_∞ ring spectra are much more amenable to homotopical analysis, E_∞ ring spectra are of course still essential to the infinite loop space level applications for which they were designed (see [37, §10-14]). In particular, there is no H_∞ analog of the recognition principle which allows one to construct E_∞ ring spectra from E_∞ ring spaces. (I must report that the passage from bipermutative categories to E_∞ ring spaces in [36, VI §4], despite being intuitively obvious, is blatantly wrong; a correct treatment will be given in [5].)

Another significant development has been the appearance of interesting examples of E_n and H_n ring spectra and of E_n ring spaces for $1 \leq n < \infty$. The definitional framework is exactly the same as when $n = \infty$, except that now G is not an E_∞ operad but an E_n operad, so that its $j^{\underline{th}}$ space has the Σ_j-equivariant homotopy type of the configuration space of j-tuples of distinct points of R^n.

Lewis [5, 28] has shown that if X is an n-fold loop space and $f:X \to BO$ is an n-fold loop map, then the resulting Thom spectrum Mf is an E_n ring spectrum; if BO is replaced by BF, one at least gets an H_n ring spectrum.

E_n ring spaces have appeared, totally unexpectedly, in connection with the analysis of the multiplicative properties of the generalized James maps

$$j_q : C_nX \to Q(C^+_{n,q} \wedge_{\Sigma_q} X^{(q)})$$

used by Cohen, Taylor and myself [9] to stably split C_nX. The product over $q \geq 0$ of the targets is an E_n ring space, and the map j with components j_q is "exponential" in the sense that it carries the additive E_n action on C_nX to the new multiplicative E_n action on the product. In principle, this completely determines the homological behavior of the James maps. I shall say more about this in [42], but it will be some time before details appear.

Another recent development concerns A_∞ ring spaces, or E_1 ring spaces in the language above. These are rings up to all higher coherence homotopies. I have constructed the algebraic K-theory of an A_∞ ring space R as follows [38] (modulo some annoying corrections necessary in the combinatories, which will be supplied in [5]). We form the space M_nR of $(n\times n)$-matrices with coefficients in R. Writing down the ordinary matrix product, but with the additions and multiplications involved parametrized by the given operad actions, we construct an A_∞ operad $\mathcal{H}_n$ which acts on M_nR. We then construct morphisms of operads $\mathcal{H}_{n+1} \to \mathcal{H}_n$ such that the usual inclusion $M_nR \to M_{n+1}R$ is an $\mathcal{H}_{n+1}$-map, where M_nR is an $\mathcal{H}_{n+1}$-space by pullback. We next form pullback diagrams of $\mathcal{H}_n$-spaces

$$\begin{array}{ccc} FM_nR & \longrightarrow & M_nR \\ \downarrow & & \downarrow \\ GL(n,\pi_0R) & \longrightarrow & M_n(\pi_0R). \end{array}$$

Thus FM_nR is the space of invertible components in M_nR. We have a classifying space functor B_n on $\mathcal{H}_n$-spaces for each n (indeed, as discussed in the previous section, a choice of equivalent functors). We let KR be the plus construction on the telescope of the spaces B_nFM_nR and define $K_*R = \pi_*KR$. Various basic properties of KR are proven in [38]; for example, if $FR = FM_1R$ is the unit space of R, then the inclusion of monomial matrices in FM_nR yields a natural map $Q_0(B_1FR \amalg \{0\}) \to KR$.

If R is a discrete ring, this is Quillen's K_*R. If R is a topological ring, it is Waldhausen's [67]. In these cases, KR is an infinite loop space [38, 10.12]. I have several more or less rigorous unpublished proofs that KR is always a first loop space, but I could easily write a disquisition on how not to prove that KR is an infinite loop space in general. The latter failures are joint work with Steiner and Thomason, but Steiner still has one promising idea that has yet to be shot down. Certainly the infinite deloopability of KR is a deep theorem if it is true.

While various other A_∞ ring spaces are known, the motivation comes from Waldhausen's work [67] connecting the Whitehead groups for stable PL concordance to algebraic K-theory. For a based space X, $Q(\Omega X \amalg \{0\})$ is an A_∞ ring space and we define $AX = KQ(\Omega X \amalg \{0\})$, this being one of Waldhausen's proposed definitions of the algebraic K-theory of a space. We also define $A(X;Z) = K\tilde{N}(\Omega X \amalg \{0\})$, where $\tilde{N}(\Omega X \amalg \{0\})$ is the free topological Abelian group generated by ΩX or, equivalently, the realization $|Z[GSX]|$ of the integral group ring of the Kan loop group on the total singular complex of X. In [38], I constructed a rational equivalence $AX \to A(X;Z)$.

Waldhausen [68] constructed another functor, call it WX, and established a natural fibration sequence with total space $WX\times Z$, fibre a homology theory (as a functor of X), and base space $Wh^{PL}(X)$. (As far as I know, proofs are not yet available. However, Steinberger and others have checked out the indications of proof in [68] and in Waldhausen's lectures. The connection with concordance groups depends on a stability claim of Hatcher [19], the published proof of which is definitely incorrect; Hatcher and Igusa (and I am told Burghelea) assure us that there is an adequate correct claim, but no proof has yet been given.) Waldhausen also claims a rational equivalence $WX \to A(X;Z)$, and it is on the basis of this claim that all calculational applications proceed. I have not yet seen or heard any convincing indications of proof. Clearly it suffices to show that WX and AX are equivalent, and this would be a deep and satisfying theorem even if an alternative argument were available. Steinberger is working towards this result and seems to be reasonably close to a proof.

There has been one other recent development of considerable interest. Woolfson has given a Segal style treatment of parts of multiplicative infinite loop space theory. His paper [72] is devoted to a theory analogous to the E_∞ ring theory summarized in [37, §12]. His paper [73] is devoted to a reformulation in his context of the orientation theory discussed in [37, §14] and to a proof of Nishida's nilpotency theorem along lines proposed by Segal [52]. (I have not read [72] or the first half of [73] for details, but the passage from particular bipermutative categories to hyper Γ-spaces sketched in [72] is unfortunately just as blatantly wrong as my passage from bipermutative categories to E_∞ ring spaces in [36]; as stated before, a correct treatment of this point will appear in [5]. The second half of [73] cannot be recommended; the proof of Theorem 2.2 is incorrect, and the argument as a whole is much harder than that based on the simpler homotopical notion of an H_∞ ring spectrum [5, 39].)

This theory raises further uniqueness questions of the sort discussed in the first section, and these have been considered by Thomason. The conclusion seems to be that there probably exists an appropriate theory but that the details would be so horrendous that it would not be worth developing unless a commanding need arose.

Incidentally, a Segal type approach to the construction and infinite delooping of KR was one of the failures mentioned above.

§3. *Descriptive analysis of infinite loop spaces*

The deepest new result under this heading is the proof of the infinite loop version of the complex Adams conjecture. When localized away from r, the composite $BU \xrightarrow{\psi^r - 1} BU \xrightarrow{j} BSF$ is not just null homotopic as a map of spaces but as a map of infinite loop spaces. That is, the associated composite map of spectra is null homotopic. This was originally announced by Friedlander and Seymour [17]. Their proposed proofs proceeded along wholly different lines. That of Seymour was based on Snaith's assertion [56, 4.1] that Seymour's bundle theoretical model [54] for the fibre JU(r) of $\psi^r - 1$ could be constructed in a more economical way. Snaith's assertion is now known to be false*, and this line of proof is moribund. (The error also makes [56, §4-7] and [57] much less interesting.) I have been carefully checking Friedlander's proof. It is an enormously impressive piece of mathematics, and I am convinced that it is correct. It will appear in [16], in due course.

The infinite loop Adams conjecture, when combined with earlier results and the uniqueness theorem for the stable classifying spaces of geometric topology discussed in section one, largely completes the program of analyzing these infinite loop spaces at odd primes. The grand conclusion is stated in the introduction of [41]. One essential

* See Seymour and Snaith, these Proceedings.

ingredient was the work of Madsen, Snaith, and Tornehave [32], and complete new proofs of their results have been given by Adams [1, §6].

There remain interesting problems at $p = 2$. Here no lifting $\alpha:BSO \to F/O$ of $\psi^r-1:BSO \to BSO$, $r \equiv \pm 3 \bmod 8$, can even be an H-map. An analysis of the homological behavior of one choice of α has been given by Brumfiel and Madsen [6] and the deviation from additivity of another choice has been studied in detail by Tornehave [67]. A provocative formulation of a possible 2-primary infinite loop version of the real Adams conjecture has been given by Miller and Priddy [46], although we have not the slightest idea of how their conjectures might be proven. Similarly, Madsen [31, 2.9] has made some very interesting conjectures about the infinite loop structure of F/Top at $p = 2$, but again there are no proofs in sight.

One very satisfying result along these lines has been given by Priddy [47]. Using the transfer and homology calculations, he has shown that, at the prime 2, SF is a direct factor (up to homotopy) in $QB(\Sigma_2 \int \Sigma_2)$ and F/O is a direct factor in QBO(2). The first assertion is a deeper multiplicative analog of the Kahn-Priddy theorem, their proof of which has just recently appeared [24,25]. That result gave that, at any prime p, Q_0S^0 is a direct factor in $QB\Sigma_p$. It is natural to conjecture that SF is also a direct factor in $QB(\Sigma_p \int \Sigma_p)$ for $p > 2$. However, because of the problems explained in [8, II §6], a proof along Priddy's lines would be much more difficult. There are three other splittings of this general nature that should be mentioned. Segal [49] proved that BU is a direct factor in QBU(1) and Becker [3] proved that BSp is a direct factor in QBSp(1) and BO is a direct factor in QBO(2). Snaith [59] rederived these last splittings and used them to deduce stable decompositions of the classifying spaces BG for G = U(n), Sp(n), or O(2n).

In my original survey, I neglected to mention Segal's paper [51]. Let $A = \{A_q | q \geq 0\}$ be a graded commutative ring. Then $\times_{q\geq 0} K(A_q,q)$ is a ring space with unit space $A_0^* \times (\times_{q\geq 1} K(A,q))$ and special unit space $\times_{q\geq 1} K(A,q)$. Segal proved that these unit spaces are infinite loop spaces. Steiner [61,62] later gave an improved argument which showed that these infinite loop structures are functorial in A and used the functoriality to prove certain splittings of these infinite loop spaces in case A is p-local, such splittings having been conjectured by Segal. (I find the earlier of Steiner's proofs the more convincing.) Snaith [58] showed that the total Stiefel-Whitney and Chern classes $\Sigma_{q\geq 1} w_q:BO \to \times_{q\geq 1} K(Z_2,q)$ and $\Sigma_{q\geq 1} c_q:BU \to \times_{q\geq 1} K(Z,2q)$ fail to commute with transfer. However, this does not disprove Segal's conjecture [51, p. 293] about these classes. Segal was quite careful to avoid such transfer pathologies by asking if the map $\coprod_{n\geq 0} BO(n) \to (\times_{q\geq 1} K(Z_2,q))\times Z$ specified by $(\sum_{q\geq 1} w_q)\times\{n\}$ on $BO(n)$ extends to an infinite loop map $BO\times Z \to (\times_{q\geq 1} K(Z_2,q))\times Z$ for a suitable infinite loop structure on the target, and similarly for the Chern classes. (I am told that this has now been proven by a student of Segal's. See [36, Remarks VIII.1.4], interpreting the remarks additively rather than multiplicatively, for a discussion of the relationship of the transfer for $BO\times Z$ to that for $BO = BO\times\{0\}$.)

The last, but by no means least, piece of progress to be reported in this area is the complete analysis by Fiedorowicz and Priddy [15] of the infinite loop spaces associated to the classical groups of finite fields and their relationship to the image of J spaces obtained as fibres of maps $\psi^r-1:BG \to BG'$ for stable classical groups G and G'. While this is an extraordinarily rich area of mathematics, the grand conclusion is that there is a one-to-one correspondence, realized by infinite loop equivalences coming from Brauer lifting of modular representations, between these two kinds of infinite loop spaces. In a sequel, Fiedorowicz [14] considers the uniqueness of the localizations at p prime to r of the infinite loop spaces $JG(r)$ obtained with $G = G'$ above being O, U, or Sp. In particular cases of geometric interest, the problem is not hard [56, §3], but the general answer is most satisfactory: $JG(r)_p$ and $JH(s)_p$ are equivalent as infinite loop spaces if and only if they have abstractly isomorphic homotopy groups.

§4. *Homological analysis of infinite loop spaces*

Probably the biggest development under this heading is again the work of Fiedorowicz and Priddy [15] just cited. They give an exhaustive analysis of the homologies, with their homology operations, of the various image of J. spaces. Amusingly, some of the most useful formulae, in particular for the real image of J spaces at the prime 2, are wholly inaccessible without the connection with finite groups. Their work also includes complete information on the homology and cohomology of all of the various classical groups of finite fields (away from the characteristic).

In [8, II §13], I used these calculations to study the Bockstein spectral sequences in the fibration sequence $B\,\mathrm{Coker}\,J \to BSF \to BJ_\otimes$

at $p = 2$. I would like to record one inconsequential error; [8, II.13.7] should read $\tilde{Q}^{2i+2} x_{(i,i)} \equiv x_{(2i+1,2i+1)}$ mod #-decomposables, the line of proof being as indicated but with due regard to the middle term of the mixed Cartan formula. In [8, II.13.8], the error term $\tilde{Q}^{4i}\beta\sigma_*\tilde{y}_{(2i,2i-1)}$ should therefore be $\sigma_*\tilde{y}_{(4i-1,4i-1)}$ rather than zero. No further changes are needed. (Another inconsequential error occurs in [8, III App]; Cohen has published the required corrections in [7, App].)

Incidentally, Madsen's assertion [31, 3.5], which is stated without proof, can be read off immediately from the calculations of [8, II §13]. This result plays a key role in Madsen's very interesting theorem that a kO-orientable spherical fibration ξ over X admits a topological reduction if and only if certain characteristic classes $\tau_i(\xi) \in H^{2^i-1}(X;Z_2)$ are zero. In other words, the obstruction to kO-orientability is not only the sole obstruction to reducibility away from 2 (as discussed in [37,§§14 and 18]), it is also a large part of the obstruction at $p = 2$.

In my original survey, I did not do justice to the work of Hodgkin and Snaith [22,55] on the mod p K-theory of infinite loop spaces in general and of those infinite loop spaces of greatest geometric interest in particular. In [37, §17], I did sketch their proof of the key fact that $K_*(\mathrm{Coker}\ J) = 0$, and they have since published a very readable account [23] of this and related calculations.

I should mention one subterranean set of calculations. A reasonably good understanding of the Adams spectral sequence converging to π_*MSTop at $p > 2$ now exists. Two preprints, by Mann and Milgram [33] and Ligaard and myself [30], gave partial and complete information respectively on $H^*(\mathrm{MSTop};Z_p)$ as an A-module. This material is also in Ligaard's thesis [29], and he did much further work with me on the calculation of E_2. In my archives, I have nearly complete information on E_2, with descriptions as matric Massey products of all generators of $E_2^{s,t}$ for $s > 0$. I also have a thorough analysis of the differentials coming from $\pi_*\mathrm{MSO} \to \pi_*\mathrm{MSTop}$ and from the Bockstein spectral sequence of BCoker J [8, II.10.7], this being an elaboration of exploratory calculations in an undistributed preprint by Mann and Milgram. Mann has in his archives a calculation of a key piece of the spectral sequence from these differentials. However, a complete calculation of all of $E_r^{s,*}$ for $s > 0$ is out of reach algebraically, and we have very little control over the huge amount of noise in $E_2^{0,*}$. Milgram has in his archives a very nice geometrical argument to show that some of this noise does in fact survive to E_∞. Altogether though, we are very far from a complete determination of π_*MSTop, and the interest of all four parties seems to have flagged.

The work reported so far was already well under way when my earlier survey was written. There are two major later homological developments to report. The first is both negative and positive. In [10], Curtis claimed to prove that the mod 2 Hurewicz homomorphism for QS^0 annihilated all elements of π_*^s except the Hopf maps and, where present, the Arf invariant maps. The assertion may or may not be true, but Wellington's careful analysis [71] makes clear that we are very far from a proof by any known techniques. On the positive side, Wellington's work gives a good hold on the global structure of the cohomology of iterated loop spaces. In principal, this is a dualization problem from the homology calculations of Cohen [8, III]. The latter give $H_*(\Omega_0^n\Sigma^n X;Z_p)$ explicitly as an algebra and with precise recursive formulae for the coproduct and action by the Steenrod algebra A (see [37, §24]). Wellington proves that $H^*(\Omega_0^n\Sigma^n X;Z_p)$ is isomorphic as an algebra to the universal enveloping algebra of a certain Abelian restricted Lie algebra M_n^*X. While M_n^*X admits an A-action with respect to which its enveloping algebra is a free A-algebra, the isomorphism does not preserve the A-module structures. With this as his starting point, Wellington gives a detailed analysis of the problem of determining the A-annihilated primitive elements in $H_*(\Omega_0^n\Sigma^n X;Z_p)$, the main technique being a method for computing Steenrod operations in M_nX by use of the differential structure of the Λ-algebra.

The last homological development I wish to report concerns the relationship between the homology of infinite loop spaces and the homology of spectra. Let $X = E_0$ be the zero$\underline{\text{th}}$ space of a spectrum $E = \{E_i\}$. In [35, p. 155-156], I pointed out that my two-sided bar construction gave spectral sequences $\{{}^iE^rX\}$ such that ${}^iE^2X$ is a well-defined computable functor of the R-algebra H_*X, where R is the Dyer-Lashof algebra, and $\{{}^iE^rX\}$ converges to H_*E_i. I specifically asked for a precise description of ${}^iE^2X$ as some homological functor of X, but I never pursued the point.

Much later, but independently, Miller [45] used resolution techniques to construct a spectral sequence $\{E^rX\}$ converging from a suitable functor of the R-algebra H_*X to H_*E. More importantly he developed techniques allowing explicit computation of E^2X in favorable cases and studied the behavior of the Steenrod operations in the spectral sequence. In particular, he showed that the spectral sequence collapses for $E = K(Z,0)$.

A little later, Kraines independently rediscovered this spectral sequence. I shall only say a little about his joint work with Lada on this subject, since their paper also appears in this volume [26]. They give a very pretty spectrum-level version of my two-sided bar construction, thus obtaining a most satisfactory geometric construction of Miller's algebraic spectral sequence. Among other things, the close connection with the geometry allows them to use the spectral sequence to disprove the long discredited conjecture that a representable functor with a transfer extends to a cohomology theory. More applications will surely appear, and further study of this spectral sequence is bound to be profitable.

§5. *Equivariant infinite loop space theory*

One of the most fashionable activities in modern topology is to take one's favorite theory, put an action of a compact Lie group G on all spaces in sight, and ask how much of the theory remains valid. Much less ambitiously, one might restrict G to be finite.

For the homotopy theorist, the first thing one wants is a thorough study of G-CW complexes. This we now have in the full generality of compact Lie groups, the relevant theory having been initiated by Matumota [34] and completed by Waner [69]. Any G-space is weakly G-equivalent to a G-CW complex and a weak G-equivalence between G-CW complexes is a G-equivalence. Actually, once ordinary CW-theory is developed properly, these and other standard results present little difficulty. Much more deeply, all of Milnor's basic theorems about spaces of the homotopy type of CW-complexes generalize to G-CW complexes; see Waner [69].

The next thing one wants is a good theory of G-bundles and G-fibrations (with some other structural group, A say, in the bundle case), including classification theorems for bundles or fibrations over G-CW complexes. This too we now have in the full generality of compact Lie groups, the bundle theory having been supplied by Segal [48], Tom Dieck [11], and Lashof and Rothenberg [27] and the fibration theory having been supplied by Waner [69], with addenda by Hauschild [21].

One is then led to ask if the resulting stable K-theories extend to cohomology theories. In the bundle case, as Segal has explained [48], one can generalize Bott periodicity. In the fibration case, and in the case of topological rather than linear bundles, one is inexorably led to develop G-infinite loop space theory.

I am quite confident that the eventual state of the art will precisely parallel the situation sketched in the first section. There will be two main approaches to the recognition principle, namely a G-Segal machine and a G-May machine, and there will be a uniqueness theorem on G-infinite loop space machines which ensures an equivalence between them. However, work in this direction is still in its infancy, and full details are not yet in place. It may well be necessary to restrict to finite groups, and we do so in the following discussion.

The present situation is this. I am in possession of three unpublished manuscripts, by Segal [53], Hauschild [21], and Waner [70], all of which I received within a month of the present writing (October 1978). In the first, Segal sketches a G-Segal machine, and I have little doubt that any missing details can be filled in. The other two give a G-May machine. In the latter approach, as I long ago explained to both authors, modulo a few technical points which turn out to be a bit tricky but not particularly difficult, it is formal to reduce the G-recognition principle to the stable G-approximation theorem.

Unstably, the G-approximation theorem asserts the existence of a natural "G-group completion" $C(V,X) \to \Omega^V\Sigma^V X$ for based G-spaces X, where V is a G-representation, Ω^V and Σ^V are the loops and suspension associated to the one-point compactification of V, and C(V,X) is the G-space of finite unordered subsets of V with labels in X. More precisely, $C(V,X) = \coprod_{j\geq 0} F(V,j)\times_{\Sigma_j} X^j/(\approx)$, where F(V,j) is the configuration space of j-tuples of distinct points of V and the equivalence relation encodes basepoint identifications. In the stable version, one takes colimits over G-representations V contained in a G-space R^∞ which contains each irreducible representation infinitely often. Hauschild [20] has published an argument for the stable theorem in the special case $X = S^0$, and Segal's manuscript [53] sketches an argument for the sharper unstable result, also for $X = S^0$. The bulk of Waner's manuscript [70] is devoted to a proof of the stable theorem for general X and the main part of Hauschild's manuscript [21] is devoted to a proof of the unstable result for general X. The various arguments are quite complicated and, at this writing, I cannot claim to fully understand any of them. However, I am reasonably sure that the union of Hauschild [20] and Waner [70] does include a complete proof of the stable theorem.

In any case, granting the stable G-approximation theorem, we have the G-recognition principle in a form applicable to $G\text{-}E_\infty$ spaces and can apply it to the classifying spaces for stable spherical G-fibrations and topological G-bundles. Thus the relevant K-theories extend to G-cohomology theories. It is to be expected that this will be a powerful tool for the study of the equivariant Adams conjecture, this application being work in progress by Waner.

In connection with the G-approximation theorem, it is worth remarking that the paper by Cohen, Taylor, and myself [9] on the splitting of spaces of the same general form as C(V,X) above applies virtually verbatim with G-actions put in everywhere. There are evident notions of G-coefficient systems $\mathcal{C}$, G-Π spaces $\underline{X}$, and a resulting general construction of $\mathcal{C}\underline{X}$ as in [9, §1-2]. The maps for the approximation theorem, but not the approximation theorem itself, can be used precisely as in [9] to obtain stable splittings of such G-spaces $\mathcal{C}\underline{X}$, provided only that each $\mathcal{C}_j$ is Σ_j-free. That is, the suspension G-spectrum of $\mathcal{C}\underline{X}$ is weakly G-equivalent to the wedge of the suspension G-spectra of the successive filtration quotients $\mathcal{C}_j^+ \wedge_{\Sigma_j} X^{(j)}$.

In fact, as we intend to make precise elsewhere, the whole argument of [9] is so formal that it can be carried out in an axiomatic setting of general topological categories with suitable extra structure. Indeed, the whole framework of definitions exploited in the study of iterated loop spaces can be set up in such a setting, and it can be expected that the resulting theory will find many future applications.

BIBLIOGRAPHY

1. J. F. Adams. Infinite loop spaces. Annals of Math. Studies No. 90. Princeton. 1978.
2. J. C. Becker. Characteristic classes and K-theory. Springer Lecture Notes in Mathematics Vol. 428, pp. 132-143. 1974.
3. J. M. Boardman and R. M. Vogt. Homotopy Invariant Algebraic Structures on Topological Spaces. Springer Lecture Notes in Mathematics Vol. 347. 1973.
4. R. Bruner. The Adams spectral sequence of H_∞ ring spectra. Thesis. Univ. of Chicago. 1977.
5. R. Bruner, G. Lewis, J. P. May, J. McClure, and M. Steinberger. H_∞ Ring Spectra and their Applications. Springer Lecture Notes in Mathematics. In preparation.
6. G. Brumfiel and I. Madsen. Evaluation of the transfer and the universal surgery classes. Inventiones math. 32 (1976), 133-169.
7. F. R. Cohen. Braid orientations and bundles with flat connections. Inventiones math. 46 (1978), 99-110.
8. F. R. Cohen, T. Lada, and J. P. May. The Homology of Iterated Loop Spaces. Springer Lecture Notes in Mathematics Vol. 533. 1976.
9. F. R. Cohen, J. P. May, and L. R. Taylor. Splitting of certain spaces CX. Math. Proc. Camb. Phil. Soc. To appear.

10. E. B. Curtis. The Dyer-Lashof algebra and the Λ-algebra. Ill. J. Math. 19 (1975), 231-246.

11. T. tom Dieck. Faserbündel mit Gruppenoperation. Archiv der Math. 20 (1969), 136-143.

12. Z. Fiedorowicz. A note on the spectra of algebraic K-theory, Topology 16 (1977), 417-422.

13. Z. Fiedorowicz. The Quillen-Grothendieck construction and extensions of pairings. Springer Lecture Notes in Mathematics Vol. 657, 163-169. 1978.

14. Z. Fiedorowicz. The primary components of the Im J spaces. Manuscript, 1978.

15. Z. Fiedorowicz and S. B. Priddy. Homology of Classical Groups over Finite Fields and their Associated Infinite Loop Spaces. Springer Lecture Notes in Mathematics Vol. 674. 1978.

16. E. Friedlander. The infinite loop Adams conjecture. Math. Proc. Camb. Phil. Soc. To appear. (Manuscript, 1978).

17. E. Friedlander and R. Seymour. Two proofs of the stable Adams conjecture. Bull. Amer. Math. Soc. 83 (1977), 1300-1302.

18. J. W. Gray. Closed categories, lax limits and homotopy limits. Preprint, 1978.

19. A. Hatcher. Higher simple homotopy theory. Annals of Math. 102 (1975), 101-137.

20. H. Hauschild. Zerspaltung äquivariante Homotopiemengen. Math. Ann. 230 (1977), 279-292.

21. H. Hauschild. Klassifizierende Räume, G-Konfigurationsräume und und äquivariante Schleifenräume. Manuscript, 1978.

22. L. Hodgkin. The K-theory of some well-known spaces I, QS^0. Topology 11 (1972), 371-375.

23. L. Hodgkin and V. P. Snaith. The K-theory of some more well-known spaces. Ill. J. Math. 22 (1978), 270-278.

24. D. S. Kahn and S. B. Priddy. On the transfer in the homology of symmetric groups. Math. Proc. Camb. Phil. Soc. 83 (1978), 91-101.

25. D. S. Kahn and S. B. Priddy. The transfer and stable homotopy theory. Math. Proc. Camb. Phil. Soc. 83 (1978), 103-111.

26. D. Kraines and T. Lada. A counterexample to the transfer conjecture. These proceedings.

27. R. Lashof and M. Rothenberg. G smoothing theory. Proc. Symp. Pure Math. Vol. 32, part I, pp. 211-266. Amer. Math. Soc. 1978.

28. G. Lewis, The stable category and generalized Thom spectra. Thesis. Univ. of Chicago. 1978.

29. H. Ligaard. On the spectral sequence for π_*(MSTop). Thesis. Univ. of Chicago. 1976.

30. H. Ligaard and J. P. May. On the Adams spectral sequence for π_*MSTop, I. Preprint, 1976.

31. I. Madsen. Remarks on normal invariants from the infinite loop space viewpoint. Proc. Symposia Pure Math. Vol. 32, part I, pp. 91-102. Amer. Math. Soc. 1978.

32. I. Madsen, V. P. Snaith, and J. Tornehave. Infinite loop maps in geometric topology. Math. Proc. Camb. Phil. Soc. 81 (1977), 399-430.

33. B. Mann and R. J. Milgram. On the action of the Steenrod algebra on H*(MSPL) at odd primes. Preprint, 1976.

34. T. Matumota. On G-CW complexes and a theorem of J.H.C. Whitehead. J. Fac. Sci. Tokyo 18 (1971-72), 363-374.

35. J. P. May. The Geometry of Iterated Loop Spaces. Springer Lecture Notes in Mathematics Vol. 271. 1972.

36. J. P. May (with contributions by F. Quinn, N. Ray, and J. Tornehave). E_∞ Ring Spaces and E_∞ Ring Spectra. Springer Lecture Notes in Mathematics Vol. 577. 1977.

37. J. P. May. Infinite loop space theory. Bull. Amer. Math. Soc. 83 (1977), 456-494.

38. J. P. May. A_∞ ring spaces and algebraic K-theory. Springer Lecture Notes in Mathematics Vol. 658, 240-315. 1978.

39. J. P. May. H_∞ ring spectra and their applications. Proc. Symp. Pure Math. Vol. 32, part 2, pp. 229-244. Amer. Math. Soc. 1978.

40. J. P. May. The spectra associated to permutative categories. Topology. To appear.

41. J. P. May. The spectra associated to $\mathcal{I}$-monoids. Math. Proc. Camb. Phil. Soc. To appear.

42. J. P. May. Applications and generalizations of the approximation theorem. Proc. conf. in Algebraic Topology. Aarhus. 1978.

43. J. P. May and R. Thomason. The uniqueness of infinite loop space machines. Topology. To appear.

44. J. McClure. H_∞^{d*} ring spectra and their cohomology operations. Thesis. Univ. of Chicago. 1978.

45. H. R Miller. A spectral sequence for the homology of an infinite delooping. Preprint, 1977.

46. H. R. Miller and S. B. Priddy. On G and the stable Adams conjecture. Springer Lecture Notes in Mathematics Vol. 658, 331-348, 1978.

47. S. B. Priddy. Homotopy splittings involving G and G/O. Comm. Math. Helv. 53 (1978), 470-484.

48. G. Segal. Equivariant K-theory. Publ. Math. I.H.E.S. Vol. 34 (1968), 129-151.

49. G. Segal. The stable homotopy of complex projective space. Quart. J. Math. Oxford (2) 24 (1973), 1-5.

50. G. Segal. Categories and cohomology theories. Topology 13 (1974), 293-312.

51. G. Segal. The multiplicative group of classical cohomology. Quart. J. Math. Oxford (3) 26 (1975), 289-293.

52. G. Segal. Power operations in stable homotopy theory. Preprint.

53. G. Segal. Some results in equivariant homotopy theory. Manuscript, 1978.

54. R. M. Seymour. Vector bundles invariant under the Adams operations. Quart. J. Math. Oxford (2) 25 (1974), 395-414.

55. V. P. Snaith. Dyer-Lashof operations in K-theory. Springer Lecture Notes in Mathematics Vol. 496, 103-294. 1975.

56. V. P. Snaith. The complex J-homomorphism. I. Proc. London Math. Soc. (3) 34 (1977), 269-302.

57. V. P. Snaith. The 2-primary J-homomorphism. Math. Proc. Camb. Phil. Soc. 82 (1977), 381-387.

58. V. P. Snaith. The total Chern and Stiefel-Whitney classes are not infinite loop maps. Illinois J. Math. 21 (1977), 300-304.

59. V. P. Snaith. Algebraic cobordism and K-theory. Memoir Amer. Math. Soc. To appear.

60. M. Steinberger. Homology operations for H_∞ ring spectra. Thesis. Univ. of Chicago. 1977.

61. R. J. Steiner. Thesis. Cambridge Univ. 1977.

62. R. J. Steiner. Spectra and products of Eilenberg-Mac Lane spaces. Preprint, 1978.

63. R. Street. Two constructions on lax functors. Cahiers de Topologie et Geometrie Differentielle 13 (1972), 217-264.

64. R. Thomason. Homotopy colimits in the category of small categories. Preprint, 1977.

65. R. Thomason. First quadrant spectral sequences in algebraic K-theory. Preprint, 1978.

66. R. Thomason. Uniqueness of delooping machines. Preprint, 1978.

67. J. Tornehave. Deviation from additivity of a solution to the Adams conjecture. To appear.

68. F. Waldhausen. Algebraic K-theory of topological spaces, I. Proc. Symp. Pure Math. Vol. 32, part 1, pp. 35-60. Amer. Math. Soc. 1978.

69. S. Waner. Equivariant classifying spaces and fibrations. Thesis. Univ. of Chicago. 1978.

70. S. Waner. Equivariant infinite loop spaces. Manuscript, 1978.

71. R. Wellington. The A-algebra $H^*\Omega_0^{n+1}\Sigma^{n+1}X$, the Dyer-Lashof algebra, and the Λ-algebra. Thesis. Univ. of Chicago. 1977.

72. R. Woolfson. Hyper-Γ-spaces and hyperspectra. Preprint, 1978.

73. R. Woolfson. Γ-spaces, orientations and cohomology operations. Preprint, 1978.

A J-homomorphism associated with a space of empty varieties

(addenda and corrigenda to two papers on the J-homomorphism)

by

Robert Seymour & Victor Snaith*

§0: Introduction

This paper concerns two previous papers by the second author [Sn 1; Sn 2]. Theorem 4.1 of [Sn 1] is wrong (see §1.2). The mistake was spotted independently by the two authors and also by Michael Crabb. Below (§1.1) we explain why the result is false and outline the effect upon the papers [Sn 1; Sn 2; F-S] in §2.

Recall the definitions of the groups $Ad_q^o(X)$ and $Pr_q(X)$.

§0.1: Let q be a prime and let X be a compact, Hausdorff space. A ψ^q - vector bundle over X is a triple (E,E',θ). Here E and E' are complex vector bundles over X. If N denotes the regular representation of Z/q on $\mathbb{C}^q$ let Z/q act on $E \oplus (E' \otimes N)$ by

$$g(e \oplus (e' \otimes n)) = e \oplus (e' \otimes gn)$$

$$(g \in Z/q,\ e \in E_x,\ e' \in E'_x,\ x \in X \text{ and } n \in N).$$

Let Z/q act on $E^{\otimes q}$ cyclically. Then

$$\theta : E^{\otimes q} \longrightarrow E \oplus (E' \otimes N)$$

is a Z/q-vector bundle isomorphism. We refer to θ as a ψ^q-isomorphism.

---

*Research partially supported by the N.R.C. of Canada

A triple (E,E',θ) is <u>proper</u> if $\theta(e^{\otimes q})$ has a non-zero E-component whenever $e \in E_x$ is non-zero $(x \in X)$. When (E,E',θ) is proper we call θ a <u>proper</u> <u>ψ^q-isomorphism</u>.

The simplest example of a proper ψ^q-vector bundle is given [Sn 1, §4.3] by the evident proper ψ^q-isomorphism

$$\theta_n : (\mathbb{C}^n)^{\otimes q} \cong \mathbb{C}^n \oplus (\mathbb{C}^d \otimes N) \tag{0.2}$$

where $qd = n^q - n$. $(\mathbb{C}^n, \mathbb{C}^d, \theta_n)$ is denoted by $\underline{n}$. There is a sum operation, $\oplus$, on ψ^q-vector bundles [Sn 1, §4.5] under which $\underline{n} = \underline{1} \oplus \underline{1} \oplus \ldots \oplus \underline{1}$ (n copies).

Two ψ^q-vector bundles (E,E',θ) and (F,F',ϕ) over X are <u>equivalent</u> if there exists vector bundle isomorphisms $\alpha : E \to F$ and $\alpha' : E' \to F'$ such that $\phi \circ \alpha^q$ is homotopic through ψ^q-isomorphisms to $\alpha \oplus (\alpha' \otimes \mathrm{id})$ where "id" is the identity map of N. Two proper ψ^q-bundles are <u>properly</u> <u>equivalent</u> if they are equivalent through a homotopy of proper ψ^q-isomorphisms. The set of equivalence classes of of ψ^q-vector bundles on X is an abelian monoid [Se 1, §1] whose associated Grothendieck group is written $Ad_q^o(X)$. Similarly proper ψ^q-vector bundles define a group $Pr_q(X)$ [Sn 1, §4.5].

There is a forgetful homomorphism

$$\Gamma_X : Pr_q(X) \to Ad_q^o(X) \tag{0.3}$$

which is claimed in [Sn 1, §4.1] to be an isomorphism. Presently we shall see this is not the case.

§1: Examples, counterexamples and ramifications concerning [Sn 1].

Before explaining why [Sn 1, §4.1] is false perhaps a few elementary examples of ψ^q-structures might help to illustrate the role of propriety in §0.1.

1.1: Examples

(a) Let PR_n be the set of proper ψ^q-structures (q a fixed prime) of the form

$$\theta : (\mathbb{C}^n)^{\otimes q} \xrightarrow{\cong} \mathbb{C}^n \oplus (\mathbb{C}^d \otimes N)$$

where $qd = n^q - n$. Forming the composition

$$\theta \circ \theta_n^{-1} \qquad (\theta_n \text{ as in } §0.2)$$

we obtain an element of $\mathrm{Aut}_{Z/q}(\mathbb{C}^n \oplus (\mathbb{C}^d \otimes N))$ and restricting to the fixed vectors, $\mathbb{C}^n \oplus \mathbb{C}^d$, we obtain a matrix, $A(\theta)$ in $GL_{n+d}\mathbb{C}$, with respect to the standard basis. Propriety is expressible in terms of the matrix, $A(\theta)$.

For example if $n = 2 = q$, $d = 1$ then propriety means precisely that the matrix $(a_{ij}) = A(\theta)$ is non-singular and that the projective variety

$$\left\{ [z_1, z_2] \in \mathbb{P}' \;\middle|\; \begin{matrix} a_{11}z_1^2 + a_{12}z_2^2 + a_{13}z_1z_2 = 0 \\ a_{21}z_1^2 + a_{22}z_2^2 + a_{23}z_1z_2 = 0 \end{matrix} \right\}$$

is empty.

In general propriety means that a similar projective variety defined in terms of the coefficients of $A(\theta)$ is the empty variety. Hence the title.

(b) A non-proper ψ^2-structure on $\mathbb{C}^2$ is given by setting the matrix $A(\theta)$ equal to

$$\begin{pmatrix} 1 & 0 & 1 \\ 0 & 1 & 1 \\ 0 & 0 & 1 \end{pmatrix}$$

(c) If we topologise PR_n as a subspace of the space of all linear maps $(\mathbb{C}^n)^{\otimes q} \to \mathbb{C}^n \oplus (\mathbb{C}^d \otimes N)$ an essential loop in the space of proper ψ^2-structures on $\mathbb{C}^2$ is given by setting $A(\theta)(w)$ $(w \in S')$ equal to

$$\begin{pmatrix} w & 0 & 2 \\ 0 & 1 & 0 \\ 1 & 0 & 1 \end{pmatrix}.$$

1.2: $Pr_q(X) \neq Ad_q^o(X)$ in general

ψ^q-vector bundles with and without the propriety condition look very different. To see $Pr_q(X) \neq Ad_q^o(X)$ consider the μ construction of [Sn 1, §5] on the space PR_n of §1.1(c). This assigns to

$$\theta : (\mathbb{C}^n)^{\otimes q} \xrightarrow{\cong} \mathbb{C}^n \oplus (\mathbb{C}^d \otimes N)$$

the map $\overline{\mu}(\theta) : \mathbb{C}^n \to \mathbb{C}^n$ given by $\overline{\mu}(\theta)(v) = \pi_1(\theta(v^{\otimes q}))$ where π_1 is the first projection. One point compactification yields $\mu : PR_n \to \varinjlim_m \Omega^m S^m = QS^o$. The μ-construction, given here for trivial bundles over a point, may be globalised to give a homomorphism [Sn 1, §5.4]

$$\mu : Pr_q(X) \to [X, QS^o][1/q].$$

We will now exhibit two proper ψ^q-vector bundles whose μ-images are different but which are equivalent (non-properly). Chose $\alpha : X \to U(n)$ whose J-homomorphism image $J(\alpha) \in [X, QS^o][1/q]$ is non-trivial. We may assume n large and make the identification

$$\mathbb{C}^n \oplus (\mathbb{C}^d \otimes N) \cong \mathbb{C}^n \oplus (\mathbb{C}^n \oplus \mathbb{C}^{d-n} \oplus (\mathbb{C}^d \otimes M))$$

where $N \underset{=}{\sim} 1 \oplus M$ as Z/q-representations. The two proper ψ^q-vector bundles are given by $(X \times \mathbb{C}^n, X \times \mathbb{C}^d, \theta(i))$ $(i = 1,2)$ where , in the terms of the above direct sum decompositions,

$$\theta(1) = (\alpha \oplus 1) \circ \theta_n$$

and

$$\theta(2) = (1 \oplus (\alpha \oplus 1 \oplus 1_{\mathbb{C}^d \otimes M})) \circ \theta_n .$$

From [Sn 1, §5.6] $\mu(\theta(1)) = J(\alpha)$ while $\mu(\theta(2)) = \mu(\text{constant map}) = 0$.

1.3: The forgetful map Γ_X

Although [Sn 1, §4.1] is false, Γ_X of (0.3) is onto when localised at an odd prime, p, different from q. That is :-

1.3.1: Theorem:

$$\Gamma_X : Pr_q(X)_{(p)} \to Ad_q^o(X)_{(p)}$$

is onto for all odd primes $p(\neq q)$, such that q generates the units mod p^2.

Here is a very brief sketch of the proof. $GL_n\mathbb{C}$ acts on Pn_n of §1.1(c) by $(\alpha \in GL_n\mathbb{C})$

$$\alpha(\theta) = (\alpha \oplus (\alpha_1 \otimes N)) \circ \alpha \circ (\alpha^{-1})^{\otimes q}$$

where α_1 is the automorphism induced by α on the $\exp(2\pi i/q)$-eigenspace of the q-cycle permutation of $(\mathbb{C}^n)^{\otimes q}$. Form

$$B(U_n, PR_n) = EU_n \times_{U_n} PR_n \qquad (1.3.2)$$

where U_n is the unitary group. Addition of $\underline{1}$ (see §0.1) yields a map $B(U_n,PR_n) \to B(U_{n+1},PR_{n+1})$ and a direct limit space

$$B(U,PR) = \varinjlim_n B(U_n,PR_n).$$

This is an H-space by [Se 1], the operation being induced by sum of ψ^q-vector bundles. It also admits a transfer. This is seen by generalising the Pr_q-theory transfer of [Sn 1, §6.3] to arbitrary finite coverings. For this purpose coverings of the form

$$Y/\Sigma_{n-1} \to Y/\Sigma_n$$

well suffice (Σ_n, the symmetric group) as explained in [K - P]. An unpublished result of Peter Eccles* [E] deduces from the transfer a map $\sigma : \Gamma^+ B(U,PR) \to B(U,PR)$ where Γ^+ is the construction of [B - E]

There is a (compatible) family of ψ^q-vector bundles θ^n over the spaces $B(U_n,PR_n)$. These may not be universal because $Pr_q(X)$ may not be representable, however pulling back $[\theta^n] - \underline{n} \in Pr_q(B(U_n,PR_n))$ yields a homomorphism

$$\lambda_X : [X,B(U,PR)] \longrightarrow Pr_q(X). \qquad (1.3.3)$$

Now let $J_{(p)}$ be the "image of J" space as in [Sn 1]. We will construct a map $f : J_{(p)} \to B(U,PR)_{(p)}$ so that the composition

$$[X,J_{(p)}] \xrightarrow{f_\#} [X,B(U,PR)_{(p)}] \xrightarrow{\lambda_X} Pr_q(X)_{(p)} \xrightarrow{\Gamma_X} Ad_q^o(X)_{(p)} \cong [X,J_{(p)}]$$

is an isomorphism when $X = S^{2p-3}$. By [Sn 1, §3.1] the above composite is then an isomorphism for all X. To construct f take the map constructed in [H - S]

---

*See also the paper of Kraines & Lada - these proceedings.

$$\tau\colon J_{(p)} \simeq (BGL\mathbb{F}_q^+)_p \to \Gamma^+BZ/p$$

and compose it with $\sigma \circ \Gamma^+ g$. Here choose $g\colon BZ/p \to B(U,PR)$, take any map such that g^* sends the reduced ("universal") ψ^q-vector bundle to the reduced, proper ψ^q-bundle over BZ/p which is described in [Sn 1, §6.8].

The sketch of the proof of 1.3.1 is now complete except for a technicality which is relegated to the final remark of the paper.

§2: Addenda to the J-homomorphism papers

2.1: The complex case

The following amendments must be made to [Sn 1] because of the falsity of [Sn 1, §4.1 and §5.1]. Firstly Theorem 4.1 may be replaced by Theorem 1.3.1 sketched above. The other amendments consist in §§5-7 of restating each result to refer to Pr_q-theory instead of Ad_q^o-theory. Unfortunately it is the latter, because of its relation to J-theory [Se 1], which interests homotopy theorists. Nevertheless Theorem 8.4 and Theorem 5.1 of [Sn 1] are true if we replace μ and τ respectively by the maps induced [Sn 1, §1.1] by the solution of the stable Adams conjecture given in [F]. This is proved by appealing to [Sn 1, Theorem 8.3] which asserts the uniqueness of transfer-commuting maps out of $J^{\oplus}_{(p)} \times Z^+$ ($Z^+ = \{0,1,2,\ldots\}$). Friedlander's solution of the stable Adams conjecture induces a map

$$J^{\oplus}_{(p)} \times Z^+ \to QS^o_{(p)}$$

in (1.1) of [Sn 1] which must be transfer-commuting as it is an infinite loop map.

Similarly [Sn 1, Theorem 9.5] remains true if we use in the proof Friedlander's map to replace f in diagram (9.1) of [Sn 1]. However this adds nothing to Tornhave's original proof (cf. J.P. May et al., Springer Lecture Notes no. 577, ch. 8).

2.2: The complex stable Adams conjecture

In [F - S] two proofs of the stable Adams conjecture were announced. Unfortunately the proof proposed by the first author rested on [Sn 1, Theorem 4.1] as it consisted of generalising $Pr_q(X)$ to a cohomology theory identifiable with J-theory. Substituting Theorem 1.3.2 above is useless for infinite loopspace purposes. However we record what the first author claims in this direction.

2.2.1: Theorem

If p is an odd prime then the p-localisation of the space B(U,PR) introduced in §§1.1(c), 1.3.1 can be given an infinite loopspace structure.

Fortunately the second proof [F-S;F] remains intact, despite its immense technical details. Nevertheless the second author would feel easier if an alternative proof existed. To this end here is simple-sounding equivalent problem which he has been encouraging people to attempt since 1974.

2.2.2: Problem: Construct an infinite loop map $\tau : BU_{(p)} \to SG/U_{(p)}$ such that in rational cohomology the following diagram commutes.

$$\begin{array}{ccc} BU_{(p)} & \xrightarrow{\tau} & (SG/U)_{(p)} \\ & {\scriptstyle \psi^q - 1} \searrow \quad \swarrow & \\ & BU_{(p)} & \end{array}$$

where p, q are odd primes and q generates $(Z/p^2)^*$.

The results of [M - S - T] imply the stable Adams conjecture from the positive solution to §2.2.2.

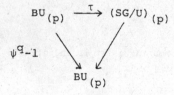

2.3: The real case

[Sn 2] treated the analogue, PR(X), of $Pr_q(X)$ for Real K-theory. Failure of [Sn 1, 4.1] superannuates this analogue and with it Theorem A of [Sn 2]. Theorems B and C concerning the deviation from additivity of the Becker-Gottlieb transfer remain true.

2.4: Remark

Ad_q-theory as described in [Se 1] does not quite give, as claimed, a model for J-theory when p-localised. The correct bundle-theoretic definition is very similar (and fortunately all published proofs apply verbatim to either definition). The correct definition is given in [Se 2]. The definition appearing in [Se 1] is elegant and does admit a nice description of products in Ad_q-theory. It does not satisfy the crucial J-theory exact sequence because it defines a strictly larger theory than J-theory. Further details will appear in [Se 2].

[B-E] M. G. Barratt and P. J. Eccles: Γ^+-structures-I: A free group functor for stable homotopy; Topology 13(1974), 24-45.

[E] P. J. Eccles: Does the transfer characterise cohomology theories?; Manchester University preprint (1975).

[F] E. M. Friedlander: The stable Adams conjecture; North-western University preprint (1977).

[F-S] E. M. Friedlander and R. M. Seymour: Two proofs of the stable Adams conjecture; Bull. A. M. Soc. 83(1977), 1300-02.

[H-S] B. Harris and G. B. Segal: K_j groups of rings of algebraic integers; Annals of Maths. 101(1)(1975), 20-33.

[K-P] D. S. Kahn and S. B. Priddy: Applications of the transfer to stable homotopy theory, Bull. Amer. Math. Soc. 78(1972), 981-987.

[M-S-T] I. Madsen, V. P. Snaith & J. Tornehave: Infinite loop maps in geometric topology, Math. Proc. Camb. Phil. Soc., 81(1977), 399-430.

[Se1] R. M. Seymour: Vector bundles invariant under the Adams operations; Quart. J. Math., Oxford (2)25(1974), 395-414.

[Se2] R. M. Seymour: A local representation of ψ^q revisited; University College preprint (1978).

[Sn1] V. P. Snaith: The Complex J-homomorphism, I; Proc. London Math. Soc. 3(34)(1977), 269-302.

[Sn2] V. P. Snaith: The 2-primary J-homomorphism; Math. Proc. Camb. Phil Soc. 82(1977), 381-387.

University College
Gower Street
London, WCIE 6BT
England

University of Western Ontario
London, Ontario
Canada, N6A 5B9

ADDRESSES OF CONTRIBUTORS

Professor Don W. Anderson
Department of Mathematics
University of California, San Diego
La Jolla, California
U.S.A. 92093

Professor Doug R. Anderson
Department of Mathematics
Syracuse University
Syracuse, New York
U.S.A. 13210

Professor A. Assadi
School of Mathematics
Institute for Advanced Studies
Princeton, New Jersey
U.S.A. 08540

Professor M. Boardman
Department of Mathematics
The Johns Hopkins University
Baltimore, Maryland
U.S.A. 21218

Professor G. Carlsson
Department of Mathematics
University of California, San Diego
La Jolla, California
U.S.A. 92093

Professor D. Davis
Department of Mathematics
Lehigh University
Bethlehem, Pennsylvania
U.S.A. 18015

Professor K. Dovermann
Department of Mathematics
University of Chicago
Chicago, Illinois
U.S.A. 60637

Professor J. Ewing
Department of Mathematics
Indiana University
Bloomington, Indiana
U.S.A. 47401

Professor I. Hambleton
Department of Mathematics
McMaster University
Hamilton, Ontario,
Canada L8S 4K1

Professor J. Harper
Department of Mathematics
University of Rochester
Rochester, New York
U.S.A. 14627

Professor R. Kane
Department of Mathematics
University of Alberta
Edmonton, Alberta
Canada T6G 2G1

Professor S. Kochman
Department of Mathematics
University of Western Ontario
London, Ontario
Canada N6A 5B9

Professor D. Kraines
Department of Mathematics
Duke University
Durham, North Carolina
U.S.A. 27706

Professor T. Lada
Department of Mathematics
North Carolina State University
Raleigh, North Carolina
U.S.A. 27650

Professor P. Landweber
Department of Mathematics
Rutgers University
New Brunswick, New Jersey
U.S.A. 08903

Professor R. Lashof
Department of Mathematics
University of Chicago
Chicago, Illinois
U.S.A. 60637

Professor A. Liulevicius
Department of Mathematics
University of Chicago
Chicago, Illinois
U.S.A. 60637

Professor P. May
Department of Mathematics
University of Chicago
Chicago, Illinois
U.S.A. 60637

Dr. J. McLeod
Trinity College
Cambridge, England

Professor J. Milgram
Department of Mathematics
Stanford University
Stanford, California
U.S.A. 94305

Professor H. Munkholm
Mathematical Department
Odense University
Niels Bohr Alle
DK 5230 Odense M.
Denmark

Professor T. Petrie
Department of Mathematics
Rutgers University
New Brunswick, New Jersey
U.S.A. 08903

Professor A. Ranicki
Department of Mathematics
Princeton University
Princeton, New Jersey
U.S.A. 08540

Professor R. Schultz
Department of Mathematics
Purdue University
West Lafayette, Indiana
U.S.A. 47907

Dr. P. Seymour
Department of Mathematics
University College
Gower St.
London WC1E 6BT England

Professor R. Sharpe
Department of Mathematics
University of Toronto
Toronto, Ontario
Canada M5S 1A1

Professor D. Sjerve
Department of Mathematics
University of British Columbia
Vancouver, British Columbia
Canada V6T 1W5

Professor V. Snaith
Department of Mathematics
University of Western Ontario
London, Ontario
Canada N6A 5B9

Professor R. Steiner
Department of Mathematics
Warwick University
Coventry, England
CV4 7AL

Professor L. Taylor
Department of Mathematics
Notre Dame University
Notre Dame, Indiana
U.S.A. 46556

Professor T. tom Dieck
Mathematical Institute
University of Gottingen
Federal Republic of Germany

Professor B. Williams
Department of Mathematics
Notre Dame University
Notre Dame, Indiana
U.S.A. 46556

LIST OF PARTICIPANTS

J. Allard
Don Anderson
Doug Anderson
D. Andrus
A. Assadi
A. Bak
A. Brender
J. Boardman
P. Booth
G. Carlsson
K. Chan
F. Connolly
D. Davis
H. Dovermann
K. Ehrlich
J. Ewing
S. Feder
P. Fillmore
W. Gilbert
D. Gottlieb
I. Hambleton
J. Harper
P. Hoffman
W. Hsiang
W. Iberkleid
D. James
D. Johnson
R. Kane
D. Kraines
S. Kochman
R. Kulkarni
T. Lada
K. Lam
P. Landweber
R. Lashof
R. Lee
N. Levitt

W.K. Li
R. Litherland
A. Liulevicius
E. Lluis
S. Lomonaco
I. Madsen
Y. Matsumoto
P. May
J. Milgram
H. Munkholm
K. Murasugi
D. Niday
T. Nishimura
S. Oldfield
W. Pardon
D. Patterson
T. Petrie
W. Ralph
A. Ranicki
M. Rothenberg
R. Schultz
R. Sharpe
G. Simons
D. Sjerve
D. Smith
V. Snaith
E. Stein
R. Steiner
N. Stoltzfus
S. Thomeier
K. Varadarajan
J. Verster
V. Vijums
C. Weibel
C. Wilkerson
B. Williams
P. Zvengrowski

Vol. 580: C. Castaing and M. Valadier, Convex Analysis and Measurable Multifunctions. VIII, 278 pages. 1977.

Vol. 581: Séminaire de Probabilités XI, Université de Strasbourg. Proceedings 1975/1976. Edité par C. Dellacherie, P. A. Meyer et M. Weil. VI, 574 pages. 1977.

Vol. 582: J. M. G. Fell, Induced Representations and Banach *-Algebraic Bundles. IV, 349 pages. 1977.

Vol. 583: W. Hirsch, C. C. Pugh and M. Shub, Invariant Manifolds. IV, 149 pages. 1977.

Vol. 584: C. Brezinski, Accélération de la Convergence en Analyse Numérique. IV, 313 pages. 1977.

Vol. 585: T. A. Springer, Invariant Theory. VI, 112 pages. 1977.

Vol. 586: Séminaire d'Algèbre Paul Dubreil, Paris 1975–1976 (29ème Année). Edited by M. P. Malliavin. VI, 188 pages. 1977.

Vol. 587: Non-Commutative Harmonic Analysis. Proceedings 1976. Edited by J. Carmona and M. Vergne. IV, 240 pages. 1977.

Vol. 588: P. Molino, Théorie des G-Structures: Le Problème d'Equivalence. VI, 163 pages. 1977.

Vol. 589: Cohomologie l-adique et Fonctions L. Séminaire de Géométrie Algébrique du Bois-Marie 1965–66, SGA 5. Edité par L. Illusie. XII, 484 pages. 1977.

Vol. 590: H. Matsumoto, Analyse Harmonique dans les Systèmes de Tits Bornologiques de Type Affine. IV, 219 pages. 1977.

Vol. 591: G. A. Anderson, Surgery with Coefficients. VIII, 157 pages. 1977.

Vol. 592: D. Voigt, Induzierte Darstellungen in der Theorie der endlichen, algebraischen Gruppen. V, 413 Seiten. 1977.

Vol. 593: K. Barbey and H. König, Abstract Analytic Function Theory and Hardy Algebras. VIII, 260 pages. 1977.

Vol. 594: Singular Perturbations and Boundary Layer Theory, Lyon 1976. Edited by C. M. Brauner, B. Gay, and J. Mathieu. VIII, 539 pages. 1977.

Vol. 595: W. Hazod, Stetige Faltungshalbgruppen von Wahrscheinlichkeitsmaßen und erzeugende Distributionen. XIII, 157 Seiten. 1977.

Vol. 596: K. Deimling, Ordinary Differential Equations in Banach Spaces. VI, 137 pages. 1977.

Vol. 597: Geometry and Topology, Rio de Janeiro, July 1976. Proceedings. Edited by J. Palis and M. do Carmo. VI, 866 pages. 1977.

Vol. 598: J. Hoffmann-Jørgensen, T. M. Liggett et J. Neveu, Ecole d'Eté de Probabilités de Saint-Flour VI – 1976. Edité par P.-L. Hennequin. XII, 447 pages. 1977.

Vol. 599: Complex Analysis, Kentucky 1976. Proceedings. Edited by J. D. Buckholtz and T. J. Suffridge. X, 159 pages. 1977.

Vol. 600: W. Stoll, Value Distribution on Parabolic Spaces. VIII, 216 pages. 1977.

Vol. 601: Modular Functions of one Variable V, Bonn 1976. Proceedings. Edited by J.-P. Serre and D. B. Zagier. VI, 294 pages. 1977.

Vol. 602: J. P. Brezin, Harmonic Analysis on Compact Solvmanifolds. VIII, 179 pages. 1977.

Vol. 603: B. Moishezon, Complex Surfaces and Connected Sums of Complex Projective Planes. IV, 234 pages. 1977.

Vol. 604: Banach Spaces of Analytic Functions, Kent, Ohio 1976. Proceedings. Edited by J. Baker, C. Cleaver and Joseph Diestel. VI, 141 pages. 1977.

Vol. 605: Sario et al., Classification Theory of Riemannian Manifolds. XX, 498 pages. 1977.

Vol. 606: Mathematical Aspects of Finite Element Methods. Proceedings 1975. Edited by I. Galligani and E. Magenes. VI, 362 pages. 1977.

Vol. 607: M. Métivier, Reelle und Vektorwertige Quasimartingale und die Theorie der Stochastischen Integration. X, 310 Seiten. 1977.

Vol. 608: Bigard et al., Groupes et Anneaux Réticulés. XIV, 334 pages. 1977.

Vol. 609: General Topology and Its Relations to Modern Analysis and Algebra IV. Proceedings 1976. Edited by J. Novák. XVIII, 225 pages. 1977.

Vol. 610: G. Jensen, Higher Order Contact of Submanifolds of Homogeneous Spaces. XII, 154 pages. 1977.

Vol. 611: M. Makkai and G. E. Reyes, First Order Categorical Logic. VIII, 301 pages. 1977.

Vol. 612: E. M. Kleinberg, Infinitary Combinatorics and the Axiom of Determinateness. VIII, 150 pages. 1977.

Vol. 613: E. Behrends et al., L^p-Structure in Real Banach Spaces. X, 108 pages. 1977.

Vol. 614: H. Yanagihara, Theory of Hopf Algebras Attached to Group Schemes. VIII, 308 pages. 1977.

Vol. 615: Turbulence Seminar, Proceedings 1976/77. Edited by P. Bernard and T. Ratiu. VI, 155 pages. 1977.

Vol. 616: Abelian Group Theory, 2nd New Mexico State University Conference, 1976. Proceedings. Edited by D. Arnold, R. Hunter and E. Walker. X, 423 pages. 1977.

Vol. 617: K. J. Devlin, The Axiom of Constructibility: A Guide for the Mathematician. VIII, 96 pages. 1977.

Vol. 618: I. I. Hirschman, Jr. and D. E. Hughes, Extreme Eigen Values of Toeplitz Operators. VI, 145 pages. 1977.

Vol. 619: Set Theory and Hierarchy Theory V, Bierutowice 1976. Edited by A. Lachlan, M. Srebrny, and A. Zarach. VIII, 358 pages. 1977.

Vol. 620: H. Popp, Moduli Theory and Classification Theory of Algebraic Varieties. VIII, 189 pages. 1977.

Vol. 621: Kauffman et al., The Deficiency Index Problem. VI, 112 pages. 1977.

Vol. 622: Combinatorial Mathematics V, Melbourne 1976. Proceedings. Edited by C. Little. VIII, 213 pages. 1977.

Vol. 623: I. Erdelyi and R. Lange, Spectral Decompositions on Banach Spaces. VIII, 122 pages. 1977.

Vol. 624: Y. Guivarc'h et al., Marches Aléatoires sur les Groupes de Lie. VIII, 292 pages. 1977.

Vol. 625: J. P. Alexander et al., Odd Order Group Actions and Witt Classification of Innerproducts. IV, 202 pages. 1977.

Vol. 626: Number Theory Day, New York 1976. Proceedings. Edited by M. B. Nathanson. VI, 241 pages. 1977.

Vol. 627: Modular Functions of One Variable VI, Bonn 1976. Proceedings. Edited by J.-P. Serre and D. B. Zagier. VI, 339 pages. 1977.

Vol. 628: H. J. Baues, Obstruction Theory on the Homotopy Classification of Maps. XII, 387 pages. 1977.

Vol. 629: W. A. Coppel, Dichotomies in Stability Theory. VI, 98 pages. 1978.

Vol. 630: Numerical Analysis, Proceedings, Biennial Conference, Dundee 1977. Edited by G. A. Watson. XII, 199 pages. 1978.

Vol. 631: Numerical Treatment of Differential Equations. Proceedings 1976. Edited by R. Bulirsch, R. D. Grigorieff, and J. Schröder. X, 219 pages. 1978.

Vol. 632: J.-F. Boutot, Schéma de Picard Local. X, 165 pages. 1978.

Vol. 633: N. R. Coleff and M. E. Herrera, Les Courants Résiduels Associés à une Forme Méromorphe. X, 211 pages. 1978.

Vol. 634: H. Kurke et al., Die Approximationseigenschaft lokaler Ringe. IV, 204 Seiten. 1978.

Vol. 635: T. Y. Lam, Serre's Conjecture. XVI, 227 pages. 1978.

Vol. 636: Journées de Statistique des Processus Stochastiques, Grenoble 1977, Proceedings. Edité par Didier Dacunha-Castelle et Bernard Van Cutsem. VII, 202 pages. 1978.

Vol. 637: W. B. Jurkat, Meromorphe Differentialgleichungen. VII, 194 Seiten. 1978.

Vol. 638: P. Shanahan, The Atiyah-Singer Index Theorem, An Introduction. V, 224 pages. 1978.

Vol. 639: N. Adasch et al., Topological Vector Spaces. V, 125 pages. 1978.

Vol. 640: J. L. Dupont, Curvature and Characteristic Classes. X, 175 pages. 1978.

Vol. 641: Séminaire d'Algèbre Paul Dubreil, Proceedings Paris 1976–1977. Edité par M. P. Malliavin. IV, 367 pages. 1978.

Vol. 642: Theory and Applications of Graphs, Proceedings, Michigan 1976. Edited by Y. Alavi and D. R. Lick. XIV, 635 pages. 1978.

Vol. 643: M. Davis, Multiaxial Actions on Manifolds. VI, 141 pages. 1978.

Vol. 644: Vector Space Measures and Applications I, Proceedings 1977. Edited by R. M. Aron and S. Dineen. VIII, 451 pages. 1978.

Vol. 645: Vector Space Measures and Applications II, Proceedings 1977. Edited by R. M. Aron and S. Dineen. VIII, 218 pages. 1978.

Vol. 646: O. Tammi, Extremum Problems for Bounded Univalent Functions. VIII, 313 pages. 1978.

Vol. 647: L. J. Ratliff, Jr., Chain Conjectures in Ring Theory. VIII, 133 pages. 1978.

Vol. 648: Nonlinear Partial Differential Equations and Applications, Proceedings, Indiana 1976–1977. Edited by J. M. Chadam. VI, 206 pages. 1978.

Vol. 649: Séminaire de Probabilités XII, Proceedings, Strasbourg, 1976–1977. Edité par C. Dellacherie, P. A. Meyer et M. Weil. VIII, 805 pages. 1978.

Vol. 650: C*-Algebras and Applications to Physics. Proceedings 1977. Edited by H. Araki and R. V. Kadison. V, 192 pages. 1978.

Vol. 651: P. W. Michor, Functors and Categories of Banach Spaces. VI, 99 pages. 1978.

Vol. 652: Differential Topology, Foliations and Gelfand-Fuks-Cohomology, Proceedings 1976. Edited by P. A. Schweitzer. XIV, 252 pages. 1978.

Vol. 653: Locally Interacting Systems and Their Application in Biology. Proceedings, 1976. Edited by R. L. Dobrushin, V. I. Kryukov and A. L. Toom. XI, 202 pages. 1978.

Vol. 654: J. P. Buhler, Icosahedral Golois Representations. III, 143 pages. 1978.

Vol. 655: R. Baeza, Quadratic Forms Over Semilocal Rings. VI, 199 pages. 1978.

Vol. 656: Probability Theory on Vector Spaces. Proceedings, 1977. Edited by A. Weron. VIII, 274 pages. 1978.

Vol. 657: Geometric Applications of Homotopy Theory I, Proceedings 1977. Edited by M. G. Barratt and M. E. Mahowald. VIII, 459 pages. 1978.

Vol. 658: Geometric Applications of Homotopy Theory II, Proceedings 1977. Edited by M. G. Barratt and M. E. Mahowald. VIII, 487 pages. 1978.

Vol. 659: Bruckner, Differentiation of Real Functions. X, 247 pages. 1978.

Vol. 660: Equations aux Dérivée Partielles. Proceedings, 1977. Edité par Pham The Lai. VI, 216 pages. 1978.

Vol. 661: P. T. Johnstone, R. Paré, R. D. Rosebrugh, D. Schumacher, R. J. Wood, and G. C. Wraith, Indexed Categories and Their Applications. VII, 260 pages. 1978.

Vol. 662: Akin, The Metric Theory of Banach Manifolds. XIX, 306 pages. 1978.

Vol. 663: J. F. Berglund, H. D. Junghenn, P. Milnes, Compact Right Topological Semigroups and Generalizations of Almost Periodicity. X, 243 pages. 1978.

Vol. 664: Algebraic and Geometric Topology, Proceedings, 1977. Edited by K. C. Millett. XI, 240 pages. 1978.

Vol. 665: Journées d'Analyse Non Linéaire. Proceedings, 1977. Edité par P. Bénilan et J. Robert. VIII, 256 pages. 1978.

Vol. 666: B. Beauzamy, Espaces d'Interpolation Réels: Topologie et Géometrie. X, 104 pages. 1978.

Vol. 667: J. Gilewicz, Approximants de Padé. XIV, 511 pages. 1978.

Vol. 668: The Structure of Attractors in Dynamical Systems. Proceedings, 1977. Edited by J. C. Martin, N. G. Markley and W. Perrizo. VI, 264 pages. 1978.

Vol. 669: Higher Set Theory. Proceedings, 1977. Edited by G. H. Müller and D. S. Scott. XII, 476 pages. 1978.

Vol. 670: Fonctions de Plusieurs Variables Complexes III, Proceedings, 1977. Edité par F. Norguet. XII, 394 pages. 1978.

Vol. 671: R. T. Smythe and J. C. Wierman, First-Passage Perculation on the Square Lattice. VIII, 196 pages. 1978.

Vol. 672: R. L. Taylor, Stochastic Convergence of Weighted Sums of Random Elements in Linear Spaces. VII, 216 pages. 1978.

Vol. 673: Algebraic Topology, Proceedings 1977. Edited by P. Hoffman, R. Piccinini and D. Sjerve. VI, 278 pages. 1978.

Vol. 674: Z. Fiedorowicz and S. Priddy, Homology of Classical Groups Over Finite Fields and Their Associated Infinite Loop Spaces. VI, 434 pages. 1978.

Vol. 675: J. Galambos and S. Kotz, Characterizations of Probability Distributions. VIII, 169 pages. 1978.

Vol. 676: Differential Geometrical Methods in Mathematical Physics II, Proceedings, 1977. Edited by K. Bleuler, H. R. Petry and A. Reetz. VI, 626 pages. 1978.

Vol. 677: Séminaire Bourbaki, vol. 1976/77, Exposés 489–506. IV, 264 pages. 1978.

Vol. 678: D. Dacunha-Castelle, H. Heyer et B. Roynette. Ecole d'Eté de Probabilités de Saint-Flour. VII-1977. Edité par P. L. Hennequin. IX, 379 pages. 1978.

Vol. 679: Numerical Treatment of Differential Equations in Applications, Proceedings, 1977. Edited by R. Ansorge and W. Törnig. IX, 163 pages. 1978.

Vol. 680: Mathematical Control Theory, Proceedings, 1977. Edited by W. A. Coppel. IX, 257 pages. 1978.

Vol. 681: Séminaire de Théorie du Potentiel Paris, No. 3, Directeurs: M. Brelot, G. Choquet et J. Deny. Rédacteurs: F. Hirsch et G. Mokobodzki. VII, 294 pages. 1978.

Vol. 682: G. D. James, The Representation Theory of the Symmetric Groups. V, 156 pages. 1978.

Vol. 683: Variétés Analytiques Compactes, Proceedings, 1977. Edité par Y. Hervier et A. Hirschowitz. V, 248 pages. 1978.

Vol. 684: E. E. Rosinger, Distributions and Nonlinear Partial Differential Equations. XI, 146 pages. 1978.

Vol. 685: Knot Theory, Proceedings, 1977. Edited by J. C. Hausmann. VII, 311 pages. 1978.

Vol. 686: Combinatorial Mathematics, Proceedings, 1977. Edited by D. A. Holton and J. Seberry. IX, 353 pages. 1978.

Vol. 687: Algebraic Geometry, Proceedings, 1977. Edited by L. D. Olson. V, 244 pages. 1978.

Vol. 688: J. Dydak and J. Segal, Shape Theory. VI, 150 pages. 1978.

Vol. 689: Cabal Seminar 76–77, Proceedings, 1976–77. Edited by A.S. Kechris and Y. N. Moschovakis. V, 282 pages. 1978.

Vol. 690: W. J. J. Rey, Robust Statistical Methods. VI, 128 pages. 1978.

Vol. 691: G. Viennot, Algèbres de Lie Libres et Monoïdes Libres. III, 124 pages. 1978.

Vol. 692: T. Husain and S. M. Khaleelulla, Barrelledness in Topological and Ordered Vector Spaces. IX, 258 pages. 1978.

Vol. 693: Hilbert Space Operators, Proceedings, 1977. Edited by J. M. Bachar Jr. and D. W. Hadwin. VIII, 184 pages. 1978.

Vol. 694: Séminaire Pierre Lelong – Henri Skoda (Analyse) Année 1976/77. VII, 334 pages. 1978.

Vol. 695: Measure Theory Applications to Stochastic Analysis, Proceedings, 1977. Edited by G. Kallianpur and D. Kölzow. XII, 261 pages. 1978.

Vol. 696: P. J. Feinsilver, Special Functions, Probability Semigroups, and Hamiltonian Flows. VI, 112 pages. 1978.

Vol. 697: Topics in Algebra, Proceedings, 1978. Edited by M. F. Newman. XI, 229 pages. 1978.

Vol. 698: E. Grosswald, Bessel Polynomials. XIV, 182 pages. 1978.

Vol. 699: R. E. Greene and H.-H. Wu, Function Theory on Manifolds Which Possess a Pole. III, 215 pages. 1979.